LONDON MATHEMATICAL SOCIETY LECTURE NOTE SERIES

Managing Editor: Professor N.J. Hitchin, Mathematical Institute,
University of Oxford, 24–29 St Giles, Oxford OX1 3LB, United Kingdom

The titles below are available from booksellers, or from Cambridge University Press at www.cambridge.org

London Mathematical Society Lecture Note Series. 300

Introduction to Möbius Differential Geometry

Udo Hertrich-Jeromin
Technische Universität Berlin

CAMBRIDGE
UNIVERSITY PRESS

CAMBRIDGE
UNIVERSITY PRESS

32 Avenue of the Americas, New York NY 10013-2473, USA

Cambridge University Press is part of the University of Cambridge.

It furthers the University's mission by disseminating knowledge in the pursuit of education, learning, and research at the highest international levels of excellence.

www.cambridge.org
Information on this title: www.cambridge.org/9780521535694

First published 2003
Reprinted 2013

A catalog record for this publication is available from the British Library.

Library of Congress Cataloging in Publication data
Hertrich-Jeromin, Udo, 1965–
Introduction to Möbius differential geometry / Udo Hertrich-Jeromin.
    p. cm. (London Mathematical Society lecture note series)
Includes bibliographical references and index
ISBN 0-521-53569-7 Paperback
1. Differential geometry. I. Title. II. Series.
QA609.H47 2003
516.3′6–dc21   2002041686

ISBN 978-0-521-53569-4 Paperback

Contents

Metric $S^n \subset \mathbb{R}^{n+1}$	Projective $S^n \subset \mathbb{R}P^{n+1}$	Minkowski $L^{n+1} \subset \mathbb{R}_1^{n+2}$
point: $p \in S^n$	point: $p \in S^n \subset \mathbb{R}P^{n+1}$	lightlike vector $p \in \mathbb{R}_1^{n+2},\ \langle p,p \rangle = 0$
sphere: $S \subset S^n$	point in "outer space": $S \in \mathbb{R}P_O^{n+1}$	spacelike vector: $S \in \mathbb{R}_1^{n+2}\ \langle S,S \rangle > 0$
incidence: $p \in S$	polarity: $S \in \mathrm{pol}[p] \cong T_p S^n$	orthogonality: $\langle p,S \rangle = 0$
orthogonal intersection: $S_1 \cap_\perp S_2$	polarity: $S_i \in \mathrm{pol}[S_j]$	orthogonality: $\langle S_1,S_2 \rangle = 0$
intersection angle: $S_1 \cap_\alpha S_2$	—	scalar product: $\cos^2 \alpha = \frac{\langle S_1,S_2 \rangle^2}{\|S_1\|^2 \|S_2\|^2}$
elliptic sphere pencil: $\{S \mid S_1 \cap S_2 \subset S\}$	line not meeting S^n: $\mathrm{inc}[S_1,S_2]$	spacelike 2-plane: $S_1 \wedge S_2, \langle S_1,S_2 \rangle = 0$
parabolic sphere pencil: $\{S \mid p \in S, T_p S = T_p S_0\}$	line touching S^n: $\mathrm{inc}[p,S_0]$	degenerate 2-plane: $p \wedge S_0, \langle p,S_0 \rangle = 0$
hyperbolic sphere pencil: $\{S \mid p_1,p_2 \in \tilde{S} \Rightarrow S \perp \tilde{S}\}$	line intersecting S^n: $\mathrm{inc}[p_1,p_2]$	Minkowski 2-plane: $p_1 \wedge p_2, p_1 \neq p_2$
elliptic sphere complex: $\{S \subset S^n \mid S \cap_\perp \mathcal{K} \cong S_\infty\}$	hyperplane intersecting S^n: $\mathrm{pol}[\mathcal{K}]$	Minkowski hyperplane: $\{\mathcal{K}\}^\perp, \langle \mathcal{K},\mathcal{K} \rangle > 0$
parabolic sphere complex: $\{S \mid S \ni \mathcal{K} \cong p_\infty\}$	hyperplane touching S^n: $\mathrm{pol}[\mathcal{K}] \cong T_{p_\infty} S^n$	degenerate hyperplane: $\{\mathcal{K}\}^\perp, \langle \mathcal{K},\mathcal{K} \rangle = 0$
hyperbolic sphere complex: $\{S \mid S^\perp \ni \mathcal{K} \cong 0 \in \mathbb{R}^{n+1}\}$	hyperplane not meeting S^n: $\mathrm{pol}[\mathcal{K}] \cong \mathbb{R}P^{n+1} \setminus \mathbb{R}^{n+1}$	spacelike hyperplane: $\{\mathcal{K}\}^\perp, \langle \mathcal{K},\mathcal{K} \rangle < 0$
m-sphere: $S : p_i \in S, \dim S = m$	$(m+1)$-plane intersecting S^n: $\mathrm{inc}[p_1,\dots,p_{m+2}]$	Minkowski $(m+2)$-space: $p_1 \wedge \dots \wedge p_{m+2}$
$(n-m)$-sphere: $S_1 \cap \dots \cap S_m$	$(m-1)$-plane in "outer space": $\mathrm{inc}[S_1,\dots,S_m]$	spacelike m-space: $S_1 \wedge \dots \wedge S_m$
inversion: $p \mapsto \frac{\sin^2(\varrho)p - 2(pm - \cos \varrho)m}{1 - 2pm \cos \varrho + \cos^2 \varrho}$	polar reflection: —	reflection: $p \mapsto p - 2\frac{\langle p,S \rangle}{\langle S,S \rangle} S$
Möbius transformation: $\mu \in M\ddot{o}b(n)$	projective transformation: $\mu \in PGl(n+2), \mu(S^n) = S^n$	Lorentz transformation: $\mu \in O_1(n+2)$

Fig. T.1. The classical model of Möbius geometry

Metric $S^n \subset \mathbb{R}^{n+1}$	**Projective** $S^n \subset \mathbb{R}P^{n+1}$	**Minkowski** $L^{n+1} \subset \mathbb{R}_1^{n+2}$
space of const. curvature: $S_\kappa^n, \mathbb{R}^3, H_\kappa^n \dot\cup H_\kappa^n = M_\kappa^n$	... the base-mf becomes... $S^n \setminus \partial_\infty M_\kappa^n$	quadric: $Q_\kappa^n \subset L^{n+1}$
sphere complex: $\{S \subset M_\kappa^n \mid I\!I \equiv 0\}$	hyperplane: $\mathrm{pol}[\mathcal{K}]$	hyperplane: $\{\mathcal{K}\}^\perp$
immersion: $f : M^m \to S^n$	proj. immersion: $\forall p, v : f(p) \neq \partial_v f(p)$	spacelike immersion: $\langle df, df \rangle > 0$
sphere congruence: $M^m \ni p \mapsto S(p) \subset S^n$	sphere congruence: $S : M^m \to \mathbb{R}P_O^{n+1}$	sphere congruence (locally): $S : M^m \to S_1^{n+1}$
f envelopes S: $f(p) \in S(p)$, $d_p f(T_p M) \subset T_{f(p)} S(p)$	f envelopes S: $T_{f(p)} f(M) \subset \mathrm{pol}[S(p)]$	strip (f, S): $f(p), d_p f(T_p M) \perp S(p)$
Möbius frame: $(S_1, \ldots, S_{n-1}, S, f, \hat{f})$	—	pseudo orthonormal frame: $F : M^m \to O_1(n+2)$

Fig. T.2. The classical model of Möbius geometry, differential geometric terms

Conformal $\mathbb{HP}^1$	Homogeneous $\mathbb{H}^2$	Minkowski $L^{n+1} \subset \mathfrak{H}(\mathbb{H}^2)$
point: $p = v\mathbb{H} \in S^4$	point: $v \in \mathbb{H}^2$	isotropic form: $S_p \in \mathfrak{H}(\mathbb{H}^2)$, $\det S_p = 0$
hypersphere: $S \subset S^4$	—	spacelike form: $S \in \mathfrak{H}(\mathbb{H}^2)$, $\det S < 0$
2-sphere: $\mathcal{S} \subset S^4$	involution: $\mathcal{S} \in \mathfrak{S}(\mathbb{H}^2)$	elliptic sphere pencil: $S, \mathcal{J}S \in \mathfrak{H}(\mathbb{H}^2)$
incidence: $p = v\mathbb{H} \in S$	isotropy: $S(v, v) = 0$	orthogonality: $\langle S_p, S \rangle = 0$
incidence: $p \in \mathcal{S}$	eigendirection: $\mathcal{S}v \parallel v$	orthogonality: $S_p \perp S, \mathcal{J}S$
intersection angle: $S_1 \cap_\alpha S_2$	—	scalar product: $\cos^2 \alpha = \frac{\{S_1, S_2\}^2}{4S_1^2 S_2^2}$
f envelopes S: $f(p) \in S(p)$, $d_p f(T_p M) \subset T_{f(p)} S(p)$	f envelopes S: $S(f, f) = 0$ $S(f, df) + S(df, f) \equiv 0$	strip (S_f, S): $S_f(p), d_p S_f(T_p M) \perp S(p)$
f envelopes $\mathcal{S}$: $f(p) \in \mathcal{S}(p)$, $d_p f(T_p M) \subset T_{f(p)} \mathcal{S}(p)$	f envelopes $\mathcal{S}$: $\mathcal{S}f \parallel f$ $d\mathcal{S} \cdot f \parallel f$	—
Möbius transformation: $\mu \in M\ddot{o}b(4)$	fractional linear: $\mu \in Sl(2, \mathbb{H})$, $v \mapsto \mu v$	Lorentz transformation: $\mu \in Sl(2, \mathbb{H})$, $S \mapsto \mu S$
stereographic projection: $v \mapsto (\nu_0 v)(\nu_\infty v)^{-1}$	affine coordinates: $v = v_0 + v_\infty \mathfrak{p}$	—
point pair map: $(f, \hat{f}) : M \to \mathfrak{P}$	Möbius frame: $F : M \to Sl(2, \mathbb{H})$	Möbius frame: $(S_1, \ldots, S_4, S_f, S_{\hat{f}})$
cross-ratio: $[p_1; p_2; p_3; p_4]$	cross-ratio: $\nu_1 v_2 \frac{1}{\nu_3 v_2} \nu_1 v_4 \frac{1}{\nu_3 v_4}$	—

Fig. T.3. A quaternionic model of Möbius geometry

Conformal S^n	Projective $S^n \subset \mathbb{R}P^{n+1}$	Clifford Algebra $L^{n+1} \subset \Lambda^1 \mathbb{R}_1^{n+2}$
point: $p \in S^n$	point: $p \in S^n$	isotropic vector $p \in \Lambda^1 \mathbb{R}_1^{n+2}$, $p^2 = 0$
hypersphere: $s \subset S^n$	point in "outer space": $s \in \mathbb{R}P_O^{n+1}$	spacelike vector: $s \in \Lambda^1 \mathbb{R}_1^{n+2}$, $s^2 < 0$
incidence: $p \in s$	polarity: $s \in \mathrm{pol}[p] \cong T_p S^n$	orthogonality: $\{p, s\} = 0$
orthogonal intersection: $s_1 \cap_\perp s_2$	polarity: $s_i \in \mathrm{pol}[s_j]$	orthogonality: $\{s_1, s_2\} = 0$
intersection angle: $s_1 \cap_\alpha s_2$	—	scalar product: $\cos^2 \alpha = \frac{\{s_1, s_2\}^2}{4 s_1^2 s_2^2}$
k-sphere: $\mathfrak{s}: p_1, \ldots, p_{k+2} \in \mathfrak{s}$	plane intersecting S^n: $\mathrm{inc}[p_1, \ldots, p_{k+2}]$	timelike pure $(k+2)$-vector: $p_1 \wedge \ldots \wedge p_{k+2} \in \Lambda^{k+2} \mathbb{R}_1^{n+2}$
k-sphere: $s_1 \cap \ldots \cap s_{n-k}$	plane in "outer space": $\mathrm{inc}[s_1, \ldots, s_{n-k}]$	spacelike pure $(n-k)$-vector: $s_1 \wedge \ldots \wedge s_{n-k} \in \Lambda^{n-k} \mathbb{R}_1^{n+2}$
incidence: $p \in \mathfrak{s}$	polarity: $p \in \mathrm{pol}[\mathfrak{s}]$	vanishing of lower grade: $\mathfrak{s}p \in \Lambda^{n-k+1} \mathbb{R}_1^{n+2}$
f envelopes $\mathfrak{s}$: $f(p) \in \mathfrak{s}(p)$, $d_p f(T_p M) \subset T_{f(p)} \mathfrak{s}(p)$	f envelopes $\mathfrak{s}$: $T_{f(p)} f(M) \subset \mathrm{pol}[\mathfrak{s}(p)]$	no lower grades: $\mathfrak{s}f, \mathfrak{s}df \mapsto \Lambda^{n-k+1} \mathbb{R}_1^{n+2}$
inversion	polar reflection	reflection: $p \mapsto \frac{1}{|s|^2} sps$
Möbius transformation: $\mathfrak{z} \in M\ddot{o}b(n)$	projective transformation: $\mathfrak{z}, \mathfrak{z}(S^n) = S^n$	spinor: $\mathfrak{z} \in Spin_1(n+2)$
cross-ratio: $[p_1; p_2; p_3; p_4]$	—	cross-ratio: $\frac{p_1 p_2 p_3 p_4 + p_4 p_3 p_2 p_1}{(p_1 p_4 + p_4 p_1)(p_2 p_3 + p_3 p_2)}$

Fig. T.4. A Clifford algebra model of Möbius geometry

Conformal S^n	Affine $S^n \cong \mathbb{R}^n \cup \{\infty\}$	Clifford Algebra $L^{n+1} \subset \Lambda^1 \mathbb{R}_1^{n+2}$
point: $\mathcal{V} \in S^n$	"vector": $\binom{v}{1}$, $v \in \mathbb{R}^n$, or $\binom{1}{0}$	isotropic vector $\left(\begin{smallmatrix} v & -v^2 \\ 1 & -v \end{smallmatrix}\right), \left(\begin{smallmatrix} 0 & 1 \\ 0 & 0 \end{smallmatrix}\right) \in \Lambda^1 \mathbb{R}_1^{n+2}$
hypersphere: $\mathcal{S} \subset S^n$	hypersphere/-plane: $\mathcal{S} \subset \mathbb{R}^n$	spacelike vector/Möbius involution: $\left(\begin{smallmatrix} m & -m^2-r^2 \\ 1 & -m \end{smallmatrix}\right), \left(\begin{smallmatrix} n & 2d \\ 0 & -n \end{smallmatrix}\right) \in \Lambda^1 \mathbb{R}_1^{n+2}$
incidence: $\mathcal{V} \in \mathcal{S}$	fixed point: $\mathcal{S}\binom{v}{1} = \binom{v}{1}a$	orthogonality: $\{\mathcal{V}, \mathcal{S}\} = 0$
orthogonal intersection: $\mathcal{S}_1 \cap_\perp \mathcal{S}_2$	—	orthogonality: $\{\mathcal{S}_1, \mathcal{S}_2\} = 0$
intersection angle: $\mathcal{S}_1 \cap_\alpha \mathcal{S}_2$	—	scalar product: $\cos^2 \alpha = \dfrac{\{\mathcal{S}_1, \mathcal{S}_2\}^2}{4\mathcal{S}_1^2 \mathcal{S}_2^2}$
k-sphere: $\int : \mathcal{V}_1, \ldots, \mathcal{V}_{k+2} \in \int$	—	timelike pure $(k+2)$-vector: $\mathcal{V}_1 \wedge \ldots \wedge \mathcal{V}_{k+2} \in \Lambda^{k+2} \mathbb{R}_1^{n+2}$
k-sphere: $\mathcal{S}_1 \cap \ldots \cap \mathcal{S}_{n-k}$	—	spacelike pure $(n-k)$-vector: $\mathcal{S}_1 \wedge \ldots \wedge \mathcal{S}_{n-k} \in \Lambda^{n-k} \mathbb{R}_1^{n+2}$
incidence: $v \in \mathcal{S}$	fixed point: $\mathcal{S}\binom{v}{1} = \binom{v}{1}a$	fixed point: $r(\mathcal{S})\mathcal{V} \parallel \mathcal{V}$
inversion	inversion: $v \mapsto m - r^2(v - m)^{-1}$	reflection: $\mathcal{V} \mapsto \frac{1}{|\mathcal{S}|^2}\mathcal{S}\mathcal{V}\mathcal{S}$
Möbius transformation: $\mu \in M\ddot{o}b(n)$	fractional linear: $v \mapsto (av + b)(cv + d)^{-1}$	spinor: $\left(\begin{smallmatrix} a & b \\ c & d \end{smallmatrix}\right) \in Pin(\mathbb{R}_1^{n+2})$
point pair map: $(f_\infty, f_0) : M \to \mathfrak{P}$	point pair map $(f_\infty, f_0), f_i : M \to \mathbb{R}^n$	Möbius frame: $\left(\begin{smallmatrix} f_\infty & f_0 \\ 1 & 1 \end{smallmatrix}\right) : M \to \Gamma(\mathbb{R}_1^{n+2})$

Fig. T.5. A Clifford algebra model: Vahlen matrices

Introduction

Over the past two decades, the geometry of surfaces and, more generally, submanifolds in Möbius geometry has (re)gained popularity. It was probably T. Willmore's 1965 conjecture [306] that stimulated this increased interest: Many geometers have worked on this conjecture, and in the course of this work it turned out (see for example [67] and [40]) that the Willmore conjecture is in fact a problem for surfaces in Möbius geometry and that the corresponding local theory was already developed by the classical geometers (cf., [218]). A crucial classical reference was [29]; however, it may not be very easy to obtain and, once found, may not be very easy to read, especially for non-German—speaking colleagues.

A similar story could be told about the recent developments on isothermic surfaces — here, it was the relation with the theory of integrable systems, first pointed out in [71], that made the topic popular again; also in this case [29] turned out to be a treasure trove, but many more results are scattered in the classical literature.

The present book has a twofold purpose:

- It aims to provide the reader with a solid background in the Möbius geometry of surfaces and, more generally, submanifolds.
- It tries to introduce the reader to the fantastically rich world of classical (Möbius) differential geometry.

The author also hopes that the book can lead a graduate student, or any newcomer to the field, to recent research results.

Before going into details, the reader's attention shall be pointed to three[1] other textbooks in the field. To the author's knowledge these are the only books that are substantially concerned with the Möbius geometry of surfaces or submanifolds:

1. W. Blaschke [29]: Currently, this book is a standard reference for the geometry of surfaces in 3-dimensional Möbius geometry where many fundamental facts (including those concerning surface classes of current interest) can be found. Möbius geometry is treated as a subgeometry of Lie sphere geometry.

2. T. Takasu [275]: Nobody seems to know this book; like [29], it is in German, and the presented results are similar to those in Blaschke's book — however, Möbius geometry is treated independently.

---

[1] Apparently there is another classical book [95] by P. C. Delens on the subject that the present author was not able to obtain so far.

3. M. Akivis and V. Goldberg [4]: This is a modern account of the theory, generalizing many results to higher dimension and/or codimension. Also, different signatures are considered, which is relevant for applications in physics. In this book the authors also discuss nonflat conformal structures; almost Grassmann structures, a certain type of Cartan geometry, are considered as a closely related topic (compare with the survey [7]).

Besides these three textbooks, there is a book with a collection of (partially introductory) articles on Möbius or conformal differential geometry [170]. Here, Möbius geometry is mainly approached from a Riemannian viewpoint (see, for example, J. Lafontaine's article [173]), which is similar to the way it is touched upon in many textbooks on differential geometry but in much greater detail and including a description of the projective model.

A more general approach to Möbius geometry than the one presented in this book may be found in [254], see also the recent paper [50]: There, Möbius geometry is treated as an example of a Cartan geometry.

I.1 Möbius geometry: models and applications

In Möbius geometry there is an angle measurement but, in contrast to Euclidean geometry, no measurement of distances. Thus Möbius geometry of surfaces can be considered as the geometry of surfaces in an ambient space that is equipped with a conformal class of metrics but does not carry a distinguished metric. Or, taking the point of view of F. Klein [160], one can describe the Möbius geometry of surfaces as the study of those properties of surfaces in the (conformal) n-sphere S^n that are invariant under Möbius (conformal) transformations of S^n. Here, "Möbius transformation" means a transformation that preserves (hyper)spheres in S^n, where a hypersphere can be understood as the (transversal) intersection of an affine hyperplane in Euclidean space $\mathbb{R}^{n+1}$ with $S^n \subset \mathbb{R}^{n+1}$.

Note that the group of Möbius transformations of $S^n \cong \mathbb{R}^n \cup \{\infty\}$ is generated by inversions, that is, by reflections in hyperspheres in S^n.

The lack of length measurement in Möbius differential geometry has interesting consequences; for example, from the point of view of Möbius geometry, the planes (as spheres that contain ∞) and (round) spheres of Euclidean 3-space are not distinguished any more.

At this point we can already see how Euclidean geometry is obtained as a subgeometry of Möbius geometry when taking the Klein point of view: The group of Möbius transformations is generated by inversions; by restricting to reflections in planes (as "special" spheres), one obtains the group of isometries of Euclidean space.

Also, the usual differential geometric invariants of surfaces in space lose their meaning; for example, the notion of an induced metric on a surface as well as the notion of curvature lose their meanings as the above example of spheres and planes in $\mathbb{R}^3$ illustrates — the reason is, of course, that these notions are defined using the metric of the ambient space. But the situation is less hopeless than one might expect at first: Besides the angle measurement (conformal structure) that a surface or submanifold inherits from the ambient space, there are conformal invariants that encode some of the curvature behavior of a surface. However, it is rather complicated and unnatural to consider submanifolds in Möbius geometry as submanifolds of a Riemannian (or Euclidean) space and then to extract those properties that remain invariant under the larger symmetry group of conformal transformations.[2]

I.1.1 Models of Möbius geometry. Instead, one can describe Möbius geometry in terms of certain "models" where hyperspheres (as a second type of "elements" in Möbius geometry, besides points) and the action of the Möbius group are described with more ease. For example, we can describe hyperspheres in $S^n \subset \mathbb{R}^{n+1}$ by linear equations instead of quadratic equations as in $S^n \cong \mathbb{R}^n \cup \{\infty\}$ — however, it is still unpleasant to describe the Möbius group acting on $S^n \subset \mathbb{R}^{n+1}$.

Let us consider another example to clarify this idea: In order to do hyperbolic geometry, it is of great help to consider a suitable model of the hyperbolic ambient space, that is, a model that is in some way adapted to the type of problems that one deals with. One possibility would be to choose the Klein model of hyperbolic space where H^n is implanted into projective n-space $\mathbb{R}P^n$; then the hyperbolic motions become projective transformations that map H^n to itself, that is, projective transformations that preserve the infinity boundary $\partial_\infty H^n$ of the hyperbolic space, and hyperplanes become the intersection of projective hyperplanes in $\mathbb{R}P^n$ with H^n. In this model, for example, it is obvious that the Euclidean Parallel Postulate does not hold in hyperbolic geometry. Another possibility would be to consider H^n as one of the connected components of the 2-sheeted hyperboloid

$$\{y \in \mathbb{R}_1^{n+1} \mid \langle y, y \rangle = -1\}$$

in Minkowski $(n+1)$-space $\mathbb{R}_1^{n+1} = (\mathbb{R}^{n+1}, \langle ., . \rangle)$. In this model[3] it is rather simple to do differential geometry. A third possibility, which we will come

---

[2] This will become obvious when we take this viewpoint in the "Preliminaries" chapter that is meant as an introduction for those readers who have a background in Riemannian geometry but are new to Möbius geometry.

[3] It is a matter of taste whether one wants to consider this as a different *model* for hyperbolic geometry: Of course there is a simple way to identify this "model" with the Klein model; it is just a convenient choice of homogeneous coordinates for the Klein model after equipping the coordinate $\mathbb{R}^{n+1}$ with a scalar product.

across in the present text more often, will be to use the Poincaré (half-space or ball) model of hyperbolic geometry, where H^n is considered as a subspace of the conformal n-sphere. In this model, the hyperbolic hyperplanes become (the intersection of $H^n \subset S^n$ with) hyperspheres that intersect the infinity boundary $\partial_\infty H^n \subset S^n$ of the hyperbolic space orthogonally and reflections in these hyperplanes are inversions.

In this book we will elaborate three and a half models[4] for Möbius geometry.

I.1.2 The projective model.   This is the approach that the classical differential geometers used, and it is still the model that is used in many modern publications in the field. In fact, it can be considered as *the* model, because all other "models" can be derived from it.[5] A comprehensive treatment of this model can be found in the book by M. Akivis and V. Goldberg [4], which the reader is also encouraged to consult; our discussions in Chapter 1 will follow instead the lines of W. Blaschke's aforementioned book [29].

To obtain this projective model for Möbius geometry the ambient (conformal) n-sphere S^n is implanted into the projective $(n+1)$-space $\mathbb{RP}^{n+1}$. In this way hyperspheres will be described as the intersection of projective hyperplanes with $S^n \subset \mathbb{RP}^{n+1}$, in a similar way as discussed above, and Möbius transformations will be projective transformations that preserve S^n as an "absolute quadric" so that, in homogeneous coordinates, the action of the Möbius group is linear. This is probably the most important reason for describing Möbius geometry as a subgeometry of projective geometry.[6]

I.1.3 The quaternionic approach.   Here the idea is to generalize the description of Möbius transformations of $S^2 \cong \mathbb{CP}^1 \cong \mathbb{C} \cup \{\infty\}$ as fractional linear transformations from complex analysis to higher dimensions. For dimensions 3 and 4, this can be done by using quaternions — this approach can be traced back to (at least) E. Study's work [262]. Recently, the use of the quaternionic model for Möbius geometry has provided remarkable progress in the global Möbius geometry of surfaces [113].

One of the main difficulties in establishing this quaternionic approach to Möbius geometry is the noncommutativity of the field of quaternions. Besides that, it is rather seamless to carry over much of the complex theory.

---

[4] As mentioned above, it is, to a certain extent, a matter of taste of what one considers as different "models" and what one considers to be just different incarnations of one model.

[5] Therefore, one could consider this model to be the only model and all other "models" to be different representations or refinements of it; however, this is not the author's viewpoint.

[6] In [160], F. Klein undertakes it to describe many geometries as subgeometries of projective geometry and, in this way, to bring order to the variety of geometries (compare with [254]).

An important issue will be the description of (hyper)spheres in this model. For this it is convenient to link the quaternionic model to the projective model by using quaternionic Hermitian forms; for the important case of 2-spheres, there will be a second description based on Möbius involutions.

I.1.4 The Clifford algebra approach. After describing Möbius geometry as a subgeometry of projective geometry it is rather natural to use the Clifford algebra of the $(n+2)$-dimensional space of homogeneous coordinates of the host projective space, equipped with a Minkowski scalar product, in order to describe geometric objects (cf., [155] and [154]) — just notice that the development of this algebra, initiated by H. Grassmann in [130] and [133] and by W. K. Clifford in [77], was originally motivated by geometry, as its original name, "geometric algebra," suggests. For example, the description of spheres of any codimension becomes extremely simple using this approach. Our discussions in Chapter 6 on this topic were motivated by an incomplete 1979 manuscript by W. Fiechte [117].

More common is an enhancement[7] of this Clifford algebra model by writing the elements of the Clifford algebra of the coordinate Minkowski space $I\!R_1^{n+2}$ of the classical model as 2×2 matrices with entries from the Clifford algebra of the Euclidean n-space $I\!R^n$. This approach can be traced back to a paper by K. Vahlen [288]; see also [2] and [3] by L. Ahlfors.

In some sense this model can also be understood as an enhancement of the quaternionic model (for 3-dimensional Möbius geometry): Möbius transformations, written as Clifford algebra 2×2 matrices, act on the conformal n-sphere $S^n \cong I\!R^n \cup \{\infty\}$ by fractional linear transformations.[8] The description of Möbius transformations by 2×2 matrices makes this approach to Möbius geometry particularly well suited for the discussion of the geometry of "point pair maps" that arise, for example, in the theory of isothermic surfaces as developed in the excellent paper by F. Burstall [47].

I.1.5 Applications. The description of each model in turn is complemented by a discussion of applications to specific problems in Möbius differential geometry. The choice of these problems is certainly influenced by the author's preferences, his interest and expertise — however, the author hopes to have chosen applications that are of interest to a wider audience and that can lead the reader to current research topics.

Conformally flat hypersurfaces are discussed in Chapter 2. Here we already touch on various topics that will reappear in another context or in more generality later; in particular, we will come across curved flats, a particularly

---

[7] Thus we count the two descriptions as one and a half "models."

[8] This matrix representation of the Clifford algebra $A I\!R_1^{n+2}$ corresponds to its "conformal split" (see [154]); this choice of splitting provides a notion of stereographic projection (cf., [47]).

simple type of integrable system, and Guichard nets, that is, a certain type of triply orthogonal system.

Willmore surfaces are touched on as a solution of a Möbius geometric problem that we will refer to as "Blaschke's problem." In this text we will cover only basic material on Willmore surfaces; in particular, we will not get into the large body of results concerning the Willmore conjecture.

Isothermic surfaces will appear as another solution of Blaschke's problem. A first encounter of the rich transformation theory of isothermic surfaces is also given in Chapter 3, but a comprehensive discussion is postponed until the quaternionic model is available in Chapter 5; in the last two sections of Chapter 8 we will reconsider isothermic surfaces and show how to generalize the results from Chapter 5 to arbitrary codimension. We will make contact with the integrable systems approach to isothermic surfaces, and we will discuss a notion of discrete isothermic nets in the last section of Chapter 5.

Orthogonal systems will be discussed in Chapter 8. We will generalize the notion of triply orthogonal systems to m-orthogonal systems in the conformal n-sphere and discuss their Ribaucour transformations, both smooth and discrete. The discussion of discrete orthogonal nets is somewhat more geometrical than that of discrete isothermic nets, and it demonstrates the interplay of geometry and the Clifford algebra formalism very nicely.

I.1.6 Integrable systems.   A referee of the present text very rightly made the remark that "the integrable systems are not far beneath the surface in the current text." This fact shall not be concealed: As already mentioned above, conformally flat hypersurfaces as well as isothermic surfaces are related to a particularly simple type of integrable system in a symmetric space, "curved flats," that were introduced by D. Ferus and F. Pedit in [112].

However, the corresponding material is scattered in the text, and we will not follow up on any implications of the respective integrable systems descriptions but content ourselves by introducing the spectral parameter that identifies the geometry as integrable. Instead we will discuss the geometry of the spectral parameter in more detail:

Conformally flat hypersurfaces in S^4 are related to curved flats in the space of circles, and the corresponding spectral parameter can already be found in C. Guichard's work [136]. In this case, a curved flat describes a circle's worth of conformally flat hypersurfaces, and the associated family of curved flats yields a 1-parameter family of such cyclic systems with conformally flat orthogonal hypersurfaces.

Isothermic surfaces are related to curved flats in the space of point pairs in the conformal 3-sphere S^3 (or, more generally, in S^n), and the existence of the corresponding spectral parameter will turn out to be intimately related to the conformal deformability of isothermic surfaces [62] and to the Calapso transformation, see [53] and [55]. In fact, the associated family of curved flats (Darboux pairs of isothermic surfaces) yields one of Bianchi's permutability theorems [20] that intertwines the Christoffel, Darboux, and Calapso transformations of an isothermic surface.

For more information about the respective integrable systems approaches, the reader shall be referred to [142] and to F. Burstall's paper [47].

I.1.7 Discrete net theory. More recently, discrete net theory has become a field of active research. In the author's opinion, discrete net theory, as it is discussed in the text, is of interest for various reasons: An obvious reason may be the application of the theory in computer graphics and experimental mathematics; however, the author thinks that it is also very interesting for methodological reasons — the proofs of "analogous" results in corresponding smooth and discrete theories are usually rather different. While proofs in (smooth) differential geometry are often very computational in nature, the proofs of the corresponding discrete results may be done by purely (elementary) geometric arguments. In this way proofs from discrete net theory sometimes resemble the proofs of the classical geometers when they applied geometric arguments to "infinitesimal" quantities (cf., [70]).

As already mentioned above, we will discuss two discrete theories, one of which is a special case — with more structure — of the other:

Discrete isothermic nets will be discussed in the last section of Chapter 5. We will see that much of the theory of smooth isothermic surfaces can be carried over to the discrete setup; in fact, many proofs can be carried over directly when using the "correct" discrete version — discrete quantities usually carry more information than their smooth versions. In this way, the analogous smooth and discrete theories can motivate and inspire each other.[9] Note that the fact that computations can be carried over from one setup to the other so seamlessly relies on using the quaternionic model for Möbius geometry.[10]

Discrete orthogonal nets will be treated more comprehensively, in Chapter 8. This topic demonstrates nicely the interplay of analytic and geometric methods in discrete net theory, as well as the interplay of algebra and geometry in the Clifford algebra model. In this way, its presentation serves a twofold purpose. A highlight of the presentation shall be the discrete analog of Bianchi's permutability theorem for the Ribaucour transformation of

---

[9] Despite the order of presentation, some of the proofs on smooth isothermic surfaces in the present text were obtained from their discrete counterparts.

[10] Or, equally, the Vahlen matrix approach should work just as well.

orthogonal systems.

I.1.8 Symmetry-breaking. Finally, the author would like to draw the reader's attention to a phenomenon that he considers to be rather interesting and that appears at various places in the text. First note that the metric (hyperbolic, Euclidean, and spherical) geometries are subgeometries of Möbius geometry.[11)] Now, imposing two Möbius geometric conditions on, say, a surface may break the symmetry of the problem and yield a characterization of the corresponding surface class in terms of a (metric) subgeometry of Möbius geometry. The most prominent example for such a symmetry-breaking may be Thomsen's theorem, which we discuss in Chapter 3: A surface that is Willmore and isothermic at the same time is (Möbius equivalent to) a minimal surface in some space of constant curvature. Other examples are Guichard cyclic systems, which turn out to come from parallel Weingarten surfaces in space forms, and isothermic or Willmore channel surfaces, which are Möbius equivalent to surfaces of revolution, cylinders, or cones in Euclidean geometry.

I.2 Philosophy and style

In the author's opinion, geometry describes certain aspects of an ideal world where geometric configurations and objects "live." In order to describe that ideal world, one needs to use some language or model — that may change even though the described objects or facts remain the same. Thus a geometer may choose from a variety of possibilities when carrying out his research or presenting his results to colleagues or students. In this choice he may be led by different motivations: sense of beauty, curiosity, pragmatism, ideology, ignorance, and so on.

I.2.1 Methodology. This book shall be an advertisement for "methodological pluralism": It will provide the reader with three and a half "models" for Möbius (differential) geometry that may be used to formulate geometric facts. In order to compare the effect of choosing different models, the reader may compare the treatment of isothermic surfaces — in terms of the classical projective model in Chapter 3, in terms of the quaternionic approach in Chapter 5, and in terms of the Vahlen matrix setup in Chapter 8. The author hopes that the chance of comparing these different descriptions of the geometry of isothermic surfaces compensates the reader for the repetitions caused by the multiple treatment.

However, we will also pursue this program of "methodological pluralism" in the details. Very often we will (because of the author's pragmatism, sense

---

[11)] For the hyperbolic geometry, we already touched on this when discussing the Poincaré model of hyperbolic geometry above.

of beauty, or ignorance) use Cartan's method of moving frames to describe the geometry of surfaces or hypersurfaces. However, we will always try to find an "adapted frame" that fits the geometric situation well. This notion of an adapted frame, which is central to many computations in the text, will change depending on the context. The discussion of isothermic and Willmore channel surfaces in the last section of Chapter 3 may serve as a typical example: Depending on the viewpoint we wish to take, we will use frames that are, to an attuned degree, adapted to the surface or to the enveloped sphere curve, respectively.

I.2.2 Style. Note that a "model" should always be distinguished from what it describes, and that it is unlikely that a description using a model will be optimal in any sense. This is even more the case as long as an author has not stopped working and learning about his topic — here is where the ignorance issue comes in. Therefore, the present text does not claim to provide the optimal description of a subject, and the reader is warmly invited to figure out better ways to think about, say, isothermic surfaces. However, there is another issue related to the aim of this book to make the classical literature more accessible to the reader: In this text we will try to adopt certain habits of the classical authors and make a compromise between modern technology and classical phrasing; in this way a reader may be better prepared to study the classical literature — which is sometimes not very easy to access.

I.2.3 Prerequisites. There is some background material that the reader is expected to be familiar with: some basics in semi-Riemannian geometry, on Lie groups and homogeneous spaces, and some vector bundle geometry. All the needed background material can be found in the excellent textbook [209]. Also, the reader shall be referred to [189], which is a treasure trove for algebraic (and historic) background material, in particular on Clifford algebras and quaternions.

In the "Preliminaries" chapter we will summarize some material on conformal differential geometry from the Riemannian point of view, mainly following ideas from [173]. It is meant to be a preparation for those readers who have some background in Riemannian geometry but who are new to conformal differential geometry; the author took this approach when giving a course for graduate students at TU Berlin in the winter of 1999—2000. This chapter also serves to collect some formulas and notions for later reference, and a discussion of certain conformal invariants of surfaces in the conformal 3-sphere is provided.[12]

---

[12] However, it shall be pointed out that this chapter is not meant as an introduction to nonflat conformal differential geometry; for this the reader is referred to [254] and [50].

I.2.4 Further reading.   At the end of the book the author gives a selection of references for further reading, with comments.

Also, the list of references contains many more references than are cited in the text: In particular, many classical references are provided to facilitate the search for additional literature — in the author's experience it can be rather difficult to locate relevant (classical) references. For the reader's convenience the coordinates of reviews are provided where the author was able to locate a review, and some cross-references to occurrences of a reference in the text are compiled into the Bibliography.[13]

I.3 Acknowledgments

Many people have contributed, in one way or another, to this book — more than I could possibly mention here. However, I would like to express my gratitude to at least some of them.

First, I would like to thank my teacher, Ulrich Pinkall, who got me interested in Möbius differential geometry in the first place. For helpful discussions or suggestions concerning the covered topics, for constructive criticism and other helpful feedback on the manuscript, or for some other kind of support I warmly thank the following friends and/or colleagues: Alexander Bobenko, Christoph Bohle, Fran Burstall, Susanne Hannappel, Gary Jensen, Catherine McCune, Patrick McDonough, Emilio Musso, Lorenzo Nicolodi, Franz Pedit, Paul Peters, Boris Springborn, Yoshihiko Suyama, Ekkehard Tjaden, Konrad Voss; as well as my wife, Heike Jeromin.

Special thanks also go to F. Burstall for many helpful discussions, in particular on isothermic surfaces, and for his suggestion to investigate the "retraction form"; Y. Suyama for providing me with a new explicit example of a generic conformally flat hypersurface and for allowing me to include it; E. Tjaden for always having the time to discuss and optimize explicit examples, as well as for his help with computing/-er questions; and K. Voss for allowing me to include joint unpublished results on Willmore channel surfaces.

I am grateful for hospitality while (and for) working on this text at the Mathematisches Forschungsinstitut Oberwolfach and the Forschungsinstitut für Mathematik at ETH Zürich. This book is based on lecture notes of a course given by the author at TU Berlin (1999—2000) and on the author's Habilitation thesis, "Models in Möbius differential geometry," TU Berlin (2002).

The text was typeset using plain TeX, the sketches were prepared using `xfig`, and the surface graphics were produced using *Mathematica*.

---

[13] Of course, this does not reflect all relevant text passages; for example, Blaschke's book [29] is not cited in every relevant paragraph, while other references only provide a technical detail.

I also would like to express my gratitude to a referee of the manuscript who provided helpful suggestions, and to Roger Astley of Cambridge University Press, who sent it to this referee and with whom it was a pleasure to work anyway. For revising my English and for help during the preparation of the final manuscript, I would like to thank Elise Oranges.

Preliminaries
The Riemannian point of view

As already mentioned before, this book grew out of the lecture notes of a course given for graduate students who had previously taken a course in Riemannian geometry. This preliminary chapter is meant to be of a didactical (and historical) nature rather than to be a modern conceptual introduction to conformal geometry; a modern treatment[1] of conformal geometry may be found in [254] or [50].

Thus, in this chapter, we will discuss basic notions and facts of conformal geometry from the point of view of Riemannian geometry: This is a point of view that most readers will be familiar with. The main goal will be to introduce the notion of the "conformal n-sphere" — this is the ambient space of the submanifolds that we are going to investigate — and to get some understanding of its geometry. In particular, we will discuss Riemannian spaces that are "conformally flat," that is, look locally like the conformal n-sphere.

In the second part of this chapter we will discuss the conformal geometry of submanifolds from the Riemannian point of view: We will deduce how the fundamental quantities of a submanifold change when the metric of the ambient space is conformally altered, and we will discuss various conformal and "Möbius invariants" that appear in the literature. Totally umbilical submanifolds will be treated in detail because they will serve as a main tool in our approach to Möbius geometry presented in the following chapters.

Much of the material in this chapter can be found in the two papers by R. Kulkarni [170] and J. Lafontaine [173]; see also the textbook [122].

The contents of this chapter are organized as follows:

Section P.1. The notions of conformal maps and conformal structures on manifolds are introduced and illustrated by various examples. The most important conformal map given in this section may be the stereographic projection. The metrics of constant curvature are discussed as representatives of the conformal structure given by the Euclidean metric. The term "conformal n-sphere" is defined.

Section P.2. In this section, the transformation formulas for the Levi-Civita connection and the curvature tensor under a conformal change of the Riemannian metric are derived.

---

[1] However, to appreciate the modern treatment of conformal geometry as a Cartan geometry, it should be rather helpful to be familiar with the classical model of Möbius differential geometry as it will be presented in Chapter 1 of the present book.

Section P.3. The Weyl and Schouten tensors are introduced via a decomposition of the Riemannian curvature tensor that conforms with the transformation behavior of the curvature tensor under conformal changes of the Riemannian metric.

Section P.4. The notion of conformal flatness is introduced and related to the existence of conformal coordinates. Then Lichtenstein's theorem on the conformal flatness of every surface (2-dimensional Riemannian manifold) is presented and two proofs are sketched. One of the proofs is based on the relation between conformal structures and complex structures on 2-dimensional manifolds.

Section P.5. The conformal flatness of higher dimensional Riemannian manifolds is discussed: The Weyl-Schouten theorem provides conditions for a Riemannian manifold of dimension $n \geq 3$ to be conformally flat. A proof of this theorem is given. As examples, spaces of constant sectional curvature and of 3-dimensional Riemannian product manifolds are given; the last example will become important when investigating conformally flat hypersurfaces in Chapter 2.

Section P.6. The geometric structures induced on a submanifold of a Riemannian manifold are introduced via the structure equations: the induced connection and the normal connection, the second fundamental form, and the Weingarten tensor field; and their transformation behavior under a conformal change of the ambient Riemannian structure is investigated. Some conformal and Möbius invariants are discussed. In particular, we arrive at Fubini's conformal fundamental forms for surfaces in the conformal 3-sphere, and we investigate Wang's "Möbius form," which turns out to be a Möbius invariant but not a conformal invariant; however, Rothe gave a conformally invariant formulation for Wang's Möbius form in the 2-dimensional case. Finally, it is shown that the notions of curvature direction and of umbilic are conformally invariant.

Section P.7. Using the conformal invariance of umbilics, we introduce hyperspheres as totally umbilic hypersurfaces that are maximal in an appropriate sense. As examples, we discuss the hyperspheres of the spaces of constant curvature — of particular interest may be the hyperspheres in hyperbolic space. Using Joachimsthal's theorem, spheres of higher codimension are then characterized in two equivalent ways. An ad-hoc definition is given for hyperspheres in the conformal 2-sphere S^2.

Section P.8. In this final section the notion of Möbius transformation is introduced, and Liouville's theorem on the relation between Möbius transformations and conformal transformations is formulated; a proof will be given later.

Remark. As indicated above, in this chapter the reader is expected to have some background in Riemannian geometry. On the other hand, most of the presented material will be familiar to the reader, so that this chapter may be omitted from a first reading; it may rather serve for reference. However, this chapter may be a good introduction when giving a course on the subject.

P.1 Conformal maps

First, we recall the notion of a conformal map between Riemannian manifolds and the notion of conformal equivalence of metrics:

P.1.1 Definition. *A map $f : (M, g) \to (\tilde{M}, \tilde{g})$ between Riemannian manifolds is called conformal[2] if the induced metric $f^*\tilde{g} = \tilde{g}(df, df) = e^{2u}g$ with some function $u : M \to \mathbb{R}$.*

Two metrics g and $\tilde{g}$ on $\tilde{M} = M$ are said to be conformally equivalent if the map $f = id$ is conformal.

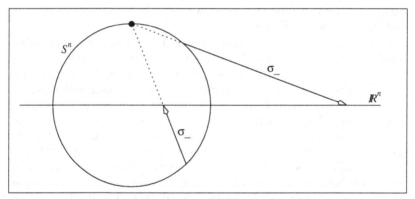

Fig. P.1. The stereographic projection

P.1.2 Note that a map $f : (M, g) \to (\tilde{M}, \tilde{g})$ is conformal iff it preserves angles iff it preserves orthogonality: Clearly, a conformal map preserves angles, and hence it preserves orthogonality. It remains to understand that f preserving orthogonality forces f to be conformal. For that purpose, let $(e_1, \ldots, e_n)$ denote an orthonormal basis of T_pM, with respect to g; then, $0 = f^*\tilde{g}_p(e_i, e_j)$ and $0 = f^*\tilde{g}_p(e_i + e_j, e_i - e_j) = f^*\tilde{g}_p(e_i, e_i) - f^*\tilde{g}_p(e_j, e_j)$ for $i \neq j$, showing that the $df(e_i)$ are orthogonal and have the same lengths. Hence f is conformal.

P.1.3 Examples. The stereographic projections $\sigma_\pm$, defined in terms of

---

[2] Note that, by definition, the conformal factor $e^{2u} > 0$ — allowing zeros of the conformal factor, the map is called "weakly conformal."

their inverses by

$$\sigma_{\pm}^{-1} : \mathbb{R}^n \to S^n \subset \mathbb{R}^{n+1} = \mathbb{R} \times \mathbb{R}^n, \quad p \mapsto \tfrac{1}{1+|p|^2}(\mp(1-|p|^2), 2p),$$

are conformal (cf., Figure P.1).

The Mercator map of the world (assuming the world is a 2-sphere), given by the parametrization $(u, v) \mapsto (\frac{1}{\cosh u} \cos v, \frac{1}{\cosh u} \sin v, \tanh u)$, is conformal;[3] the Archimedes-Lambert projection (projection of the sphere to the cylinder along lines perpendicular to the axis) is area-preserving but not conformal. Note that the system of meridians and parallels of latitude still forms an orthogonal net (see [260]).

The metrics $g_k|_p = \frac{4 dp^2}{(1+k|p|^2)^2}$, $k \geq k_0$ for some $k_0 \in \mathbb{R}$, are conformally equivalent on $\{p \in \mathbb{R}^n \mid 1 + k_0 |p|^2 > 0\}$; in particular, the standard metrics of constant curvature[4] $k \geq -\frac{1}{r^2}$ (cf., [122]) are conformally equivalent on the ball $B(r) := \{p \mid |p| < r\}$. The Poincaré ball model of hyperbolic space of curvature $k = -\frac{1}{r^2}$ is obtained as the only complete space in the family.

The metric $\frac{1}{p_0^2}(dp_0^2 + dp^2)$ on $(0, \infty) \times \mathbb{R}^n$ is clearly conformally equivalent to the standard Euclidean metric on that set as a subset of $\mathbb{R}^{n+1}$. This metric has constant sectional curvature -1 (cf., [162]) as one either computes directly or one concludes by defining an isometry onto the Poincaré ball model $(B(1), g_{-1})$ given above; this can be done by using a suitable "inversion" (we will learn about inversions later). This is the Poincaré half-space model of hyperbolic space.

P.1.4 Definition. *A conformal equivalence class of metrics on a manifold M is called a conformal structure on M.*

P.1.5 Clearly, "conformally equivalent" defines an equivalence relation for Riemannian metrics on a given manifold M. Thus, this definition makes sense.

Note that in [170] a slightly more general definition is given: There, a conformal structure is defined via locally defined metrics; however, any conformal structure in this wider sense contains a globally defined representative by a partition of the unity argument (see [170]).

One particular example of such a conformal structure on a specific manifold will become very important to us — as the ambient space of the submanifolds or hypersurfaces that we are going to examine:

---

[3] Note that the lines $v = const$ give unit speed geodesics in the Poincaré half-plane model (see the next example); the sphere is then isometrically parametrized as a surface in the space $S^1 \times H^2$, which is conformally equivalent to $\mathbb{R}^3 \setminus \{(0, 0, t)\}$ equipped with the standard Euclidean metric.

[4] We will see later how to describe these models via "generalized stereographic projections."

P.1.6 Definition. *The n-sphere S^n equipped with its standard conformal structure will be called the conformal n-sphere.*

P.2 Transformation formulas

Our next goal is to give the transformation formulas for the Levi-Civita connection and the curvature tensor of a Riemannian manifold under conformal changes of the metric.

P.2.1 Lemma. *If $\tilde{g} = e^{2u}g$, then $\tilde{\nabla} = \nabla + B$ with the symmetric $(2,1)$-tensor field $B(x,y) := du(x)y + du(y)x - g(x,y)\mathrm{grad}\, u$.*

P.2.2 *Proof.* Obviously, $\tilde{\nabla} := \nabla + B$ defines a torsion-free connection as B is symmetric. Since

$$(\tilde{\nabla}_x \tilde{g})(y,z) = e^{2u}[2du(x)g(y,z) - g(B(x,y),z) - g(y,B(x,z))] = 0,$$

this is the Levi-Civita connection for $\tilde{g}$.   ◁

The next definition is of a rather technical nature but allows us to write the transformation formulas for the curvature tensor in a much more condensed form (cf., [173]):

P.2.3 Definition. *Let $b_1, b_2 : V \times V \to \mathbb{R}$ be two symmetric bilinear forms on a vector space. The Kulkarni-Nomizu product of b_1 and b_2 is defined as*

$$(b_1 \wedge b_2)(x,y,z,w) := \begin{vmatrix} b_1(x,z) & b_1(x,w) \\ b_2(y,z) & b_2(y,w) \end{vmatrix} + \begin{vmatrix} b_2(x,z) & b_2(x,w) \\ b_1(y,z) & b_1(y,w) \end{vmatrix}.$$

P.2.4 Note that the Kulkarni-Nomizu product is bilinear (thus it qualifies as a "product") and symmetric, $b_1 \wedge b_2 = b_2 \wedge b_1$, and that it satisfies the same (algebraic) identities as a curvature tensor:
1. $b_1 \wedge b_2(x,y,w,z) = b_1 \wedge b_2(y,x,z,w) = -b_1 \wedge b_2(x,y,z,w)$,
2. $b_1 \wedge b_2(x,y,z,w) = b_1 \wedge b_2(z,w,x,y)$,
3. $b_1 \wedge b_2(x,y,z,w) + b_1 \wedge b_2(y,z,x,w) + b_1 \wedge b_2(z,x,y,w) = 0$.

These statements are proven by straightforward computation.

With this, the transformation formula for the curvature tensor reads

P.2.5 Lemma. *If $\tilde{g} = e^{2u}g$, then the $(4,0)$-curvature tensor of $\tilde{g}$ is given by $\tilde{r} = e^{2u}(r - b_u \wedge g)$, with[5] the symmetric bilinear form*

$$b_u(x,y) := \mathrm{hess}\, u(x,y) - du(x)du(y) + \tfrac{1}{2}g(\mathrm{grad}\, u, \mathrm{grad}\, u)g(x,y).$$

---

[5] We use the following sign convention: $R(x,y)z = -[\nabla_x\nabla_y z - \nabla_y\nabla_x z - \nabla_{[x,y]}z]$.
Also note that, $\tilde{b}_{-u} = -b_u$ so that changing the metric conformally and then changing it back has no effect on the curvature tensor — as it should be.

P.2.6 *Proof.* Write $B = \tilde{\nabla} - \nabla$; then,

$$\tilde{R}(x,y)z - R(x,y)z$$
$$= -(\nabla_x B)(y,z) + (\nabla_y B)(x,z) - B(x, B(y,z)) + B(y, B(x,z)).$$

With $B(y,z) = du(y)z + du(z)y - g(y,z)\mathrm{grad}\, u$,

$$g((\nabla_x B)(y,z), w)$$
$$= \mathrm{hess}\, u(x,y)g(z,w) + \mathrm{hess}\, u(x,z)g(y,w) - \mathrm{hess}\, u(x,w)g(y,z),$$

$$g(B(x, B(y,z)), w)$$
$$= \quad [du(x)du(y)g(z,w) - g(x,y)du(z)du(w)]$$
$$\quad + [du(x)du(z)g(y,w) + 2g(x,w)du(y)du(z) - g(x,z)du(y)du(w)]$$
$$\quad - g(x,w)g(y,z)g(\mathrm{grad}\, u, \mathrm{grad}\, u).$$

Adding the respective terms up, the claim follows.   ◁

P.2.7 Corollary. *For the sectional curvatures, a conformal change of the metric $g \to \tilde{g} = e^{2u}g$ yields*[6] $\tilde{K}(x \wedge y) = e^{-2u}(K(x \wedge y) - \mathrm{tr}\, b_u|_{x \wedge y})$.
If $\dim M = 2$, *then* $\tilde{K} = e^{-2u}(K - \Delta u)$.

P.2.8 *Proof.* With an orthonormal basis (e_1, e_2) of $x \wedge y$ with respect to g, we have

$$
\begin{aligned}
\tilde{K}(x \wedge y) &= \frac{\tilde{r}(e_1, e_2, e_1, e_2)}{\tilde{g}(e_1, e_1)\tilde{g}(e_2, e_2)} \\
&= e^{-2u}\big(r(e_1, e_2, e_1, e_2) - b_u(e_1, e_1) - b_u(e_2, e_2)\big) \\
&= e^{-2u}\big(K(x \wedge y) - \mathrm{tr}\, b_u|_{x \wedge y}\big).
\end{aligned}
$$

If $\dim M = 2$, then $\mathrm{tr}\, b_u = \mathrm{tr}\, \mathrm{hess}\, u = \Delta u$.   ◁

P.3 The Weyl and Schouten tensors

From the transformation formula for the $(4,0)$-curvature tensor we see that a conformal change of metric effects only the "trace part" of the curvature tensor, while the trace-free part is just scaled. This suggests a decomposition of the curvature tensor into a "trace part" and a "trace-free part" that is adapted to conformal geometry (cf., [122] or [173]).

P.3.1 Definition. *Let (M,g) be an n-dimensional Riemannian manifold. Then, the Weyl and Schouten tensors are defined as*

$$
\begin{aligned}
s &:= \tfrac{1}{n-2}\left(ric - \tfrac{scal}{2(n-1)}g\right) &&\text{(Schouten tensor),} \\
w &:= r - s \wedge g &&\text{(Weyl tensor).}
\end{aligned}
$$

---

[6] For the moment, "$x \wedge y$" just denotes the 2-plane spanned by the two vectors x and y.

P.3.2 The Weyl tensor is "trace-free": With an orthonormal basis e_i at p, we have $\sum_i w(x, e_i, y, e_i) = ric(x, y) - (n-2)s(x, y) - \operatorname{tr} s \cdot g(x, y) = 0$ indeed. On the other hand, if $r = w + s \wedge g$ with a "trace-free" tensor w, then

$$
\begin{aligned}
ric &= (n-2)s + \operatorname{tr} s \cdot g \\
scal &= 2(n-1)\operatorname{tr} s,
\end{aligned}
$$

which implies that s is the Schouten tensor and w is the Weyl tensor (cf., [122]).

The transformation behavior of the curvature tensor is now easily described by that of the Weyl and Schouten tensors:

P.3.3 Lemma. *Under a conformal change $g \to \tilde{g} = e^{2u}g$ of the metric, the Weyl and Schouten tensors transform as follows:*

$$
\begin{aligned}
w &\to \tilde{w} = e^{2u}w\,, \\
s &\to \tilde{s} = s - b_u.
\end{aligned}
$$

In particular, the $(3,1)$-Weyl tensor W, $w = g(W., .)$, is invariant under conformal changes of the metric; therefore, it is also called the "conformal curvature tensor."

P.4 Conformal flatness

As a first application of our notions, we want to discuss conformal flatness of a Riemannian manifold and give, in the following section, a criterion for conformal flatness in terms of the Weyl and Schouten tensors.

P.4.1 Definition. *A Riemannian manifold (M, g) is called conformally flat if, for any point $p \in M$, there is a neighbourhood U of p and some function $u : U \to \mathbb{R}$ so that the (local) metric $\tilde{g} = e^{2u}g$ is flat on U.*

P.4.2 Note that some authors define conformal flatness globally; that is, the conformal structure associated with the given metric is required to contain a (global) flat representative.

P.4.3 Lemma. *A manifold (M, g) is conformally flat if and only if, around each point $p \in M$, there exist conformal coordinates; that is, there is a coordinate map $x : M \supset U \to \mathbb{R}^n$ and a function $u : U \to \mathbb{R}$ such that the metric $g|_U = e^{2u} \sum_{i=1}^{n} dx_i^2$.*

P.4.4 *Proof.* Obviously, a metric of the form $g = e^{2u} \sum_{i=1}^{n} dx_i^2$ is conformally flat; on the other hand, to a flat metric $e^{-2u}g$, there always exist local coordinates x such that $e^{-2u}g = \sum_{i=1}^{n} dx_i^2$ (any flat manifold is locally isometric to $\mathbb{R}^n$, via the exponential map; cf., [122]). ◁

P.4.5 Any 1-dimensional Riemannian manifold (M^1, g) is conformally flat (actually: flat — there exist arc length parameters).

P.4.6 Theorem (Lichtenstein [183]). *Any 2-dimensional Riemannian manifold (M^2, g) is conformally flat.*[7]

P.4.7 *Proof.* According to the transformation formula for the Gauss curvature from §P.2.7, we have to (locally) solve the partial differential equation $K = \Delta u$ for u. Then the (locally defined) metric $\tilde{g} := e^{2u} g$ will be flat.

P.4.8 Here is a sketch of a proof for the existence of local solutions:

On a 2-torus (T^2, g), the total curvature $\langle 1, K \rangle = \int_{T^2} 1 \cdot K \, dA = 0$ (Gauss-Bonnet theorem), that is, $K \in \ker^\perp \Delta = \{u : T^2 \to \mathbb{R} \,|\, u \equiv const\}$. By the Hodge decomposition theorem [298], $\ker^\perp \Delta = \mathrm{im}\Delta$ in $C^\infty(T^2, \mathbb{R})$; thus there exists a (global!) solution $u \in C^\infty(T^2)$ of $\Delta u = K$.

On an arbitrary (M^2, g) choose coordinates $x : M \supset U \to \mathbb{R}^2$ around a point $p \in M$ with $x(U) \subset (0, 1)^2$ and introduce a metric $\hat{g}$ on $T^2 = \mathbb{R}^2/\mathbb{Z}^2$ such that $x^* \hat{g} = g$ on some neighborhood $\hat{U} \subset U$ of p (partition of unity). By the above argument, there exists a function $\hat{u} : T^2 \to \mathbb{R}$ such that $e^{2\hat{u}} \hat{g}$ is flat on T^2; since $x : (\hat{U}, g) \to (x(\hat{U}), \hat{g})$ is an isometry, the metric $e^{2u} g|_{\hat{U}}$, with $u = \hat{u} \circ x$, is then flat on $\hat{U}$. ◁

P.4.9 The theorem of §P.4.6 establishes a relation between 2-dimensional manifolds equipped with a conformal structure and Riemann surfaces, that is, with 1-dimensional complex manifolds: Given isothermal (conformal) coordinates (x, y) around some $p \in M$ on (M^2, g), complex coordinates can be defined by $z := x + iy$. The transition functions of such complex coordinates are (as angle-preserving maps) either holomorphic or antiholomorphic; restricting to holomorphic (orientation preserving: M has to be orientable) transition functions, an atlas of complex coordinates is obtained. This identifies 2-dimensional (orientable) Riemannian manifolds as complex curves.

On the other hand, the pullback of the multiplication by i,

$$J_p : T_p M \to T_p M, \quad d_p z \circ J_p = i \cdot d_p z,$$

provides, if it can be extended to a global tensor field (that is, if M is orientable), 90° rotations (isometries with $J_p^2 = -id$) on each tangent space: an "almost complex structure." The above theorem states that any almost complex structure J on M (as it defines a conformal structure) comes from complex coordinate charts (that is, it is a "complex structure").

---

[7] Compare also [182]. Concerning the realization of (compact) Riemann surfaces as submanifolds of Euclidean 3-space, see [124] and [125].

P.4.10 By using the concept of an almost complex structure, another (more constructive) proof for the existence of isothermal coordinates can be given (cf., [260]):

Suppose we are given[8] a harmonic function $x_1 : M^2 \supset U \to \mathbb{R}$ on a simply connected neighborhood U around $p \in M$ such that $dx_1 \neq 0$; then we find[9]

$$d(\star dx_1) := d(dx_1 \circ J) = dg(\operatorname{grad} x_1, J.) = -\Delta x_1 \, dA = 0,$$

so that there is a function $x_2 : U \to \mathbb{R}$ with $dx_2 = \star dx_1$, the "conjugate harmonic function." Then, $(x_1, x_2) : U \to \mathbb{R}^2$ define isothermal coordinates since $\frac{\partial}{\partial x_2} = -J\frac{\partial}{\partial x_1}$ yields $0 = g(\frac{\partial}{\partial x_1}, \frac{\partial}{\partial x_2})$ and $g(\frac{\partial}{\partial x_2}, \frac{\partial}{\partial x_2}) = g(\frac{\partial}{\partial x_1}, \frac{\partial}{\partial x_1})$.

P.5 The Weyl-Schouten theorem

In higher dimensions, the Weyl-Schouten theorem gives a characterization of conformally flat Riemannian manifolds (cf., [81], [82], [300], [301], [252]):

P.5.1 Theorem (Weyl-Schouten). *A Riemannian manifold (M^n, g) of dimension $n \geq 3$ is conformally flat if and only if*
- *the Schouten tensor is a Codazzi tensor, $(\nabla_x s)(y, z) = (\nabla_y s)(x, z)$, in the case $n = 3$; and*
- *the Weyl tensor vanishes, $w \equiv 0$, in the case $n > 3$.*

P.5.2 *Proof.* (cf., [173]). By the transformation formula for the curvature tensor from §P.2.5, (M, g) is conformally flat if and only if $w = 0$ and there exists (locally) a function u with $s = b_u$.

We divide the proof into three steps:

P.5.3 Step 1. If $n = 3$, then $w = 0$, by algebra.

Let (e_1, e_2, e_3) be an orthonormal basis of some T_pM; then[10]

$$r(e_i, e_j, e_i, e_j) = ric(e_i, e_i) + ric(e_j, e_j) - \tfrac{1}{2} scal = (s \wedge g)(e_i, e_j, e_i, e_j)$$

---

[8] Of course, to obtain a complete proof, one would have to show that such nonconstant harmonic functions (locally) always exist.

[9] Note that $\nabla J = 0$ because, for any vector field v, $g((\nabla J)v, v) = 0$ and $g((\nabla J)v, Jv) = 0$.

[10] The computation is best done from right to left.
Another, more conceptual proof was pointed out by Konrad Voss: Let $Sym(V)$ denote the space of symmetric bilinear forms on a vector space V, and note that $Sym(\Lambda^2 V)$ is the space of algebraic curvature tensors on V if $\dim V = 3$ (the first Bianchi identity follows from the other symmetries in this case). Now, observe that

$$Sym(V) \ni s \mapsto r(s) = s \wedge g \in Sym(\Lambda^2 V) \quad \text{and} \quad Sym(\Lambda^2 V) \ni r \mapsto s(r) \in Sym(V)$$

are linear maps. Since $Sym(V) \ni s \mapsto s(r(s)) = s \in Sym(V)$ is the identity, the linear map $s \mapsto r(s)$ injects and hence is an isomorphism since $\dim Sym(V) = \dim Sym(\Lambda^2 V)$ in the case $\dim V = 3$.

for $i \neq j$, showing that w vanishes as all its sectional curvatures do.

P.5.4 Step 2. If $n > 3$, then $w = 0$ implies $(\nabla_x s)(y, z) = (\nabla_y s)(x, z)$; this is a consequence of the second Bianchi identity.

If $r = s \wedge g$, then $(\nabla_x r) = (\nabla_x s) \wedge g$; thus,

$$0 = [(\nabla_x s) \wedge g](y, z, v, w) + [(\nabla_y s) \wedge g](z, x, v, w) + [(\nabla_z s) \wedge g](x, y, v, w).$$

Contraction with respect to x and w yields

$$
\begin{aligned}
0 &= [(\nabla_z s)(y, v) - (\nabla_y s)(z, v) + g(y, v)(\text{div } s)(z) - g(z, v)(\text{div } s)(y)] \\
&\quad + [(n-2)(\nabla_y s)(z, v) + \text{tr}(\nabla_y s)g(z, v)] \\
&\quad - [(n-2)(\nabla_z s)(y, v) + \text{tr}(\nabla_z s)g(y, v)] \\
&= (n-3)[(\nabla_z s)(y, v) - (\nabla_y s)(z, v)],
\end{aligned}
$$

since[11] $\text{div } s - d(\text{tr} s) = \frac{1}{n-2}(\text{div } ric - \frac{1}{2}d\, scal) = 0$.

P.5.5 Step 3. $(\nabla_x s)(y, z) = (\nabla_y s)(x, z)$ is the integrability condition of the partial differential equation $b_u = s$ for u when $w = 0$.

With the $(1, 1)$-Schouten tensor S, $s = g(., S.)$, and the ansatz $v = \text{grad} u$, the equation $b_u = s$ is equivalent to the system

$$
\begin{aligned}
du &= g(v, .), \\
\nabla v &= S + g(v, .)v - \tfrac{1}{2}g(v, v)\, id.
\end{aligned}
$$

Because $A := S + g(v, .)v - \frac{1}{2}g(v, v)\, id$ is symmetric, any solution v of the second equation is locally a gradient.

The integrability condition for the second equation[12] reads

$$
\begin{aligned}
0 &= (\nabla_x A)(y) - (\nabla_y A)(x) + R(x, y)v \\
&= (\nabla_x S)(y) - (\nabla_y S)(x) + W(x, y)v,
\end{aligned}
$$

since $W(x, y)v = R(x, y)v - (g(Sx, v)y - g(y, v)Sx) + (g(Sy, v)x - g(x, v)Sy)$. Thus, assuming $W = 0$, the claim follows. ◁

P.5.6 Example. The spaces of constant curvature $K \equiv c = const$ are conformally flat: Here, the curvature tensor $R(x, y)z = c[g(z, x)y - g(z, y)x]$, that is, $r = \frac{c}{2}g \wedge g$. Hence, the Weyl tensor vanishes, $w = 0$, and the Schouten tensor $s = \frac{c}{2}g$ clearly is a Codazzi tensor.

---

[11] This can also be obtained by contracting again with respect to z and v.

[12] In local coordinates, the integrability condition for the second equation $\partial_i v^k = A_i^k - \Gamma_{im}^k v^m$ reads $0 = (\partial_i \partial_j - \partial_j \partial_i)v^k = (\partial_i A_j^k - \partial_j A_i^k + \Gamma_{im}^k A_j^m - \Gamma_{jm}^k A_i^m) + R_{mij}^k v^m$.

P.5.7 The Weyl-Schouten theorem shows that the condition of being conformally flat is a second-order condition on the metric in the case $n > 3$, but it is a third-order condition in the case $n = 3$. This makes the case $n = 3$ different from higher dimensions, and it becomes more complicated in some respect: For example, a Riemannian manifold of dimension $n > 3$ is conformally flat if and only if to any orthogonal basis of a tangent space there exists a (local) orthogonal coordinate system with the Gaussian basis fields at the point pointing along the prescribed directions,[13] that is, if and only if there is (locally) an n-fold orthogonal system of hypersurfaces in any position. On the other hand, such orthogonal coordinate systems in arbitrary positions exist on any 3-dimensional Riemannian manifold (see [252] or [200]).

Another example is the study of conformally flat hypersurfaces in the conformal $(n + 1)$-sphere S^{n+1}: These are well understood in the case of dimension $n > 3$ (see [61], [102], [222]), while 3-dimensional conformally flat hypersurfaces in S^4 are still not fully classified. We will come back to this point later; there, the following example will be important.

P.5.8 Example. Consider a 3-dimensional product manifold $\hat{M}^3 = M^2 \times I$ with a product metric $g + ds^2$, where (M, g) is a Riemannian surface with Gaussian curvature K and I is an interval with parameter s. Because covariant differentiation on $(\hat{M}, g + ds^2)$ splits and is trivial on the 1-dimensional component, we have

$$\hat{r} = \pi^* r = \pi^* \tfrac{1}{2} K g \wedge g = \tfrac{1}{2} K (g - ds^2) \wedge (g + ds^2),$$

where $\pi : M \times I \to M$ denotes the projection onto the first component. The last equality follows from $ds^2 \wedge ds^2 = 0$ and the symmetry of the Kulkarni-Nomizu product (see §P.2.3). Consequently, $\hat{s} = \tfrac{1}{2} K (g - ds^2)$ is the Schouten tensor of $\hat{M}^3$. Since $(g - ds^2)$ is clearly parallel, $\hat{\nabla}(g - ds^2) = 0$, the covariant derivative of $\hat{s}$ is given by $(\hat{\nabla}_{x + \frac{\partial}{\partial s}} \hat{s}) = \tfrac{1}{2} dK(x) \cdot (g - ds^2)$, where $x \in TM$. Thus, by the Weyl-Schouten theorem, $(M \times I, g + ds^2)$ is conformally flat if and only if $K \equiv const$:

$$0 = (\hat{\nabla}_{\frac{\partial}{\partial s}} \hat{s})(x, \tfrac{\partial}{\partial s}) - (\hat{\nabla}_x \hat{s})(\tfrac{\partial}{\partial s}, \tfrac{\partial}{\partial s}) = \tfrac{1}{2} dK(x).$$

---

[13] One direction of the proof is trivial; the other relies on the fact that the Weyl tensor vanishes if and only if $r(e_1, e_2, e_3, e_4) = 0$ for four (different) orthogonal vectors, and on the integrability of the coordinate hypersurfaces.

P.6 Submanifolds

For a submanifold $M \subset \hat{M}$ of a Riemannian manifold $(\hat{M}, \hat{g})$, the induced covariant derivative ∇ (which turns out to be the Levi-Civita connection of the induced metric $I = g = \hat{g}|_{TM \times TM}$), the second fundamental form $I\!I$, the Weingarten tensor fields A_n, and the normal connection $\nabla^{\perp}$ are defined through the orthogonal decomposition $T_p\hat{M} = T_pM \oplus N_pM$:

$$
\begin{aligned}
\hat{\nabla}_x y &= \nabla_x y &+& \quad I\!I(x,y), \\
\hat{\nabla}_x n &= -A_n x &+& \quad \nabla^{\perp}_x n.
\end{aligned}
$$

With these notations, the compatibility conditions (Gauss-Codazzi equations) read as follows:

$$
\begin{aligned}
[\hat{R}(x,y)z]^T &= R(x,y)z + [A_{I\!I(y,z)}x - A_{I\!I(x,z)}y] && \text{(Gauss eqn)}, \\
[\hat{R}(x,y)z]^{\perp} &= -[(\nabla^{\perp}_x I\!I)(y,z) - (\nabla^{\perp}_y I\!I)(x,z)] && \text{(Codazzi eqn)}, \\
[\hat{R}(x,y)n]^T &= [(\nabla_x A)_n y - (\nabla_y A)_n x] && \text{(Codazzi eqn)}, \\
[\hat{R}(x,y)n]^{\perp} &= R^{\perp}(x,y)n - [I\!I(A_n x, y) - I\!I(x, A_n y)] && \text{(Ricci eqn)}.
\end{aligned}
$$

Our next task is to determine the transformation behavior of these quantities when the ambient space's metric changes conformally:

P.6.1 Lemma. *Under a conformal change $\hat{g} \to \tilde{g} = e^{2u}\hat{g}$ of the ambient manifold's metric, the second fundamental form, the Weingarten tensor, and the normal connection change as follows:*

$$
\begin{aligned}
I\!I(x,y) &\to \tilde{I\!I}(x,y) &=& \quad I\!I(x,y) &-& \quad I(x,y)\widehat{[\text{grad } u]}^{\perp}, \\
A_n x &\to \tilde{A}_n x &=& \quad A_n x &-& \quad du(n)x, \\
\nabla^{\perp}_x n &\to \tilde{\nabla}^{\perp}_x n &=& \quad \nabla^{\perp}_x n &+& \quad du(x)n.
\end{aligned}
$$

Moreover, the normal curvature $R^{\perp}(x,y)n$ is invariant.

P.6.2 Proof. The formulas follow by a straightforward computation from the transformation formula for the ambient Levi-Civita connection $\hat{\nabla}$ (see §P.2.1).

For the last statement, we employ the Ricci-equation: With

$$(b_u \wedge \hat{g})(x,y,n,\hat{n}) = 0$$

and

$$
\begin{aligned}
&\tilde{g}(\tilde{A}_n x, \tilde{A}_{\hat{n}} y) \\
&= e^{2u}\{I(A_n x, A_{\hat{n}} y) - I(A_n x, y)\partial_{\hat{n}} u - I(x, A_{\hat{n}} y)\partial_n u + I(x,y)\partial_n u \, \partial_{\hat{n}} u\},
\end{aligned}
$$

we deduce

$$
\begin{aligned}
\tilde{r}^{\perp}(x,y,n,\hat{n}) &= \tilde{r}(x,y,n,\hat{n}) + [\tilde{g}(\tilde{A}_n x, \tilde{A}_{\hat{n}} y) - \tilde{g}(\tilde{A}_{\hat{n}} x, \tilde{A}_n y)] \\
&= e^{2u}\{\hat{r}(x,y,n,\hat{n}) + [I(A_n x, A_{\hat{n}} y) - I(A_{\hat{n}} x, A_n y)]\} \\
&= e^{2u} r^{\perp}(x,y,n,\hat{n}),
\end{aligned}
$$

where we use $\hat{g}(I\!I(A_n x, z), \hat{n}) = g(A_n x, A_{\hat{n}} z)$. This proves the claim.   ◁

P.6.3 At this point it is straightforward to derive various conformal invariants for submanifolds. Later, we will occasionally be concerned with certain "simple" conformal (or Möbius) invariants.[14] However, here we do *not* attempt to study invariants of submanifolds systematically or to determine a (complete) set of invariants that describes a submanifold uniquely in the geometry considered (that is, up to conformal or Möbius transformations in our case) in the sense of a Bonnet-type "fundamental theorem" for submanifolds in conformal or Möbius geometry. But, we will recover some of the discussed conformal invariants discussed in the following paragraphs by geometric considerations later — which will in some cases naturally lead to certain special classes of surfaces in Möbius geometry as, for example, Willmore surfaces as Möbius analogs of Euclidean minimal surfaces.

The systematic study of conformal (or Möbius) invariants has a long tradition in differential geometry, and for readers interested in getting more information on the subject some references may prove useful, even if the list is far from being complete: For surfaces in a 3-dimensional (conformally flat or flat) ambient space, various results were obtained around 1900 by A. Tresse [283], P. Calapso [54] (compare also [53], [55], and the survey [59] by his son, R. Calapso), R. Rothe [247], and K. Ogura [210], as well as G. Fubini [120] (cf., [121]). A short survey on the state of the art around that time was given by L. Berwald [14]. Relevant work by G. Thomsen [277], [278] and by É. Cartan [61], [62] will be considered in the course of the present text. In 1944, A. Fialkov [115] gave a rather exhaustive treatment of conformal invariants of submanifolds from the Riemannian point of view; for the special cases of curves and (2-dimensional) surfaces, the reader may also consider the two papers [114] and [116]. More recently, there is work

---

[14] We distinguish between "conformal invariants" and "Möbius invariants": The latter are invariant under Möbius (or conformal) transformations, that is, under those conformal changes of the (ambient) metric that are caused by pulling back the ambient metric with a Möbius transformation, whereas the former are invariant under arbitrary conformal changes. For example, as we will see in the following section, the "distance spheres" on S^n are Möbius invariant — to any (hyper)sphere of S^n there is a "center"; however, when changing the metric on (part of) S^n conformally (so that it becomes flat, for example) not all hyperspheres will remain distance spheres, as some of them will lose the property of having a center. Thus, the notion of a distance sphere is not conformally invariant in general. Below we shall discuss briefly another example of a Möbius invariant for submanifolds that is not invariant under general conformal changes of the ambient metric.

by R. Sulanke [265], [266], M. Akivis and V. Goldberg [5] (see also [4]), G. Preissler [225], and C.-P. Wang [296], [297], as well as by F. Burstall, F. Pedit, and U. Pinkall [48].

P.6.4 Thus let us derive some simple *conformal invariants*. Besides the induced conformal structure $[I]$ on M, which is clearly a conformal invariant, the most prominent conformal invariant is the (normal bundle—valued) trace-free second fundamental form:

$$I\!\!I - H \cdot I \quad \text{where} \quad H := \tfrac{1}{m}\mathrm{tr}_I I\!\!I = \tfrac{1}{m}\sum_{i=1}^m I\!\!I(e_i, e_i)$$

denotes the mean curvature vector field; herein, $(e_1, \ldots, e_m)$ denotes an orthonormal basis. Accordingly, the normalized length h of the trace-free second fundamental form,

$$\begin{aligned} h^2 &:= \tfrac{1}{m}\hat{g}(I\!\!I - HI, I\!\!I - HI) \\ &= \tfrac{1}{m}\sum_{i=1}^m \hat{g}((I\!\!I - HI)(e_i, e_j), (I\!\!I - HI)(e_i, e_j)), \end{aligned}$$

changes opposite to the induced metric: If $\hat{g} \to e^{2u}\hat{g}$, then $h^2 \to e^{-2u}h^2$. Therefore, as soon as $I\!\!I - HI \neq 0$, so that $h \neq 0$, an invariant representative of the induced conformal structure $[I]$ (the "conformal (surface) measure tensor" in [115]) can be defined by[15]

$$I_{conf} := h^2 \cdot I.$$

In case $M \subset \hat{M}$ is a hypersurface, one usually considers the second fundamental form (and hence the mean curvature) to be real-valued,

$$I\!\!I_{\mathbb{R}} = \hat{g}(I\!\!I, n) \quad \text{and} \quad H_{\mathbb{R}} = \hat{g}(H, n),$$

with a unit normal field n. As $n \to \tilde{n} = e^{-u}n$ when $\hat{g} \to \tilde{g} = e^{2u}\hat{g}$, a conformally invariant normal field may be defined by

$$n_{conf} := h^{-1}n,$$

with a metric unit normal field n; hence a real-valued "conformal second fundamental form" will be given by[16]

$$\begin{aligned} I\!\!I - HI &= (I\!\!I - HI)_{conf} \cdot n_{conf} \\ \Leftrightarrow \quad (I\!\!I - HI)_{conf} &= h^2\hat{g}(I\!\!I - HI, n_{conf}) = h\hat{g}(I\!\!I - HI, n). \end{aligned}$$

---

[15] In fact, h can be used to define an invariant representative $h^2\hat{g}$ of the ambient space's conformal structure $[\hat{g}]$ "along" $M \subset \hat{M}$ (cf., [115]): this "conformal ambient metric along M" ("conformal (space) measure tensor" in [115]) will prove useful below.

[16] Here we use the "conformal ambient metric along M" as discussed in the previous footnote.

With this we recover Fubini's [120] conformally invariant fundamental forms for surfaces: As $h^2 = (H_{\mathbb{R}}^2 - K)^2$ with the (real) mean and (extrinsic) Gauss curvature $H_{\mathbb{R}}$ and K of $M^2 \subset \hat{M}$, we find

$$I_{conf} = (H_{\mathbb{R}}^2 - K)\, I \quad \text{and} \quad (I\!I - HI)_{conf} = \sqrt{H_{\mathbb{R}}^2 - K}\,(I\!I_{\mathbb{R}} - H_{\mathbb{R}}I).$$

We will eventually come across these two invariant "conformal fundamental forms" by means of a different geometric construction later; in particular, the "conformal metric" I_{conf} will play a distinguished role in various contexts.

Clearly, other conformal invariants can then be derived from the above conformal invariants; for example, from the conformally invariant metric I_{conf} on M one can derive a conformally invariant Levi-Civita connection[17] as well as a related curvature tensor (cf., [115]).

P.6.5 To demonstrate the difference between *Möbius invariants* and conformal invariants, we briefly discuss Wang's "Möbius form" (see [297] or [181]): for a submanifold in S^n (or any other space of constant curvature) with $h \neq 0$, this normal bundle—valued 1-form can be defined by[18]

$$\omega(x) := \tfrac{1}{h^2}(\nabla_x^\perp H + (I\!I - HI)(\operatorname{grad}\ln h, x)).$$

A lengthy but straightforward computation[19] then shows that, for $\xi \in NM$,

$$h^2 \hat{g}(\omega(x) - \tilde{\omega}(x), \xi) = b_u(x, \xi)$$

when changing the ambient metric conformally, $\hat{g} \to \tilde{g} = e^{2u}\hat{g}$. Thus ω stays invariant under conformal changes of the ambient metric for which the normal and tangent bundles of the submanifold are orthogonal with respect to the symmetric bilinear form b_u. However, it will not be invariant under arbitrary conformal changes of the ambient metric.

Now we consider the conformal change $\hat{g} \to \tilde{g} = e^{2u}\hat{g} = \mu^*\hat{g}'$ induced by a conformal transformation μ of the space $(\hat{M}, \hat{g})$ of constant curvature c into a space $(\hat{M}', \hat{g}')$ of (possibly different) constant curvature $\tilde{c}$: Considering μ as an isometry, $\mu : (\hat{M}, \tilde{g}) \to (\hat{M}', \hat{g}')$, we infer that

$$e^{2u}(\tfrac{c}{2}\hat{g} - b_u) \wedge \hat{g} = e^{2u}(\hat{r} - b_u \wedge \hat{g}) = \tilde{r} = \mu^*\hat{r}' = \tfrac{\tilde{c}}{2}\mu^*\hat{g}' \wedge \mu^*\hat{g}' = \tfrac{\tilde{c}}{2}e^{4u}\hat{g} \wedge \hat{g},$$

---

[17] However, even though the conformal metric can be "extended" to a conformal ambient metric along M, this connection does not extend because we are lacking the derivative of h in normal directions.

[18] The form given here differs from the one given by Wang by a constant.

[19] We use the conformally invariant "conformal ambient metric" $h^2\hat{g}$ in this computation.

so that $b_u = \frac{c-c'e^{2u}}{2}\hat{g} = \frac{1}{2}(c\hat{g} - \tilde{c}\hat{g})$. Hence Wang's Möbius form is indeed invariant under such conformal changes of the ambient metric, showing two things: First, ω is a Möbius invariant, and, second, it does not depend on the value of the constant curvature of the ambient space. In particular, it does not matter whether we compute ω for a submanifold in the sphere S^n or for, say, its stereographic projection in Euclidean ambient space $\mathbb{R}^n$.

For hypersurfaces Wang's Möbius form can be expressed in terms of a real-valued quantity by using the conformal normal field introduced in §P.6.4:

$$\omega =: n_{conf}\omega_{\mathbb{R}} = \frac{n}{h}\left\{\frac{dH_{\mathbb{R}}}{h} + \frac{1}{h^2}(I\!I_{\mathbb{R}} - H_{\mathbb{R}}I)(\mathrm{grad}\,h,.)\right\};$$

in the case of a surface, we introduce curvature line coordinates[20] (x_1, x_2) so that (here we use the notation of the classical authors)

$$I = Edx_1^2 + Gdx_2^2 \quad \text{and} \quad I\!I_{\mathbb{R}} = k_1Edx_1^2 + k_2Gdx_2^2,$$

and the real Möbius form $\omega_{\mathbb{R}}$ reduces[21] to

$$\omega_{\mathbb{R}} = \frac{1}{h}\left\{\frac{\partial k_1}{\partial x_1}dx_1 + \frac{\partial k_2}{\partial x_2}dx_2\right\}.$$

This form of $\omega_{\mathbb{R}}$ reveals a close relationship with another Möbius invariant for surfaces in Euclidean 3-space that was first given[22] by Weingarten [299] (see also [88] and [247]):

$$\Omega := \frac{dh}{h} + \frac{1}{h^2}(I\!I_{\mathbb{R}} - H_{\mathbb{R}}I)(\mathrm{grad}\,H_{\mathbb{R}},.) = \frac{1}{h}\left\{\frac{\partial k_1}{\partial x_1}dx_1 - \frac{\partial k_2}{\partial x_2}dx_2\right\}.$$

In fact, the two forms $\omega_{\mathbb{R}}$ and Ω give rise to a whole pencil $t\omega_{\mathbb{R}} + (1-t)\Omega$, $t \in \mathbb{R}$, of Möbius invariant 1-forms. To see the Möbius invariance of Ω we can directly use the Möbius invariance of $\omega_{\mathbb{R}}$ and the conformal invariance of the curvature directions, given below: Since

$$\Omega(\tfrac{\partial}{\partial x_1}) = \omega_{\mathbb{R}}(\tfrac{\partial}{\partial x_1}) \quad \text{and} \quad \Omega(\tfrac{\partial}{\partial x_2}) = -\omega_{\mathbb{R}}(\tfrac{\partial}{\partial x_2}),$$

---

[20] Note that higher dimensional and/or codimensional submanifolds do not carry curvature line coordinates in general.

[21] This form of $\omega_{\mathbb{R}}$ lets us recover a classification result of Wang [297]: A surface in 3-space with vanishing Möbius form, $\omega = 0$, is a Dupin cyclide. We will come across Dupin cyclides various times in this text; they will appear as double channel surfaces in the following chapter, that is, as surfaces with two families of circular curvature lines.

[22] The form presented here differs from those given in the literature by constants. This 1-form appeared first in the context of isothermic surfaces — this will be discussed in greater detail later. Compare [65], [299], [164], [244], and [246].

and since our curvature line coordinates (x_1, x_2) are conformally invariant, we have

$$(\Omega - \tilde{\Omega})(\tfrac{\partial}{\partial x_1}) \;=\; (\omega_{\mathbb{R}} - \tilde{\omega}_{\mathbb{R}})(\tfrac{\partial}{\partial x_1}) \;=\; b_u(\tfrac{\partial}{\partial x_1}, n_{conf}) \;=\; 0,$$
$$(\Omega - \tilde{\Omega})(\tfrac{\partial}{\partial x_2}) \;=\; -(\omega_{\mathbb{R}} - \tilde{\omega}_{\mathbb{R}})(\tfrac{\partial}{\partial x_2}) \;=\; -b_u(\tfrac{\partial}{\partial x_2}, n_{conf}) \;=\; 0.$$

However, Rothe [247] derived a form of Ω that is indeed conformally invariant: By employing the Codazzi equations,

$$\tfrac{\partial k_1}{\partial x_2} + (k_1 - k_2)\tfrac{\partial}{\partial x_2}\ln\sqrt{E} = 0 \quad\text{and}\quad \tfrac{\partial k_2}{\partial x_1} - (k_1 - k_2)\tfrac{\partial}{\partial x_1}\ln\sqrt{G} = 0,$$

and $k_1 - k_2 = \pm 2h$ so that (formally) $d\ln h = d\ln(k_1 - k_2)$, we obtain

$$\pm\tfrac{1}{2}\Omega = d\ln(h\sqrt[4]{EG}) + \tfrac{1}{4}(\tfrac{\partial}{\partial x_1}\ln\tfrac{G}{E})dx_1 + \tfrac{1}{4}(\tfrac{\partial}{\partial x_2}\ln\tfrac{E}{G})dx_2.$$

P.6.6 As we will build our approach to Möbius geometry on the notion of sphere and sphere-preserving transformations (rather than conformal transformations) we will need the following lemma. Besides showing that the curvature lines, or curvature directions of a submanifold, (if they exist) are conformally invariant, it provides the basis for a discussion of spheres from a local, conformal geometry point of view — it will enable us to define the notion of a sphere without referring to a "center." And, in fact, we will obtain a more general notion of a sphere than the notion of a "distance sphere."

The following lemma follows directly from §P.6.1:

P.6.7 Lemma. *For fixed* $n \in N_pM$, *the eigenspaces of the Weingarten tensor* $A_n : T_pM \to T_pM$ *are conformally invariant. In particular, the notion of an umbilic[23]* $p \in M$, *that is,* $A_n = \lambda(n)\,id_{T_pM}$ *with a (linear) functional* $\lambda : N_pM \to \mathbb{R}$, *is invariant.*

P.7 Spheres

Using this last lemma, we can give an obviously conformally invariant definition of a (hyper)sphere; because completeness is not a conformal invariant, we have to encode the desired "maximality" of a sphere a little differently:

P.7.1 Definition. *A hypersphere* $S \subset \hat{M}$ *in a Riemannian manifold* $(\hat{M}, \hat{g})$ *of dimension* $n \geq 3$ *is a connected, totally umbilic hypersurface that is*

---

[23] Equivalently, an umbilic can be characterized by $h(p) = 0$ since $I\!I = \Sigma_j I(., A_{n_j}.)n_j$ for an orthonormal basis $(n_1, \dots, n_{n-m})$ of N_pM: Thus $I\!I_p = H(p)I_p$ for some vector $H(p) \in N_pM$ if and only if $A_n = \hat{g}(H(p), n)\,id_{T_pM} = \lambda(n)\,id_{T_pM}$.

"*maximal*" *in the sense that there is no strictly larger connected, totally umbilic hypersurface containing* S.

P.7.2 In particular:
  (i) this notion of "hyperspheres" in (M, g) depends only on the conformal structure $[g]$ defined by g; and,
  (ii) if $f : (M, g) \to (\tilde{M}, \tilde{g})$ is a conformal diffeomorphism, then it maps hyperspheres to hyperspheres.

P.7.3 Examples. On $S^n \subset \mathbb{R}^{n+1}$, the hyperspheres are the intersections of S^n with affine hyperplanes $H \subset \mathbb{R}^{n+1}$, $dist(H, 0) < 1$.

In $\mathbb{R}^n$, a "hypersphere" is either a hypersphere (in the usual sense) or a hyperplane: The stereographic projection (cf., §P.1.3) $\sigma(S) \subset \mathbb{R}^n$ of a hypersphere $S \subset S^n$ is a hypersphere or a hyperplane (if S contains the center of projection).

In hyperbolic space $H^n = (B(1), \frac{4}{(1-|p|^2)^2} dp^2)$ (see §P.1.3) the hyperspheres are the intersections of $B(1)$ with hyperspheres in $\mathbb{R}^n$. Note that (as in Euclidean space) not all hyperspheres in H^n are distance spheres: If $S \subset H^n$ touches the infinity boundary $S^{n-1} = \partial H^n$, then the orthogonal geodesics meet on the infinity boundary; if S intersects ∂H^n, then the orthogonal geodesics do not intersect at all. Also note that the hyperspheres intersecting $S^{n-1} = \partial H^n$ orthogonally are the hyperplanes of H^n.

P.7.4 It can be shown that a Riemannian manifold of dimension $n > 3$ has totally umbilic hypersurfaces in any position (that is, for any prescribed vector $v \in TM$ there exists locally an umbilic hypersurface orthogonal to that vector) if and only if the manifold is conformally flat[24] (see [252]). This might serve as a motivation why we will restrict to the conformal n-sphere as an ambient space soon, since "Möbius geometry" has the hypersphere-preserving transformations as its symmetry group.

P.7.5 Later we will wish to describe spheres of higher codimension as intersections of hyperspheres. Here, a key tool is the following well-known generalization of a theorem dealing with the intersection of two surfaces in Euclidean 3-space (cf., [260]):

P.7.6 Theorem (Joachimsthal). *Let* $M := M_1 \cap M_2$ *be the transversal intersection of two hypersurfaces of a Riemannian manifold* $(\hat{M}, \hat{g})$; *then two of the following imply the third:*
  (i) $\forall p \in M, n_1 \in N_p M_1 : A_{n_1}^{M_1} T_p M \subset T_p M$;
  (ii) $\forall p \in M, n_2 \in N_p M_2 : A_{n_2}^{M_2} T_p M \subset T_p M$;
  (iii) M_1 *and* M_2 *intersect at a constant angle.*

---

[24] This is not true in dimension $n = 3$, where totally umbilic (hyper)surfaces always exist.

P.7.7 *Proof.* $0 = \partial_x \hat{g}(n_1, n_2) + \hat{g}(n_1, A^{M_2}_{n_2} x) + \hat{g}(A^{M_1}_{n_1} x, n_2)$ for $x \in T_p M$ and unit normal fields n_i of M_i, defined near p. $\triangleleft$

P.7.8 As an immediate consequence, hyperspheres intersects at a constant angle. In particular, if a family of hyperspheres $S_1, \ldots, S_k$ intersect transversally at *one* point $p \in M := \bigcap_{i=1}^{k} S_i$, then it intersects transversally along the intersection M, which is consequently ensured to be a submanifold.

P.7.9 Lemma. *The transversal intersection $M := \bigcap_{i=1}^{k} S_i$ of hyperspheres S_i, $i = 1, \ldots, k$, is a connected, totally umbilic submanifold with flat normal bundle.*

P.7.10 *Proof.* Let n_i denote (local) unit normal fields of S_i. Then, since the S_i's are totally umbilic, $\hat{\nabla}_x n_i = -\lambda_i x$ for $x \in T_p M = \bigcap_{i=1}^{k} T_p S_i$, with suitable functions λ_i. In particular, the n_i's are parallel (local) sections of the normal bundle NM.

Since[25] $N_p M = \mathrm{span}\{n_i\}$, any normal field can be written in terms of the n_i's, $n = \sum_{i=1}^{k} a_i n_i$. Then $\hat{\nabla}_x n = -\lambda(n)x + \sum_{i=1}^{k} (\partial_x a_i) n_i$ with the 1-form $\lambda(n) := \sum_{i=1}^{k} \lambda_i a_i$.

Thus, M is totally umbilic and, because it has a (local) basis of parallel normal fields, has flat normal bundle. $\triangleleft$

P.7.11 In Euclidean space, it is an easy exercise to prove the converse: Any totally umbilic submanifold $M \subset \mathbb{R}^n$, $m := \dim M > 1$, with flat normal bundle lies in the transversal intersection of $n - m$ hyperspheres (or hyperplanes, which are also hyperspheres in our more general sense). Namely, using the Codazzi equation, $(\nabla_x A)_n y = (\nabla_y A)_n x$, we obtain $(\nabla_x^\perp \lambda)(n) \cdot y = (\nabla_y^\perp \lambda)(n) \cdot x$, where λ denotes the 1-form on NM given by $A_n x = \lambda(n) \cdot x$. Consequently, for any parallel normal field n, the function $\lambda(n)$ is constant. Now, either $\lambda(n) = 0$ and n is constant (in which case M lies in the hyperplane with normal n) or $\lambda(n) \neq 0$ and $p \mapsto p + \frac{1}{\lambda(n_p)} n_p$ is constant, giving the center of a hypersphere that contains M. Since M has flat normal bundle, there is (locally) a basis of parallel normal fields to get $n - m$ hyperspheres that contain the submanifold, because the constancy of n or $p \mapsto p + \frac{1}{\lambda(n_p)} n_p$, respectively, extends to all of M if M is connected.

P.7.12 By stereographic projection, this result carries over to submanifolds of S^n since the notions of flatness of the normal bundle and total umbilicity are conformally invariant. Note that, in S^n, the distinction between hyperspheres and hyperplanes becomes obsolete. By conformal invariance again,

---

[25] At this point, we use the transversality assumption, so that the n_i's form a local basis.

the result is valid for all ambient manifolds that are (globally) conformally equivalent to a subset of S^n equipped with its standard metric, that is, those that are conformally equivalent to a subset of the conformal n-sphere:

P.7.13 Lemma. *Any "maximal," connected, totally umbilic submanifold M^m, $m > 1$, with flat normal bundle in a conformal manifold $(\hat{M}^n, [g])$, which conformally embeds into the conformal n-sphere, is the intersection of $n - m$ hyperspheres.*

P.7.14 Note that, for dimension $\dim M = 1$, the above definition of "hypersphere" is not exactly what we want: Any "maximal" curve would be a hypersphere.[26] We will leave this situation as it is; however, as we will abandon the Riemannian picture at this point anyway, we give the following brute-force definition:

P.7.15 Definition. *On (subsets of) S^2 equipped with a metric that is conformally equivalent to the standard metric, a hypersphere (circle) is a maximal (in the above sense) curve that has constant geodesic curvature with respect to the standard S^2-metric.*

P.8 Möbius transformations and Liouville's theorem

We now turn away from the Riemannian setup and restrict our attention to (subsets of) the conformal n-sphere[27] as the ambient manifold. In this setup, we will develop the Möbius geometry of submanifolds, that is, the aspects of their geometry that are invariant under the group of Möbius transformations, or conformal transformations, as we will learn from Liouville's theorem:

P.8.1 Definition. *A diffeomorphism $\mu : S^n \to S^n$ on S^n that preserves hyperspheres (maps hyperspheres to hyperspheres) is called a Möbius transformation; the group of Möbius transformations of S^n is called the Möbius group, denoted by $M\ddot{o}b(n)$.*

P.8.2 Any conformal diffeomorphism $\mu : S^n \to S^n$, $n \geq 2$, is a Möbius transformation: For $n > 2$, this is a consequence of our previous developments; for $n = 2$, this is a case of the uniqueness part of the uniformization theorem (see [257]). However, in the case $n > 2$, we have an even stronger

---

[26] This is less absurd than it appears at first: Any (real) analytic embedding $c : S^1 \to \mathbb{C}$ of the unit circle extends to an analytic (conformal) diffeomorphism defined on an annulus about S^1.

[27] In higher dimensions, only the conformal n-sphere provides "enough" hyperspheres to be preserved by Möbius transformations (see §P.7.4). On any other Riemannian manifold the conformal group (group of Möbius transformations) is inessential (see [119] or [111]).

statement that ties down a conformal diffeomorphism even if it is only known locally:

P.8.3 Theorem (Liouville). *Let $f : U \to V$ be a conformal diffeomorphism between connected open subsets $U, V \subset S^n$, $n \geq 3$; then, there is a unique Möbius transformation of S^n, $\mu \in M\ddot{o}b(n)$, such that $f = \mu|_U$.*

P.8.4 *Proof.* See [170]; another proof will be given later.   ◁

Chapter 1

The projective model

Following Klein's Erlanger program [160], a geometry can be described in terms of a transformation group acting on an underlying space (see also [254]): Two geometric configurations are considered equivalent if one can be mapped onto the other by an element of the transformation group under consideration. Thus, we describe Möbius geometry as the geometry given by the Möbius group $M\ddot{o}b(n)$ acting on the conformal n-sphere S^n. The goal of this chapter is to describe Möbius geometry as a subgeometry of projective geometry, as the classical geometers did; see [239], [86], and [88], or [161], where interesting historical remarks also are to be found. This description provides a linearization of Möbius geometry: Spheres will be described by certain linear subspaces and Möbius transformations by linear transformations.

After the description of this model of Möbius geometry, culminating in a "translation table" for the occurring geometric objects and relations (see Figure T.1), we will discuss the metric geometries (hyperbolic, Euclidean, and spherical geometry) as subgeometries of Möbius geometry — and give the announced proof of Liouville's theorem by making use of the subgeometry description of hyperbolic (resp. Euclidean) geometry.

Turning to differential geometry, we will introduce the concept of (hyper)sphere congruences and their envelopes, which is central to many constructions in Möbius differential geometry (cf., Figure T.2). We will set up frame formulas for certain kinds of adapted frames for later reference. There are many possibilities to carry out computations; for reasons of personal taste of the author, we will often use Cartan's method of moving frames. However, in contrast to a common approach, we will *not* introduce a "canonical frame." We will also frequently switch between the moving frame method and the invariant formalism; to make the relation between the two formalisms more transparent, we will discuss the geometry contained in the frame equations after stating them.

In this chapter, we mainly follow the presentations in W. Blaschke's book [29] (which was actually written by G. Thomsen, as Blaschke mentions in the Preface) and in G. Thomsen's papers [277], [278]. The interested reader may also want to check the book [275] by T. Takasu, which appeared around the same time as Blaschke's book, and the more recent book [4] by M. Akivis and V. Goldberg; the reader shall also be referred to the very readable introduction in T. Cecil's book [66].

The contents of this chapter are organized as follows:

Section 1.1. In this section the basic geometric constructions are presented. These constructions yield a description of points of the conformal n-sphere and of hyperspheres in S^n as points in a projective space with an absolute quadric. Introducing a Minkowski scalar product on the space of homogeneous coordinates the incidence of a point and a sphere, and the intersection angle of hyperspheres can be described conveniently.

Section 1.2. Hypersphere pencils and hypersphere complexes are introduced as configurations of hyperspheres that are linear in our projective model space, and their geometry is discussed. These configurations will, on one hand, become important in our description of the Möbius group and, on the other hand, for the description of spheres of arbitrary dimension.

Section 1.3. In this section we provide the relation between Möbius transformations, projective transformations of our projective model space that preserve the absolute quadric, and Lorentz transformations of the coordinate Minkowski space. Inversions are discussed as an important example. We will also see how Möbius geometry arises as the boundary geometry of hyperbolic geometry.

Section 1.4. Hyperbolic, Euclidean, and spherical geometry are characterized as subgeometries of Möbius geometry. In fact, we will learn how the space forms arise as quadrics of constant curvature in the coordinate Minkowski space of Möbius geometry. A generalized notion of stereographic projection from (part of) the conformal n-sphere onto a space of constant curvature is introduced. The groups of motions or similarities are characterized as subgroups of the Möbius group.

Section 1.5. In this section the announced proof of Liouville's theorem is given by making use of the description of hyperbolic geometry as a subgeometry of Möbius geometry. The (global) conformal diffeomorphisms of Euclidean and hyperbolic space are discussed.

Section 1.6. The notions of sphere congruence and of an envelope of a sphere congruence, which are central to many constructions in Möbius differential geometry, are introduced and a convenient formulation for the enveloping condition is given. Conditions on a (hyper)sphere congruence to have one or two envelopes are derived, and the notion of a "strip" is defined.

Section 1.7. The notion of an adapted frame — adapted to a hypersphere congruence or one of its envelopes — of a strip is introduced, and the corresponding structure equations and compatibility conditions are derived. The geometry of the occurring forms is discussed from different points of view. The notion of a curvature sphere congruence of an immersion is introduced.

Section 1.8. As a somewhat degenerate example, channel hypersurfaces are discussed: This example will become important in the context of conformally flat hypersurfaces. As an application, the Möbius geometric classification of Dupin cyclides, that is, surfaces in S^3 that are channel surfaces in two ways, is presented.

Remark. Some basic knowledge of projective geometry is required in this chapter. However, it should not be difficult to extrapolate the required facts from projective geometry from the present text; the reader may also wish to have a look at [162]. An understanding of the topics covered in this chapter is absolutely necessary for everything that follows. And for the reasons indicated above, it may be of profit, even to a reader familiar with the classical setup for Möbius geometry, to flip briefly through this chapter. However, it should do no harm to postpone reading the more technical part concerning frames.

1.1 Penta- and polyspherical coordinates

In §P.8.1, we defined Möbius transformations as those transformations of (the conformal) S^n that preserve hyperspheres. Thus, we will consider two kinds of "elements of Möbius geometry": the points of the conformal n-sphere and the hyperspheres $S \subset S^n$. Möbius transformations act on both types of elements[1] simultaneously. We start by giving a description[2] of these "elements" and their interrelations.

1.1.1 Remember, from §P.7.3, that we considered a hypersphere $S \subset S^n$ as the intersection $S = E^n \cap S^n$ of an n-dimensional hyperplane $E^n \subset \mathbb{R}^{n+1}$ with $S^n \subset \mathbb{R}^{n+1}$ embedded into $\mathbb{R}^{n+1}$ as the standard unit sphere. Now, consider the cone $C \subset \mathbb{R}^{n+1}$ that touches S^n along a hypersphere S in S^n: If S is not a great sphere, then it can be identified with the vertex of C, that is, with the "pole[3]" of the (projective) hyperplane that cuts out S from

---

[1] Not intertwining them, in contrast to Lie transformations (cf., [29] or [66]).

[2] *Caution:* In this section, we introduce certain identifications (as we will later also do for the other models), and soon we will stop (notationally) distinguishing between the identified objects (unless explicitly stated). This will be unfamiliar to many readers; thus, we will emphasize these identifications several times, at the beginning. However, subsequently, we will leave it to the reader to detect the correct setting from the context — here the "translation table" in Figure T.1 may be helpful.

[3] For our purposes, this construction shall serve as a definition of the relation between a "pole" and its "polar hyperplane." Another possible way to characterize this relation is provided by the observation that the two intersection points of a line through the pole separate the pole and the intersection point of the line with the polar hyperplane harmonically. In a moment, we will give a more technical characterization of "polarity," using a scalar product on the space of $\mathbb{R}^{n+2}$ of homogeneous coordinates of $\mathbb{R}P^{n+1}$. Note that this relation of polarity can be extended to a relation between m- and $(n - m)$-planes in $\mathbb{R}P^{n+1}$.

S^n. By embedding $\mathbb{R}^{n+1} \hookrightarrow \mathbb{R}P^{n+1}$ (by "adding" the infinity hyperplane), this identification is extended to great spheres, where the vertex of the cone C (that degenerates to a cylinder) lies in the infinity hyperplane.

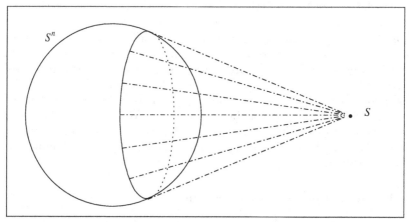

Fig. 1.1. Hyperspheres in the classical model of Möbius geometry

Subsequently, we will not distinguish between a hypersphere $S \subset S^n$ and the corresponding point in the "outer space[4])" of $S^n \subset \mathbb{R}P^{n+1}$.

Thus, the elements of Möbius geometry are described in a unified manner: The points of S^n are the points of the "absolute quadric" $S^n \subset \mathbb{R}P^{n+1}$ while hyperspheres become points in $\mathbb{R}P^{n+1}$ "outside" this absolute[5]) quadric. Observe that a hypersphere, or a point, is described by $n+2$ coordinates in this model, that is, by 5 coordinates if we consider S^3. This might explain the term "pentaspherical coordinates" — see also the explanation that Klein gives in [161].

In the future, we will (notationally) not distinguish between a point in projective $(n+1)$-space and its homogeneous coordinates, nor will we distinguish between the object in S^n and its representative in $\mathbb{R}P^{n+1}$: We may, at the same time, write $S \subset S^n$, $S \in \mathbb{R}P^{n+1}$, and $S \in \mathbb{R}^{n+2}$ or $p \in S^n \subset \mathbb{R}^{n+1}$, $p \in \mathbb{R}P^{n+1}$, and $p \in \mathbb{R}^{n+2}$ as we identify $p \simeq [(1,p)] \simeq (1,p)$. Which interpretation has to be used will be clear from the context.

1.1.2 On the space $\mathbb{R}^{n+2}$ of homogeneous coordinates for points in $\mathbb{R}P^{n+1}$, we introduce a scalar product $\langle .,. \rangle$ such that its null lines are the points $p \in S^n$: If $\mathbb{R}^{n+1} = \{x \in \mathbb{R}^{n+2} \,|\, x_0 = 1\}$, then $\langle .,. \rangle = -dx_0^2 + \sum_{i=1}^{n+1} dx_i^2$

---

[4]) The complement $\mathbb{R}P^{n+1} \setminus S^n$ consists of two connected components: One of them is the "outer space" containing those points that correspond to spheres, i.e., that lie on (real) tangent lines to S^n.

[5]) The term "absolute" refers to the fact that polarity (see below for an algebraic description) is with respect to that quadric.

yields such a scalar product; up to nonzero multiples, this scalar product is unique.[6] Given a hypersphere $S \subset S^n$ in terms of a center[7] $m \in S^n$ and the corresponding radius $\varrho \in (0, \pi)$, we have $S = [(\cos \varrho, m)] \in \mathbb{R}P^{n+1}$. Note that, with the above choice of scaling, $\langle S, S \rangle = \sin^2 \varrho > 0$.

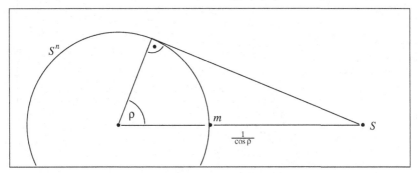

Fig. 1.2. A hypersphere in terms of a "center" and its radius

Subsequently, we let $\langle ., . \rangle$ be a Minkowski product on $\mathbb{R}^{n+2} = \mathbb{R}_1^{n+2}$, that is, we rule out negative scalings of $-dx_0^2 + \sum dx_i^2$ in the above setup. Thus the space of hyperspheres $S \subset S^n$ becomes the space of spacelike lines in $\mathbb{R}_1^{n+2}$, as the outer space of $S^n \subset \mathbb{R}P^{n+1}$ can now be described in this way (cf., Figure T.1).

1.1.3 Later on, we will need to provide the space of hyperspheres with a metric: Similar to the way the Fubini-Study metric is introduced on a real projective space by considering the (metric) sphere as a Riemannian double cover, we obtain a metric on

$$\mathbb{R}P_O^{n+1} := \{\mathbb{R}v \in \mathbb{R}P^{n+1} \mid v \in \mathbb{R}_1^{n+2}, \langle v, v \rangle > 0\}$$

by considering the Lorentz sphere

$$S_1^{n+1} := \{v \in \mathbb{R}_1^{n+2} \mid \langle v, v \rangle = 1\}$$

as a Lorentzian double cover of $\mathbb{R}P_O^{n+1}$. Thus, by denoting the light cone in our coordinate Minkowski space by

$$L^{n+1} := \{v \in \mathbb{R}_1^{n+2} \mid \langle v, v \rangle = 0\},$$

we obtain the identifications

$$S^n \cong L^{n+1}/\mathbb{R} \quad \text{and} \quad \mathbb{R}P_O^{n+1} \cong S_1^{n+1}/\pm 1$$

---

[6] ... since $0 = \langle \frac{\partial}{\partial x_0} \pm \frac{\partial}{\partial x_i}, \frac{\partial}{\partial x_0} \pm \frac{\partial}{\partial x_i} \rangle = \langle \frac{\partial}{\partial x_0} + \frac{1}{\sqrt{2}}(\frac{\partial}{\partial x_i} \pm \frac{\partial}{\partial x_j}), \frac{\partial}{\partial x_0} + \frac{1}{\sqrt{2}}(\frac{\partial}{\partial x_i} \pm \frac{\partial}{\partial x_j}) \rangle$ ...

[7] Note that, with m, $-m$ is a center for the same sphere (with radius $\pi - \varrho$), too.

for the n-sphere S^n and the space[8] $\mathbb{RP}_O^{n+1}$ of hyperspheres in S^n.

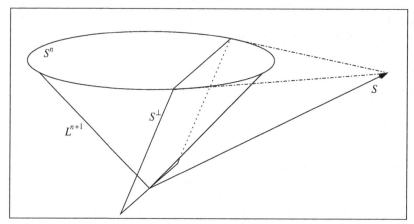

Fig. 1.3. Hyperspheres in S^n as lines in Minkowski $\mathbb{R}_1^{n+2}$

1.1.4 Having established this description for hyperspheres in S^n, we now turn to the interrelations between the elements of Möbius geometry.

First, we discuss the incidence relation between points and hyperspheres: a point $p \in S^n$ lies on a hypersphere $S \subset S^n$ if and only if $\langle p, S \rangle = 0$.

Namely, if $c : (-\varepsilon, \varepsilon) \to \mathbb{R}_1^{n+2}$ are homogeneous coordinates of a curve in $S^n \subset \mathbb{RP}^{n+1}$, then $(c \wedge c')(0)$ defines its tangent line[9] at $c(0)$. Because c is isotropic, $\langle c, c \rangle \equiv 0$, the tangent hyperplane (not the tangent *space*) of S^n at $c(0)$ is therefore given by $\{c(0)\}^\perp$ — and, by construction, $p \in S$ if and only if S lies in the tangent hyperplane[10] of $S^n \subset \mathbb{RP}^{n+1}$ at p (see Figure 1.1).

Another proof can be obtained by bare-hands computation: Obviously, the condition $\langle p, S \rangle = 0$ is independent of the choice of homogeneous co-ordinates (as it has to be); by describing $S = (\cos \varrho, m)$ in terms of m

---

[8] Using the two representatives in S_1^{n+1} of a hypersphere to distinguish the two possible orientations of the hypersphere, the Lorentz sphere can be considered as the space of *oriented hyperspheres*. For example, by using the "inner" center $m \in S^n \subset \mathbb{R}^{n+1}$ to describe an oriented hypersphere of radius ϱ in S^n, a change of orientation, $(m, \varrho) \mapsto (-m, \pi - \varrho)$, yields a change of sign for the S_1^{n+1}-representative: $\frac{1}{\sin \varrho}(\cos \varrho, m) \mapsto \frac{1}{\sin \varrho}(-\cos \varrho, -m)$.

[9] $(c \wedge c')(0) \in \Lambda^2(\mathbb{R}_1^{n+2})$ determines the 2-plane spanned by $c(0)$ and $c'(0)$ that, in the projective picture, defines a line — the line tangent to the curve at $[c(0)] \in S^n$. We will often use the $\wedge$-product of vectors in $\mathbb{R}_1^{n+2}$ to indicate the subspace spanned by the vectors (if they are linearly independent); this will be discussed in greater detail later, in the context of the Clifford algebra model. Note that c' does depend on the choice of homogeneous coordinates of the curve — thus, it makes no sense to consider c' as a tangent *vector* in projective geometry: $c'(0)$ just defines some point (depending on the choice of homogeneous coordinates for c) on the tangent line.

[10] Note that the tangent hyperplane of S^n at $p \in S^n$ is exactly the polar hyperplane of p.

and ϱ, as above, we find[11] that the point $p \simeq (1,p) \in S$ if and only if $p \cdot m = \cos \varrho$, that is, if and only if the projection of p onto m has length $\cos \varrho$ (see Figure 1.2).

1.1.5 Next, we want to examine the interaction of two hyperspheres: Two hyperspheres $S_1, S_2 \subset S^n$ intersect orthogonally if and only if $\langle S_1, S_2 \rangle = 0$; moreover, their angle of intersection,[12] $\alpha \in [0, \frac{\pi}{2}]$, is given by

$$\langle S_1, S_2 \rangle^2 = |S_1|^2 |S_2|^2 \cos^2 \alpha.$$

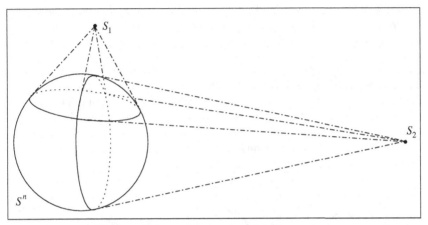

Fig. 1.4. Orthogonal intersection of two hyperspheres

Namely, the scalar product $\langle S_1, S_2 \rangle = 0$ vanishes if and only if S_1 lies in the polar hyperplane of S_2, if and only if the two spheres intersect and the generators of the cones C_1 and C_2, which meet at an intersection point $p \in S_1 \cap S_2$, intersect orthogonally (see Figure 1.4).

To find the angle formula, suppose $p \in S_1 \cap S_2$ and the S_i's are given in terms of their centers $m_i \in S^n$ and their radii $\varrho_i \in (0, \pi)$, that is, we have $S_i = (\cos \varrho_i, m_i)$. Then, the intersection angle of S_1 and S_2 at $p \simeq (1, p)$ is the angle (in $\mathbb{R}^{n+1}$) between the vectors $m_i - \cos \varrho_i\, p$. Since $p \cdot m_i = \cos \varrho_i$, we find

$$\begin{aligned} \langle S_1, S_2 \rangle &= (m_1 - \cos \varrho_1\, p) \cdot (m_2 - \cos \varrho_2\, p) \quad \text{and} \\ \langle S_i, S_i \rangle &= (m_i - \cos \varrho_i\, p)^2, \end{aligned}$$

which proves the claim.

---

[11] We denote the scalar product of $p, m \in \mathbb{R}^{n+1}$ by $p \cdot m$, to distinguish it from the Minkowski product $\langle ., . \rangle$ on $\mathbb{R}^{n+2}_1$.

[12] We consider the *unoriented* angle of intersection.

1.2 Sphere pencils and sphere complexes

As a next step, we want to examine more complicated configurations of hyperspheres and points that correspond to m-planes in $\mathbb{R}P^{n+1}$. Here, of particular interest are "sphere pencils," which correspond to lines in $\mathbb{R}P^{n+1}$, and "sphere complexes," which correspond to hyperplanes in $\mathbb{R}P^{n+1}$.

To start, we first prove the following

1.2.1 Lemma. *Let $S_1, S_2 \in \mathbb{R}P_O^{n+1}$ denote two hyperspheres, $S_1 \neq S_2$. Then:*

* *If $|S_1|^2 |S_2|^2 - \langle S_1, S_2 \rangle^2 > 0$, then S_1 and S_2 intersect transversally.*
* *If $|S_1|^2 |S_2|^2 - \langle S_1, S_2 \rangle^2 = 0$, then they touch at one point.*
* *If $|S_1|^2 |S_2|^2 - \langle S_1, S_2 \rangle^2 < 0$, then they have no points in common.*

1.2.2 *Proof.* The sign of $|S_1|^2 |S_2|^2 - \langle S_1, S_2 \rangle^2$ determines the signature of the induced metric on the plane $S_1 \wedge S_2 \subset \mathbb{R}_1^{n+2}$: Depending on whether this metric is positive definite, degenerate, or Minkowski, its orthogonal complement intersects the light cone in an $(n-2)$-sphere, in a point, or not at all. Moreover, in the second case, we see that S_1 and S_2 "intersect" at an angle $\alpha = 0$, that is, they touch.   ◁

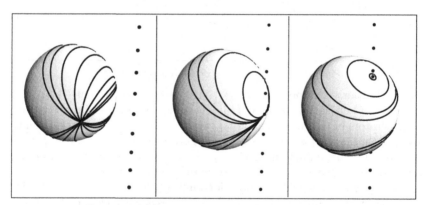

Fig. 1.5. Elliptic, parabolic and hyperbolic sphere pencils

1.2.3 Definition. *Thus, in the projective picture, we get the following:*

* *A line that does not meet $S^n \subset \mathbb{R}P^{n+1}$ can be identified with an $(n-2)$-dimensional sphere in S^n; such a line of spheres will be called[13] an "elliptic sphere pencil."*

---

[13] The notions of elliptic versus hyperbolic pencils are not used consistently in the literature. Here we consider a pencil elliptic when it carries an elliptic measure (the intersection angle), whereas it is hyperbolic when it carries a hyperbolic measure (with two limit points) (cf., [162]).

* *A line that touches $S^n \subset \mathbb{R}P^{n+1}$ can be identified with a 1-parameter family of spheres that touch each other in a point (the point p of contact of the line with S^n); such a line is called a "parabolic sphere pencil," or "contact element" on S^n (cf., [66]).*

* *A line that intersects $S^n \subset \mathbb{R}P^{n+1}$ can be identified with the point pair (0-sphere) that is the intersection of the line with S^n or, by polarity, with the set of all hyperspheres containing these two intersection points; such a line shall be called a "hyperbolic sphere pencil."*

1.2.4 Earlier (see §P.7.5ff) we agreed on describing m-dimensional spheres in S^n as the intersection of $n - m$ hyperspheres. Now, we see that an m-sphere can be identified with

* an $(m + 1)$-plane $P^{m+1} \subset \mathbb{R}P^{n+1}$ that intersects S^n transversally, that is, with a Minkowski $(m + 2)$-subspace in the coordinate Minkowski space $\mathbb{R}_1^{n+2}$;

or, via orthogonality[14] (polarity $P^{m+1} \mapsto P^{n-m-1} = \mathrm{pol}P^{m+1}$) with

* an $(n - m - 1)$-plane $P^{n-m-1} \subset \mathbb{R}P^{n+1}$ that does not meet S^n, that is, with a spacelike $(n - m)$-subspace in $\mathbb{R}_1^{n+2}$.

The first characterization is directly obtained by describing an m-sphere in S^n as the intersection of an $(m + 1)$-plane with S^n. In particular, an m-sphere is determined by $m + 2$ points $p_1, \ldots, p_{m+2} \in S^n$ in general position.

On the other hand, by choosing $n - m$ hyperspheres $S_1, \ldots, S_{n-m}$ in general position on the polar $(n - m - 1)$-plane, the same m-sphere is equivalently obtained[15] as the transversal intersection of the S_i, $1 \le i \le n - m$. This brings us back to our original description of m-spheres.

Note how hyperbolic and elliptic sphere pencils occur as the special cases of 0-spheres (point pairs) and $(n - 2)$-spheres.

1.2.5 In the previous paragraph, we established a Möbius geometric interpretation for $(m + 1)$-planes in $\mathbb{R}P^{n+1}$ that intersect S^n, and for $(n - m - 1)$-planes that do not intersect. We want to conclude this section by giving a characterization of the set of all spheres that lie in a hyperplane in $\mathbb{R}P^{n+1}$ that may intersect S^n, touch, or not intersect S^n at all:

1.2.6 Definition. *Given $\mathcal{K} \in \mathbb{R}P^{n+1}$, the set $\{S \subset S^n \,|\, \langle S, \mathcal{K} \rangle = 0\}$ of hyperspheres is called a "linear sphere complex."*

---

[14] $\mathrm{pol}P^{m+1}$ can be obtained as the projectivization of the orthogonal complement of the coordinate $(m + 2)$-subspace of P^{m+1} in $\mathbb{R}_1^{n+2}$. Readers without background in projective geometry may take this as the definition of "polarity."

[15] It shall be a small exercise for the reader to verify that any point in the intersection of the "spanning" spheres S_i also lies on any sphere in the span $\mathrm{inc}[S_1, \ldots, S_{n-m}]$ of the S_i. Here, "$\mathrm{inc}[S_1, \ldots, S_{n-m}]$" denotes the $(n - m - 1)$-plane containing all S_i, that is, it is the projectivization of $\mathrm{span}\{S_1, \ldots, S_{n-m}\} \subset \mathbb{R}_1^{n+2}$.

Again, three types of sphere complexes occur, depending on whether $\mathcal{K}$ lies in the outer space of $S^n \subset \mathbb{RP}^{n+1}$, on S^n, or in the "inner space" (the other connected component of $\mathbb{RP}^{n+1} \setminus S^n$).

* *If $\langle \mathcal{K}, \mathcal{K} \rangle > 0$, then $\mathcal{K}$ defines[16] an "elliptic sphere complex"; in this case, $\mathcal{K} = S_0$ can be identified with a hypersphere, and the sphere complex consists of all hyperspheres that intersect S_0 orthogonally.*

  Interpreting S_0 as the infinity boundary of a hyperbolic space, an elliptic sphere complex becomes the system of hyperplanes of that hyperbolic space.

* *If $\langle \mathcal{K}, \mathcal{K} \rangle = 0$, then the sphere complex is called a "parabolic sphere complex"; here, $\mathcal{K} = p_\infty$ can be considered as a point, identifying the sphere complex as the system of all spheres containing this point.*

  Interpreting p_∞ as the point at infinity of $\mathbb{R}^n \subset S^n$, a parabolic sphere complex becomes the system of hyperplanes in that Euclidean space.

* *If $\langle \mathcal{K}, \mathcal{K} \rangle < 0$, then $\mathcal{K}$ defines a "hyperbolic sphere complex"; considering all lines (hyperbolic sphere pencils) that contain $\mathcal{K}$ (as an "inner point" of $S^n \subset \mathbb{RP}^{n+1}$), we find that $\mathcal{K}$ defines a notion of "antipodal points."[17]*

  Interpreting $\mathcal{K}$ as the center of $S^n \subset \mathbb{R}^{n+1} = \mathbb{RP}^{n+1} \setminus \text{pol}[\mathcal{K}]$, we obtain the system of hyperplanes (great spheres) in $S^n \subset \mathbb{R}^{n+1}$; note that this reverses our initial reasoning to embed $\mathbb{R}^{n+1} \hookrightarrow \mathbb{RP}^{n+1}$.

1.2.7 A particular interest in linear sphere complexes is triggered by the fact that they single out the "hyperplanes" in the subgeometries of Möbius geometry, as we will discuss in more detail later.

We will identify a sphere complex with the point $\mathcal{K} \in \mathbb{RP}^{n+1}$ defining it.

1.2.8 Summarizing. In Figure T.1 we have collected various identifications and relations discussed in this chapter, for the reader's reference. In that table, we denote by $\text{inc}[y_1, \ldots, y_m]$ the unique $(m-1)$-plane in $\mathbb{RP}^{n+1}$ that contains the m point $y_1, \ldots, y_m$, the "incidence plane of the points"; otherwise said, it is the projectivization of $\text{span}\{y_1, \ldots, y_m\} \subset \mathbb{R}_1^{n+2}$. Note the twofold interpretation of the y_i: as points in $\mathbb{RP}^{n+1}$ in the first formulation and as vectors in $\mathbb{R}_1^{n+2}$ in the second.

---

[16] Again, the notions of elliptic versus hyperbolic are not consistent in the literature. We follow [29]. Note that, in the case $n = 1$, this leads to the following curiosity: A hyperbolic sphere pencil becomes an elliptic sphere complex, and vice versa.

[17] We will come back to this point in the next section, to see how hyperbolic sphere pencils relate to certain Möbius transformations that can be considered as "antipodal maps."

1.3 The Möbius group

In this section we will see that the Möbius transformations of the conformal n-sphere are exactly the projective transformations of $\mathbb{RP}^{n+1}$ that preserve the absolute quadric $S^n \subset \mathbb{RP}^{n+1}$. Further, we will see that the Lorentz group $O_1(n+2)$ is a (trivial) double cover of the Möbius group, providing a linear representation of the Möbius group. We start with the easy part:

1.3.1 Lemma. *Any Lorentz transformation $\mu \in O_1(n+2)$ descends to a projective transformation[18] $\mu : \mathbb{RP}^{n+1} \to \mathbb{RP}^{n+1}$ that preserves S^n (as the absolute quadric).*

Any projective transformation on $\mathbb{RP}^{n+1}$ that preserves S^n gives rise to a Möbius transformation $\mu : S^n \to S^n$, with its action $\mu : \mathbb{RP}^{n+1}_O \to \mathbb{RP}^{n+1}_O$ on the outer space being exactly the action of the Möbius transformation on the hyperspheres of S^n.

1.3.2 *Proof.* As a linear transformation of $\mathbb{R}^{n+2}$, a Lorentz transformation maps lines in $\mathbb{RP}^{n+1}$ to lines, and it preserves $S^n \subset \mathbb{RP}^{n+1}$ as it maps the light cone $L^{n+1} \subset \mathbb{R}^{n+2}_1$ onto itself.

A projective transformation also maps (hyper)planes in $\mathbb{RP}^{n+1}$ to (hyper)planes, and as it maps S^n onto itself it induces a Möbius transformation on S^n; moreover, since polarity is preserved, its restriction to $\mathbb{RP}^{n+1}_O$ yields the action of this Möbius transformation on hyperspheres.   ◁

1.3.3 Example. Given a point (sphere) $S \in \mathbb{RP}^{n+1}_O$ of the outer space, the reflection $p \mapsto \mu(p) := p - 2\frac{\langle p, S \rangle}{\langle S, S \rangle} S$ is a "polar reflection" on $\mathbb{RP}^{n+1}$. Namely, given S and any point $p \neq S$, the line through S and p is parametrized by $t \mapsto q + tS$, where $q = p - \langle p, S \rangle S$ is the intersection point of this line with the polar hyperplane $\{S\}^\perp$ of S; since, with this parametrization of the line, $S \simeq \infty$ and $q \simeq 0$, we find that $p \simeq \langle p, S \rangle$ and $\mu(p) \simeq -\langle p, S \rangle$ separate S and q harmonically.[19]

Moreover, the restriction of μ to S^n is exactly the "inversion" of S^n at the hypersphere $S \subset S^n$: When describing $S = (\cos \varrho, m)$ in terms of a center $m \in S^n$ and the corresponding radius $\varrho \in (0, \pi)$, we obtain

$$S^n \ni p \simeq (1, p) \quad \mapsto \quad \left(\frac{1 + \cos^2(\varrho) - 2p \cdot m \cos \varrho}{\sin^2 \varrho}, \frac{\sin^2(\varrho)p - 2(p \cdot m - \cos \varrho)m}{\sin^2 \varrho} \right)$$

$$\simeq \frac{\sin^2(\varrho)p - 2(p \cdot m - \cos \varrho)m}{1 + \cos^2 \varrho - 2p \cdot m \cos \varrho} \in S^n.$$

---

[18] Or "collineation": a diffeomorphism of projective space that maps lines to lines.

[19] This argument also applies to points S in the inner space, showing that the resulting map is a polar reflection in that case, too. Note that we just proved the aforementioned characterization of "polarity" (see Footnote 1.3) — this is a common way to introduce polarity in projective geometry.

1.3.4 Now we turn to the more difficult part of the description of Möbius transformations in our model: We have to show that a Möbius transformation of the conformal n-sphere $S^n \subset \mathbb{RP}^{n+1}$ (uniquely) extends to a projective transformation on $\mathbb{RP}^{n+1}$. For that purpose, we discuss the effect that a Möbius transformation has on sphere pencils and sphere complexes.

1.3.5 Lemma.   *Möbius transformations preserve sphere pencils of each type, as well as the orthogonal intersection of hyperspheres (polarity).*

1.3.6 *Proof.* Clearly, Möbius transformations do not change the type of a sphere pencil. Because elliptic and parabolic sphere pencils are characterized in terms of incidence and contact (cf., Figure T.1), these are preserved by Möbius transformations. The spheres of hyperbolic sphere pencils can be characterized in terms of the incidence and orthogonal intersection of hyperspheres; namely, the spheres of a hyperbolic sphere pencil are orthogonal to all spheres that contain the point pair of the pencil. Thus, hyperbolic sphere pencils are preserved by Möbius transformations as soon as orthogonal intersection is preserved.

The invariance of orthogonal intersection depends on a curious ("merkwürdige") property of parabolic sphere pencils: On one hand, a parabolic sphere pencil consists of all spheres S that have first-order contact in the point p of contact with S^n while, on the other hand, it can be considered as the tangent line of any curve intersecting these spheres orthogonally in p. Thus, two spheres S_1 and S_2 intersect orthogonally[20] in p if the two parabolic pencils $\mathrm{inc}[p, S_1] \subset T_p S_2$ and, conversely, $\mathrm{inc}[p, S_2] \subset T_p S_1$. Consequently, the orthogonal intersection of two hyperspheres is preserved by Möbius transformations.   ◁

1.3.7 Lemma.   *Möbius transformations preserve linear sphere complexes.*

1.3.8 *Proof.* Elliptic sphere complexes can be identified with spheres via orthogonality (cf., Figure T.1), and parabolic sphere complexes can be characterized in terms of incidence; thus, these are preserved.

To show that Möbius transformations preserve hyperbolic sphere complexes, we use induction: Suppose $(k-1)$-planes $P^{k-1} \subset \mathbb{RP}^{n+1}$ that do not meet S^n are preserved by the action of a Möbius transformation $\mu : \mathbb{RP}^{n+1}_O \to \mathbb{RP}^{n+1}_O$ on the space of spheres.[21] Now, choose $k+1$ orthogonally intersecting spheres S_j that determine a k-plane[22] $P^k \subset \mathbb{RP}^{n+1}_O$, and let $S \in P^k$. Then, the line $\mathrm{inc}[S, S_{k+1}]$ intersects the $(k-1)$-plane

---

[20] Note that this, in fact, was how we came by our characterization of orthogonal intersection.

[21] For $k \leq n$, this is just the fact that Möbius transformations preserve $(n-k)$-spheres.

[22] This fails for $k > n$; for $k \leq n$ it works since the k-plane does not meet S^n.

$P^{k-1} := \text{inc}[S_1, \ldots, S_k]$ in a sphere $\tilde{S} \neq S_{k+1}$. By assumption $\mu\tilde{S} \in \mu P^{k-1}$, so that the line $\text{inc}[\mu\tilde{S}, \mu S_{k+1}] \subset \text{inc}[\mu S_1, \ldots, \mu S_{k+1}]$ lies in the k-plane determined by the μS_j's. Because Möbius transformations preserve orthogonal intersection, the μS_j's span a k-plane. Then, since elliptic sphere pencils are preserved, $\mu S \in \text{inc}[\mu\tilde{S}, \mu S_{k+1}]$, and, because Möbius transformations act bijectively on the space of spheres, $\mu P^k = \text{inc}[\mu S_1, \ldots, \mu S_{k+1}]$.   ◁

1.3.9 Theorem. *Any Möbius transformation $\mu \in M\ddot{o}b(n)$ naturally extends to a projective transformation $\mathbb{R}P^{n+1} \to \mathbb{R}P^{n+1}$ that maps S^n onto itself.*

1.3.10 *Proof.* A Möbius transformation $\mu \in M\ddot{o}b(n)$ naturally acts on the elements of Möbius geometry, points and spheres, $S^n \cup \mathbb{R}P_O^{n+1} \subset \mathbb{R}P^{n+1}$. To extend the action to all of $\mathbb{R}P^{n+1}$, we consider its action on linear sphere complexes — since the identification of elliptic and parabolic sphere complexes with spheres and points is compatible with the action of Möbius transformations, this yields nothing new on $S^n \cup \mathbb{R}P_O^{n+1}$. If we happen to show that this action on sphere complexes preserves lines, then the claim follows by:

1.3.11 Fundamental theorem of projective geometry (§50 in [29]). *Any bijection $\mu : \mathbb{R}P^{n+1} \to \mathbb{R}P^{n+1}$ that preserves lines and the incidence of points and lines comes from a linear map on the space of homogeneous coordinates, $\mu \in PGl(n+2)$.*

1.3.12 We already know that Möbius transformations preserve sphere pencils. Thus the only thing left is to show that they also preserve the incidence of inner points of $S^n \subset \mathbb{R}P^{n+1}$ (hyperbolic sphere complexes) with lines that intersect S^n (hyperbolic sphere pencils). But, an inner point $\mathcal{K}$, $\langle \mathcal{K}, \mathcal{K} \rangle < 0$, lies on a line $\text{inc}[p_1, p_2]$ intersecting S^n if and only if $\mathcal{K}^\perp \supset \{p_1, p_2\}^\perp$, that is, if and only if the sphere complex $\mathcal{K}$ contains all spheres $S \in T_{p_1}S^n \cap T_{p_2}S^n$ that contain both points p_1 and p_2. This characterization is clearly invariant under Möbius transformations.   ◁

1.3.13 Remark. Our approach was not the most efficient one; indeed, it would have been much easier to define projective transformations as linear transformations on the space of homogeneous coordinates and Möbius transformations to be restrictions of projective transformations to the n-sphere $S^n \subset \mathbb{R}P^{n+1}$ (cf., [161]). However, the approach we took seemed to be the more intuitive.

Now, obtaining the last identification of our table Figure T.1 at this point, it is not hard to show that

1.3.14 Theorem. *Any Möbius transformation $\mu \in M\ddot{o}b(n)$ comes from a*

Lorentz transformation $\mu \in O_1(n+2)$; this Lorentz transformation is unique up to sign.

1.3.15 *Proof.* As we already saw, any Möbius transformation naturally gives rise to a unique projective transformation $\mathbb{R}\mu \in PGl(n+2)$. Since $\mu \in Gl(n+2)$ is a transformation that preserves the light cone of $\mathbb{R}_1^{n+2}$, we have[23] that, for an orthonormal basis $(e_0, e_1, \ldots, e_{n+1})$, the vectors $\mu(e_0 \pm e_i)$ and $\mu(e_0 + \frac{1}{\sqrt{2}}(e_i \pm e_j))$ are isotropic. Hence, the $\mu \, e_i$'s form, up to a common scale, an orthonormal basis; that is, μ is a multiple of a Lorentz transformation.   ◁

1.3.16 As a consequence, proved by linear algebra, we find that any Möbius transformation can be decomposed into inversions, that is, polar reflections with poles in the outer space $\mathbb{R}P_O^{n+1}$. First note that any pseudo orthogonal transformation $\mu \in O_k(n)$, $0 \leq k \leq n$, is generated by at most n reflections in nondegenerate hyperplanes (see [66]). Since, moreover, a reflection in a spacelike hyperplane in $\mathbb{R}_1^{n+2}$ can be written as the composition of $n-1$ reflections in Minkowski hyperplanes and $-id$, we obtain the result:

1.3.17 Corollary. *Any Möbius transformation is generated by inversions.*

1.3.18 Before turning to the metric subgeometries, we want to examine a class of Möbius transformations that contains the inversions in hyperspheres as a special case: the Möbius involutions. These transformations can be used to define the "tangent planes" of the hyperbolic and elliptic subgeometries of Möbius geometry, as we will see later, and, their fixed point sets yield m-spheres in S^n.

1.3.19 Lemma. *Let $\mu \in O_1(n+2)$ be an involutive Möbius transformation, that is, $\mu^2 = \pm id$; then, μ is the composition of at most $n+1$ inversions in orthogonal hyperspheres.*

1.3.20 *Proof.* We have to consider two cases: $\mu^2 = -id$ and $\mu^2 = id$.

$\mu^2 = -id$: We would have $\langle v, \mu(v) \rangle = -\langle \mu^2(v), \mu(v) \rangle = -\langle \mu(v), v \rangle$; thus $\mu(v) \perp v$ for any $v \in \mathbb{R}_1^{n+2}$. In particular, if $|v|^2 = -1$, then v and $\mu(v)$ would be two orthogonal timelike vectors, which is not possible. Thus, this case does not occur.

$\mu^2 = +id$: Here, *any* vector $v \in \mathbb{R}_1^{n+2}$ decomposes into orthogonal eigenvectors of μ, $2v = (v + \mu v) + (v - \mu v)$. Thus, $\mathbb{R}_1^{n+2}$ is the direct and orthogonal sum of the $+1$- and the -1-eigenspaces of μ, $\mathbb{R}_1^{n+2} = E_{+1} \oplus_\perp E_{-1}$. Since the sum is orthogonal and direct, one of the eigenspaces is Minkowski

---

[23] The same argument proves the uniqueness up to scale of a Lorentz product when its light cone is prescribed.

(w.l.o.g. the $+1$-eigenspace, since we are only interested in the Möbius transformation $\pm\mu$), whereas the other is spacelike. By choosing an orthonormal basis of $\mathbb{R}_1^{n+2}$ compatible with this decomposition, we can write μ as the composition of $m = \dim E_{-1}$ reflections in Minkowski hyperplanes. ◁

1.3.21 If the dimension m of the spacelike eigenspace E_{-1} of a Möbius involution μ is less than n, $m \leq n$, then E_{-1} (resp. $E_{+1} = E_{-1}^{\perp}$) defines an $(n-m)$-sphere as the fixed point set of μ. Thus we may consider μ as the inversion in that $(n-m)$-sphere. Clearly the converse is true: Any $(n-m)$-sphere defines a Möbius involution with this $(n-m)$-sphere as its fixed point set.

Now consider a Möbius involution of S^n with no fixed points; that is, the dimension of the spacelike eigenspace E_{-1} of μ is $m = \dim E_{-1} = n + 1$. Then μ defines a hyperbolic sphere complex (via its unique timelike eigendirection), and the spheres of the complex are exactly those hyperspheres that are mapped onto themselves by the involution. Such an involution is the composition of $n+1$ inversions in orthogonal hyperspheres of the complex.

1.3.22 Finally, we note that n-dimensional Möbius geometry appears as the infinity boundary geometry of $(n+1)$-dimensional hyperbolic geometry: The projective transformations of $\mathbb{RP}^{n+1}$ that preserve S^n act as hyperbolic motions on the Klein model of hyperbolic space, given by the inner part of $\mathbb{RP}^{n+1} \setminus S^n$, where (hyperbolic) straight lines are segments of the lines of the ambient projective space (cf., [29]).

1.4 The metric subgeometries

Clearly, the isometries of the (metric) n-sphere $S^n \subset \mathbb{R}^{n+1}$ are Möbius transformations, because they preserve hyperspheres.

The Euclidean space $\mathbb{R}^n$ is conformally equivalent to the n-sphere minus a point, via stereographic projection $\sigma : S^n \setminus \{\infty\} \to \mathbb{R}^n$; thus it can be considered as a subspace of the conformal n-sphere. In this picture, the isometries of $\mathbb{R}^n$ yield conformal, that is, hypersphere-preserving diffeomorphisms; we will see that these are the Möbius transformations that fix the "point at infinity."

Considering the Poincaré model of hyperbolic space (cf., §P.1.3), we see that it is conformally embedded into Euclidean space, and hence into the conformal n-sphere. Again, the hyperbolic isometries yield hypersphere-preserving diffeomorphisms. We want to see that they extend (uniquely) to (special) Möbius transformations of the conformal n-sphere — characterizing hyperbolic geometry as a subgeometry of Möbius geometry also.

However, we will approach the description of the metric subgeometries as subgeometries of Möbius geometry in a slightly different way, using our

recently developed model. Our first objective will be to develop suitable
models for the spherical, Euclidean, and hyperbolic geometries (cf., [29]).

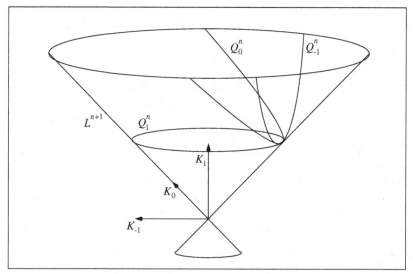

Fig. 1.6. The spaces of constant curvature as affine light cone quadrics

1.4.1 Lemma. *Let* $\mathcal{K} \in \mathbb{R}_1^{n+2} \setminus \{0\}$ *and let* $\kappa := -|\mathcal{K}|^2$. *Then the n-dimensional affine quadric*

$$Q_\kappa^n := \{p \in L^{n+1} \,|\, \langle \mathcal{K}, p \rangle = -1\},$$

equipped with the induced metric $\langle .,. \rangle$, *is a space of constant curvature* κ.

If $\kappa < 0$, *then* Q_κ^n *consists of two connected components that can be (isometrically) identified via the inversion*

$$i_\infty : Q_\kappa^n \to Q_\kappa^n, \quad p \mapsto i_\infty(p) = -(p + \tfrac{2}{\kappa}\langle p, \mathcal{K} \rangle \mathcal{K})$$

at the "infinity boundary" $S_\infty \simeq \mathcal{K}$ of Q_κ^n.

1.4.2 *Proof.* We consider the cases $\kappa = 0$ and $\kappa \neq 0$ separately.[24]

If $\kappa \neq 0$, then the affine hyperplane $E_\mathcal{K}^{n+1} := \{y \in \mathbb{R}_1^{n+2} \,|\, \langle y, \mathcal{K} \rangle = -1\}$ becomes an $(n+1)$-dimensional Euclidean (if $\kappa > 0$) or Minkowski (if $\kappa < 0$) space — with $0 \simeq \tfrac{1}{\kappa}\mathcal{K}$ — that contains Q_κ^n as the standard round sphere or the standard 2-sheeted hyperboloid of "radius" $\dfrac{1}{\sqrt{|\kappa|}}$, respectively:

$$\forall y = \tfrac{1}{\kappa}\mathcal{K} + y^\perp : \langle y, y \rangle = 0 \Leftrightarrow \langle y^\perp, y^\perp \rangle = \tfrac{1}{\kappa}.$$

---

[24] Indeed, these two cases are qualitatively slightly different — this is related to the fact that Euclidean space is the only space of constant curvature that allows homotheties, that is, constant scaling of the metric without changing the curvature.

If $\kappa < 0$, then the map i_∞ clearly preserves $E_{\mathcal{K}}^{n+1}$, and since $i_\infty|_{E_{\mathcal{K}}^{n+1}} = -id$ (where we think of $E_{\mathcal{K}}^{n+1}$ as the linear space of the $y^\perp$'s, with $0 \simeq \frac{1}{\kappa}\mathcal{K}$), it interchanges the two sheets of the hyperboloid Q_κ^n.

In the case $\kappa = 0$, we may assume that $\mathcal{K} = (1, 0, -1)$ in terms of some orthonormal basis $(e_0, \dots, e_{n+1})$ of $\mathbb{R}_1^{n+2} = \mathbb{R} \times \mathbb{R}^n \times \mathbb{R}$. Then the map

$$\mathbb{R}^n \ni x \mapsto \left(\tfrac{1+x\cdot x}{2}, x, \tfrac{1-x\cdot x}{2}\right) \in L^{n+1}$$

is an isometry parametrizing Q_0^n. $\quad\triangleleft$

1.4.3 Identifying, as above, $\mathbb{R}^n \ni x \leftrightarrow \left(\frac{(1+|x|^2)}{2}, x, \frac{(1-|x|^2)}{2}\right) \in Q_0^n$ and the round sphere $\mathbb{R}^{n+1} \supset S^n \ni p \leftrightarrow (1, p) \in Q_1^n$, we recover the inverse of a stereographic projection (cf., §P.1.3) as a simple rescaling,

$$Q_0^n \ni \left(\tfrac{1+|x|^2}{2}, x, \tfrac{1-|x|^2}{2}\right) \mapsto \left(1, \tfrac{2x}{1+|x|^2}, \tfrac{1-|x|^2}{1+|x|^2}\right) \in Q_1^n,$$

that is, as a central projection of Q_0^n to Q_1^n along the generators of the light cone. This may motivate the following generalization of the stereographic projection:

1.4.4 Definition. *Any choice of representative* $\sigma : S^n \setminus \{\mathcal{K}\}^\perp \to Q_\kappa^n$ *for the inclusion map* $S^n \hookrightarrow \mathbb{RP}^{n+1}$ *of the conformal n-sphere into the light cone that takes values in a quadric* Q_κ^n *of constant curvature will be called a stereographic projection.*

1.4.5 Note that, for $-\langle \mathcal{K}, \mathcal{K}\rangle = \kappa > 0$, such a stereographic projection is globally defined while, for $\kappa \leq 0$, it ceases to be defined:
– at a point (the point $p_\infty = \mathcal{K}$ at infinity) if $\kappa = 0$, and
– on a hypersphere (the infinity boundary $S_\infty = \frac{1}{\sqrt{-\kappa}}\mathcal{K}$) if $\kappa < 0$.
To understand that these "stereographic projections" are conformal, we prove the following:

1.4.6 Lemma. *Let* $f_1, f_2 : M^m \to \mathbb{R}_1^{n+2}$ *denote two representatives of the same immersion* $f : M^m \to S^n \subset \mathbb{RP}^{n+1}$. *Then their induced metrics* $g_i = \langle df_i, df_i \rangle$ *are conformally equivalent.*

1.4.7 *Proof.* First note that, if $f_i : M^m \to L^{n+1}$ is a representative of an immersion $f : M^m \to S^n$, then g_i is a (positive definite) metric indeed: Suppose $g_i|_p(v, v) = \langle \partial_v f_i(p), \partial_v f_i(p)\rangle = 0$ for some $p \in M$ and $v \in T_pM$; then $\partial_v f_i(p) \perp f_i(p), \partial_v f_i(p)$, that is, $\partial_v f_i(p) \parallel f_i(p)$, which contradicts the assumption of $f : M^m \to S^n$ being an immersion if $v \neq 0$.

Now, f_i are two representatives of the same immersion $f : M \to S^n$ if and only if $f_2 = \pm e^u f_1$ with some function $u : M \to \mathbb{R}$. Then, since $|f_1|^2 \equiv 0$, we find $g_2 = \langle d(e^u f_1), d(e^u f_1)\rangle = e^{2u}\langle df_1, df_1\rangle = e^{2u}g_1$. $\quad\triangleleft$

The conformality of the above stereographic projections follows now from the following more general corollary, which finally justifies the term "conformal n-sphere" for our inclusion map $S^n \hookrightarrow \mathbb{R}P^{n+1}$:

1.4.8 Corollary. *Any choice of representative* $i : S^n \rightarrow \mathbb{R}_1^{n+2}$ *of the inclusion map* $S^n \hookrightarrow \mathbb{R}P^{n+1}$ *induces the standard conformal structure.*

1.4.9 *Proof.* With the scalar product $\langle . , . \rangle = -dx_0^2 + \sum_{i=1}^{n+1} dx_i^2$ on $\mathbb{R}_1^{n+2}$, the representative $\mathbb{R}^{n+1} \supset S^n \ni p \mapsto i(p) := (1, p) \in L^{n+1}$ is an isometry. ◁

1.4.10 Next we want to determine those hyperspheres that yield the hyperplanes (complete, totally geodesic[25]) hypersurfaces) in our spaces Q_κ^n of constant curvature.

For $\kappa \neq 0$, the orthogonal hyperplane $\{S\}^\perp$ of a spacelike vector $S \in \mathcal{K}^\perp$ (hypersphere of the linear complex $\mathcal{K}$) contains the point

$$\tfrac{1}{\kappa}\mathcal{K} \simeq 0 \in E_\mathcal{K}^{n+1} = \{y \in \mathbb{R}_1^{n+2} \mid \langle y, \mathcal{K} \rangle = -1\}.$$

Consequently, its intersection with $Q_\kappa^n \subset E_\mathcal{K}^{n+1}$ (the set of points of S that lie in the quadric Q_κ^n) is a hyperplane in Q_κ^n (cf., [122]).

Note that any sphere $S \in \mathcal{K}^\perp$ of the linear sphere complex $\mathcal{K}$ is invariant under the Möbius involution[26]

$$i_\infty : p \mapsto -(p + \tfrac{2}{\kappa}\langle p, \mathcal{K} \rangle \mathcal{K},$$

since $i_\infty(\tfrac{1}{\kappa}\mathcal{K} + y^\perp) = \tfrac{1}{\kappa}\mathcal{K} - y^\perp$ for $\tfrac{1}{\kappa}\mathcal{K} + y^\perp \in E_\mathcal{K}^{n+1}$, that is, $i_\infty|_{E_\mathcal{K}^{n+1}}$ is a point reflection of $E_\mathcal{K}^{n+1}$ at $\tfrac{1}{\kappa}\mathcal{K} \simeq 0 \in E_\mathcal{K}^{n+1}$.

If $\kappa = 0$, then the spheres S of the linear complex $\mathcal{K} \simeq p_\infty$ are just the spheres that contain the point p_∞ at infinity of $\mathbb{R}^n \cong Q_0^n = S^n \setminus \{p_\infty\}$.

Thus, we have obtained the following

1.4.11 Lemma. *Given* $\mathcal{K} \in \mathbb{R}_1^{n+2}$, *the spheres of the linear sphere complex* $\mathcal{K}$ *are the hyperplanes of* $Q_\kappa^n = \{p \in L^{n+1} \mid \langle p, \mathcal{K} \rangle = -1\}$.

If $\kappa < 0$, *then these are the hyperplanes that are invariant under the inversion* i_∞, *that is, that smoothly extend through the infinity boundary* $S_\infty \simeq \mathcal{K}$ *of* Q_κ^n.

1.4.12 Going back to F. Klein's characterization of a geometry [160], we are left to show that the motion groups of our constant curvature quadrics $Q_\kappa^n \subset S^n$ are subgroups of the Möbius group $O_1(n+2)/\pm$ — then we have

---

[25] Remember that a hypersurface $M \subset \tilde{M}$ is called "totally geodesic" if its second fundamental form vanishes identically.

[26] If $\kappa < 0$, this involution is the reflection in the infinity sphere $S_\infty \simeq \mathcal{K}$ of Q_κ^n that identifies the two copies of hyperbolic space (see §1.4.1). If $\kappa > 0$, then i_∞ is the antipodal map.

established the spherical, Euclidean, and hyperbolic geometries as subgeometries of Möbius geometry. However, in the hyperbolic case, Q_κ^n consists of two components, each of which is a hyperbolic space and which we identify via the inversion i_∞ at the infinity boundary; we will show that the group of motions that commute with i_∞ is a subgroup of the Möbius group in this case.

1.4.13 Lemma. *Let $\mathcal{K} \in \mathbb{R}_1^{n+2} \setminus \{0\}$. Then the Lorentz transformations $\mu \in O_1(n+2)$ that fix $\mathcal{K}$, $\mu(\mathcal{K}) = \mathcal{K}$, are isometries of Q_κ^n.*

If $\kappa \geq 0$, then all isometries occur in this way.

If $\kappa < 0$, then these are the isometries that commute with the inversion i_∞ at $S_\infty \simeq \mathcal{K}$, that is, those isometries that smoothly extend through the infinity boundary of Q_κ^n.

1.4.14 *Proof.* Any Lorentz transformation $\mu \in O_1(n+2)$ with $\mu(\mathcal{K}) = \mathcal{K}$ induces an isometry on $E_\mathcal{K}^{n+1}$, and hence on Q_κ^n.

On the other hand, for $\kappa > 0$, the isometries of Q_κ^n are the linear isometries of the Euclidean space $E_\mathcal{K}^{n+1}$, where we identify $0 \simeq \frac{1}{\kappa}\mathcal{K}$ (for the linearity); and, for $\kappa < 0$, the isometries of Q_κ^n that commute with i_∞ are the Lorentz transformations of the Minkowski space $E_\mathcal{K}^{n+1}$ with $0 \simeq \frac{1}{\kappa}\mathcal{K}$. Thus, in both cases, an isometry of Q_κ^n (that commutes with i_∞ if $\kappa < 0$) extends uniquely to a Lorentz transformation with $\mu(\mathcal{K}) = \mathcal{K}$.

Finally, for $\kappa = 0$, any isometry of Q_0^n is a Möbius transformation that preserves the point $p_\infty = \mathcal{K}$; that is, it comes from a Lorentz transformation $\mu \in O_1(n+2)$ with $\mu(\mathcal{K}) = r\mathcal{K}$ for some $r \in \mathbb{R} \setminus \{0\}$. Projecting back onto Q_0^n, $p \mapsto \frac{1}{r}\mu p$, the resulting map becomes an isometry if and only if we have $r^2 = 1$. ◁

1.4.15 Note that, in the projective setting, the Q_κ^n's cannot be defined any more (since the defining equation depends on a choice of homogeneous coordinates). However, the corresponding linear sphere complex $\mathcal{K}$ is still well defined, providing a notion of "hyperplanes": In the case of a parabolic sphere complex this yields the geometry of similarities on Euclidean space; in the case of an elliptic or hyperbolic sphere complex, the group of hyperplane-preserving Möbius transformations is isomorphic to the isometry group of hyperbolic or elliptic geometry, respectively.

In fact, by identifying points $p \in S^n$ via the polar reflection defined by the linear sphere complex, we obtain the classical hyperbolic and elliptic geometries: the metric geometries of H^n (rather than on two copies of H^n) and of $\mathbb{R}P^n \cong S^n/\pm 1$ equipped with the Fubini-Study metric (see [29]).

1.4.16 Thus we may also break the Möbius symmetry by factorizing the conformal S^n by a Möbius involution (see §1.3.21) and considering the group of Möbius transformations that commute with the involution. If the fixed point set of the involution[27] $S^n \to S^n$ is a hypersphere, we obtain hyperbolic geometry, and if its fixed point set is empty,[28] we obtain the classical elliptic geometry; in both cases the fixed point set of the involution (now considered as an involution on the projective ambient space $\mathbb{RP}^{n+1}$ of S^n) determines a sphere complex that consists of the hyperplanes of the geometry.

1.5 Liouville's theorem revisited

At this point, we have collected the necessary material to prove Liouville's theorem (see §P.8.3) in our setup. Remember that, in §P.8.2, we already saw that a globally defined conformal diffeomorphism $S^n \to S^n$ is a Möbius transformation (this is true even in case $n = 2$). First note that the converse of this statement is easily derived from our model of Möbius geometry:

1.5.1 Theorem. *Möbius transformations are conformal.*

1.5.2 *Proof.* (Compare with the proof given in [170].) Since Möbius transformations $\mu \in M\ddot{o}b(n)$ are represented by isometries $\mu \in O_1(n + 2)$, and any change of representative for the inclusion $S^n \hookrightarrow \mathbb{RP}^{n+1}$ is conformal (see §1.4.8), the statement follows immediately.   ◁

1.5.3 Now, Liouville's theorem states that even local conformal diffeomorphisms come from Möbius transformations if the dimension $n \geq 3$, showing that any "conformal geometry" as determined (following Klein) by a base manifold M^n that is globally conformally equivalent to a subset of S^n acted on by the group of conformal diffeomorphisms is actually a subgeometry of Möbius geometry:

1.5.4 Theorem (Liouville). *Let $f : U \to V$ be a conformal diffeomorphism between connected open subsets $U, V \subset S^n \subset \mathbb{R}^{n+1}$, $n \geq 3$. Then there is a unique Möbius transformation $\mu \in M\ddot{o}b(n)$ such that $f = \mu|_U$.*

1.5.5 *Proof.* We show that the restriction of f to any ball in U is the restriction of a unique Möbius transformation. Thus, let $B \subset U$ be an open ball with boundary $\partial B = S_\infty \subset U$. Since conformal diffeomorphisms map hyperspheres to hyperspheres,[29] $f(S_\infty) \subset V$ is a hypersphere. By possibly postcomposing f with a Möbius transformation, we may assume that $f(S_\infty) = S_\infty$ and $f(B) = B$.

---

[27] It should be an interesting exercise to study the geometries that are obtained from an involution with an $(m-1)$-sphere, $1 \leq m \leq n-1$, as its fixed point set (cf., §1.3.21).

[28] In this case it seems reasonable to call the involution an "antipodal map."

[29] At this point, we need $n \geq 3$.

Now we equip $B \cong H^n$ with a hyperbolic metric g: Consider B as one of the connected components of Q^n_{-1} where $\mathcal{K} = S_\infty$; without loss of generality we assume that $|S_\infty|^2 = 1$. Then f preserves the straight lines of this hyperbolic space — circles that intersect S_∞ orthogonally. Thus, for the covariant derivative $\tilde{\nabla}$ of the induced metric $\tilde{g} := f^*g$, we infer that

$$\tilde{\nabla}_x y - \nabla_x y = \omega(x)y + \omega(y)x,$$

with a suitable 1-form ω since ∇ and $\tilde{\nabla}$ have the same pregeodesics (this is Weyl's theorem; cf., [179]).

On the other hand, $\tilde{g} = e^{2u}g$ and g are conformally equivalent since f is conformal. Comparison with the transformation formula for the covariant derivative (see §P.2.1) shows that $\text{grad}\, u = 0$, that is, $u \equiv const$ is constant. Since $\tilde{g}$ has — as the pullback of the hyperbolic metric g — constant curvature -1, we finally learn that $\tilde{g} = g$; for example, by employing the transformation formula for the sectional curvature from §P.2.7.

Consequently, f is an isometry of one of the connected components of Q^n_{-1}, and, therefore, there is a unique Lorentz transformation $\mu \in O_1(n+2)$ with $\mu(S_\infty) = S_\infty$ and $f|_B = \mu|_B$. By postcomposing both f as well as μ with the inverse of the Möbius transformation that we may have used earlier to achieve that $f(B) = B$, we derive that $f|_B : B \to f(B) \subset V$ extends to a unique Möbius transformation on S^n.

Finally, because B was an arbitrary open ball in U, we conclude that $f : U \to V$ extends to a unique Möbius transformation. ◁

1.5.6 Note that this result obviously does not hold in dimension $n = 2$: Here, *any* biholomorphic function $f : U \to V$ of subsets $U, V \subset \mathbb{C}$ provides a conformal diffeomorphism.

As a consequence of Liouville's theorem we obtain the following corollary characterizing (global) conformal diffeomorphisms of the Euclidean and hyperbolic spaces:

1.5.7 Corollary. *Any (global) conformal diffeomorphism $f : \mathbb{R}^n \to \mathbb{R}^n$ of Euclidean space is a similarity, and any (global) conformal diffeomorphism $f : H^n \to H^n$ of hyperbolic space is an isometry.*

1.6 Sphere congruences and their envelopes

Now, we turn to a key concept in Möbius differential geometry: The concept of sphere congruences and their envelopes. In some sense, a sphere congruence enveloping a hypersurface replaces the Euclidean concept of the Gauss map (therefore, a certain sphere congruence that can be uniquely associated to a surface in S^3 is called the "conformal Gauss map"). However, in certain contexts, the sphere congruence is considered to be the primary object

and its envelope(s) take the role of the Gauss map. In the following two
chapters on conformally flat hypersurfaces and on Willmore and isothermic
surfaces we will get to discuss each of these two viewpoints in detail. These
remarks will become clearer below.

We start by giving the following

1.6.1 Definition. *A differentiable m-parameter family* $S : M^m \to \mathbb{RP}_O^{n+1}$
of spheres is called a sphere congruence.

A differentiable map $f : M^m \to S^n$ *is called an envelope of a sphere
congruence* S *if, for all* $p \in M$,

$$f(p) \in S(p) \qquad \text{and} \qquad T_{f(p)} f(M) \subset T_{f(p)} S(p) \,.$$

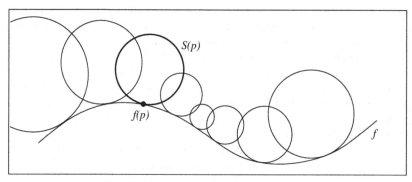

Fig. 1.7. An envelope of a sphere congruence

1.6.2 Note that, again, we consider the tangent spaces of $S(p)$ and $f(M)$
at $f(p)$ as planes in projective space $\mathbb{RP}^{n+1}$, that is, we do not consider
their linear structure. Also, we do not require f (or S) to be an immersion,
in general: Usually, we will require one of them to be regular, depending on
the viewpoint we wish to take — the object of primary interest will usually
be assumed to be immersed.

The following lemma gives a useful way to handle the congruence—
envelope relation formally. Note that the conditions are independent of
the choices of homogeneous coordinates for f and S.

1.6.3 Lemma. *A differentiable map* $f : M^m \to S^n$ *envelopes a sphere
congruence* $S : M^m \to \mathbb{RP}_O^{n+1}$ *if and only if*

$$\langle f, S \rangle = 0 \qquad \text{and} \qquad \langle df, S \rangle \equiv 0 \,.$$

1.6.4 *Proof.* We already saw (see Figure T.1) that $f(p) \in S(p)$ if and only
if $\langle f(p), S(p) \rangle = 0$. The tangent plane $T_{f(p)} f(M) \subset T_{f(p)} S(p)$ if and only if

any tangent line of f at $f(p)$, $\mathrm{inc}[f(p), \partial_v f(p)]$, lies in the polar hyperplane $\mathrm{pol}[S(p)]$ of $S(p)$, that is, $\mathrm{span}\{f(p), \partial_v f(p)\} \subset \{S(p)\}^\perp$. ◁

1.6.5 This characterization shows that a sphere congruence S and an envelope f of S can, if both are regular, be considered as an "immersion—normal field pair" in $\mathbb{R}_1^{n+2}$ in two ways: S as a spacelike normal field of f, or f as an isotropic normal field of S.

In the following, we want to cultivate this viewpoint a little further, after discussing some obstructions for a sphere congruence to have envelopes.

1.6.6 Remember that we can equip the space $\mathbb{R}P_O^{n+1}$ of spheres with a spherical metric by considering the Lorentz sphere S_1^{n+1} as a double cover (the $\pm$-ambiguity can be interpreted as the orientation) (see §1.1.3). Thus we can (locally) choose homogeneous coordinates for a sphere congruence such that $|S|^2 \equiv 1$, that is, so that $S : M^m \to S_1^{n+1}$ takes values in the Lorentz sphere.

Subsequently, we will always choose (local) homogeneous coordinates for a sphere congruence to take values in the Lorentz sphere, $S : M^m \to S_1^{n+1}$.

In this way it makes sense[30] to consider the induced metric $\langle dS, dS \rangle$ of a sphere congruence S, as in the following two lemmas:

1.6.7 Lemma. *If a regular sphere congruence $S : M^m \to S_1^{n+1}$ has an envelope, then $\partial_v S$ never becomes timelike.*

If S has two envelopes, then S induces a Riemannian metric on M.

1.6.8 *Proof.* If $d_p S(T_p M)$ would contain a timelike vector, then the corresponding normal space $\{S(p), d_p S(T_p M)\}^\perp \subset T_{S(p)} S_1^{n+1}$ would be spacelike, that is, there would be no isotropic normal vectors. Hence there would be no envelope because an envelope can be interpreted as an isotropic normal field of the sphere congruence.

If S has two (different) envelopes, then each normal space has to be a Minkowski space, and hence the tangent spaces have to be Euclidean. ◁

1.6.9 Lemma. *A regular hypersphere congruence $S : M^{n-1} \to S_1^{n+1}$ has two envelopes if and only if S induces a Riemannian metric $\langle dS, dS \rangle$ on M.*

1.6.10 *Proof.* The "only—if" part is a special case of the previous lemma.

For the "if" part, note that the normal bundle of $S : M^{n-1} \to S_1^{n+1}$ as an immersion into the Lorentz sphere is 2-dimensional. Thus, if S induces a Riemannian metric, then the normal planes are Minkowski, containing at

---

[30] Note that the pullback of the Lorentz metric on $\mathbb{R}_1^{n+2}$ is by no means (half) invariant under changes of the homogeneous coordinates of a sphere congruence, in contrast to the pullback by a light cone map or the pullback by a sphere congruence in Lie sphere geometry, where only the induced conformal structure is well defined.

each point exactly two (distinct) null lines (that smoothly depend on the point); these provide two envelopes of S.   ◁

1.6.11 From these considerations we learn that, in particular, a sphere congruence can have at most two envelopes.

We close with a definition that emphasizes the "immersion—normal field" viewpoint on a sphere congruence with an envelope, treating the two maps symmetrically: Note that $\langle f, S \rangle = 0$ implies $\langle df, S \rangle \equiv 0 \Leftrightarrow \langle f, dS \rangle \equiv 0$.

1.6.12 Definition. *A pair of smooth maps $(S, f) : M^m \to S_1^{n+1} \times L^{n+1}$ that satisfies the enveloping conditions $\langle f, S \rangle = 0$ and $\langle df, S \rangle \equiv 0$ will be called a strip.*

1.7 Frame formulas and compatibility conditions

There are various possibilities to carry out computations in our model of Möbius (differential) geometry. Working with $O_1(n + 2)$ acting on $\mathbb{R}_1^{n+2}$, we will often use Cartan's method of moving frames. However, we will not (as it is often done) stick to some "canonical frame," but we will rather try to choose our frames to be adapted to each problem in order to simplify computations and arguments as much as possible in this framework. In fact, the definition of "adapted frames" in this section should be understood as a temporary definition only; it will be convenient in the following two chapters. However, for other situations other notions may be more suitable.

Thus a central notion will be that of an "adapted frame" for a strip:

1.7.1 Definition. *Let $(e_0, e_1, \ldots, e_n, e_\infty)$ be a pseudo-orthonormal basis of $\mathbb{R}_1^{n+2}$, that is, $\langle e_i, e_j \rangle = \delta_{ij}$ and $e_0, e_\infty \in \{e_1, \ldots, e_n\}^\perp$ are isotropic and scaled so that $\langle e_0, e_\infty \rangle = 1$. Further, let $(S, f) : M^{n-1} \to S_1^{n+1} \times L^{n+1}$ be a strip. Then a smooth map $F : M^{n-1} \to O_1(n + 2)$ will be called a frame for the strip (S, f) if*

$$S = Fe_n \quad \text{and} \quad f = Fe_0.$$

Moreover, F is called[31]
  * *S-adapted if $\forall p \in M : d_p S(T_p M) = \mathrm{span}\{F(p)e_i \,|\, 1 \le i \le n - 1\}$, and*
  * *f-adapted if $\forall p \in M : d_p f(T_p M) = \mathrm{span}\{F(p)e_i \,|\, 1 \le i \le n - 1\}$.*

1.7.2 Note that our definition of an (adapted) frame depends on various choices: First, it depends on our model for Möbius geometry, that is, on the representation of the Möbius group $M\ddot{o}b(n)$ as (doubly covered by) the

---

[31] Note that these conditions force S resp. f to be immersed and of "good" geometric behavior: S will have two envelopes, or f will be an immersion into the conformal S^n, respectively.

Lorentz group $O_1(n+2)$. In Chapters 4, 6, and 7, other types of frames will be used as we will use different representations of the Möbius group. Second, it depends on our choice of using a pseudo-orthonormal basis as the (fixed) "reference frame": This is convenient but not necessary; in particular, our choice for the relative scaling of e_0 and e_∞ will not always be the best. In Chapter 4 we will naturally come across a different relative scaling that will cause the structure equations to look slightly different.

Finally note that we assume the domain manifold (usually just an open piece of $\mathbb{R}^{n-1}$) to be $(n-1)$-dimensional. This is a restriction that we impose here since, in the following two chapters, we are interested only in treating hypersurfaces. Remember that, in this text, frames are understood as a convenient tool to carry out computations — nothing more — and that we will change our view of what an (adapted!) frame is whenever it seems convenient.

1.7.3 However, there is some geometry attached to a frame that we do not want to conceal: The maps $S_i = Fe_i : M^{n-1} \to S_1^{n+1}$, $i = 1,\dots,n-1$, are congruences of spheres that intersect S orthogonally in f, and $\hat{f} := Fe_\infty$ is the map of the second intersection points of the spheres S_i and S.

If the frame is f-adapted, then these spheres S_i, and so their second intersection points $\hat{f}$, depend[32] on the choice of lift $f : M^{n-1} \to L^{n+1}$, and the S_i's also depend on the "tangential gauge" of the frame.

If, however, F is S-adapted, then $\hat{f}$ is the second envelope of S.

In Figure T.2 we have condensed some of the developed differential geometric terms, extending our table from Figure T.1 in this way.

1.7.4 Compatibility Conditions. Let F be a frame for a strip (S, f); denote by $\Phi := F^{-1}dF$ its connection form,

$$
\Phi = \begin{pmatrix}
\omega_{11} & \cdots & \omega_{1,n-1} & -\psi_1 & \omega_1 & \chi_1 \\
\vdots & \ddots & \vdots & \vdots & \vdots & \vdots \\
\omega_{n-1,1} & \cdots & \omega_{n-1,n-1} & -\psi_{n-1} & \omega_{n-1} & \chi_{n-1} \\
\psi_1 & \cdots & \psi_{n-1} & 0 & \omega & \chi \\
-\chi_1 & \cdots & -\chi_{n-1} & -\chi & \nu & 0 \\
-\omega_1 & \cdots & -\omega_{n-1} & -\omega & 0 & -\nu
\end{pmatrix}, \qquad (1.1)
$$

where $\omega_{ij} + \omega_{ji} = 0$ since $\Phi \in \mathfrak{o}_1(n+2)$, and $\omega = 0$ since (f, S) is a strip.

---

[32] It should be an instructive exercise for the reader to investigate this dependence; to determine the effect on the connection form of the frame is a nightmare computation that will provide some additional insight, though — the reader should compare his result with our transformation formulas from Section P.2.

Then the Maurer-Cartan equation[33] $0 = d\Phi + \frac{1}{2}[\Phi \wedge \Phi]$ is equivalent to
(i) the *Gauss equations*

$$\varrho_{ij} := d\omega_{ij} + \sum_k \omega_{ik} \wedge \omega_{kj} \;=\; \psi_i \wedge \psi_j + \omega_i \wedge \chi_j + \chi_i \wedge \omega_j, \quad (1.2)$$

(ii) the *Codazzi equations*

$$\begin{aligned} d\psi_i + \sum_j \omega_{ij} \wedge \psi_j &= \chi \wedge \omega_i \\ d\omega_i + \sum_j \omega_{ij} \wedge \omega_j &= \nu \wedge \omega_i \\ d\chi_i + \sum_j \omega_{ij} \wedge \chi_j &= \psi_i \wedge \chi + \chi_i \wedge \nu, \end{aligned} \quad (1.3)$$

(iii) and the *Ricci equations*

$$\begin{aligned} 0 &= \sum_i \psi_i \wedge \omega_i \\ d\chi &= \sum_i \chi_i \wedge \psi_i - \nu \wedge \chi \\ d\nu &= \sum_i \chi_i \wedge \omega_i. \end{aligned} \quad (1.4)$$

1.7.5 The equivalence is verified by straightforward computation.

In the case where F is an adapted frame (adapted to S or f) we want to give some geometric interpretations, for the coefficient forms of Φ as well as for the Gauss-Codazzi-Ricci equations; it should be a worthwhile exercise for the reader to verify these geometric interpretations:

1.7.6 S-adapted frame ($\chi = 0$). The induced metric and second fundamental form of the sphere congruence $S : M^{n-1} \to S_1^{n+1}$ are given by

$$I = \langle dS, dS \rangle = \sum_i \psi_i^2 \quad \text{and} \quad I\!I = (\sum_i \omega_i \psi_i)\hat{f} + (\sum_i \chi_i \psi_i)f. \quad (1.5)$$

Introducing the dual vector fields v_i of the first fundamental forms ψ_i, that is, $\psi_i(v_j) = \delta_{ij}$ or, equivalently, $dS(v_i) = -S_i$, the Weingarten tensor fields of S become

$$A_f = \sum_i v_i \omega_i \quad \text{and} \quad A_{\hat{f}} = \sum_i v_i \chi_i. \quad (1.6)$$

The first and second Ricci equations (1.4) then show that A_f and $A_{\hat{f}}$ are self-adjoint tensor fields — and are, therefore, diagonalizable by orthonormal vector fields. The induced covariant derivative is given by

$$\nabla v_i = -\sum_j \omega_{ij} v_j, \quad (1.7)$$

---

[33] For two Lie algebra—valued 1-forms, we define $[\Phi \wedge \Psi](x,y) := [\Phi(x), \Psi(y)] - [\Phi(y), \Psi(x)]$; note that $[\Phi \wedge \Psi]$ is symmetric in Φ and Ψ since $[.,.]$ is skew. In the case of a matrix Lie algebra, where $[.,.]$ is the commutator, we get $[\Phi \wedge \Phi](x,y) = 2\Phi \wedge \Phi(x,y)$, where the second $\wedge$ means matrix multiplication with $\wedge$ for the components.

and the forms ϱ_{ij} provide its curvature tensor,

$$R(.,.)v_i = \sum_j \varrho_{ij} v_j . \tag{1.8}$$

The normal connection $\nabla^\perp$ of S is given by

$$\nabla^\perp f = \nu f \quad \text{and} \quad \nabla^\perp \hat{f} = -\nu \hat{f} , \tag{1.9}$$

and the normal curvature becomes

$$R^\perp(.,.)f = -d\nu\, f \quad \text{and} \quad R^\perp(.,.)\hat{f} = d\nu\, \hat{f} . \tag{1.10}$$

The third Ricci equation (1.4) then shows that f and $\hat{f}$ can be normalized to be parallel normal fields if and only if $d\nu = 0$, that is, if and only if the Weingarten tensor fields A_f and $A_{\hat{f}}$ can be diagonalized simultaneously.

1.7.7 f-adapted frame ($\nu = 0$). Here, the induced metric and second fundamental form of the immersion $f : M^{n-1} \to \mathbb{R}_1^{n+2}$ into the light cone (note that we consider f as an immersion into the ambient Minkowski space in order to make the normal bundle a nondegenerate metric vector bundle) read

$$I = \sum_i \omega_i^2 \quad \text{and} \quad I\!I = (\sum_i \omega_i \psi_i)S - (\sum_i \omega_i \chi_i)f - I\,\hat{f} . \tag{1.11}$$

Taking v_i to be the dual vector fields of the first fundamental forms ω_i, that is, $df(v_i) = S_i$, the corresponding Weingarten tensor fields become

$$A_f = -\sum_i v_i \omega_i = -id, \quad A_{\hat{f}} = -\sum_i v_i \chi_i, \quad \text{and} \quad A_S = \sum_i v_i \psi_i , \tag{1.12}$$

all three being self-adjoint, as the first and third Ricci equations show. As before, the ω_{ij}'s and ϱ_{ij}'s define the covariant derivative and curvature tensor of the induced metric; as we will study conformal flatness of hypersurfaces using this framework, it will be convenient to express the Weyl and Schouten tensors, $\eta_{ij} = w(.,.,v_i,v_j)$ and $s(.,v_i) = \sigma_i$, in terms of the occuring forms by

$$\sigma_i = \tfrac{1}{n-2}(\varrho_i - \tfrac{\varrho}{2(n-1)}\omega_i) \quad \text{and} \quad \eta_{ij} = \varrho_{ij} - (\sigma_i \wedge \omega_j + \omega_i \wedge \sigma_j) , \tag{1.13}$$

where $\varrho_i = \sum_j \varrho_{ij}(.,v_j)$ and $\varrho = \sum_i \varrho_i(v_i)$ give the Ricci tensor and scalar curvature. The normal connection of f is given by

$$\nabla^\perp S = -\chi f, \quad \nabla^\perp f = 0, \quad \text{and} \quad \nabla^\perp \hat{f} = \chi S . \tag{1.14}$$

The second Ricci equation then shows that the normal bundle of f is flat ($d\chi = 0$) if and only if the Weingarten tensors A_S and $A_{\hat{f}}$ commute, that is, if and only if they can be diagonalized simultaneously; in that case, the normal gauge transformation

$$S \to S + a\,f \quad \text{and} \quad \hat{f} \to \hat{f} - a\,S - \tfrac{1}{2}a^2\,f\,, \tag{1.15}$$

where $da = \chi$, yields a parallel normal frame for f.

1.7.8 Now suppose that $f : M^{n-1} \to Q^n_\kappa$ is an immersion into a quadric of constant curvature, $\langle \mathcal{K}, f \rangle \equiv -1$. A canonical choice of an enveloped sphere congruence is its tangent plane congruence, $S : M^{n-1} \to S^{n+1}_1 \cap \{\mathcal{K}\}^\perp$. Since $df, dS \perp \mathcal{K}, S, f$, we learn that an adapted frame F will be adapted to both f and S at the same time — if f and S are immersed. In particular,

$$\hat{f} = \tfrac{\kappa}{2}f - \mathcal{K} \simeq f - 2\tfrac{\langle f, \mathcal{K} \rangle}{\langle \mathcal{K}, \mathcal{K} \rangle}\mathcal{K}$$

is Möbius equivalent to f (if $\kappa \neq 0$) or constant (if $\kappa = 0$).

Conversely, if there is a $\kappa \in \mathbb{R}$ such that $\tfrac{\kappa}{2}df - d\hat{f} = 0$, then f takes values in the space Q^n_κ of constant curvature κ defined by $\mathcal{K} := \tfrac{\kappa}{2}f - \hat{f}$.

1.7.9 With this construction in mind, we motivate the definition of a curvature sphere of a hypersurface in the conformal n-sphere:

Let $f : M^{n-1} \to Q^n_\kappa$ be a hypersurface in a quadric of constant curvature, and let $S : M^{n-1} \to S^{n+1}_1 \cap \{\mathcal{K}\}^\perp$ denote its tangent plane congruence. Note that S can be interpreted as the Gauss map of f as a hypersurface in Q^n_κ since $S(p) \in T_{f(p)}Q^n_\kappa = \{f(p), \mathcal{K}\}^\perp$ for $p \in M^{n-1}$. Then[34]

$$I = \langle df, df \rangle \quad \text{and} \quad I\!I_S = -\langle df, dS \rangle = I(., A_S.)$$

yield the first and second fundamental forms of f as an immersion into Q^n_κ.

Now denote by a_i the eigenvalues of the Weingarten tensor A_S, that is, the a_i's are the principal curvatures of $f : M^{n-1} \to Q^n_\kappa$, and let v_i be the corresponding principal vector fields, so that[35] (Rodrigues' formula)

$$(dS + a_i df)(v_i) = df(a_i v_i - A_S v_i) = 0.$$

Then the sphere congruences $S_i := S + a_i f$ consist of touching spheres that have the same (principal) curvature[36] $a_i = -\langle S_i, \mathcal{K} \rangle$ as the hypersurface in

---

[34] Compare this with (1.11): Since f takes values in Q^n_κ, we have $\chi_i = \tfrac{\kappa}{2}\omega_i$, and its (vectorial) second fundamental form becomes $I\!I = I\!I_S S - I \cdot (\hat{f} + \tfrac{\kappa}{2}f) \perp \mathcal{K}$.

[35] Here we use that $df, dS \perp f, S, \mathcal{K}$ (see §1.7.8).

[36] This formula for the curvature of a hypersphere can be checked by assuming that f locally parametrizes the (for that purpose constant) hypersphere in Q^n_κ: Let $\tilde{S}$ be a constant sphere that is parametrized by f; then $\tilde{S} = S + a f$ as f envelopes $\tilde{S}$, and $0 = \langle d\tilde{S}, df \rangle = a I - I\!I_S$ shows that $a = -\langle \tilde{S}, \mathcal{K} \rangle$ is the (constant) curvature of $\tilde{S} \subset Q^n_\kappa$.

the corresponding direction: The $S_i(p)$ are the "curvature spheres" of f at $f(p)$, and the $S_i : M^{n-1} \to S_1^{n+1}$ are the "curvature sphere congruences" of f.

A characteristic property of these curvature sphere congruences S_i is that

$$\partial_{v_i} S_i = dS_i(v_i) = (dS + a_i df)(v_i) + da_i(v_i) \cdot f = \partial_{v_i} a_i \cdot f$$

becomes isotropic (in fact, parallel to f) for an eigendirection v_i of the Weingarten operator A_S that corresponds to the principal curvature a_i.

To see the converse, let $S : M^{n-1} \to S_1^{n+1}$ be a sphere congruence with a regular envelope $f : M^{n-1} \to S^n$ so that $dS(v) \parallel f$ for some vector field v on M^{n-1}. Then we may choose any $\mathcal{K} \in \mathbb{R}_1^{n+2} \setminus \{0\}$ and consider homogeneous coordinates $f : M^{n-1} \to Q_\mathcal{K}^n$ into the corresponding quadric of constant curvature. Now let $T : M^{n-1} \to S_1^{n+1}$ denote the tangent plane map of f and write $S = T + a\,f$: From

$$dS = (dT + a\,df)(v) + da(v)\,f \parallel f$$

we deduce that $(dT + a\,df)(v) \perp df$, so that a is a principal curvature of f (eigenvalue of A_T) by Rodrigues' formula.

1.7.10 Definition. *A sphere congruence $S : M^{n-1} \to \mathbb{RP}_O^{n+1}$ with a regular envelope $f : M^{n-1} \to S^n$ is a* curvature sphere congruence *of f if, at each point $p \in M$, there is a direction $v \in T_p M$ so that $d_p S(v) \in \mathrm{inc}[f(p), S(p)]$.*

1.7.11 Clearly, this definition does not depend on any choices of homogeneous coordinates for f or S, and the curvature spheres of a hypersurface (as well as the corresponding curvature directions; cf., §P.6.7) are invariant under Möbius transformations.[37]

By choosing homogeneous coordinates $S : M^{n-1} \to S_1^{n+1}$ of the sphere congruence to take values in the Lorentz sphere, we have $dS \perp S$, so that the above condition reads $dS(v) \parallel f(p)$. Since also $dS \perp f$ (the enveloping condition), the sphere congruence S becomes a curvature sphere congruence as soon as its induced metric has, at every point $p \in M^{n-1}$, some null direction: If v is a null vector, $|d_p S(v)|^2 = 0$, then we have $d_p S(v) \parallel f(p)$, by the enveloping condition.

Usually, we will avoid this situation and assume that the considered sphere congruences have a nondegenerate (positive definite) induced metric, as discussed above.

---

[37] Actually, the curvature spheres of a hypersurface are invariant under Lie sphere transformations (see [66]).

1.8 Channel hypersurfaces

To conclude this chapter, we discuss channel hypersurfaces as an example[38]:
a "channel hypersurface" is the envelope $f : I \times S^{n-2} \to S^n$ of a 1-parameter
family $S : I \to \mathbb{RP}_O^{n+1}$ of hyperspheres in S^n.

1.8.1 We start by discussing channel hypersurfaces from the point of view
of a "sphere curve." We explicitly do not assume that its envelope is immersed. At the end of the section, we will switch viewpoints and discuss
the notion of a channel hypersurface by considering the (regular) hypersurface as the primary object of interest, and we will introduce the notion
of a "branched channel hypersurface" that will become important in the
following chapter.

1.8.2 Not every 1-parameter family of spheres possesses such an envelope.
Assuming that $f(\{t\} \times S^{n-2})$ is an $(n-2)$-sphere (as it generally should be),
a necessary condition — since f envelopes S — is that $S(t)$ and $S'(t)$ are
both perpendicular to the Minkowski n-space that defines this $(n-2)$-sphere
(see Figure T.1). In particular, $(S \wedge S')(t)$ has to be spacelike, that is, the
tangent line $\mathrm{inc}[S(t), S'(t)]$ of S at $S(t)$ should not intersect $S^n \subset \mathbb{RP}^{n+1}$.

If we (locally) normalize S to take values in the Lorentz sphere, this is
equivalent to the sphere curve being spacelike, $|S'| > 0$, since $S \perp S'$.

The next lemma shows how to construct the envelope f from the (regular)
sphere curve S and gives an additional condition on S under which the
envelope f is regular:

1.8.3 Lemma. *Let $S : I \to S_1^{n+1}$ be a spacelike sphere curve, $|S'|^2 > 0$,
and choose an orthonormal[39] parallel normal frame $(S_0, S_1, \ldots, S_{n-1})$ for
the sphere curve. Then*

$$f : I \times S^{n-2} \to L^{n+1}, \quad (t, \vartheta) \mapsto f(t, \vartheta) := S_0(t) + \sum_{i=1}^{n-1} \vartheta_i S_i(t)$$

*envelopes S, and $f : \{t\} \times S^{n-2} \to S(t) \cap S'(t)$ is a diffeomorphism onto
the "generating spheres" $S(t) \cap S'(t)$ for any fixed $t \in I$.*

*f is an immersion if and only if all osculating planes $\mathrm{inc}[S(t), S'(t), S''(t)]$
of S intersect S^n transversally.[40]*

1.8.4 *Proof.* f envelopes S since $\langle f, S \rangle = 0$ and $\langle df, S \rangle = -\langle f, S'dt \rangle \equiv 0$.

---

[38] This will become important, for example, in the context of our discussion of conformally
flat hypersurfaces in the next chapter (compare [61]). See also [124]. A more general notion of
channel hypersurfaces than the one we are going to discuss here was investigated in [6] (cf., [66]).

[39] Remember that the scalar products of parallel vector fields along a curve are constant.
"Orthonormal" means: $1 = -|S_0|^2 = |S_i|^2$ and $S_i \perp S_j$ for $i \neq j$.

[40] In particular, for f to be regular, the sphere curve has to be curved, that is, $S(t)$, $S'(t)$,
and $S''(t)$ have to be linearly independent.

For the rest of the proof we assume, without loss of generality, that S is parametrized by arc length, $|S'|^2 \equiv 1$.

Introducing the curvatures $\kappa_i := \langle S'', S_i \rangle$ of the sphere curve S, the induced metric of f becomes

$$\langle d_{(t,\vartheta)}f, d_{(t,\vartheta)}f \rangle = (\kappa_0(t) + \textstyle\sum_{i=1}^{n-1} \kappa_i(t)\vartheta_i)^2 \, dt^2 + d\vartheta^2 \,.$$

Thus, for fixed $t \in I$, $f(t,.) : S^{n-2} \to \{S(t), S'(t)\}^{\perp}$ is an isometry and hence a regular parametrization of the generating $(n-2)$-sphere $S(t) \cap S'(t)$.

However, f may fail to be an immersion (into S^n) because $\frac{\partial}{\partial t} f(t, \vartheta)$ may become isotropic for some (t, ϑ). By interpreting the forefactor of dt^2 in the above formula as the squared inner product of the curvature vector $(-\kappa_0, \dots, \kappa_{n-1})$ and the isotropic vector $(1, \vartheta_1, \dots, \vartheta_{n-1})$ in the Minkowski space $\mathbb{R}_1^n$, we learn that f induces a positive definite metric (is an immersion) if and only if the curvature vector is timelike, that is, if and only if $|(S + S'')|^2 < 0$ for all t. Thus, in the projective setting, f is an immersion if and only if the osculating planes $\mathrm{inc}[S, S', S''](t)$ of S all intersect $S^n \subset \mathbb{RP}^{n+1}$ transversally. ◁

1.8.5 Example. We consider the case of a "planar" curve $S : I \to S_1^4$ of spheres in S^3, that is, the osculating planes $\mathrm{inc}[S, S', S''] \subset \mathbb{RP}^4$ are a fixed plane. There are three cases to consider, depending on the signature of the plane of the curve:

(i) $\mathrm{inc}[S, S', S'']$ intersects S^3 transversally (that is, f is an immersion). Then this intersection yields a fixed circle; by describing that circle in terms of two fixed, orthogonally intersecting spheres $S_1, S_2 \in S_1^4$, we realize that all spheres $S(t)$ intersect S_1 and S_2, and hence this circle, orthogonally. Stereographically projecting $S^3 \to \mathbb{R}^3$ from a point $p_\infty \in S_1 \cap S_2$, this circle becomes a straight line in $\mathbb{R}^3$ that intersects the spheres S of the 1-parameter family orthogonally: The immersion f becomes a surface of revolution about that circle as an axis of rotation.

(ii) $\mathrm{inc}[S, S', S'']$ touches S^3 in a point p_∞. In this case, all spheres $S(t)$ contain that point; choosing a (any) fixed sphere $\tilde{S} \in \mathrm{pol}[S, S', S'']$ in the polar line of the osculating plane yields another sphere that contains p_∞ and intersects the spheres $S(t)$ orthogonally. Stereographically projecting $S^3 \to \mathbb{R}^3$ from p_∞, we obtain a 1-parameter family of planes $t \to S(t)$ that intersect a fixed plane $\tilde{S}$ orthogonally: f becomes a cylinder — that fails to be immersed at the point p_∞ at infinity.

(iii) $\mathrm{inc}[S, S', S'']$ does not meet S^3. Here we consider the two points p_0 and p_∞ of the intersection of the polar line $\mathrm{pol}[S, S', S'']$ of the osculating plane with S^3: The spheres $S(t)$ of our 1-parameter family all contain both of these points. Stereographically projecting $S^3 \to \mathbb{R}^3$ from p_∞

yields a family $t \rightarrow S(t)$ of planes that contain p_0 and intersect all spheres of the pencil $\mathrm{inc}[p_0, p_\infty]$ orthogonally; those become concentric spheres with a common center p_0, and f becomes a cone with vertex p_0 — failing to be immersed at p_0 and p_∞.

1.8.6 Reverting these arguments, we see that any surface of revolution, cylinder, or cone is the envelope of a planar sphere curve.[41]

Note that this example can, with minor reformulations, be generalized to channel hypersurfaces in any S^n; we leave this as an exercise to the reader.

1.8.7 As an application of the just-derived classification of channel surfaces with a planar sphere curve we obtain a Möbius geometric classification of Dupin cyclides[42] (cf., [217] or [66]), that is, of surfaces that are channel in two different ways[43]:

1.8.8 Classification of Dupin Cyclides. *If a surfaces in S^3 is a channel surface in two different ways, then it is Möbius equivalent to (part of) a circular torus of revolution, a circular cylinder, or a circular cone.*

1.8.9 *Proof.* Let $S_i : I_i \rightarrow S_1^4$, $i = 1, 2$, denote the two (different) 1-parameter families of spheres, and let $f : I_1 \times I_2 \rightarrow S^3$ be their common envelope. We *assume* the sphere curves to be curved, that is, S_i, S_i', and S_i'' are linearly independent: If one of them was not curved, then it would describe an elliptic sphere pencil, that is, a fixed circle, so that f would be highly degenerate.

Since $S_1(s)$ and $S_2(t)$ touch at $f(s, t)$, the spheres $S_1(s)$ and $S_2(t)$ belong to one parabolic sphere pencil, and $S_1(s)$, $S_2(t)$, and $f(s, t)$ are (as points in projective space) collinear:

$$\langle S_1(s), S_2(t) \rangle \equiv 1 \quad \text{and} \quad f(s, t) = S_1(s) - S_2(t),$$

up to scale. Consequently, by taking derivatives of $\langle S_1(s), S_2(t) \rangle$ with respect to s and t, respectively, we find $S_i', S_i'' \perp S_j, S_j', S_j''$ for $i \neq j$. In particular, $S_1' \perp S_2'$, so that the generating circles $S_1(s) \cap S_1'(s)$ and $S_2(t) \cap S_2'(t)$ intersect orthogonally in $f(s, t)$ for any choice of $(s, t) \in I_1 \times I_2$. Moreover, because we required the sphere curves S_i to be curved so that S_i' and S_i'' are linearly independent, we have

$$\mathrm{inc}[S_1, S_1', S_1''](s) = \mathrm{pol}[S_2', S_2''](t) \quad \text{and} \quad \mathrm{inc}[S_2, S_2', S_2''](t) = \mathrm{pol}[S_1', S_1''](s)$$

---

[41] Compare [8] for an investigation of Möbius equivariant surfaces; see also [267] and [268].

[42] From a Lie geometric point of view, all Dupin cyclides are the same (Lie equivalent).

[43] Note that a surface in Euclidean 3-space qualifies as a channel surface if and only if one of its principal curvatures is constant along its curvature lines, as we will discuss at the end of this section. Thus a surface in Euclidean space is a Dupin cyclide if and only if Weingarten's invariant 1-form $\Omega = 0$ (see §P.6.5) vanishes.

for all $(s,t) \in I_1 \times I_2$. As a consequence, the osculating planes $\mathrm{inc}[S_i, S_i', S_i'']$ of the two sphere curves are fixed.

Now, three cases can occur:

1.8.10 If $\mathrm{inc}[S_1, S_1', S_1'']$ does not intersect $S^3 \subset \mathbb{RP}^4$, then f stereographically projects to a cone with the circles $S_1 \cap S_1'$ as its generators; since the spheres S_2' belong to the hyperbolic pencil $\mathrm{pol}[S_1, S_1', S_1''] = \mathrm{inc}[p_0, p_\infty]$ of spheres with center p_0 in $\mathbb{R}^3 \cong S^3 \setminus \{p_\infty\}$ — which contain the other family of generating circles $S_2 \cap S_2'$ — it is a circular cone.

In homogeneous coordinates: Denote $\mathrm{span}\{S_i, S_i', S_i''\} =: E_i$ and choose a pseudo-orthonormal basis $(e_1, e_2, e_3, p_0, p_\infty)$ of $\mathbb{R}_1^5$, with $\langle p_0, p_\infty \rangle = 1$, so that $E_1 = \mathrm{span}\{e_1, e_2, e_3\}$ and $E_2 = \mathrm{span}\{e_3, p_0, p_\infty\}$. Since $S_i' \perp E_j$ for $i \neq j$, we find, up to reparametrization,

$$
\begin{aligned}
S_1(s) &= \sin\alpha \cdot [\cos(s)e_1 + \sin(s)e_2] + \cos\alpha \cdot e_3, \\
S_2(t) &= \sec\alpha \cdot e_3 + \tfrac{1}{\sqrt{2}} \tan\alpha \cdot [e^t p_0 - e^{-t} p_\infty],
\end{aligned}
$$

with some constant α; consequently, $f(s,t) = S_1(s) - S_2(t)$ becomes a cone in $\mathbb{R}^3 \cong Q_0^3$ with p_∞ as the point at infinity:

$$
f(s,t) \simeq e^{-t}\sqrt{2}[\cos\alpha \cdot (\cos(s)e_1 + \sin(s)e_2) - \sin\alpha \cdot e_3] - p_0 + e^{-2t}p_\infty.
$$

1.8.11 If $\mathrm{inc}[S_1, S_1', S_1'']$ meets S^3 in a point $p_\infty \in S^3$, then f yields a cylinder in $\mathbb{R}^3 = S^3 \setminus \{p_\infty\}$ and its generators $S_1 \cap S_1'$ all contain p_∞; since all spheres S_2' project to planes (spheres containing p_∞) that intersect this cylinder orthogonally and contain the other family of generating circles $S_2 \cap S_2'$, it becomes a circular cylinder.

In homogeneous coordinates: Here $E_1 = \mathrm{span}\{e_1, e_2, p_\infty\}$, with the above notations; $S_1(s) = \cos(s)e_1 + \sin(s)e_2 + \alpha(s)\,p_\infty$ with some function α; and $S_2(t) = c_1 e_1 + c_2 e_2 + (c_3 + t)e_3 + c_0 p_0 + \tfrac{1}{2}(t^2 + c_\infty)p_\infty$, without loss of generality, since $S_2' \perp E_1$; $|S_2|^2 \equiv 1$ and $\langle \dot{S}_1, S_2 \rangle \equiv 1$ then yield $c_3 = 0$, $c_0 = -1$, $c_\infty = c_1^2 + c_2^2 - 1$, and $\alpha(s) = c_1 \cos s + c_2 \sin s - 1$:

$$
\begin{aligned}
S_1(s) &= \cos(s)e_1 + \sin(s)e_2 + (c_1 \cos s + c_2 \sin s - 1)\,p_\infty, \\
S_2(t) &= c_1 e_1 + c_2 e_2 + t\,e_3 - p_0 + \tfrac{1}{2}(c_1^2 + c_2^2 + t^2 - 1)p_\infty.
\end{aligned}
$$

Thus $f(s,t) = S_1(s) - S_2(t)$ becomes a circular cylinder in $\mathbb{R}^3 \cong Q_0^3$:

$$
\begin{aligned}
f(s,t) \simeq &-[(\cos s - c_1)e_1 + (\sin s - c_2)e_2 - t\,e_3] \\
&-p_0 + \tfrac{1}{2}((\cos s - c_1)^2 + (\sin s - c_2)^2 + t^2)p_\infty.
\end{aligned}
$$

1.8.12 If $\mathrm{inc}[S_1, S_1', S_1'']$ intersects $S^3 \subset \mathbb{RP}^4$ transversally, we may restrict ourselves to the case where $\mathrm{inc}[S_2, S_2', S_2'']$ also intersects S^3 transversally (otherwise we are back to one of the previous cases with the roles of S_1 and S_2 interchanged). Here, f projects to a surface of revolution in $\mathbb{R}^3$ in two ways, where the the intersection of the osculating planes $\mathrm{inc}[S_i, S_i', S_i'']$ of S_i with S^3 provides the respective axis of revolution; since all spheres S_j' contain the axis given by the osculating plane of S_i, $i \neq j$, f becomes a circular torus of revolution in either way.

In homogeneous coordinates: Here, we choose an honest orthonormal basis $(e_0, e_1, \ldots, e_4)$ of $\mathbb{R}_1^5$, $1 = -|e_0|^2 = |e_i|^2$ and $e_i \perp e_j$ for $i \neq j$, such that we have $E_1 = \mathrm{span}\{e_0, e_1, e_2\}$ and $E_2 = \mathrm{span}\{e_0, e_3, e_4\}$. Then

$$
\begin{aligned}
S_1(s) &= \sinh\alpha \cdot e_0 + \cosh\alpha \cdot [\cos(s)e_1 + \sin(s)e_2], \\
S_2(t) &= -\tfrac{1}{\sinh\alpha}\, e_0 + \tfrac{\cosh\alpha}{\sinh\alpha}\, [\cos(t)e_3 + \sin(t)e_4],
\end{aligned}
$$

with a constant α, and

$$
f(s,t) \simeq e_0 + \tfrac{\sinh\alpha}{\cosh\alpha}[\cos(s)e_1 + \sin(s)e_2] - \tfrac{1}{\cosh\alpha}[\cos(t)e_3 + \sin(t)e_4]
$$

becomes a Clifford torus in $S^3 \cong Q_1^3$ that stereographically projects to a circular torus of revolution in $\mathbb{R}^3$ in two ways.   ◁

1.8.13 Note how, in the cases of circular cones and cylinders, the osculating plane of one of the enveloped sphere curves does not intersect S^3 transversally: This corresponds to the envelope being singular at two or one point(s), that is, two or one of the generating circles degenerates to a point.

1.8.14 So far we have considered "sphere curves," dealing with their envelopes as derived objects that can be constructed from the curve under weak regularity assumptions. In the next chapter we will reconsider channel hypersurfaces in a different context, where the hypersurface will be of primary interest and the enveloped sphere congruence will be a derived object: There, we will not be able (and it will not be desirable, cf., [102]) to keep the regularity requirements on the enveloped sphere curve. Thus we will weaken the regularity assumption on the enveloped sphere congruence while requiring the envelope to be an immersion:

1.8.15 Definition. *A regular map $f : M^{n-1} \to S^n$ is called a branched channel hypersurface if it envelopes a sphere congruence S with* $\mathrm{rk}\,dS \leq 1$.

1.8.16 To conclude this chapter, we want to prove that a branched channel hypersurface can be characterized by a slightly weaker condition than the one given in the definition when the dimension is high enough. This characterization[44] will turn out to be useful later:

---

[44] Compare [6] or [66].

1.8.17 Lemma. *A spacelike submanifold* $f : M^{n-1} \to L^{n+1}$, $n \geq 4$, *of the lightcone defines a branched channel hypersurface if and only if its shape operators* A_S *with respect to some spacelike normal field* $S : M^{n-1} \to S_1^{n+1}$ *have* $\mathrm{rk}\, A_S \leq 1$.

1.8.18 *Proof.* Clearly, if S is a spacelike normal field (defining a sphere congruence) with $\mathrm{rk}\, dS \leq 1$, then $\mathrm{rk}\, A_S \leq 1$.

The converse is a consequence of the Codazzi equations (1.3): Using an f-adapted frame (cf., §1.7.7), the equations for the ψ_i read

$$(\nabla_x A_S - \chi(x))y = (\nabla_y A_S - \chi(y))x$$

when written invariantly. Now, choosing local orthonormal vector fields x, y in the $\ker A_S$ distribution, we obtain

$$dS(y) = -\chi(y)g(x,x)f = g((\nabla_x A_S)y - (\nabla_y A_S)x - \chi(x)y, x)f = 0,$$

showing that we have $dS(y) = 0$, for $y \in \ker A_S$. ◁

1.8.19 A change $S \mapsto S + af$ of the normal field (enveloped sphere congruence) with a function a results in a change $A_S \mapsto A_S - a \cdot id$ of the corresponding Weingarten tensor field: $\langle dS, df \rangle \mapsto \langle dS + a\, df, df \rangle$.

Thus, a hypersurface in S^n, $n \geq 4$, is a branched channel hypersurface as soon as it is "quasi-umbilic," that is, as soon as the Weingarten tensor field A_S — with respect to *any* enveloped sphere congruence S — has an eigenvalue of multiplicity $n - 2$.

1.8.20 Note that the above lemma does not hold in dimension $n = 3$: There, the Weingarten tensor fields with respect to a congruence of curvature spheres S (see §1.7.10) have $\mathrm{rk}\, A_S \leq n - 1 - 1 = 1$ for any surface. The reason why the above proof fails is the same as the one that made our definition of hyperspheres fail in dimension $n = 2$ (cf., §P.7.15): The Codazzi equations cannot be applied to prove that a "totally umbilic" curve is a circle — because "totally umbilic" is not a condition on curves.

However, it remains true that a surface in the conformal 3-sphere is a channel surface if it carries one family of circular curvature lines.[45] Namely, by Joachimsthal's theorem from §P.7.6, any sphere of the elliptic pencil (cf., Figure T.1) that contains the circular curvature line intersects the surface at a constant angle. In particular, if the sphere touches the surface in one point of the circle, then it touches along the circle, and hence we find a 1-parameter family of spheres that is enveloped by the surface.

---

[45] If we consider the surface in Euclidean ambient space $\mathbb{R}^3$, then it has circular curvature lines if and only if one of its principal curvatures is constant along its curvature lines: Suppose that the principal curvature a_i of the surface f, with unit normal field n, does not vanish; then the focal point $f + \frac{1}{a_i} n$ of f is constant along the corresponding curvature line if and only if a_i is.

Chapter 2

Application: Conformally flat hypersurfaces

As a first application of the just-elaborated classical model of Möbius geometry we want to study conformally flat hypersurfaces in the conformal n-sphere. Here, we will consider the (regular) hypersurface as the primary object of interest and an enveloped (1-parameter) family of spheres as its "Gauss map," in contrast to the way we considered channel hypersurfaces in most of Section 1.8.

Conformally flat hypersurfaces have been a topic of interest for some time. In [61], É. Cartan gave a complete local classification of conformally flat hypersurfaces[1] in S^n when $n \geq 5$, that is, in the case where conformal flatness is detected by the vanishing of the Weyl tensor (cf., §P.5.1). Cartan's result was reproven several times; we will present a proof close to Cartan's original proof to show that any conformally flat hypersurface in dimension $n \geq 5$ is (a piece of) a branched channel hypersurface, using the model for Möbius geometry developed in the previous chapter. Later, the global structure of compact conformally flat (submanifolds and) hypersurfaces was analyzed, showing that the conformal structure has, in general, no global flat representative (see [199]) deriving restrictions on the intrinsic structure (cf., [198], [222], [194]), and giving a complete geometric description of their shape (see [102]; cf., [173]). However, in contrast to the 2-dimensional case (cf., [124] and [125]), the question of whether a given "classical Schottky manifold[2]" can be immersed conformally seems to be still open (cf., [231] and [271]).

As indicated earlier, in §P.5.7, the situation seems more complicated in the case of 3-dimensional conformally flat hypersurfaces in the conformal 4-sphere. Here, the condition for conformal flatness is a third-order system of partial differential equations for the metric,[3] as opposed to the higher dimensional case where the condition is of second-order (see §P.5.1). It is not hard to see that branched channel hypersurfaces are conformally flat. Thus, the remaining (local) problem is to characterize generic conformally flat hypersurfaces, that is, conformally flat hypersurfaces with three distinct

---

[1] For related results, see also [208], [190], and [195]; for a survey, see [292].

[2] This is a conformal manifold that is obtained by removing $2k$ disjoint round balls from the round n-sphere and then identifying their boundaries pairwise by means of Möbius transformations [222].

[3] Most approaches to the classification problem are of a local nature; for an entirely different approach, see [68] and [69].

principal curvatures.[4] Cartan [61] gave a rather mysterious characterization in terms of the integrability of the "umbilic distributions." Below we present his characterization, along with a relation with certain maps into the symmetric space of circles, "curved flats" (see [112]), which might make Cartan's characterization less mysterious since his umbilic distributions turn out to be exactly the "root distributions" of the curved flat [145]. In fact, we are going to relate Cartan's result to the integrability of the "conformal fundamental forms" of a generic hypersurface — which leads to the notion of "Guichard nets" [143], a special type of triply orthogonal system.

By constructing the cone, cylinder, or hypersurface of revolution over a surface of constant Gauss curvature in S^3, $\mathbb{R}^3$, or H^3, respectively, we obtain examples of generic conformally flat hypersurfaces in the conformal 4-sphere (see [173]). We will discuss the corresponding Guichard nets and see that they are rather special — showing that these examples do not provide all conformally flat hypersurfaces in the conformal 4-sphere with three distinct principal curvatures (see [143]). More explicit examples of generic conformally flat hypersurfaces that are not of the above type are given by Y. Suyama in [273].

We will conclude the chapter by discussing a certain class of Guichard nets, "cyclic Guichard nets," that lead to families of parallel linear Weingarten surfaces in spaces of constant curvature. An interesting feature of this point of view on linear Weingarten surfaces is that it provides a unified framework to prove analogs of Bonnet's theorem on parallel constant mean and constant Gauss curvature surfaces in space forms (cf., [214] and [146]).

The contents of this chapter are organized as follows:

Section 2.1. We prove Cartan's theorem that conformally flat hypersurfaces in ambient dimensions $n \geq 5$ are branched channel hypersurfaces. A notion of a conformal metric for 3-dimensional hypersurfaces in S^4 is introduced.

Section 2.2. The notion of a curved flat is introduced, and basic properties of curved flats are discussed. A relation between conformally flat hypersurfaces in S^4, cyclic systems, and curved flats in the symmetric space of circles is established. The aforementioned conformal metric for 3-dimensional hypersurfaces is shown to arise naturally from the curved flat description of conformally flat hypersurfaces.

Section 2.3. The conformal fundamental forms of a 3-dimensional hypersurface are introduced, and their integrability is shown to be equivalent to the conformal flatness of the hypersurface. We discuss to what extent the conformal fundamental forms determine a conformally flat hypersurface.

---

[4] For an early result, see [118]. Interesting classes of examples may be found in [190].

Finally, the relation of the conformal fundamental forms with Cartan's umbilic distributions is given, and Cartan's respective characterization of conformal flatness is discussed.

Section 2.4. Guichard nets are introduced as certain triply orthogonal systems in (Euclidean) 3-space. Their relation with conformally flat hypersurfaces is revealed. Dupin's classical theorem is obtained as a by-product (a more general version of this theorem will be proved in Chapter 8), and Lamé's equations are derived for later use. Various examples of 3-dimensional conformally flat hypersurfaces are given, and the geometry of their Guichard nets is investigated.

Section 2.5. The geometry of cyclic systems (normal congruences of circles) and of the triply orthogonal system given by a cyclic system are investigated. It is shown that any Guichard cyclic system is a normal line congruence in some space of constant curvature.

Section 2.6. Continuing the line of thought from the previous section, it is proved here that the orthogonal surfaces of a Guichard cyclic system are parallel linear Weingarten surfaces in a space form and, conversely, that any such family of surfaces gives rise to a Guichard cyclic system.

Section 2.7. Bonnet's classical theorem on parallel surfaces of constant Gaussian and constant mean curvature in Euclidean space is obtained as a consequence of the previous discussions on Guichard cyclic systems.

Remark. The reading of this chapter requires familiarity with the classical model of Möbius geometry presented in Chapter 1.

2.1 Cartan's theorem

In this section we will present Cartan's local classification result for conformally flat hypersurfaces in S^n, $n \geq 5$. Our proof will be very similar to the original proof by Cartan [61]. Because we are interested in the geometry of the hypersurface, we will work with an f-adapted frame of a hypersurface $f : M^{n-1} \to S^n$ (cf., §1.7.7).

We start by discussing the conformal flatness of channel hypersurfaces:

2.1.1 Example. Any branched channel hypersurface is conformally flat.

2.1.2 *Proof.* Denote by S the enveloped sphere congruence that characterizes f as a branched channel hypersurface, that is, $\mathrm{rk}\, dS \leq 1$. Further, let $F : M^{n-1} \to O_1(n+2)$ be an f-adapted frame for the strip (S, f). For the components of the connection form (1.1) of F, the assumption $\mathrm{rk}\, dS \leq 1$ implies that $\psi_i \wedge \psi_j = 0$ and $\psi_i \wedge \chi = 0$ for all i, j (recall that, for an f-adapted frame, $\omega = \nu = 0$). Consequently, the Gauss equations (1.2) reduce to $\varrho_{ij} = \chi_i \wedge \omega_j + \omega_i \wedge \chi_j$, that is, the Schouten forms, $\sigma_i = \chi_i$ (cf., §P.3.2), and the Weyl tensor vanishes, $\eta_{ij} = 0$.

Thus, for $n \geq 5$, we are done: A branched channel hypersurface is conformally flat by the Weyl-Schouten theorem §P.5.1.

For $n = 4$ (the case $n = 3$ is not interesting, anyway; see §P.4.6), the Weyl tensor always vanishes, and the condition for conformal flatness becomes that the Schouten tensor be a Codazzi tensor (see §P.5.1). Here, we employ the Codazzi equations (1.3) for χ_i to obtain $d\sigma_i + \sum_j \omega_{ij} \wedge \sigma_j = 0$, that is, conformal flatness in this case also. ◁

2.1.3 In order to prove the converse, we first state a lemma — that is the key to other interesting results, too. Cartan's classification will then be an immediate consequence.

2.1.4 Lemma. *If $f : M^{n-1} \to \mathbb{R}_1^{n+2}$, $n \geq 4$, is a flat spacelike immersion with values in the light cone — a flat lift of a conformally flat hypersurface in S^n — then f has flat normal bundle (as an immersion into $\mathbb{R}_1^{n+2}$).*

2.1.5 *Proof.* Let F be an f-adapted frame with connection form Φ. Suppose (cf., §1.7.7) that the dual vector fields v_i of ω_i diagonalize the Weingarten tensor field A_S, that is, that we have $\psi_1 = a_i\omega_i$ with suitable functions a_i. With the ansatz $\chi_i = -\sum_j b_{ij}\omega_j$, the Gauss equation (1.2) then yields

$$0 = a_i a_j - b_{ii} - b_{jj} \quad \text{and} \quad 0 = b_{ij} \tag{2.1}$$

for $i \neq j$. In particular, $\chi_i = -b_{ii}\omega_i$, so that $A_{\hat{f}}$ is also diagonal; hence the second Ricci equation (1.4) shows that the normal bundle of f is flat. ◁

2.1.6 Now let i, j, k, l be pairwise distinct indices. From (2.1) we deduce

$$(a_i - a_k)(a_j - a_l) = (b_{ii} + b_{jj}) - (b_{ii} + b_{ll}) - (b_{kk} + b_{jj}) + (b_{kk} + b_{ll}) = 0.$$

Hence $n - 2$ of the a_i's coincide: $a_2 = \ldots = a_{n-1}$, without loss of generality. Then $\operatorname{rk} A_{\tilde{S}} \leq 1$ for the sphere congruence $\tilde{S} := S + a_2 f$ (cf., (1.15)), so that our lemma from §1.8.17 tells us that f is a branched channel hypersurface.

Since we can have four pairwise distinct indices $i, j, k, l \in \{1, \ldots, n-1\}$ as soon as $n \geq 5$, we have just proved a theorem by Cartan [61] (see also [252]):

2.1.7 Theorem (Cartan). *If $f : M^{n-1} \to S^n$, $n \geq 5$, is a conformally flat immersion, then f is a branched channel hypersurface.*

2.1.8 For $n = 4$, Cartan [61] introduces the tensor $B := \sum_{i=1}^3 v_i(\sigma_i - \chi_i)$. As the Weyl tensor vanishes, $\varrho_{ij} = \sigma_i \wedge \omega_j + \omega_i \wedge \sigma_j$, the Gauss equations (1.2) then imply that the coefficients b_{ij} of B satisfy (2.1) when working with a principal f-adapted frame, that is, $\psi_i = a_i\omega_i$, as above. Consequently, $2b_{ii} = a_i a_j + a_i a_k - a_j a_k$ for pairwise distinct indices $i, j, k \in \{1, 2, 3\}$ or, written in an invariant way,

$$B = (\operatorname{tr} A_S \cdot A_S - A_S^2) - \tfrac{1}{4}((\operatorname{tr} A_S)^2 - \operatorname{tr} A_S^2)\, id. \tag{2.2}$$

This formula now holds for *any* light cone lift (choice of homogeneous coordinates) of an immersion $f : M^3 \to S^4$ because the induced curvature has been built in — in contrast to the above reasoning, where we chose a flat lift.

2.1.9 Obviously, the Cartan tensor B has the same eigendirections as A_S, and the induced metric of f is conformally flat, that is, the Schouten tensor is a Codazzi tensor, if and only if B satisfies

$$
\begin{aligned}
& (\nabla_x B - \chi(x) A_S) y = (\nabla_y B - \chi(y) A_S) x \\
\Leftrightarrow \quad & (\nabla_x B - A_S \nabla_x A_S) y = (\nabla_y B - A_S \nabla_y A_S) x,
\end{aligned}
\tag{2.3}
$$

since $d\sigma_i + \sum_j \omega_{ij} \wedge \sigma_j = 0 \Leftrightarrow d(\sigma_i - \chi_i) + \sum_j \omega_{ij} \wedge (\sigma_j - \chi_i) = \chi \wedge \psi_i$ by (1.3) for an f-adapted frame F; the equivalence with the second equation is a consequence of the Codazzi equations, again.

For example, for an immersion $f : M^3 \to Q_\kappa^4$ into a space of constant curvature, $\chi = 0$ so that the Schouten tensor is Codazzi if and only if the Cartan tensor is.

2.1.10 Also note that the Cartan tensor is not invariant under "admissible changes" $(f, S) \mapsto (e^u f, S + af)$ of the strip (f, S). By (2.2), the transformation behavior of B is determined by that of A_S. Namely,

$$
\langle df, df \rangle \mapsto e^{2u} \langle df, df \rangle \quad \text{and} \quad \langle df, dS \rangle \mapsto e^u \langle df, dS + a \, df \rangle
$$

implies that the Weingarten tensor

$$
A_S \mapsto e^{-u} (A_S - a \cdot id) \quad \Rightarrow \quad B \mapsto e^{-2u} (B - a \cdot A_S + \tfrac{1}{2} a^2 \cdot id).
\tag{2.4}
$$

Thus $B - \tfrac{1}{2} A_S^2 \mapsto e^{-2u} (B - \tfrac{1}{2} A_S^2)$, so that $I(., (B - \tfrac{1}{2} A_S^2).)$ defines a scalar product that is invariant under such admissible changes:

2.1.11 Definition. *The scalar product $C := I(., (2B - A_S^2).)$ will be called the conformal metric of an immersion $f : M^3 \to S^4$ into the conformal 4-sphere.*

2.1.12 By choosing $f : M^3 \to Q_\kappa^4$ to take values in a space of constant curvature, S can be chosen to be the tangent plane map of f, that is, its unit normal field in Q_κ^4, so that A_S will be the usual Weingarten tensor field of a hypersurface in a space form.

Before analyzing this conformal metric C further, let us see how it arises in another context.

2.2 Curved flats

In this section we want to examine some consequences of the lemma in §2.1.4 for 3-dimensional conformally flat hypersurfaces. Thus, let $f : M^3 \to L^5$ be a (local: remember that such flat lifts usually do not exist globally [199]) flat lift of a conformally flat hypersurface in the conformal 4-sphere. Then we learned that, besides the tangent bundle, the normal bundle of $f : M^3 \to \mathbb{R}_1^6$ is a flat subbundle of $f^\star T\mathbb{R}_1^6$ also.

2.2.1 The "generalized Gauss map" γ of any immersion $f : M^3 \to L^5$ with spacelike tangent planes,

$$\gamma : \quad M^3 \quad \to \quad G_+(6,3), \\ p \quad \mapsto \quad \gamma(p) := \operatorname{span}\{d_p f(T_p M)\}, \tag{2.5}$$

provides a map[5] into the Grassmannian of spacelike 3-planes in $\mathbb{R}_1^6$, that is, into the space of circles in S^4 (see Figure T.1). Its target space

$$G_+(6,3) = O_1(6)/K, \quad \text{where} \quad K := O(3) \times O_1(3),$$

is a reductive homogeneous space (cf., [209]): The Lie algebra $\mathfrak{o}_1(6)$ of $O_1(6)$ has a symmetric decomposition $\mathfrak{o}_1(6) = \mathfrak{k} \oplus \mathfrak{p}$, where $\mathfrak{k} = \mathfrak{o}(3) \oplus \mathfrak{o}_1(3)$ is the isotropy algebra and $\mathfrak{p}$ is a complementary subspace, so that

$$[\mathfrak{k}, \mathfrak{k}] \subset \mathfrak{k}, \quad [\mathfrak{k}, \mathfrak{p}] \subset \mathfrak{p}, \quad \text{and} \quad [\mathfrak{p}, \mathfrak{p}] \subset \mathfrak{k}. \tag{2.6}$$

2.2.2 Now, by taking γ to be the generalized Gauss map of a flat spacelike immersion $f : M^3 \to L^5$ and $F : M^3 \to O_1(6)$ a lift of γ with connection form[6] $\Phi = \Phi_{\mathfrak{k}} + \Phi_{\mathfrak{p}}$, the Gauss and Ricci equations, (1.2) and (1.4), decompose — since both bundles, $\gamma(M) \to M$ as well as $\gamma^\perp(M) \to M$, are flat, we have in addition to the Gauss-Ricci and Codazzi equations,[7]

$$0 = d\Phi_{\mathfrak{k}} + \tfrac{1}{2}[\Phi_{\mathfrak{k}} \wedge \Phi_{\mathfrak{k}}] + \tfrac{1}{2}[\Phi_{\mathfrak{p}} \wedge \Phi_{\mathfrak{p}}] \\ 0 = d\Phi_{\mathfrak{p}} + [\Phi_{\mathfrak{k}} \wedge \Phi_{\mathfrak{p}}], \tag{2.7}$$

the integrability[8] of the form $\Phi_{\mathfrak{k}}$: $0 = d\Phi_{\mathfrak{k}} + \tfrac{1}{2}[\Phi_{\mathfrak{k}} \wedge \Phi_{\mathfrak{k}}]$ or, equivalently,

$$0 = [\Phi_{\mathfrak{p}} \wedge \Phi_{\mathfrak{p}}]. \tag{2.8}$$

---

[5] Note that this generalized Gauss map *does* depend on the lift of an immersion $f : M^3 \to S^4$ into the conformal 4-sphere: It is not a well-defined notion in Möbius geometry.

[6] In our setup from §1.7.4, this symmetric decomposition of Φ will be decomposition into diagonal and off-diagonal blocks in (1.1).

[7] For the definition of the product $[\Phi \wedge \Phi]$, see Footnote 1.33; note that, by (2.6), the Maurer-Cartan equation splits into a $\mathfrak{k}$-part and a $\mathfrak{p}$-part.

[8] This means that F can be chosen to provide parallel basis fields for both bundles: If H is a solution of $H^{-1}dH = \Phi_{\mathfrak{k}}$, then the gauge transformation $F \mapsto FH^{-1}$ causes $\Phi_{\mathfrak{k}} \mapsto 0$ to vanish, that is, the derivatives of sections Fe_i of one bundle take values in the other bundle.

This last condition makes the generalized Gauss map γ of our flat light cone immersion qualify as a "curved flat":

2.2.3 Definition. *A map* $\gamma : M^m \to G/K$ *into a reductive homogeneous space* G/K *is called a curved flat if* $[\Phi_{\mathfrak{p}} \wedge \Phi_{\mathfrak{p}}] = 0$, *where* $F : M^m \to G$ *is any lift of* γ *and*

$$F^{-1}dF = \Phi = \Phi_{\mathfrak{k}} + \Phi_{\mathfrak{p}} : TM \to \mathfrak{g} = \mathfrak{k} \oplus \mathfrak{p}$$

is the $\mathfrak{k} \oplus \mathfrak{p}$-*decomposition of its connection form.*

2.2.4 Note that the condition $[\Phi_{\mathfrak{p}} \wedge \Phi_{\mathfrak{p}}] = 0$ to be a curved flat does not independ on the lift (frame) $F : M^m \to G$ of $\gamma : M^m \to G/K$. When computing the effect of a gauge transformation (change of frame) $F \mapsto FH$ on the connection form $\Phi = F^{-1}dF$, where $H : M^m \to K$ a is a map into the isotropy group K, we find

$$\Phi_{\mathfrak{k}} \mapsto \mathrm{Ad}_H \Phi_{\mathfrak{k}} + H^{-1}dH \quad \text{and} \quad \Phi_{\mathfrak{p}} \mapsto \mathrm{Ad}_H \Phi_{\mathfrak{p}}.$$

Hence, the notion of curved flat is a well-defined notion for maps into the reductive homogeneous space.

2.2.5 Geometrically, the curved flat condition says that $\Phi_{\mathfrak{p}}(T_pM)$ is, at every point $p \in M$, an abelian subalgebra of $\mathfrak{p}$. Identifying $d\gamma$ with $\Phi_{\mathfrak{p}}$ (cf., [44]), this means that, in case the ambient space G/K is a symmetric space, γ is tangent to a "flat," that is, to a totally geodesic flat submanifold of the ambient space (see [140]).

Otherwise said, γ is an envelope of a congruence of flats (cf., [112]).

2.2.6 Clearly, by the Gauss equations (1.2), the induced metric of a space-like light cone immersion $f : M^3 \to L^5$ is flat as soon as its generalized Gauss map γ is a curved flat (in the space of circles). Together with §2.1.4, this gives us the converse of our above assertion, and we have the following lemma[9]:

2.2.7 Lemma. *A light cone representative* $f : M^3 \to L^5$ *of an immersion into the conformal 4-sphere is flat if and only if its generalized Gauss map is a curved flat.*

2.2.8 Trying to reconstruct the light cone map $f : M^3 \to L^5$ from the associated curved flat γ, we encounter a problem: *Any* parallel isotropic section $\tilde{f} : M^3 \to L^5$ of the bundle $\gamma^{\perp}(M) \to M$ has γ as its generalized

---

[9] That, in view of the lemma in §2.1.4, is true in any dimension (cf., [145]); however, here we restrict our attention to the 3-dimensional case $n = 4$.

Gauss map. In terms of parallel, pseudo-orthonormal[10] basis fields $(S, f, \hat{f})$ of the bundle, we obtain a 2-parameter family

$$(r, t) \mapsto f_{r,t} := r \cdot \left(\tfrac{1}{\sqrt{2}}(f - \hat{f}) + \tfrac{1}{\sqrt{2}}(f + \hat{f}) \cos t + S \sin t \right)$$

of parallel isotropic sections. Here, the first parameter r is just (constant) scaling of the section and, as such, geometrically not interesting. Thus, we are left with essentially a circle's worth of light cone maps that share (if they are immersed) the generalized Gauss map γ.

2.2.9 And indeed, in terms of Möbius geometry, a map $\gamma : M^3 \to G_+(6, 3)$ into $G_+(6, 3)$ can be interpreted as a congruence of circles: At any point, $\gamma(p)$ is a circle in S^4 (see Figure T.1). Moreover, each circle $\gamma(p)$ intersects $f(M)$ in $f(p)$ orthogonally: When choosing an f-adapted frame $F = (S_1, S_2, S_3, S, f, \hat{f})$, any circle $\gamma(p)$ is the intersection of the spheres $S_1(p)$, $S_2(p)$, and $S_3(p)$ that intersect $S(p)$ orthogonally in $f(p)$ (and $\hat{f}(p)$), as discussed in §1.7.3; but, $S(p)$ touches $f(M)$ in $f(p)$ by the enveloping condition, so that the circle $\gamma(p)$ intersects $f(M)$ orthogonally in $f(p)$.

2.2.10 Moreover, using the above parametrization for the circles, it is an easy exercise to show that, when S, f, and $\hat{f}$ are *parallel* sections of $\gamma^{\perp}$, the maps

$$f_t = \tfrac{1}{\sqrt{2}}(f - \hat{f}) + \tfrac{1}{\sqrt{2}}(f + \hat{f}) \cos t + S \sin t \qquad (2.9)$$

envelope the sphere congruences

$$S_t = -\tfrac{1}{\sqrt{2}}(f + \hat{f}) \sin t + S \cos t. \qquad (2.10)$$

Since all spheres $S_t(p)$ intersect the circles $\gamma(p)$ orthogonally in $f_t(p)$, the maps f_t represent (if they do not degenerate) hypersurfaces in the conformal 4-sphere that intersect the circles of the congruence γ orthogonally.[11]

Hence we have identified the congruence $\gamma : M^3 \to G_+(6, 3)$ of circles as a "cyclic system" (cf., [79]):

2.2.11 Definition. *A map $\gamma : M^{n-1} \to G_+(n + 2, n - 1)$ into the space of circles in the conformal S^n is called a* cyclic system, *or* normal congruence of circles, *if there is a 1-parameter family of hypersurfaces $f_t : M^{n-1} \to S^n$ that intersect all circles orthogonally.*

---

[10] Remember that the scalar products of parallel sections are constant as soon as the connection on the bundle is compatible with the metric.

[11] Note that this is a Möbius geometric generalization of the family of parallel hypersurfaces in the metric geometries.

2.2.12 Thus, given a curved flat $\gamma : M^3 \to G_+(6,3)$, there is a 1-parameter family of light cone maps[12] $f_t : M^3 \to L^5$ that have, if they are immersed, γ as their generalized Gauss map and induce a flat metric: This is an immediate consequence of the lemma in §2.2.7.

We summarize our discussions in the following theorem (cf., [39]):

2.2.13 Theorem. *Given a conformally flat hypersurface, the generalized Gauss map of any flat light cone lift is a curved flat.*

Conversely, any curved flat gives rise to light cone lifts of a 1-parameter family of maps into the conformal 4-sphere; when such a map is immersed[13] it is a flat lift of a conformally flat hypersurface in S^4 with the curved flat as its generalized Gauss map.

2.2.14 Note that the curved flat that we can associate to a conformally flat hypersurface is only locally defined (as flat light cone lifts usually do only exist locally [199]), and it is not unique. Consequently, the Möbius geometric construction that comes from it also is local and not unique (cf., [145]):

2.2.15 Corollary. *Any conformally flat hypersurface is (locally) an orthogonal hypersurface to a cyclic system with all of its orthogonal hypersurfaces being conformally flat; this cyclic system is not unique.*

2.2.16 Example. From §2.1.1, we know that (branched) channel hypersurfaces are conformally flat. Thus, suppose $f : M^3 \to L^5$ is a flat lift of a channel hypersurface and $S : M^3 \to S_1^5$ is the enveloped 1-parameter family of spheres — for simplicity, we exclude umbilics on the hypersurface, so that $\mathrm{rk}\,dS = 1$. Consequently, the corresponding Weingarten tensor A_S has a 2-dimensional kernel, and we may (locally) choose an f-adapted frame such that, in the connection form (1.1), $\psi_2 = \psi_3 = 0$. As usual, we analyze the integrability conditions and then interpret the results geometrically:

By the Gauss equations (1.2), $\chi_i = 0$, $i = 1, 2, 3$, and the Ricci equations (1.4) yield $\psi_1 = a\omega_1$ with a nonvanishing function $a \neq 0$, and $d\chi = 0$. Further, $0 = a\omega_1 \wedge \chi$ and $d(a\omega_1) = \chi \wedge \omega_1 = 0$ by the Codazzi equations (1.3). This last assertion gives us the integrability of the distribution $\omega_1 = 0$, that is, locally $a\omega_1 = ds$; moreover, $d\chi = 0$ and $\chi \wedge ds = 0$ tell us that $\chi = db$ with a function $b = b(s)$.

---

[12] As for parallel hypersurfaces in the metric geometries, regularity of the f_t's depends on "curvature quantities," that is, on the quantities arising in the connection form (1.1). However, if $f = f_0$ is regular, then the maps f_t are (locally) regular for small t.

[13] By using the normal form for Φ_p given by the Cartan-Moore lemma [197] it can be shown that, under certain regularity conditions on the curved flat, there are always immersions among these maps; cf. [112] and [142], [145].

Since $d\hat{f} = S \cdot \chi$ and $\chi \wedge \omega_1 = 0$, the point map $\hat{f}$ is constant along the foliating 2-sphere (pieces) $\omega_1 = 0$ of the channel hypersurface — just as the enveloped 1-parameter family of S hyperspheres is. Hence, all circles of the normal congruence (given by the curved flat associated to the flat lift f) that intersect in a fixed foliating 2-sphere are perpendicular to all spheres of the parabolic pencil given by $S(s)$ and $\hat{f}(s) \in S(s)$. In particular, all circles pass through the fixed point $\hat{f}(s)$. In order to obtain the family $(f_t)_t$ of conformally flat hypersurfaces orthogonal to this cyclic system, we apply a normal gauge transformation

$$S \to S + bf \quad \text{and} \quad \hat{f} \to \hat{f} - bS - \tfrac{1}{2}b^2 f$$

(cf., (1.15): this yields a parallel normal frame for the cyclic system) and use the parametrization (2.9): With $g_t(b) := (1 + \cos t) + b \sin t + \tfrac{1}{2}b^2(1 - \cos t)$, we obtain

$$f_t = \tfrac{1}{\sqrt{2}} g_t(b)\, f + \tfrac{1}{\sqrt{2}} g_t'(b)\, S - \tfrac{1}{\sqrt{2}} g_t''(b)\, \hat{f}.$$

It is then easily checked that $S_t := S - \frac{g_t'(b)}{g_t(b)} \hat{f}$ provides, for each t, a 1-parameter family of spheres that, of course, belong to the parabolic pencils spanned by S and $\hat{f}$ and that is enveloped by f_t: By the definition of S_t we have $\langle S_t, f_t \rangle = 0$, and the enveloping condition $\langle S_t', f_t \rangle = 0$ follows with $\langle S, \hat{f}' \rangle = -\langle S', \hat{f} \rangle = b'$ since $g_t'^2(b) = 2g_t''(b)g_t(b)$, so that $(\frac{g_t'(b)}{g_t(b)})' = -\frac{g_t''(b)}{g_t(b)} b'$.

Consequently, all of the orthogonal hypersurfaces of our cyclic system are channel hypersurfaces also.

2.2.17 Finally, we want to see how the conformal metric C that we defined before, in §2.1.11, arises in the curved flat theory.

Supplying the space $G_+(6,3)$ of circles with the Killing metric, it becomes a pseudo-Riemannian symmetric space with $O_1(6)$ acting on $G_+(6,3)$ by isometries. The Killing metric on $G_+(6,3) = \frac{O_1(6)}{O(3) \times O_1(3)}$ is just a multiple of the trace form $\mathfrak{p} \times \mathfrak{p} \ni (X,Y) \mapsto \operatorname{tr} XY \in \mathbb{R}$ (see [140]); hence the metric induced by γ becomes a multiple of $\operatorname{tr} \Phi_{\mathfrak{p}}^2$.

Now, let γ be the generalized Gauss map of a flat light cone lift of a conformally flat hypersurface in S^4. Assuming that the lift F of γ is an f-adapted frame for the flat light cone map $f : M^3 \to L^5$ such that the Weingarten tensor field A_S with respect to $S = Fe_4$ is diagonal, $\psi_i = a_i \omega_i$ (cf., §1.7.7), we saw in (2.1) that $\chi_i = -b_i \omega_i$ with $b_i + b_j = a_i a_j$. Thus we recover a multiple of our conformal metric from §2.1.11:

$$\tfrac{1}{2} \operatorname{tr} \Phi_{\mathfrak{p}}^2 = \sum_i (2b_i - a_i^2) \omega_i^2 = C.$$

2.3 The conformal fundamental forms

Next we want to analyze the condition (2.3), on the Cartan tensor B, for conformal flatness further. We are interested in "generic hypersurfaces" of the conformal 4-sphere, that is, hypersurfaces with three distinct principal curvatures — if two principal curvatures coincide, then the hypersurface is a channel hypersurface (cf., §1.8.17f) that we already detected to be conformally flat in §2.1.1. Thus we will restrict our attention to the case where the Weingarten operator of any lift $f : M^3 \to L^5$ of a hypersurface with respect to a (any) enveloped sphere congruence $S : M^3 \to S_1^5$ has three distinct eigenvalues.

2.3.1 Let $F : M^3 \to O_1(6)$ denote an f-adapted frame for a strip (f, S), where f is generic. Then, we can (locally) choose the frame to diagonalize the Weingarten tensor field A_S of f with respect to S, that is, $\psi_i = a_i \omega_i$. Thus, with the dual vector fields v_i of the ω_i, the Weingarten and Cartan tensors take the forms

$$A_S = \sum_i v_i a_i \omega_i \quad \text{and} \quad B = \sum_i v_i b_i \omega_i,$$

where, as in §2.1.8, we have $2b_i = a_i a_j + a_i a_k - a_j a_k$ for pairwise distinct indices i, j, k and the a_i's are the eigenvalues of the Weingarten tensor A_S.

2.3.2 Now the second equation in (2.3) detecting conformal flatness reads

$$
\begin{aligned}
0 &= d(b_i \omega_i) + \sum_j \omega_{ij} \wedge (b_j \omega_j) - a_i [d(a_i \omega_i) + \sum_j \omega_{ij} \wedge (a_j \omega_j)] \\
&= (db_i - a_i da_i) \wedge \omega_i + \sum_i (-2b_i + a_i^2 + b_i + b_j - a_i a_j) \omega_{ij} \wedge \omega_j \\
&= \sqrt{2b_i - a_i^2} \cdot d(\sqrt{2b_i - a_i^2}\, \omega_i),
\end{aligned}
$$

$$(2.11)$$

where the second equality relies on the second of the Codazzi equations (1.3) and the last equality uses (2.2) in addition.

Hence, we have proved the following:

2.3.3 Lemma and Definition. *A generic hypersurface $f : M^3 \to S^4$ is conformally flat if and only if the 1-forms*[14]

$$
\gamma_i := \sqrt{2b_i - a_i^2}\, \omega_i = \begin{cases} \sqrt{(a_3 - a_1)}\sqrt{(a_1 - a_2)}\,\omega_1 \\ \sqrt{(a_1 - a_2)}\sqrt{(a_2 - a_3)}\,\omega_2 \\ \sqrt{(a_2 - a_3)}\sqrt{(a_3 - a_1)}\,\omega_3 \end{cases} \qquad (2.12)
$$

are closed.

We call the forms γ_i the "conformal fundamental forms" of the hypersurface.

---

[14] There is always a choice of signs for the roots that makes the second equality hold; see the following paragraph.

2.3.4 These conformal fundamental forms are, up to sign, well-defined geometric objects: For example, because we have $\sum_i \gamma_i^2 = I(., (2B - A_S^2).)$, they can be described as the dual forms of the principal vector fields v_i on M that are normalized to be unit vector fields with respect to the conformal metric C of f (see §2.1.11).

Clearly, if two principal curvatures of f coincide, then two of the conformal fundamental forms vanish; on the other hand, if f is generic, then the conformal fundamental forms do not vanish and the conformal metric is nondegenerate.

However, note that (independent of the order of the a_i) only one of the conformal fundamental forms, say γ_2, is real-valued, whereas the other two are imaginary, $\gamma_3, \gamma_1 : TM \to i\mathbb{R}$. Hence the conformal metric C of a generic hypersurface $f : M^3 \to S^4$ has the signature $(+ - -)$.

2.3.5 Our conformal fundamental forms are a variant of Wang's invariant differential forms θ_i [296]: Introducing the functions l_i by $\omega_i = l_i \gamma_i$, we have

$$\theta_1 = -\tfrac{l_2}{l_3}\gamma_1, \quad \theta_2 = -\tfrac{l_3}{l_1}\gamma_2, \quad \text{and} \quad \theta_3 = -\tfrac{l_1}{l_2}\gamma_3.$$

Since two of the forms γ_i take values in the imaginary complex numbers so do two of the functions l_i; as a consequence, the forms θ_i are real-valued. An easy computation yields $l_1^2 \tilde{S}_1 + l_2^2 \tilde{S}_2 + l_3^2 \tilde{S}_3 = 0$, where $\tilde{S}_i = S + a_i f$ denotes the curvature spheres of f (cf., §1.7.10). Multiplication with $-\tfrac{1}{l_3^2}$ then gives

$$W\tilde{S}_1 + (1 - W)\tilde{S}_2 - \tilde{S}_3 = 0, \quad \text{where} \quad W := -\frac{l_1^2}{l_3^2} = \frac{a_2 - a_3}{a_2 - a_1}$$

denotes Wang's "Möbius curvature" — which is conformally invariant up to rearranging indices. Note that the ratios of the functions l_i are invariant, whereas the functions themselves are not. In his paper [296], Wang gives a fundamental theorem: The invariants W and θ_i, $i = 1, 2, 3$, form a complete set of invariants, that is, they determine a 3-dimensional generic hypersurface uniquely up to a Möbius transformation.[15]

2.3.6 However, we are interested in a weaker version of such a fundamental theorem — restricting attention to conformally flat hypersurfaces (cf., [143]).

First, we want to determine how far a flat light cone lift $f : M^3 \to L^5$ of a generic conformally flat hypersurface is determined by its conformal fundamental forms and the functions[16] l_i or, equivalently, by its induced

---

[15] In fact, he also formulates an existence theorem: Given W and θ_i, $i = 1, 2, 3$, that satisfy certain equations, there exists a generic hypersurface with these invariants.

[16] Note that the Möbius curvature W determines the functions l_i up to a common factor, that is, it determines a conformal class of metrics for f.

metric $I = \sum_i l_i^2 \gamma_i^2$. To do so, we will study how unique the connection form Φ of an f-adapted frame in terms of principal directions is, using the integrability conditions.

Then, we will extract the remaining integrability conditions in terms of the functions l_i. This will provide us with necessary and sufficient conditions on a set of (closed) forms γ_i and a set of functions l_i, defined on some manifold M^3, to (locally) define a flat light cone immersion $f : M^3 \to L^5$ with the γ_i's as conformal fundamental forms and $\sum_i l_i^2 \gamma_i^2$ as its induced metric.

Later, in the next section, we will then use these more technical results to formulate a fundamental-like theorem for generic conformally flat hypersurfaces in the conformal 4-sphere.

2.3.7 Thus, let $\Phi : TM^3 \to \mathfrak{o}_1(6)$ be the connection form of an f-adapted frame $F : M^3 \to O_1(6)$ for a given flat light cone lift $f : M^3 \to L^5$ of a conformally flat hypersurface with conformal fundamental forms γ_i. In order to make the frame F unique, we require the sphere congruence $S = Fe_4$ to be the central sphere congruence of f, that is, we require $\operatorname{tr} A_S = 0$. Moreover, because we restrict our attention to generic hypersurfaces, we may assume that F is a principal frame for f, so that the $S_i = Fe_i$, $i = 1, 2, 3$, yield principal directions for the strip (f, S), that is, A_S is diagonal: $\psi_i = a_i \omega_i$ with the "principal curvatures" a_i of f with respect to S. Let l_i, $i = 1, 2, 3$, be functions with $\omega_i = l_i \gamma_i$.

Since the γ_i's ought to be the conformal fundamental forms of our conformally flat hypersurface, we have to have (2.12), which gives us

$$a_1 - a_2 = \tfrac{l_3}{l_1 l_2}, \quad a_2 - a_3 = \tfrac{l_1}{l_2 l_3}, \quad \text{and} \quad a_3 - a_1 = \tfrac{l_2}{l_3 l_1}. \tag{2.13}$$

This determines[17] the a_i up to an additive function a. Investing the assumption that S be the central sphere congruence of f, $a_1 + a_2 + a_3 = 0$, we obtain:

$$a_1 = \tfrac{1}{3}\big(\tfrac{l_3}{l_1 l_2} - \tfrac{l_2}{l_3 l_1}\big), \quad a_2 = \tfrac{1}{3}\big(\tfrac{l_1}{l_2 l_3} - \tfrac{l_3}{l_1 l_2}\big), \quad a_3 = \tfrac{1}{3}\big(\tfrac{l_2}{l_3 l_1} - \tfrac{l_1}{l_2 l_3}\big). \tag{2.14}$$

Flatness of the induced metric then allows us to uniquely determine the forms χ_i: As in (2.1), the Gauss equation (1.2) yields

$$\chi_i = -b_i \omega_i \quad \text{with} \quad b_i = \tfrac{1}{2}(a_i a_j + a_i a_k - a_j a_k), \tag{2.15}$$

where i, j, k are pairwise distinct indices. Using the fact that the conformal fundamental forms are closed, $d\gamma_i = 0$, the components ω_{ij} of the covariant

---

[17] Here, we want to disregard the sign ambiguity of the conformal fundamental forms γ_i; this effects the sign of the l_i and, consequently, the *simultaneous* sign of the a_i.

derivative can be determined from the second of the Codazzi equations (1.3), for the $\omega_i = l_i \gamma_i$. We find

$$\omega_{ij} = k_{ij}\omega_j - k_{ji}\omega_i \quad \text{with} \quad k_{ij} = -\frac{\partial_i l_j}{l_j}, \tag{2.16}$$

where ∂_i denotes the directional derivatives, $\partial_i u = du(v_i)$, with the dual vector fields v_i of the ω_i when u denotes some function. Then the Codazzi equations for the $\psi_i = a_i \omega_i$ read

$$\chi \wedge \omega_i = d(a_i\omega_i) + \sum_j \omega_{ij} \wedge (a_j\omega_j) = [da_i - \sum_j k_{ji}(a_i - a_j)\omega_j] \wedge \omega_i; \tag{2.17}$$

this (over)determines the form χ: By writing χ in terms of the basis forms ω_i, we obtain two equations for each coefficient.

To summarize, we formulate the following:

2.3.8 Lemma. *A flat light cone representative of a generic conformally flat hypersurface is uniquely determined (up to Lorentz transformation) by the conformal fundamental forms and the function triplet (l_1, l_2, l_3) that relates the conformal metric C to the (flat) induced metric I.*

In particular, the corresponding generic conformally flat hypersurface is determined up to Möbius transformation.

2.3.9 Now we turn to the converse: Suppose we are given three closed, linearly independent forms γ_i, $i = 1, 2, 3$, two of which take values in $i\mathbb{R}$ and the remaining takes values in $\mathbb{R}$, along with three functions l_i, so that:
(i) the three forms $\omega_i := l_i \gamma_i$ are real-valued,
(ii) $\sum_i l_i^2 = 0$, and
(iii) the metric $I := \sum_i \omega_i^2$ is flat.
The problem is to determine a flat light cone lift, with induced metric I, of a generic conformally flat hypersurface such that the γ_i's are its conformal fundamental forms.

The idea is to *define* the connection form Φ of the previously discussed f-adapted frame F and to determine conditions on the l_i's and, possibly, the forms γ_i that ensure that Φ is integrable. Thus, let the components (cf., (1.1)) of Φ be defined by the above equations (2.14), (2.15), and (2.16), and let χ be defined through (2.17). However, for the definition of χ, we have to check that the two equations for each coefficient are compatible, that is,

$$\begin{array}{rlcl}
\chi(v_1): & \partial_1 a_2 - k_{12}(a_2 - a_1) & = & \partial_1 a_3 - k_{13}(a_3 - a_1), \\
\chi(v_2): & \partial_2 a_3 - k_{23}(a_3 - a_2) & = & \partial_2 a_1 - k_{21}(a_1 - a_2), \\
\chi(v_3): & \partial_3 a_1 - k_{31}(a_1 - a_3) & = & \partial_3 a_2 - k_{32}(a_2 - a_3).
\end{array} \tag{2.18}$$

But, using (2.13) — which follows from (2.14) — the set of three equations is shown to be equivalent to $\frac{1}{l_1 l_2 l_3}(l_1 dl_1 + l_2 dl_2 + l_3 dl_3) = 0$, which holds since $\sum_i l_i^2 \equiv 0$. Hence χ can consistently be defined by (2.17).

Now we check the integrability conditions:

- The Codazzi equations (1.3) for ω_i hold by definition (2.16) of the forms ω_{ij} to provide the components of the Levi-Civita connection of I.
- The Codazzi equations for ψ_i hold by the definition of χ, as we just discussed.
- The Codazzi equations for χ_i reduce, by the computation (2.11) (which uses the definition of the χ_i and the Codazzi equations for the ω_i), to the Codazzi equations for the ψ_i together with $d\gamma_i = 0$.
- The Gauss equation (1.2) holds, on one hand because of the definition of the χ_i and, on the other hand since we required the metric $I = \sum_i \omega_i^2$ to be flat, $\varrho_{ij} = 0$ (cf., §1.7.7: note that the forms ω_{ij} do indeed provide the components of the Levi-Civita connection of I by their definition[18]).
- The Ricci equations (1.4) hold as soon as $d\chi = 0$: This will be a straightforward computation — after some preparation:

2.3.10 First, we will need the Lie brackets of the dual vector fields v_i of the basis forms ω_i: $\omega_k([v_i, v_j]) = -d\omega_k(v_i, v_j) = k_{ik}\omega_k(v_j) - k_{jk}\omega_k(v_i)$ by (1.3), so that we obtain

$$[v_i, v_j] \;=\; k_{ij}v_j - k_{ji}v_i. \tag{2.19}$$

As in §2.1.4, we will also use the flatness of the induced metric. To make this fact accessible to us, we compute $0 = \varrho_{ij} = d\omega_{ij} + \omega_{ik} \wedge \omega_{kj}$ for pairwise distinct indices i, j, k (cf., §1.7.7). This yields Lamé's equations (cf., [248]):

$$
\begin{aligned}
\partial_3 k_{12} &= k_{32}(k_{12} - k_{13}), & \partial_3 k_{21} &= k_{31}(k_{21} - k_{23}), \\
\partial_1 k_{23} &= k_{13}(k_{23} - k_{21}), & \partial_1 k_{32} &= k_{12}(k_{32} - k_{31}), \\
\partial_2 k_{31} &= k_{21}(k_{31} - k_{32}), & \partial_2 k_{13} &= k_{23}(k_{13} - k_{12}),
\end{aligned} \tag{2.20}
$$

$$
\begin{aligned}
\partial_1 k_{12} + \partial_2 k_{21} &= k_{12}^2 + k_{21}^2 + k_{31}k_{32}, \\
\partial_2 k_{23} + \partial_3 k_{32} &= k_{23}^2 + k_{32}^2 + k_{12}k_{13}, \\
\partial_3 k_{31} + \partial_1 k_{13} &= k_{31}^2 + k_{13}^2 + k_{23}k_{21}.
\end{aligned}
$$

2.3.11 Another formula that will prove to be useful is obtained by using $\sum_j l_j^2 = 0$: For pairwise distinct indices i, j, k, we derive

$$d\left(\frac{l_i}{l_j l_k}\right) \;=\; \frac{l_j}{l_i l_k}\frac{dl_k}{l_k} + \frac{l_k}{l_i l_j}\frac{dl_j}{l_j}. \tag{2.21}$$

---

[18] Defining a connection by $\nabla v_i = -\Sigma_j \omega_{ij} v_j$, the skew-symmetry $\omega_{ij} + \omega_{ji} = 0$ provides compatibility with the metric, and the Codazzi equations (1.3) for the ω_i yield that the connection is torsion-free; compare the following paragraph.

Now we are prepared to check $d\chi = 0$. By cross differentiating the terms from (2.18), we compute

$$
\begin{aligned}
\partial_1 \chi(v_2) - \partial_2 \chi(v_1) &= \partial_1 \partial_2 a_3 + (a_2 - a_3)\partial_1 k_{23} + k_{23}\partial_1(a_2 - a_3) \\
&\quad -\partial_2 \partial_1 a_3 + (a_3 - a_1)\partial_2 k_{13} + k_{13}\partial_2(a_3 - a_1) \\
&= k_{12}\partial_2 a_3 - k_{21}\partial_1 a_3 \\
&\quad +(a_2 - a_3)k_{13}(k_{23} - k_{21}) \\
&\quad -k_{23}[(a_3 - a_1)k_{13} + (a_1 - a_2)k_{12}] \\
&\quad +(a_3 - a_1)k_{23}(k_{13} - k_{12}) \\
&\quad -k_{13}[(a_1 - a_2)k_{21} + (a_2 - a_3)k_{23}] \\
&= k_{12}[\partial_2 a_3 + k_{23}(a_2 - a_3)] - k_{21}[\partial_1 a_3 - k_{13}(a_3 - a_1)] \\
&= k_{12}\chi(v_2) - k_{21}\chi(v_1) \\
&= \chi([v_1, v_2]),
\end{aligned}
$$

where we use (2.19) for the second and last equations, and the second also uses (2.20) and (2.21). The other two terms are obtained by cyclic rotation of indices because all of the facts that we used remain valid under cyclic permutations of indices. Thus we have proved:

2.3.12 Lemma. *Given three closed, linearly independent forms γ_i, one with values in $\mathbb{R}$ and two with values in $i\mathbb{R}$, and three functions l_i such that*

$$
\omega_i := l_i \gamma_i : TM \to \mathbb{R}, \quad \sum_{j=1}^3 l_j^2 = 0, \quad \text{and} \quad I := \sum_{j=1}^3 l_j^2 \gamma_j^2 \text{ is flat,}
$$

there exists a flat light cone lift, with induced metric I, of a generic conformally flat hypersurface such that the γ_i's are its conformal fundamental forms.

2.3.13 Before turning to a geometric application of the two lemmas that we just proved, let us mention another observation, which links the conformal fundamental forms §2.3.3 to the geometry of a hypersurface: The six distributions $\gamma_i \pm \gamma_j = 0$ are exactly Cartan's "umbilic distributions" (see [61]). Namely, for example, for $i = 1$ and $j = 2$, let

$$
v := v_3 \quad \text{and} \quad w := i\left(\frac{\sqrt{a_2 - a_3}}{\sqrt{a_2 - a_1}} v_1 \mp \frac{\sqrt{a_3 - a_1}}{\sqrt{a_2 - a_1}} v_2\right),
$$

where, as before, the v_i's denote the dual basis fields of the ω_i, that is, the v_i's are principal basis fields that are orthonormal with respect to the induced metric I of $f : M^3 \to L^5$. A straightforward computation shows that v and w form orthonormal (with respect to I) basis fields for the distribution $\gamma_1 \pm \gamma_2 = 0$ and that[19] $I\!I_S(v, v) = I\!I_S(w, w) = a_3$. Hence, the

---

[19] Here $I\!I_S = -\langle dS, df \rangle$ denotes the (real-valued) second fundamental form of f with respect to S as a unit normal field (cf., §1.7.7).

planes $\gamma_1 \pm \gamma_2 = 0$ intersect the Dupin indicatrix (curvature ellipsoid) of f (with respect to S) in circles.

2.3.14 Consequently, from the lemma in §2.3.3, half of Cartan's characterization of 3-dimensional conformally flat hypersurfaces [61] follows: If a generic hypersurface is conformally flat, then the six umbilic distributions are integrable.

To see the converse, first note that the integrability of the umbilic distributions implies the integrability of the distributions $\gamma_i = 0$: If

$$0 = d(\gamma_i \pm \gamma_j) \wedge (\gamma_i \pm \gamma_j) = (d\gamma_i \wedge \gamma_i + d\gamma_j \wedge \gamma_j) \pm (d\gamma_i \wedge \gamma_j + d\gamma_j \wedge \gamma_i)$$

for $i \neq j$, then $d\gamma_i \wedge \gamma_i = 0$ for $i = 1, 2, 3$. Thus $d\gamma_i \wedge \gamma_j + d\gamma_j \wedge \gamma_i = 0$ for *any* choice of indices $i, j \in \{1, 2, 3\}$. By employing the Codazzi equations (1.3) for the forms $\omega_i = l_i \gamma_i$, we have $l_i d\gamma_i = -\frac{dl_i}{l_i} \wedge \omega_i - \sum_j \omega_{ij} \wedge \omega_j$. When writing the ω_{ij}'s in terms of the basis forms ω_i, $\omega_{ij} = k_{ij}\omega_j - k_{ji}\omega_i + s_{ij}\omega_k$ with $k \neq i, j$, we obtain

$$l_1 d\gamma_1 = \quad (s_{12} - s_{13})\omega_2 \wedge \omega_3 - (\tfrac{\partial_3 l_1}{l_1} + k_{31})\omega_3 \wedge \omega_1 + (\tfrac{\partial_2 l_1}{l_1} + k_{21})\omega_1 \wedge \omega_2,$$

$$l_2 d\gamma_2 = \quad (\tfrac{\partial_3 l_2}{l_2} + k_{32})\omega_2 \wedge \omega_3 + (s_{23} - s_{21})\omega_3 \wedge \omega_1 - (\tfrac{\partial_1 l_2}{l_2} + k_{12})\omega_1 \wedge \omega_2,$$

$$l_3 d\gamma_3 = -(\tfrac{\partial_2 l_3}{l_3} + k_{23})\omega_2 \wedge \omega_3 + (\tfrac{\partial_1 l_3}{l_3} + k_{13})\omega_3 \wedge \omega_1 + (s_{31} - s_{32})\omega_1 \wedge \omega_2.$$

Now $d\gamma_i \wedge \gamma_i = 0$ implies $s_{12} = s_{13} = -s_{31} = -s_{32} = s_{23} = s_{21} = -s_{12}$ and, consequently, $s_{ij} = 0$. From $d\gamma_i \wedge \gamma_j + d\gamma_j \wedge \gamma_i = 0$ we then deduce three equations:

$$\tfrac{\partial_1 l_2}{l_2} + k_{12} = \tfrac{\partial_1 l_3}{l_3} + k_{13}, \quad \tfrac{\partial_2 l_3}{l_3} + k_{23} = \tfrac{\partial_2 l_1}{l_1} + k_{21}, \quad \tfrac{\partial_3 l_1}{l_1} + k_{31} = \tfrac{\partial_3 l_2}{l_2} + k_{32}.$$

On the other hand, the Codazzi equations for the ψ_i, written in the form (2.17), give us the three equations (2.18). Expressing the differences of the a_i in terms of the functions l_i (see (2.13)) and using (2.21), those become

$$0 = \tfrac{l_3}{l_1 l_2}(\tfrac{\partial_1 l_2}{l_2} + k_{12}) + \tfrac{l_2}{l_3 l_1}(\tfrac{\partial_1 l_3}{l_3} + k_{13}),$$

$$0 = \tfrac{l_1}{l_2 l_3}(\tfrac{\partial_2 l_3}{l_3} + k_{23}) + \tfrac{l_3}{l_1 l_2}(\tfrac{\partial_2 l_1}{l_1} + k_{21}),$$

$$0 = \tfrac{l_2}{l_3 l_1}(\tfrac{\partial_3 l_1}{l_1} + k_{31}) + \tfrac{l_1}{l_2 l_3}(\tfrac{\partial_3 l_2}{l_2} + k_{32}),$$

which finally yield, together with the previous set of equations, $k_{ij} = -\frac{\partial_i l_j}{l_j}$, since $\frac{l_2}{l_3 l_1} + \frac{l_3}{l_1 l_2} = -\frac{l_1}{l_2 l_3} \neq 0$, and so on. Consequently, $d\gamma_i = 0$ and the hypersurface is conformally flat, according to our lemma in §2.3.3:

2.3.15 Theorem (Cartan). *A generic hypersurface in S^4 is conformally flat if and only if its six umbilic distributions are integrable.*

2.4 Guichard nets

By §2.3.3, the conformal fundamental forms of a conformally flat hypersurface $f : M^3 \to S^4$ in the conformal 4-sphere are closed. Hence they can locally be integrated to obtain a coordinate system

$$x = (x_1, x_2, x_3) : M^3 \supset U \to \mathbb{R}^3_2, \quad dx_i = \gamma_i,$$

in case f is generic since, in that case, the conformal fundamental forms are linearly independent. Here we identify $\mathbb{R}^3_2 \cong \mathbb{R} \times i\mathbb{R} \times i\mathbb{R}$. This is the starting observation that will lead to the announced "fundamental-like" theorem.

2.4.1 On the other hand, because f is conformally flat, there exist local conformal coordinates $y : M^3 \supset V \to \mathbb{R}^3$ with respect to the induced conformal structure, by §P.4.3. Then the map $x \circ y^{-1} : \mathbb{R}^3 \supset y(V \cap U) \to \mathbb{R}^3_2$ defines a "triply orthogonal system" of surfaces in $\mathbb{R}^3$: Since, by construction, (x_1, x_2, x_3) are curvature line coordinates, any coordinate surfaces from different families, $x_i = const$ and $x_j = const$ with $i \neq j$, intersect orthogonally.

2.4.2 Choosing different conformal coordinates $\tilde{y}$, the transition map $\tilde{y} \circ y^{-1}$ yields a conformal map between open sets in $\mathbb{R}^3$. Hence it is the restriction of a Möbius transformation of $S^3 \cong \mathbb{R}^3 \cup \{\infty\}$ by Liouville's theorem (see §1.5.4). On the other hand, any Möbius transformation $\mu \in M\ddot{o}b(3)$ (that does not map an image point of y to ∞) gives rise to a choice of conformal coordinates $\tilde{y} = \mu \circ y$.

Thus the triply orthogonal system $x \circ y^{-1} : \mathbb{R}^3 \supset y(V) \to \mathbb{R}^3_2$ is unique up to Möbius transformation. In order to simplify notations we will, in what follows, assume $V = U = M^3 \subset \mathbb{R}^3$ and $y = id$. In this way $x : U \to \mathbb{R}^3_2$ itself defines the triply orthogonal system under investigation.

2.4.3 Clearly, the Gaussian basis fields $\frac{\partial}{\partial x_i}$ of x are orthogonal, and their lengths l_i, $l_i^2 = I(\frac{\partial}{\partial x_i}, \frac{\partial}{\partial x_i})$, where I denotes the Euclidean metric U, are well-defined functions up to sign: Since two of the coordinate functions x_i are imaginary, two of the functions l_i are also, along with their corresponding Gaussian basis fields $\frac{\partial}{\partial x_i}$ that live in the complexified tangent bundle.

Introducing the orthonormal vector fields $v_i := \frac{1}{l_i} \frac{\partial}{\partial x_i}$, along with their dual forms given by $\omega_i = I(v_i, .) = l_i dx_i$, that is, $I = \sum_i \omega_i^2$, we are back to the picture of a flat light cone lift of a conformally flat hypersurface in S^4, intrinsically. In particular, the "Lamé functions" l_i of the triply orthogonal system x satisfy $\sum_i l_i^2 = 0$ because they relate a (flat) representative I of the induced conformal structure to the induced conformal metric C (cf., §2.1.11). Thus we have identified the triply orthogonal system $x : \mathbb{R}^3 \supset U \to \mathbb{R}_2^3$ as a Guichard net[20]:

2.4.4 Definition. *A triply orthogonal system* $x : \mathbb{R}^3 \supset U \to \mathbb{R}_2^3$ *will be called a Guichard net if its Lamé functions* l_i, *which are uniquely defined up to sign by* $I = \sum_i l_i^2 dx_i^2$, *satisfy* $\sum_i l_i^2 = 0$.

2.4.5 Note that the Guichard condition $\sum_i l_i^2 = 0$ is invariant under conformal (Möbius) transformations of the ambient $\mathbb{R}^3 \cup \{\infty\}$ or, more general, under conformal changes $I \to e^{2u} I$ of the ambient space's metric I.

If we consider a triply orthogonal system $x : \mathbb{R}^3 \to \mathbb{R}^3$ given by *real* coordinate functions, then the Guichard condition reads $l_1^2 + l_2^2 = l_3^2$, after suitable choice of indices. On occasion, we will prefer to give Guichard nets a "real" treatment, as, for example, in one of the examples below.

Merging this terminology with our lemmas in §2.3.8 and §2.3.12 in the previous section, we obtain the announced fundamental-like theorem:

2.4.6 Theorem. *Any generic conformally flat hypersurface* $f : M^3 \to S^4$ *in the conformal 4-sphere gives rise to a Guichard net* $x : \mathbb{R}^3 \supset U \to \mathbb{R}_2^3$; *this Guichard net is unique up to Möbius transformation.*

Conversely, given a Guichard net $x : U \to \mathbb{R}_2^3$, *there exists locally a generic conformally flat hypersurface in* S^4 *with conformal fundamental forms* $\gamma_i = dx_i$; *this conformally flat hypersurface is unique up to Möbius transformation.*

2.4.7 As an application of this 1-to-1 correspondence between (Möbius equivalence classes of) generic conformally flat hypersurfaces in S^4 and (Möbius equivalence classes of) Guichard nets, we want to study how the geometry of the Guichard net is related to the geometry of the conformally flat hypersurface, in a special case.

For this purpose, we first analyze the extrinsic geometry of the coordinate surfaces $x_i = const$ of our Guichard net: Clearly, $v_i = \frac{1}{l_i} \frac{\partial}{\partial x_i}$ yields a unit normal field for the surfaces $x_i = const$. To determine the curvatures of these coordinate surfaces, we compute its derivative $\nabla_{v_j} v_i$ in a tangential direction v_j, $j \neq i$. From Koszul's formula,

$$2I(\nabla_{\frac{\partial}{\partial x_j}} \frac{\partial}{\partial x_i}, \frac{\partial}{\partial x_k}) = \frac{\partial}{\partial x_i} I(\frac{\partial}{\partial x_j}, \frac{\partial}{\partial x_k}) + \frac{\partial}{\partial x_j} I(\frac{\partial}{\partial x_i}, \frac{\partial}{\partial x_k}) - \frac{\partial}{\partial x_k} I(\frac{\partial}{\partial x_i}, \frac{\partial}{\partial x_j}),$$

we deduce

$$\nabla_{\frac{\partial}{\partial x_j}} \frac{\partial}{\partial x_i} = \frac{1}{l_j} \frac{\partial l_j}{\partial x_i} \frac{\partial}{\partial x_j} + \frac{1}{l_i} \frac{\partial l_i}{\partial x_j} \frac{\partial}{\partial x_i}, \quad \text{or} \quad \nabla_{v_j} v_i = -k_{ij} v_j,$$

---

[20] These Guichard nets were (first?) considered by Guichard in [136], where he also derived a relation with 3-dimensional submanifolds in a 6-dimensional space, in particular, with 3-dimensional submanifolds of a light cone.

where $k_{ij} = -\frac{\partial_i l_j}{l_j} = -\frac{1}{l_i l_j}\frac{\partial}{\partial x_i}l_j$, as in (2.16). Consequently, k_{ij} is the principal curvature of a surface $x_i = const$ in the (principal) direction v_j. Note that we did not use the Guichard property; thus, as a by-product, we have just proved Dupin's theorem (see [248]):

2.4.8 Theorem (Dupin). *Any two surfaces of a triply orthogonal system that are from different families intersect along curvature lines.*

2.4.9 Also note that we did not use flatness of the metric $I = \sum_i l_i^2 dx_i^2$: Thus, the above statement about k_{ij} being the principal curvature of a coordinate surfaces $x_i = const$ in the principal direction v_j stays true for any conformally equivalent metric $\tilde I = e^{2u}I$ with

$$\tilde k_{ij} = -\frac{1}{\tilde l_i \tilde l_j}\frac{\partial}{\partial x_i}\tilde l_j = -e^{-u}\frac{1}{l_i l_j}\frac{\partial}{\partial x_i}(l_j + u).$$

For later reference, we formulate this fact as a lemma, while also collecting a previous result on detecting flatness of the metric I:

2.4.10 Lemma. *Given a triply orthogonal system $x : \mathbb{R}^3 \supset U \to \mathbb{R}^3$ with Lamé functions l_i, where $I = \sum_i l_i^2 dx_i^2$ is any metric that is conformally equivalent to the Euclidean metric on U, $k_{ij} = -\frac{1}{l_i l_j}\frac{\partial}{\partial x_i}l_j$ is the principal curvature of a surface $x_i = const$ in the principal direction $v_j = \frac{1}{l_j}\frac{\partial}{\partial x_j}$.*

The metric I is flat if and only if the k_{ij}'s satisfy Lamé's equations (2.20), or the following equivalent version of the Lamé equations:

$$\begin{aligned}
\frac{\partial}{\partial x_1}\frac{\partial}{\partial x_2}l_3 &= \frac{1}{l_1}\frac{\partial l_3}{\partial x_1}\frac{\partial l_1}{\partial x_2} + \frac{1}{l_2}\frac{\partial l_3}{\partial x_2}\frac{\partial l_2}{\partial x_1},\\
\frac{\partial}{\partial x_2}\frac{\partial}{\partial x_3}l_1 &= \frac{1}{l_2}\frac{\partial l_1}{\partial x_2}\frac{\partial l_2}{\partial x_3} + \frac{1}{l_3}\frac{\partial l_1}{\partial x_3}\frac{\partial l_3}{\partial x_2},\\
\frac{\partial}{\partial x_3}\frac{\partial}{\partial x_1}l_2 &= \frac{1}{l_3}\frac{\partial l_2}{\partial x_3}\frac{\partial l_3}{\partial x_1} + \frac{1}{l_1}\frac{\partial l_2}{\partial x_1}\frac{\partial l_1}{\partial x_3},\\
0 &= \frac{\partial}{\partial x_2}\Big(\frac{1}{l_2}\frac{\partial l_1}{\partial x_2}\Big) + \frac{\partial}{\partial x_1}\Big(\frac{1}{l_1}\frac{\partial l_2}{\partial x_1}\Big) + \frac{1}{l_3^2}\frac{\partial l_1}{\partial x_3}\frac{\partial l_2}{\partial x_3},\\
0 &= \frac{\partial}{\partial x_3}\Big(\frac{1}{l_3}\frac{\partial l_2}{\partial x_3}\Big) + \frac{\partial}{\partial x_2}\Big(\frac{1}{l_2}\frac{\partial l_3}{\partial x_2}\Big) + \frac{1}{l_1^2}\frac{\partial l_2}{\partial x_1}\frac{\partial l_3}{\partial x_1},\\
0 &= \frac{\partial}{\partial x_1}\Big(\frac{1}{l_1}\frac{\partial l_3}{\partial x_1}\Big) + \frac{\partial}{\partial x_3}\Big(\frac{1}{l_3}\frac{\partial l_1}{\partial x_3}\Big) + \frac{1}{l_2^2}\frac{\partial l_3}{\partial x_2}\frac{\partial l_1}{\partial x_2}.
\end{aligned} \tag{2.22}$$

2.4.11 *Proof.* The only thing left is to check the second statement: For $i \neq j \neq k$ we have $\omega_{ij} = I(v_i, \nabla_{v_k}v_j) = -k_{jk}\delta_{ik}$; hence $\omega_{ij} = k_{ij}\omega_j - k_{ji}\omega_i$. This implies that the computation that provided (2.20) applies to give a flatness criterion.

Using $\partial_i = \frac{1}{l_i}\frac{\partial}{\partial x_i}$ and $k_{ij} = -\frac{\partial_i l_j}{l_j}$, it is a lengthy but straightforward calculation to show the equivalence of (2.20) and (2.22) (cf., [248]). ◁

2.4.12 Example. Now, we want to turn to some special (generic) conformally flat hypersurfaces, the conformally flat "conformal product hypersurfaces." By §P.5.8, a 3-dimensional product manifold $(M \times I, g + ds^2)$

is conformally flat if and only if the 2-dimensional part (M, g) has constant Gauss curvature K. Using this fact, we can construct three types of conformally flat hypersurfaces in (part of) the conformal 4-sphere:

(i) If $f : M^2 \to \mathbb{R}^3$ is a surface of constant Gauss curvature K, then the cylinder

$$M^2 \times \mathbb{R} \ni (p, s) \mapsto (s, f(p)) \in \mathbb{R} \times \mathbb{R}^3 = \mathbb{R}^4$$

over $f(M^2)$ defines a conformally flat hypersurface.

(ii) If $f : M^2 \to S^3 \subset \mathbb{R}^4$ is a surface of constant Gauss curvature K, then

$$M^2 \times \mathbb{R} \ni (p, s) \mapsto e^s f(p) \in \mathbb{R} \times S^3 = (\mathbb{R}^4 \setminus \{0\}, \tfrac{dy^2}{|y|^2})$$

defines a conformally flat hypersurface in the conformal 4-sphere: The metric induced from $\mathbb{R} \times S^3$ is easily seen to be the product metric $df^2 + ds^2$. Note that the metric of $\mathbb{R} \times S^3$ is conformally equivalent to the Euclidean metric on $\mathbb{R}^4$; hence $\mathbb{R} \times S^3$ can be considered as part of the conformal 4-sphere.

(iii) If $f : M^2 \to H^3 = (\{(y_1, y_2, y_3) \mid y_1 > 0\}, \tfrac{1}{y_1^2}(dy_1^2 + dy_2^2 + dy_3^2))$ is a surface of constant Gauss curvature K in hyperbolic 3-space, thought of as a surface in the Poincaré half-space model (cf., §P.1.3), then

$$
\begin{array}{ccc}
M^2 \times S^1 & \to & S^1 \times H^3 = (\mathbb{R}^4 \setminus \{y_0 = y_1 = 0\}, \tfrac{1}{y_0^2 + y_1^2} \sum_i dy_i^2) \\
(p, s) & \mapsto & (f_1(p) \cdot \sin s, f_1(p) \cdot \cos s, f_2(p), f_3(p))
\end{array}
$$

again defines a conformally flat hypersurface as a computation of the metric $f_1^{-2}(df_1^2 + df_2^2 + df_3^2) + ds^2$ induced from the ambient $S^1 \times H^3$ shows. Note that, again, the metric of $S^1 \times H^3$ is conformally equivalent to the Euclidean metric on $\mathbb{R}^4$, so that the hypersurface's ambient space can be considered part of the conformal 4-sphere.

Because all of these hypersurfaces have an extrinsic product structure, their principal directions are the principal directions v_1 and v_2 of the 2-dimensional part together with $v_3 = \frac{\partial}{\partial s}$; and, since the induced metric is a product metric, the induced covariant differentiation splits. In particular, $\nabla_{v_i} \frac{\partial}{\partial s} = -k_{3i} v_i = 0$, so that in the induced metric the coordinate surfaces $s = const$ are planar. Hence, choosing any conformal coordinates, the surfaces $s = const$ are mapped to totally umbilic surfaces in Euclidean 3-space — otherwise said, if the hypersurface is generic, then the corresponding Guichard net has to contain one family of sphere pieces.

Conversely, by §2.3.12, any Guichard net containing one family of sphere pieces, say $k_{31} = k_{32} =: k$, gives rise to a generic conformally flat hypersurface in the conformal 4-sphere, which is unique up to Möbius transformation by §2.3.8. A careful case-by-case study then recovers the above "conformally flat conformal product hypersurfaces" (see [142]) — a key step is to show that the orthogonal trajectories of the sphere pieces consist of circles: With $\partial_1 k = \partial_2 k = 0$, as derived from the Lamé equations (2.20), and the Guichard property $\sum_i l_i^2 = 0$, one computes $0 = d^2 \frac{1}{l_3} (\frac{\partial}{\partial x_3}, \frac{\partial}{\partial x_i}) = l_i \partial_3 k_{i3}$ for $i = 1, 2$. As a consequence, the sphere pieces belong to a sphere pencil. This yields six cases to consider, depending on the type of the sphere pencil and on whether the corresponding coordinate function x_3 is real or imaginary.

2.4.13 Another example. Motivated by the above observation, that the orthogonal trajectories of the sphere pieces are circles, we might ask whether there are more general Guichard nets with the property that one family of curvature lines consists of circles. Explicit examples of conformally flat hypersurfaces carrying such Guichard nets are given by Y. Suyama in [273]; we will present one of his examples below. A classification of these "cyclic Guichard nets" was given in [146]; later we will present this classification, leading to an interesting application of Möbius geometric arguments in its metric subgeometries. For the moment though, we shall just discuss an example that is rather typical: If we additionally assume that all circles of a cyclic Guichard net pass through one fixed point ∞, then the triply orthogonal system will comprise a family of parallel surfaces in Euclidean space. In what follows we shall see how the parallel surfaces of Dini's helices yield a cyclic Guichard net.

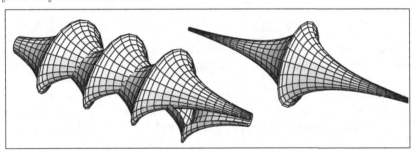

Fig. 2.1. Dini's helix and the pseudosphere

Identifying $\mathbb{R}^3 \cong \mathbb{R} \times \mathbb{C}$, a curvature line parametrization for Dini's helices (cf., [17]) is given by

$$f(x_1, x_2) := (x_1 - a \tanh g(x_1, x_2), e^{ix_2} \frac{a}{\cosh g(x_1, x_2)}),$$

where $g(x_1, x_2) := g(x_1, x_2) := \frac{1}{a}(x_1 + bx_2)$ and $a, b \in \mathbb{R}$ satisfy $a^2 + b^2 = 1$. Computing derivatives we find

$$\frac{\partial}{\partial x_1} f = \tanh g \cdot (\tanh g, -e^{ix_2} \frac{1}{\cosh g}),$$
$$\frac{\partial}{\partial x_2} f = \frac{1}{\cosh g} \cdot \{-b(\frac{1}{\cosh g}, e^{ix_2} \tanh g) + a(0, ie^{ix_2})\}.$$

A unit normal field n of f is then

$$n(x_1, x_2) := a(\frac{1}{\cosh g(x_1, x_2)}, e^{ix_2} \tanh g(x_1, x_2)) + b(0, ie^{ix_2}).$$

It is straightforward to see that (x_1, x_2) are indeed curvature line parameters, and to compute the principal curvatures of f:

$$\frac{\partial}{\partial x_1} n \;=\; -\frac{1}{\sinh g} \cdot \frac{\partial}{\partial x_1} f \quad \text{and} \quad \frac{\partial}{\partial x_2} n \;=\; \sinh g \cdot \frac{\partial}{\partial x_2} f.$$

Now we parametrize the family of parallel surfaces of f by

$$h(x_1, x_2, x_3) := f(x_1, x_2) + \sinh x_3 \cdot n(x_1, x_2).$$

Then the Lamé functions of the triply orthogonal system formed by the coordinate surfaces $x_i = const$ are given (up to sign) by

$$
\begin{aligned}
l_1^2(x_1, x_2, x_3) &= \left|\tfrac{\partial}{\partial x_1} h\right|^2(x_1, x_2, x_3) &= \frac{(\sinh g(x_1,x_2) - \sinh x_3)^2}{\cosh^2 g(x_1,x_2)} \\
l_2^2(x_1, x_2, x_3) &= \left|\tfrac{\partial}{\partial x_2} h\right|^2(x_1, x_2, x_3) &= \frac{(1 + \sinh g(x_1,x_2)\sinh x_3)^2}{\cosh^2 g(x_1,x_2)} \\
l_3^2(x_1, x_2, x_3) &= \left|\tfrac{\partial}{\partial x_3} h\right|^2(x_1, x_2, x_3) &= \cosh^2 x_3.
\end{aligned}
$$

We derive $l_1^2 + l_2^2 = l_3^2$ — this is (up to permutation of the indices) the Guichard condition when calculating in the reals. Moreover, the principal curvatures[21] of the coordinate surfaces $x_i = const$ become

$$
\begin{aligned}
k_{12} &= \tfrac{-1}{l_1 l_2}\tfrac{\partial}{\partial x_1} l_2 = \tfrac{1}{a(1+\sinh g \sinh x_3)}, & k_{13} &= \tfrac{-1}{l_1 l_3}\tfrac{\partial}{\partial x_1} l_3 = 0; \\
k_{21} &= \tfrac{-1}{l_2 l_1}\tfrac{\partial}{\partial x_2} l_1 = -\tfrac{b}{a(\sinh g - \sinh x_3)}, & k_{23} &= \tfrac{-1}{l_2 l_3}\tfrac{\partial}{\partial x_2} l_3 = 0; \\
k_{31} &= \tfrac{-1}{l_3 l_1}\tfrac{\partial}{\partial x_3} l_1 = \tfrac{1}{\sinh g - \sinh x_3}, & k_{32} &= \tfrac{-1}{l_3 l_2}\tfrac{\partial}{\partial x_3} l_2 = -\tfrac{\sinh g}{1+\sinh g \sinh x_3}.
\end{aligned}
$$

This shows that none of the families of surfaces $x_i = const$ consists of sphere pieces, unless $b = 0$; in that case, the Dini helix becomes the pseudosphere, and the surfaces $x_2 = const$ become the planes of its meridian curves.

Consequently, for $b \neq 0$, these Guichard nets correspond to generic conformally flat hypersurfaces of the conformal 4-sphere that are not among the conformal product hypersurfaces considered in the preceding example (cf., [143]).

2.4.14 Suyama's example. A more explicit construction of conformally flat hypersurfaces that are not among the conformally flat conformal product hypersurfaces from §2.4.12 was communicated to the author by Y. Suyama (see [273]). In a certain sense it "intertwines" the construction from the preceding example and that of the generic conformally flat cylinders in §2.4.12(i). Namely, given a surface $f : M^2 \to \mathbb{R}^3$ of constant Gauss curvature, a hypersurface is defined by

$$(u, f + vn) : M \times I \to \mathbb{R} \times \mathbb{R}^3 = \mathbb{R}^4,$$

---

[21] Remember that a choice of sign for $l_i = \sqrt{l_i^2}$ results in a simultaneous choice of sign for both curvatures k_{ij} of the surfaces $x_i = const$, that is, to an orientation for these surfaces, while the other curvatures do not depend on that choice.

where $u, v : I \to \mathbb{R}$ are suitable functions defined on some interval I.

As an example we will reconsider the Dini helices from the preceding example (cf., [273]). Thus let f be a Dini's helix as given above, and let

$$\hat{f} := (u, f + vn) : \quad (x_1, x_2, s) \mapsto (u(s), f(x_1, x_2) + v(s)\, n(x_1, x_2)) \in \mathbb{R}^4.$$

Further, we define a function $\phi = \phi(s)$ by[22)]

$$\tanh \phi(s) = p\, \mathrm{sn}_p(s) \quad \Leftrightarrow \quad \phi' = +\sqrt{1 - q^2 \cosh^2 \phi}, \quad \phi(0) = 0,$$

where $p^2 + q^2 = 1$ and $\mathrm{sn}_p(s) = \sin(\mathrm{am}_p(s))$ denotes the Jacobi elliptic function sn_p with modulus p. We note that, with this notation,

$$\cosh \phi = \tfrac{1}{\mathrm{dn}_p} \quad \text{and} \quad \sinh \phi = p \tfrac{\mathrm{sn}_p}{\mathrm{dn}_p}$$

are also Jacobi elliptic functions. Now we let

$$v(s) := \sinh \phi(s) \quad \text{and} \quad u(s) := q \int_0^s \cosh^2 \phi(s) ds = q \int_0^s \tfrac{ds}{\mathrm{dn}_p^2(s)}.$$

Then, with $\hat{n} := (-\phi', q \cosh \phi\, n) = (-\tfrac{p\,\mathrm{cn}_p}{\mathrm{dn}_p}, \tfrac{q}{\mathrm{dn}_p} n)$, we compute[23)]

$$\begin{array}{llll}
\frac{\partial}{\partial x_1}\hat{f} = & \frac{\sinh g - \sinh \phi}{\sinh g} \cdot (0, \frac{\partial}{\partial x_1}f), & \frac{\partial}{\partial x_1}\hat{n} = & -q\frac{\cosh \phi}{\sinh g} \cdot (0, \frac{\partial}{\partial x_1}f); \\[2mm]
\frac{\partial}{\partial x_2}\hat{f} = & (1 + \sinh g \sinh \phi) \cdot (0, \frac{\partial}{\partial x_2}f), & \frac{\partial}{\partial x_2}\hat{n} = & q \cosh \phi \sinh g \cdot (0, \frac{\partial}{\partial x_2}f); \\[2mm]
\frac{\partial}{\partial s}\hat{f} = & \cosh \phi \cdot (q \cosh \phi, \phi' n), & \frac{\partial}{\partial s}\hat{n} = & q \sinh \phi \cdot (q \cosh \phi, \phi' n),
\end{array}$$

showing that $\hat{n}$ is a (unit) normal field of $\hat{f}$ and that (x_1, x_2, s) are principal curvature line coordinates for $\hat{f}$ — its principal curvatures are given by

$$a_1 = q\frac{\cosh \phi}{\sinh g - \sinh \phi}, \quad a_2 = -q\frac{\cosh \phi \sinh g}{1 + \sinh g \sinh \phi}, \quad a_3 = -q \tanh \phi.$$

The induced metric of $\hat{f}$ is given by[24)]

$$I = \frac{(\sinh g - \sinh \phi)^2}{\cosh^2 g} dx_1^2 + \frac{(1 + \sinh g \sinh \phi)^2}{\cosh^2 g} dx_2^2 + \cosh^2 \phi\, ds^2.$$

[22)] The equivalence can be verified by using the Pythagorean rules $1 = \mathrm{cn}_p^2 + \mathrm{sn}_p^2 = \mathrm{dn}_p^2 + p^2 \mathrm{sn}_p^2$ and $\mathrm{sn}_p' = \mathrm{cn}_p \mathrm{dn}_p$, so that $\mathrm{sn}_p'^2 = (1 - \mathrm{sn}_p^2)(1 - p^2 \mathrm{sn}_p^2)$. For what follows it may also be useful to note that $\mathrm{cn}_p' = -\mathrm{sn}_p \mathrm{dn}_p$ and $\mathrm{dn}_p' = -p^2 \mathrm{sn}_p \mathrm{cn}_p$, which can also be deduced from the Pythagorean rules and the derivative of sn_p.

[23)] $g = g(x_1, x_2) = \frac{1}{a}(x_1 + bx_2)$, as in the preceding example.

[24)] Note the similarity with the metric induced by h in the preceding example.

Having determined the induced metric in terms of the curvature line coordinates and the principal curvatures of $\hat{f}$, it is straightforward to determine the conformal metric C of $\hat{f}$ that we considered in §2.1.11:

$$
\begin{aligned}
C &= (a_3 - a_1)(a_1 - a_2)\omega_1^2 + (a_1 - a_2)(a_2 - a_3)\omega_2^2 + (a_2 - a_3)(a_3 - a_1)\omega_3^2 \\
&= q^2 \cdot \{-dx_1^2 - dx_2^2 + ds^2\}.
\end{aligned}
$$

To check now that $\hat{f}$ is conformally flat, we employ the criterion we derived in §2.3.3: From C we obtain

$$
\gamma_1 = iq\,dx_1, \quad \gamma_2 = iq\,dx_2, \quad \text{and} \quad \gamma_3 = q\,ds
$$

as the conformal fundamental forms of $\hat{f}$ — since those forms are clearly closed, the induced metric I of $\hat{f}$ is conformally flat.

Finally, to see that $\hat{f}$ is not Möbius equivalent to one of the hypersurfaces discussed in §2.4.12, we use the same argument as in the preceding example. With the (real-valued) Lamé functions[25]

$$
\begin{aligned}
l_1^2(x_1, x_2, s) &= |\tfrac{\partial}{\partial x_1}\hat{f}|^2(x_1, x_2, s) &= \tfrac{(\sinh g(x_1,x_2) - \sinh \phi(s))^2}{\cosh^2 g(x_1,x_2)} \\
l_2^2(x_1, x_2, s) &= |\tfrac{\partial}{\partial x_2}\hat{f}|^2(x_1, x_2, s) &= \tfrac{(1+\sinh g(x_1,x_2)\sinh \phi(s))^2}{\cosh^2 g(x_1,x_2)} \\
l_3^2(x_1, x_2, s) &= |\tfrac{\partial}{\partial s}\hat{f}|^2(x_1, x_2, s) &= \cosh^2 \phi(s)
\end{aligned}
$$

we determine[26] the principal curvatures k_{ij} of the surfaces $x_i = const$ as surfaces "inside" $\hat{f}(M \times I)$, that is, with respect to the induced metric I:

$$
\begin{aligned}
k_{12} &= \tfrac{-1}{l_1 l_2}\tfrac{\partial}{\partial x_1}l_2 = \tfrac{1}{a(1+\sinh g \sinh \phi)}, & k_{13} &= \tfrac{-1}{l_1 l_3}\tfrac{\partial}{\partial x_1}l_3 = 0; \\
k_{21} &= \tfrac{-1}{l_2 l_1}\tfrac{\partial}{\partial x_2}l_1 = -\tfrac{b}{a(\sinh g - \sinh \phi)}, & k_{23} &= \tfrac{-1}{l_2 l_3}\tfrac{\partial}{\partial x_2}l_3 = 0; \\
k_{31} &= \tfrac{-1}{l_3 l_1}\tfrac{\partial}{\partial x_3}l_1 = \tfrac{1}{\sinh g - \sinh \phi}\phi', & k_{32} &= \tfrac{-1}{l_3 l_2}\tfrac{\partial}{\partial x_3}l_2 = -\tfrac{\sinh g}{1+\sinh g \sinh \phi}\phi'.
\end{aligned}
$$

Just as before, we see that none of the families of coordinate surfaces consists of totally umbilic surfaces unless we started with the pseudosphere — or we have $q = 1$ and $\phi \equiv 0$, in which case[27] we are obviously back to one of the hypersurfaces from §2.4.12(i).

---

[25] Any simultaneous rescalings of the Lamé functions will not affect the result of our quest.

[26] This is just the very same computation as in the preceding example.

[27] Note that this is the limiting case $p \to 0$, where $\mathrm{sn}_p \to \sin$ and $\mathrm{dn}_p \to 1$; in the other limiting case $p \to 1$, where $\mathrm{sn}_p \to \tanh$, we are back to the example of the preceding paragraph — the hypersurface $\hat{f}$ then becomes totally umbilic, which is reflected by the simultaneous vanishing of all conformal fundamental forms.

2.5 Cyclic systems

Generalizing the examples of Guichard nets from the preceding section, we want to study Guichard nets where one family of curvature lines — the orthogonal trajectories to one family of surfaces of the net — consists of circles. Obviously, such a triply orthogonal system gives rise to a cyclic system (see §2.2.11). The first thing we want to do is to see that, conversely, any cyclic system $\gamma : M^2 \to G_+(5,2)$ in the conformal 3-sphere gives rise to a triply orthogonal system,[28] consisting of the orthogonal surfaces of the cyclic system and two families of surfaces given by collecting all the circles along a curvature line of an orthogonal surface.

2.5.1 Thus, let $\gamma : M^{n-1} \to G_+(n+2, n-1)$ denote a congruence of circles (see Figure T.1). In the context of §2.2.11, we already discussed that any parallel isotropic section of the normal bundle $\gamma^\perp(p) \mapsto p$ provides, if it is immersed, an orthogonal hypersurface of γ. Hence, if there exist $3 = \mathrm{rk}\,\gamma^\perp$ parallel basis fields for $\gamma^\perp$, then γ is a cyclic system; locally this is the case if and only if the bundle $\gamma^\perp$ is flat. We want to elaborate on this characterization of cyclic systems a little further.

2.5.2 Choosing a basis for $\gamma^\perp$ that consists of two light cone maps f and $\hat{f}$ and a perpendicular unit section S, the circles of the congruence can be parametrized by (2.9). However, for our present argument, it seems more convenient to parametrize them by

$$t \mapsto f_t = f + tS - \tfrac{1}{2}t^2 \hat{f}. \tag{2.23}$$

Then, similar to (2.10), we can parametrize a 1-parameter family of hyper-sphere congruences S_t by

$$S_t = S - t\,\hat{f} \tag{2.24}$$

such that[29] each sphere $S_t(p)$ intersects the circle $\gamma(p)$ orthogonally in $f_t(p)$: We have $S_t(p) \perp \gamma(p), f_t(p)$ for each $p \in M$. Now we seek $t = t(p)$ such that f_t envelopes S_t. This yields a differential equation

$$0 = \langle df_t, S_t \rangle = dt + \tfrac{1}{2}t^2 \chi - t\nu + \omega \tag{2.25}$$

---

[28] Note that the notion of a triply orthogonal system is naturally a conformal notion because it only refers to the intersection angles of the surfaces. For example, Dupin's theorem in §2.4.8 remains valid in the conformal 3-sphere because it only refers to conformally invariant notions; in contrast, Lamé's equations (2.20) or (2.22) are not valid in the conformal setting because they express the flatness of the ambient metric.

[29] Note that these sphere congruences are by no means unique; they just have to belong to the parabolic pencils spanned by $S_t(p)$ and $f_t(p)$ for each $t \in \mathbb{R}$ and $p \in M$.

for t, where ω, χ, and ν denote the connection forms of the bundle $\gamma^\perp$ relative to the chosen basis $(S, f, \hat{f})$, as in (1.1). The integrability condition of (2.25) reads

$$0 = \tfrac{1}{2}t^2(d\chi + \nu \wedge \chi) - t(d\nu + \omega \wedge \chi) + (d\omega + \omega \wedge \nu), \qquad (2.26)$$

which, for fixed $p \in M$, is a quadratic polynomial in t. Consequently, there are more than two solutions t of the differential equation, that is, the circle congruence γ has (locally) more than two orthogonal hypersurfaces if and only if all three coefficients in (2.26) vanish. But, these coefficients are just the curvature forms of the induced connection $\nabla^\perp$ on the vector subbundle $\gamma^\perp(p) \mapsto p$ of $M \times \mathbb{R}_1^{n+2}$ (cf., §1.7.4ff).

Summarizing, we obtain the following lemma (cf., [78] or [79]):

2.5.3 Lemma. *A circle congruence* $\gamma : M^{n-1} \to G_+(n+2, n-1)$ *is (locally) normal if and only if the normal bundle* $\gamma^\perp(M) \to M$ *is flat.*

This is the case as soon as there are three orthogonal hypersurfaces of the circle congruence.

2.5.4 Thus, the orthogonal hypersurfaces of a cyclic system γ can (at least locally) always be described by parallel isotropic sections of the (normal) bundle $\gamma^\perp(M) \to M$. If we choose, as before, basis fields $(S, f, \hat{f})$ for $\gamma^\perp$, assuming now that they are parallel, the family of orthogonal hypersurfaces can be parametrized by (2.9), or (2.23) if we do without $\hat{f}$, where t is a real parameter (not a function).

Extending $(S, f, \hat{f})$ to a frame F, we obtain a frame that is adapted to S and f (as well as $\hat{f}$) at the same time (cf., §1.7.6 and §1.7.7). If we assume, for a moment, that all three maps, S, f, and $\hat{f}$, are immersions, then we learn from the Ricci equation (1.4) that the Weingarten tensors A_f and $A_{\hat{f}}$ of S with respect to f and $\hat{f}$ as its isotropic normal fields commute (see §1.7.6). Hence, they simultaneously diagonalize. Reinterpreting the equations in terms of f or $\hat{f}$ (see §1.7.7), we find that the Weingarten tensors A_f^{-1} of f and $A_{\hat{f}}^{-1}$ of $\hat{f}$ (with respect to S as a normal field) also commute and, therefore, have the same eigenspaces. Because these yield exactly the (conformally invariant) principal directions of $f, \hat{f} : M^{n-1} \to S^n$ as hypersurfaces into the conformal n-sphere, and since f and $\hat{f}$ were arbitrary orthogonal hypersurfaces of the cyclic system, we see that mapping one orthogonal hypersurface onto another along the circles of the congruence preserves the principal directions:

2.5.5 Lemma. *The principal directions on any two (immersed) orthogonal hypersurfaces of a cyclic system do correspond.*

2.5.6 As a consequence, any cyclic system in the conformal 3-sphere[30] gives rise to a triply orthogonal system where the surfaces orthogonal to the orthogonal surfaces of the cyclic system are assembled by collecting all circles of the cyclic system along a curvature line of a (any) orthogonal surface. By construction, these surfaces intersect the orthogonal surfaces of the cyclic system orthogonally; the fact that curvature lines intersect orthogonally ensures that any surfaces from the other two systems are also perpendicular.

By Dupin's theorem in §2.4.8, these surfaces assembled from the circles of the cyclic system have those circles as curvature lines. Hence, from §1.8.20, we know that the corresponding triply orthogonal system contains two families of channel surfaces. This observation gives rise to a characterization of cyclic systems as special triply orthogonal systems[31]:

2.5.7 Lemma. *A triply orthogonal system comes from a cyclic system if and only if two of the 1-parameter families of orthogonal surfaces consist of channel surfaces.*

2.5.8 Now, we want to study the effect of the Guichard condition in §2.4.4 on a cyclic system. Again, it seems preferable to use the parametrization (2.23) for a cyclic Guichard system, where f, S, and $\hat{f}$ now depend on the common curvature line parameters (x_1, x_2) of all orthogonal surfaces of the cyclic system and the family parameter $t = t(x_3)$ is a function of the third "Guichard parameter." Without loss of generality, we assume $t(0) = 0$ and $t'(0) = 1$. This last condition is achieved by possibly rescaling f by the constant $\frac{1}{t'(0)}$. Since (x_1, x_2) are principal coordinates for all f_t, we can extend the (parallel) normal frame $(S, f, \hat{f})$ to a frame $F = (S_1, S_2, S, f, \hat{f})$ such that the connection form (1.1) satisfies

$$\psi_i \wedge \omega_i = \chi_i \wedge \omega_i = 0,$$

that is, the ψ_i's and χ_i's are multiples of the metric fundamental forms ω_i of our chosen "base surface" $f = f_0$. In this setup, the Guichard condition reads (cf., §2.4.5)

$$t'^2 \;=\; |\tfrac{\partial}{\partial x_3} f_t|^2 \;=\; |\tfrac{\partial}{\partial x_1} f_t|^2 \pm |\tfrac{\partial}{\partial x_2} f_t|^2. \qquad (2.27)$$

2.5.9 For $x_3 = 0$, (2.27) becomes $1 = |\tfrac{\partial}{\partial x_1} f|^2 \pm |\tfrac{\partial}{\partial x_2} f|^2$. Thus we may assume

$$\omega_1 = \cos u \, dx_1 \quad \text{and} \quad \omega_2 = \sin u \, dx_2,$$

---

[30] With minor modification the following assertions remain true in any dimension. We leave this as an exercise for the reader.

[31] By Dupin's theorem again, it actually suffices that *one* family of surfaces consists of channel surfaces in order to conclude that a triply orthogonal system comes from a cyclic system.

where[32] $u = u(x_1, x_2)$ denotes a suitable function (that takes values in $\mathbb{R}$ or in $i\mathbb{R}$, depending on $\pm$ in the equation (2.27)). With the ansatz

$$\begin{aligned} \psi_1 &= (a_1 \cos u - a_2 \sin u)\, dx_1, \\ \psi_2 &= (a_1 \sin u + a_2 \cos u)\, dx_2, \end{aligned} \quad \text{and} \quad \begin{aligned} \chi_1 &= -(b_1 \cos u - b_2 \sin u)\, dx_1, \\ \chi_2 &= -(b_1 \sin u + b_2 \cos u)\, dx_2, \end{aligned}$$

$$(2.28)$$

where a_i, b_i are suitable functions of x_1, x_2, (2.27) then becomes

$$\begin{aligned} t'^2 &= (1 - ta_1 + \tfrac{1}{2}t^2 b_1)^2 + (ta_2 - \tfrac{1}{2}t^2 b_2)^2 \\ &= (1 - (a_1 + ia_2)\, t + \tfrac{1}{2}(b_1 + ib_1)\, t^2)(1 - (a_1 - ia_2)\, t + \tfrac{1}{2}(b_1 - ib_1)\, t^2). \end{aligned}$$

This shows that the parametrizing function t of the family of orthogonal surfaces of a cyclic system satisfying the Guichard condition has to be an elliptic function. Since this function $t = t(x_3)$ is independent of the parameters x_1 and x_2 of the orthogonal surfaces of the circle congruence, so are its branch values $t' = 0$, and, therefore,[33] the a_i's and b_i's are constant.

2.5.10 As a consequence, $\mathcal{K} := b_2 S + (a_1 b_2 - a_2 b_1)f - a_2 \hat{f}$ is a constant vector,

$$d\mathcal{K} = d[b_2 S + (a_1 b_2 - a_2 b_1)f - a_2 \hat{f}] = 0.$$

If $\mathcal{K} = 0$, then $a_2 = b_2 = 0$ and all the surfaces $x_3 = const$ are sphere pieces.

If, on the other hand, $\mathcal{K} \neq 0$, then it defines a quadric Q_κ^3 of constant sectional curvature $\kappa = -|\mathcal{K}|^2$ and a linear sphere complex that consists of the hyperplanes in Q_κ^3 (see Section 1.4). Since $S_i \perp \mathcal{K}$ for $i = 1, 2$, the spheres S_1 and S_2 are planes in the subgeometry given by $\mathcal{K}$, and hence the circles $S_1 \cap S_2$ of the cyclic Guichard system become straight lines:

2.5.11 Lemma. *Any cyclic system that defines a Guichard net is a normal line congruence in some space of constant curvature.*

2.5.12 To begin with we had chosen our parallel normal frame $(S, f, \hat{f})$ for the cyclic Guichard net γ under consideration rather arbitrarily; now, because we know that the cyclic system is a normal line congruence in some space of constant curvature, there are more natural choices: We may normalize the "initial surface" $f = f_0$ to take values in the quadric Q_κ^3 defined by $\mathcal{K}$ and the sphere congruence S to be its tangent plane congruence. Thus, we fix the normal frame $(S, f, \hat{f})$ to satisfy (cf., §1.7.8)

$$\langle f, \mathcal{K} \rangle \equiv -1 \quad \text{and} \quad \langle S, \mathcal{K} \rangle \equiv 0, \quad \text{hence} \quad \hat{f} = \tfrac{\kappa}{2} f - \mathcal{K}$$

---

[32] Remember that the forms ω_i, ψ_i, and χ_i depend, with the frame F, only on x_1 and x_2.

[33] Another option is to take the x_i-derivative, $i = 1, 2$, on both sides of the equation to obtain a vanishing polynomial on the right side. Its coefficients, which contain the derivatives of the a_i's and b_i's, then have to vanish, showing that the a_i's and b_i's have to be constant.

with $\kappa = -|\mathcal{K}|^2$. Instead of choosing some parametrization for the circles of the cyclic system directly, we now choose a simple parametrization for the family of sphere congruences S_t, requiring all of them to be plane congruences:

$$S_t := \frac{1}{\sqrt{1+\kappa t^2}}(S + t[\tfrac{\kappa}{2}f + \hat{f}]). \tag{2.29}$$

The surfaces f_t, which we normalize to take values in Q_κ^3, are then obtained as one of the intersection points of S_t with the circles of the cyclic system,

$$f_t = \frac{1}{\sqrt{1+\kappa t^2}}(f - t[S + \frac{1}{1+\sqrt{1+\kappa t^2}}\mathcal{K}]), \tag{2.30}$$

whereas the other $\hat{f} = \frac{\kappa}{2}f - \mathcal{K}$ is just its mirror image or the point at infinity in the case $\kappa = 0$ (see §1.7.8). Note that, in (2.30) as well as in (2.29), the range of t is restricted by the condition $1 + \kappa t^2 > 0$: If $\kappa < 0$, the maps

$$\sqrt{1+\kappa t^2} \cdot f_t \to f + \tfrac{1}{\kappa}\mathcal{K} \pm \frac{1}{\sqrt{-\kappa}}S \perp \mathcal{K} \quad \text{as} \quad t \to \pm\frac{1}{\sqrt{-\kappa}},$$

that is, f_t approaches the intersection points of the circles of the cyclic system with the infinity boundary ∂Q_κ^3.

With this new parametrization the Guichard condition (2.27) changes shape: Using (2.28), where now $b_1 = -\frac{\kappa}{2}$ and $b_2 = 0$, we find

$$\frac{t'^2}{(1+\kappa t^2)^2} = \frac{(1+a_1 t)^2 + a_2^2 t^2}{1+\kappa t^2} \quad \Leftrightarrow \quad t'^2 = (1+\kappa t^2)(1 + 2a_1 t + (a_1^2 + a_2^2)t^2), \tag{2.31}$$

with the a_i being constants, as before.

2.6 Linear Weingarten surfaces in space forms

Since the principal curvatures of the surfaces f_t in Q_κ^3 are obtained from expressions comprising the forms in (2.28), they only depend on one function u. Thus the principal curvatures of the surfaces f_t are (functionally) dependent, that is, the surfaces f_t are all Weingarten surfaces.

We want to make this statement more explicit.

2.6.1 Since S_t are the tangent plane congruences of the surfaces f_t in Q_κ^3, $\langle S_t, \mathcal{K} \rangle = 0$, the maps S_t can be interpreted as unit normal fields for f_t within the quadrics Q_κ^3 of constant curvature: $S_t(p) \in T_{f_t(p)}Q_\kappa^3$. Hence the first and second fundamental forms of the $f_t : (x_1, x_2) \mapsto f_t(x_1, x_2) \in Q_\kappa^3$ are given by

$$I_t = \langle df_t, df_t \rangle \quad \text{and} \quad II_t = -\langle df_t, dS_t \rangle.$$

In this way we obtain the principal curvatures of the f_t:

$$\frac{(a_1 - \kappa t)\cos u - a_2 \sin u}{(1+a_1 t)\cos u - a_2 t \sin u} \quad \text{and} \quad \frac{(a_1 - \kappa t)\sin u + a_2 \cos u}{(1+a_1 t)\sin u + a_2 t \cos u}. \tag{2.32}$$

With this it is now pure algebra to verify that the (extrinsic) Gauss curvatures K_t and mean curvatures H_t of the surfaces f_t in Q_κ^3 satisfy an affine relation

$$0 = c_K(t)K_t + 2c_H(t)H_t + c(t) \tag{2.33}$$

with

$$\begin{aligned}
c_K(t) &= (a_1^2 + a_2^2)t^2 + 2a_1t + 1, \\
c_H(t) &= \kappa a_1t^2 + (\kappa - (a_1^2 + a_2^2))t - a_1, \\
c(t) &= \kappa^2 t^2 - 2\kappa a_1 t + (a_1^2 + a_2^2).
\end{aligned} \tag{2.34}$$

Note that the sign of $c_K(t)c(t) - c_H^2(t) = a_2^2(1 + \kappa t^2)^2$ is an invariant of the family of parallel Weingarten surfaces: It determines whether we have $+$ or $-$ in (2.27), that is, whether a_2 is, with the function u, real or imaginary.

Thus we have proved the following theorem (cf., [146]):

2.6.2 Theorem. *The orthogonal surfaces of a cyclic system that defines a Guichard net are parallel linear Weingarten surfaces in some space Q_κ^3 of constant sectional curvature; that is, their (extrinsic) Gauss and mean curvatures K_t and H_t satisfy, with suitable functions c_K, c_H, and c of the family parameter, an affine relation*

$$0 = c_K(t)K_t + 2c_H(t)H_t + c(t).$$

2.6.3 The surfaces of constant Gauss or mean curvature among the surfaces f_t of a family are characterized by the (distinct) zeros[34] of c_H and c_K, respectively. By (2.31) we have $t'^2 = (1 + \kappa t^2)c_K(t)$, so that the Guichard net degenerates exactly at the surfaces of constant mean curvature in the family — and at the infinity boundary of Q_κ^3 in the case $\kappa < 0$.

We will follow up on the study of these special surfaces among the surfaces of a family $(f_t)_t$ in a moment; first we want to understand the converse of the above theorem though.

2.6.4 Does any given linear Weingarten surface $f : M^2 \to Q_\kappa^3$ give rise to a cyclic Guichard system via its normal line congruence? Obviously, by the previous paragraph, we have to exclude surfaces of constant mean curvature since they would be singular for the Guichard net. Thus, assume f is a linear Weingarten surface, satisfying $0 = c_K K + 2c_H H + c$ in some quadric Q_κ^3 of constant curvature κ, with nonconstant mean curvature. In this way $c_K \neq 0$ and we may assume $c_K = 1$, without loss of generality. Excluding also surfaces with a constant principal curvature, that is, $c = c_H^2$, where the Weingarten condition factorizes, we make the ansatz

$$a_1 - a_2 \tan u \quad \text{and} \quad a_1 + a_2 \cot u$$

---

[34] Note that $c_K(t)c(t) - c_H^2(t) = a_2^2(1 + \kappa t^2)^2 \neq 0$, because we excluded the case $a_2 = 0$ of sphere pencils, so that c_H and c_K cannot vanish simultaneously.

for the principal curvatures of f, where $a_1 := -c_H$, $a_2 := \sqrt{c - c_H^2}$, and u denotes a suitable function. Using a principal frame we consequently have

$$\psi_1 = (a_1 - a_2 \tan u)\,\omega_1 \quad \text{and} \quad \psi_2 = (a_1 + a_2 \cot u)\,\omega_2$$

in (1.1). Combining the Codazzi equations (1.3) for the ψ_i's and ω_i's to eliminate the connection forms ω_{ij}, we obtain

$$
\begin{aligned}
0 &= -a_2\left(\frac{d\omega_1}{\cos u \sin u} + \frac{du \wedge \omega_1}{\cos^2 u}\right) = -\frac{a_2}{\sin u}d\left(\frac{\omega_1}{\cos u}\right)\\
0 &= a_2\left(\frac{d\omega_2}{\cos u \sin u} - \frac{du \wedge \omega_2}{\sin^2 u}\right) = \frac{a_2}{\sin u}d\left(\frac{\omega_2}{\sin u}\right),
\end{aligned}
$$

showing that principal curvature line coordinates (x_1, x_2) can be chosen such that $\omega_1 = \cos u\, dx_1$ and $\omega_2 = \sin u\, dx_2$, that is, $|\frac{\partial}{\partial x_1}f|^2 \pm |\frac{\partial}{\partial x_2}f|^2 = 1$, where the sign depends on whether u is real- or imaginary-valued.

At this point, our previous computations show that parametrizing the parallel surfaces of f by (2.30) with the correct speed (2.31) provides us with a Guichard net. Thus we have proved the converse of the above theorem.

2.6.5 Theorem. *Given a linear Weingarten surface with nonconstant mean or principal curvatures, in a space of constant curvature its normal line congruence gives rise to a cyclic Guichard net.*

2.7 Bonnet's theorem

Now, we want to resume our above discussion about surfaces of constant mean or constant Gaussian curvature among the surfaces of a family of linear Weingarten surfaces. By the theorem we just stated, any linear Weingarten surface with nonconstant mean or principal curvatures is part of a cyclic Guichard system. However, by dropping the Guichard condition on the family, we may include these surfaces. For that purpose we will redo some of our computations, not assuming the special form for the principal curvatures of the surface $f = f_0$ given by (2.28).

2.7.1 Thus, let $f : M^2 \to Q_\kappa^3$ denote a linear Weingarten surface, its Gauss and mean curvatures satisfying $c_K K + 2c_H H + c = 0$ with suitable constants c, c_H, and c_K. Further, let f_t denote the parallel surfaces of $f = f_0$, parametrized as in (2.30), whereas S_t shall denote the tangent plane congruences (unit normal fields in Q_κ^3) of f_t. A similar computation as the one providing (2.32) before now yields

$$H_t = \frac{H + (K - \kappa)t - \kappa H t^2}{1 + 2Ht + Kt^2} \quad \text{and} \quad K_t = \frac{K - 2\kappa H t + \kappa^2 t^2}{1 + 2Ht + Kt^2}.$$

Straightforward computation then proves that the parallel surfaces f_t of f are linear Weingarten again, that is, there are functions

$$
\begin{aligned}
c_K(t) &= ct^2 - 2c_H t + c_K,\\
c_H(t) &= -\kappa c_H t^2 + (\kappa c_K - c)t + c_H, \quad \text{and}\\
c(t) &= \kappa^2 c_K t^2 + 2\kappa c_H t + c
\end{aligned}
$$

such that $c_K(t)K_t + 2c_H(t)H_t + c(t) = 0$, similar to (2.34) providing (2.33). We state this as an intermediate result:

2.7.2 Lemma.   *The parallel surfaces of a linear Weingarten surface in a space of constant curvature are linear Weingarten.*

2.7.3 Among the surfaces of such a family of parallel linear Weingarten surfaces there are special ones that are distinguished by the vanishing of one of the coefficient functions c_K, c_H, or c: If $c_K(t) = 0$, then H_t is constant; if $c_H(t) = 0$, then K_t is constant — as we already discussed in §2.6.3 — and if $c(t) = 0$, then the ratio $\frac{H_t}{K_t}$ is constant, that is, the sum of the principal curvature radii is constant.

This provides a simple tool for studying analogs of Bonnet's theorem about parallel constant mean and constant Gauss curvature surfaces in Euclidean space (cf., [214]). The only thing that is not reduced to the study of quadratic equations in this way is the computation of the (geodesic) distance between the distinguished surfaces of the family — this is obtained as the arc length integral $\int |\frac{\partial}{\partial t} f_t| dt = \int \frac{dt}{1+\kappa t^2}$ along the lines (geodesics) between the surfaces under investigation.

2.7.4 Example (Bonnet's theorem). We want to conclude these considerations deriving Bonnet's classical theorem in this context, leaving a detailed analysis of the other cases as an exercise for the reader (cf., [146]).

Thus, we are considering Euclidean ambient space Q_0^3, that is, $\kappa = 0$. In this case we have $c(t) = c$ for all t, and $c_H(t) = c_H - ct$ becomes (at most) linear. Now there are two different cases to consider:

$c \neq 0$. Here the function $t \mapsto c_H(t)$ is an honest linear function with exactly one zero at $t_0 = \frac{c_H}{c}$, so that our family of linear Weingarten surfaces contains one surface of constant Gauss curvature[35]

$$K_{t_0} = -\frac{c}{c_K(t_0)} = \frac{c^2}{c_H^2 - c_K c}.$$

The zeros $t_\pm = \frac{c_H \pm \sqrt{c_H^2 - c_K c}}{c}$ of the quadratic polynomial $t \mapsto c_K(t)$ become real if and only if $K_{t_0} > 0$: Then there are two parallel constant mean curvature surfaces at $t_\pm$ with mean curvature

$$H_{t_\pm} = -\frac{c}{2c_H(t_\pm)} = \pm\frac{1}{2\sqrt{K_{t_0}}}.$$

The two constant mean curvature surfaces lie symmetrically to the constant Gauss curvature surface in distance

$$|\int_{t_0}^{t_\pm} dt| = |t_\pm - t_0| = \frac{1}{\sqrt{K_{t_0}}} = \frac{1}{2|H_{t_\pm}|},$$

---

[35] Note that $c_K(t_0)K_{t_0} = -c \neq 0$ forces $c_K(t_0) \neq 0$ and hence $c_H^2 - c_K c \neq 0$.

having themselves distance $\frac{1}{|H_{t_\pm}|}$. Thus, we obtained Bonnet's classical assertion.

$c = 0$. In this case we can further distinguish two cases: If also $c_H = 0$, then both functions, $t \mapsto c(t)$ as well as $t \mapsto c_H(t)$, vanish identically, so that we must have $K_t = 0$ for all t; if, on the other hand, $c_H \neq 0$, then $c_H(t) = c_H$ for all t, whereas $t \mapsto c_K(t)$ becomes an honest linear function with exactly one zero at $t_0 = \frac{c_K}{c_H}$. Consequently, we obtain one surface of constant mean curvature $H_{t_0} = \frac{c}{c_H} = 0$ while all its parallel surfaces have constant sum of their curvature radii.

Chapter 3

Application: Isothermic and Willmore surfaces

As a second application of the classical model of Möbius geometry we will discuss some aspects of isothermic surfaces and Willmore surfaces. As unrelated as the two surface classes seem to be at first, they occur as two (*the* two nontrivial) solutions of the same problem that we will refer to[1] as "Blaschke's problem," since it arises in [29]. The contents of this chapter are loosely arranged around the solution of this problem, collecting many of Blaschke's key results in surface theory in the conformal[2] 3-sphere. Isothermic as well as Willmore surfaces are a field of current interest after they have been neglected for quite some time — Willmore surfaces even seem to have been forgotten and had a different name in classical times. The study of both surface classes traces back to the first half of the nineteenth century; however, the origins remain obscure to the author. It is also true for both surface classes that physical considerations play a role — as indicated for isothermic surfaces by their name.

"Willmore surfaces" were apparently (re-) discovered at least three times.

First, in the early nineteenth century, Germain [126] (cf., [127]) studied "elastic surfaces" — analogous to the elastic curves discussed in the works of Bernoulli and Euler [185]. Her main contribution seems to be "her hypothesis" that the "elastic force" of a thin plate is proportional to its mean curvature. As a consequence, the "elastic energy" is then the total squared mean curvature, the "Willmore functional." For a detailed discussion of Germain's work as well as related work by, for example, Poisson, see [83]. It may also be remarkable that this aspect of Willmore surfaces is known in physics [176] and of interest in current research [289]. As was already indicated, the mean curvature of a surface plays a central role in the theory of Willmore surfaces. It seems worth mentioning that Germain studied the mean curvature of surfaces (see also the colloquial presentation of the history of differential geometry [261] for an interesting remark) and introduced the "mean curvature sphere" [128], a concept that became central in Möbius differential geometry: Around the turn of the century, the 2-parameter family of the mean curvature spheres was known as the "central sphere congruence" of the surface [29]; nowadays, after Bryant's paper [40], it goes by the name "conformal Gauss map."

---

[1] This problem is also referred to as "Ribaucour's problem" (see [80]).

[2] Of course, we do not cover any results in Lie sphere geometry that comprise the main part of the book.

This brings us to the second, Möbius geometric approach to Willmore surfaces, which we will follow: The central sphere congruence can be defined in a conformally invariant way. One possible characterization is that its induced metric is conformally equivalent to the (conformal class of) metrics induced on the surface, justifying the notion of the conformal Gauss map. Following the classical authors (see [29]) we call a surface "conformally minimal" if its central sphere congruence is minimal, that is, if it minimizes the area for *all* variations through sphere congruences — as the "simplest variational problem that can be invariantly associated with a surface in Möbius geometry." Following [110], we will then show that the conformally minimal surfaces are exactly the Willmore surfaces, a fact that is already stated in [29].

The third approach was by T. Willmore in 1965 [306], after whom these surfaces are usually named in our days. He considers the "total squared mean curvature" of a closed surface as an analog for the total Gaussian curvature — which is a topological invariant. Then he notices that the infimum value of this "Willmore functional" characterizes the round sphere. As a natural next step, he asks for the infimum value and a minimizer in the class of (topological) tori.[3] The conjecture is that a stereographic projection of the minimal Clifford torus in S^3 (and its Möbius transforms) minimize with value $2\pi^2$ of the total squared mean curvature. Since then many mathematicians, too numerous to cite here (for some of the better known papers, check the list of references), have worked on this conjecture and the characterization of the corresponding critical surfaces, using methods that range from Euclidean and Möbius differential geometry to global analysis, calculus of variations, and integrable systems methods. See [307] for an inspiring survey.

"Isothermic surfaces," on the other hand, have a less inconsistent history, even though the origins are as obscure as in the Willmore case, and they were pretty much off the main stream after a period of intensive research at the end of the nineteenth century. It was only recently that isothermic surfaces regained interest because of their integrable systems description, initiated in [71].

It seems that the notion of "isothermal lines" was motivated by their physical interpretation as lines of equal temperature [184], and it appears as early as 1833 in a work by Lamé (cf., [175]; see also [274]). He was apparently led to the notion of "isothermic surfaces," that is, surfaces with isothermal lines of curvature, by considering triply orthogonal systems constructed from surfaces of equal temperature in a (homogeneous) body with a stationary, nonconstant temperature [191], [237]. Thus it seems that, in the beginning,

---

[3] Candidates for minimizers among surfaces of other genus may be found in [172].

the study of isothermic surfaces was closely related to the study of triply orthogonal systems, and that it was only later that they were studied on their own,[4] as, for example, in [64] or [299].

Among the first papers that deal with isothermic surfaces from the point of view of surface geometry is Christoffel's paper [70], where a problem very similar to Blaschke's problem is treated in Euclidean geometry. The solution leads to a transformation, or "duality," for isothermic surfaces that is at the basis of an exceptionally rich transformation theory. Later, a second type of transformation for isothermic surfaces was introduced by Darboux [93], which will appear from the solution of Blaschke's problem in a very similar way as Christoffel's transformation appears from Christoffel's problem. As in the Willmore case, a special type of sphere congruence occurs in this context, "Ribaucour sphere congruences," and Darboux's transformation is a special type of the corresponding notion of "Ribaucour transformations." The Darboux transformation is closely related to the integrable systems approach to isothermic surfaces (cf., [46]) as well as to the theory of "discrete isothermic nets" [30]. Finally, a third type of transformation is found in papers by Bianchi [21], [20], and P. Calapso [53], [55]. This last transformation turns out to be again closely related to the integrable systems approach to isothermic surfaces (see, for example, [153]) as well as to rigidity questions in Möbius geometry (see [62] and [203]).

However, this transformation theory of isothermic surfaces is much more smoothly dealt with in the quaternionic or the Clifford algebra setting. Therefore, we only touch on it in this chapter, in order to fix some basic notions and ideas, and postpone a detailed analysis.

After completing the discussion of Blaschke's problem and related topics, we will discuss the "intersection" of both surface classes considered, giving a proof of Thomsen's theorem [29]. Here an interesting phenomenon occurs: Imposing these two Möbius geometric properties on a surface causes symmetry breaking, so that the surface naturally lies in a metric subgeometry of Möbius geometry, similar to §2.6.2. To conclude, we will reconsider channel surfaces as an example, experiencing another symmetry-breaking (see [276] and [144]).

The contents of this chapter are organized as follows:

Section 3.1. We state Blaschke's problem, and we introduce the notion of Ribaucour sphere congruence — that will be central to various later discussions. One (trivial) class of solutions of Blaschke's problem is given.

---

[4] Also of interest is the relation with Bonnet surfaces (cf., [18], [157], and [201]).

Section 3.2. The notions of isothermic surfaces in the conformal 3-sphere and of Darboux pairs of isothermic surfaces are defined. A second (nontrivial) solution to Blaschke's problem is given: Darboux pairs of isothermic surfaces.

Section 3.3. A first encounter of the transformation theory of isothermic surfaces in S^3 is given — a more complete (and more general) treatment is to be found in Chapter 5 or 8. We discuss the relation between Darboux pairs and curved flats, and introduce the spectral transformation of a Darboux pair. As an example, Lawson's correspondence for constant mean curvature surfaces is discussed.

Then the limit behavior of the spectral transformation is used to motivate Christoffel's problem, which can be considered as the Euclidean analog of Blaschke's problem. The solutions to Christoffel's problem are given: This leads to the notion of Christoffel pairs of isothermic surfaces and to families of minimal surfaces in Euclidean 3-space that generalize the associated family of minimal surfaces. Again, constant mean curvature surfaces in $\mathbb{R}^3$ are discussed as an example.

Section 3.4. The central sphere congruence of a surface is introduced, and its characterizations as the mean curvature sphere congruence and as the conformal Gauss map are given. The notion of a conformally minimal surface is defined and a third class of solution of Blaschke's problem is established: dual pairs of conformally minimal surfaces. The (quadratic) Clifford torus is discussed as an example of a "self-dual" conformally minimal surface.

The transformation formula for the Laplacian under a conformal change of the metric on the domain is given, and conformal minimality is proven to be equivalent to the harmonicity of the conformal Gauss map.

Section 3.5. The notion of Willmore surface is introduced. The first variation formula for the Willmore functional is derived, and it is shown that Willmore surfaces are the same as conformally minimal surfaces. The Euler-Lagrange equation for Willmore surfaces in a space of constant curvature is formulated.

Section 3.6. It is shown that a surface is isothermic if and only if its central sphere congruence is Ribaucour. Thomsen's theorem is proved: Isothermic Willmore surfaces are minimal surfaces in some space of constant curvature.

Section 3.7. A theorem by Vessiot is presented: Isothermic channel surfaces are (part of) cones, cylinders, or surfaces of revolution. It is shown that Willmore channel surfaces are isothermic, and further analysis shows that planar free elastic curves are involved. It is shown that a Willmore channel surface is an equivariant minimal surface in a certain space form.

The Willmore functional of a channel surface is computed in terms of the enveloped sphere curve, and it is shown that the Willmore functional can be lowered by replacing the channel surface by a surface of revolution. It is discussed how this fact can be used to prove the Willmore conjecture for (umbilic-free) channel tori.

Remark. In this chapter the reader is expected to be familiar with the classical description of Möbius geometry, as provided in Chapter 1.

3.1 Blaschke's problem

In §78 of [29], the following problem[5] is considered: Let $S : M^2 \to \mathbb{RP}^4_O$ be a sphere congruence with two envelopes $f, \hat{f} : M^2 \to S^3$, such that the two envelopes induce the same (possibly degenerate) conformal structure on M; characterize the sphere congruence and its envelopes.

It will turn out that three cases can occur. In this section we will provide the setup to solve the problem, and we will describe the first (trivial) case. The two remaining (interesting) cases will be treated in the following two sections, respectively.

3.1.1 We will consider $S : M^2 \to S^4_1$ as a regular, spacelike map into the Lorentz sphere[6] — remember that a sphere congruence has two envelopes if and only if the induced metric of S is positively definite (see §1.6.9). Any lifts $f, \hat{f} : M^2 \to \mathbb{R}^5_1$ of the envelopes of S can then be interpreted as two isotropic normal fields (see §1.6.3) of $S : M^2 \to S^4_1$.

Fixing some scale for f and $\hat{f}$, the induced metrics I_f and $I_{\hat{f}}$ become the third fundamental forms of S with respect to f and $\hat{f}$ (cf., §1.7.6):

$$I_f = I(A., A.) \quad \text{and} \quad I_{\hat{f}} = I(\hat{A}., \hat{A}.),$$

where $A = A_f$ and $\hat{A} = A_{\hat{f}}$ denote the Weingarten tensors of S with respect to f and $\hat{f}$, respectively.

3.1.2 Now f and $\hat{f}$ induce the same conformal structure on M^2 if and only if the induced metrics I_f and $I_{\hat{f}}$ are conformally equivalent, $\hat{\lambda}^2 I_f = \lambda^2 I_{\hat{f}}$. Since the Weingarten operators are self-adjoint by the Ricci equations (1.4), the induced metrics are conformally equivalent if and only if $\hat{\lambda}^2 A^2 = \lambda^2 \hat{A}^2$.

By using the Cayley-Hamilton identity, $A^2 = \text{tr}A \cdot A - \det A \cdot id$ and similarly for $\hat{A}$, we see that (pointwise) two different cases can be considered:

1. A, $\hat{A}$, and id are linearly dependent — in this case, A and $\hat{A}$ commute, so that they can simultaneously be diagonalized; denoting their

---

[5] Compare also [80], [93], [98], and [99].

[6] In this way we exclude nonorientable surfaces, but our analysis will be local anyway.

eigenvalues by a_i and $\hat{a}_i$, respectively, $\hat{\lambda}^2 A^2 = \lambda^2 \hat{A}^2$ yields $\hat{\lambda} a_i = \pm\lambda \hat{a}_i$, $i = 1, 2$. Thus we further distinguish two subcases:

a. $\hat{\lambda}^2 \det A = \lambda^2 \det \hat{A}$, and

b. $\hat{\lambda}^2 \det A = -\lambda^2 \det \hat{A}$.

2. A, $\hat{A}$, and id are linearly independent — in this case, we must have $\text{tr} A = \text{tr} \hat{A} = 0$.

3.1.3 Regularity assumption. To analyze the problem we will *assume* that these cases do not mix, that is, at all points $p \in M^2$ the same case applies. Moreover, we will assume that at least one of the envelopes, say f, is immersed, that is, I_f is positive definite, $\det A \neq 0$.

3.1.4 We start by collecting some general facts about the first case.

Consider the bilinear form $I(A., \hat{A}.)$. Choosing an S-adapted frame F for S (cf., §1.7.6), we have $I(A., \hat{A}.) = \sum_i \omega_i \chi_i$: Thus $I(A., \hat{A}.)$ is symmetric if and only if A and $\hat{A}$ commute, $[A, \hat{A}] = 0$, if and only if the normal curvature $d\nu = \sum \chi_i \wedge \omega_i = 0$ of S vanishes, by the third of the Ricci equations (1.4).

3.1.5 Moreover, if both envelopes $f, \hat{f} : M^2 \to S^3$ are immersions, then both Weingarten tensors A and $\hat{A}$ are invertible, and their inverses, A^{-1} and $\hat{A}^{-1}$, are the Weingarten tensors of f and $\hat{f}$ with respect to S as a unit spacelike normal field. This gives the reversed viewpoint with f or $\hat{f}$ as the primary object of interest (see §1.7.7). Since $[A, \hat{A}] = 0$ if and only if $[A^{-1}, \hat{A}^{-1}] = 0$, we find that $I(A., \hat{A}.)$ is symmetric if and only if f and $\hat{f}$ have the same curvature directions.[7]

Thus, we have proved the following.[8]

3.1.6 Lemma. *Let $S : M^{n-1} \to S_1^{n+1}$ be an immersed sphere congruence with two envelopes f and $\hat{f}$. Then the corresponding Weingarten tensors A and $\hat{A}$ commute, $[A, \hat{A}] = 0$, if and only if the normal bundle of S is flat.*

When both f and $\hat{f}$ are regular, this is the case if and only if the curvature directions of f and $\hat{f}$ correspond.

3.1.7 This type of sphere congruences, where the curvature lines on both envelopes correspond, has been studied extensively by the classical geometers and goes by the name "Ribaucour congruence." This classical charac-

---

[7] Remember (§P.6.7) that the notion of "curvature direction" makes sense in the conformal n-sphere. Moreover (§1.7.7), the curvature directions of an immersion f are the eigendirections of the Weingarten tensor of f with respect to *any* enveloped sphere congruence S, viewed as a normal field of f: Any change $S \to S + af$ of the enveloped sphere congruence yields a change $A \to A - a\,id$ of the corresponding Weingarten tensor field that does not effect its eigendirections. And, by having f take values in one of the constant curvature quadrics, the Weingarten tensor of f with respect to its tangent plane congruence gives the Weingarten tensor of f in that space of constant curvature.

[8] Note that we never used $n = 3$ in the proof.

terization, however, causes problems when one of the envelopes is degenerate — and therefore has no curvature lines. On the other hand, we also wish to have a definition that applies to the case where the sphere congruence is not immersed. Thus, we give a definition using no regularity assumptions:

3.1.8 Definition. *A sphere congruence $S : M^{n-1} \to \mathbb{RP}^{n+1}_O$ with two envelopes $f, \hat{f} : M^{n-1} \to S^n$ is called Ribaucour (sphere) congruence if the normal 2-plane bundle spanned by any[9] lifts $f, \hat{f} : M^{n-1} \to \mathbb{R}^{n+2}_1$ is flat.*

3.1.9 We are now prepared to analyze the situation of case 1a in §3.1.2.

We use an S-adapted frame such that, without loss of generality,[10] f and $\hat{f}$ are parallel sections of the normal bundle of S. Thus, $\lambda df = \pm \lambda d\hat{f}$. Since f was assumed to be immersed, $\lambda \neq 0$. Now let $\kappa := \pm 2\frac{\lambda}{\lambda}$, that is, $0 = \frac{\kappa}{2} df - d\hat{f}$. The Codazzi equations then yield

$$0 = d(\tfrac{\kappa}{2}\omega_i - \chi_i) = \tfrac{1}{2}d\kappa \wedge \omega_i - \sum_j \omega_{ij} \wedge (\tfrac{\kappa}{2}\omega_j - \chi_j) = \tfrac{1}{2}d\kappa \wedge \omega_i.$$

Hence $d\kappa = 0$ because the ω_i's are linearly independent, and $\mathcal{K} := \frac{\kappa}{2}f - \hat{f}$ is a constant vector. Consequently, S is the tangent plane congruence of a surface $f : M^2 \to Q^3_\kappa$ into the space Q^3_κ of constant curvature defined by $\mathcal{K}$ (see Figure T.2). Moreover, the second envelope $\hat{f}$ is either constant (if $\kappa = 0$) or $\hat{f} = \frac{\kappa}{2}(f - 2\frac{\langle f, \mathcal{K} \rangle}{\langle \mathcal{K}, \mathcal{K} \rangle}\mathcal{K})$ is a Möbius transform of f (cf., §1.7.8).

3.1.10 The converse clearly holds: Given any surface $f : M^2 \to Q^3_\kappa$ in a space of constant curvature, its tangent plane congruence is Ribaucour and the second envelope of the tangent plane congruence is either constant (the point at infinity of Q^3_0) or a Möbius transform of f. Thus the tangent plane congruence of any surface in a space of constant curvature gives a (Ribaucour) sphere congruence whose envelopes induce the same conformal structure on M.

Hence we have obtained a first classification result:

3.1.11 Case 1a. *The sphere congruence $S : M^2 \to \mathbb{RP}^4_O$ is the common tangent plane congruence of its two (Möbius equivalent) envelopes in some quadric Q^3_κ of constant curvature.*

Conversely, the tangent plane congruence of any surface in a space Q^3_κ of constant curvature provides a Ribaucour sphere congruence whose envelopes induce (weakly) conformally equivalent metrics.

---

[9] Note that this 2-plane bundle does not depend on the choice of lifts for f and $\hat{f}$.

[10] Since $d\nu = 0$, locally $\nu = du$; then $f \to e^{-u}f$ yields a parallel rescaling of f.

3.2 Darboux pairs of isothermic surfaces

The second case, 1b, of §3.1.2 is more interesting. Clearly, as discussed in the previous section, the sphere congruence under consideration is Ribaucour in this case, too. The condition 1b, $\hat{\lambda}^2 \det A = -\lambda^2 \det \hat{A}$, is distinguished from 1a only where $\det \hat{A} \neq 0$, that is, where $\hat{f}$ is immersed. Thus we restrict attention to (the open set of) points where $\det \hat{A} \neq 0$; the results then extend to all of M^2 by continuity — as long as we prohibit the singular set of $\hat{A}$ from having interior points. But such open singular set would fall into case 1a, and we assumed that the cases should not mix.

3.2.1 Thus A and $\hat{A}$ simultaneously diagonalize in exactly *one* way,[11] such that their eigenvalues satisfy $\hat{\lambda} a_1 = \lambda \hat{a}_1$ and $\hat{\lambda} a_2 = -\lambda \hat{a}_2$ with *positive* functions $\lambda, \hat{\lambda}$. Let $u := \frac{\ln \lambda - \ln \hat{\lambda}}{2}$. We choose an S-adapted frame F that diagonalizes A and $\hat{A}$, that is, $\omega_i = a_i \psi_i$ and $\chi_i = \hat{a}_i \psi_i$, and such that f and $\hat{f}$ are parallel sections of the normal bundle of S. In particular, we have $\chi_1 = e^{-2u} \omega_1$ and $\chi_2 = -e^{-2u} \omega_2$. Now the Codazzi equations (1.3) yield

$$
\begin{aligned}
d(e^{-u}\omega_1) &= & d(e^u \chi_1) &= & e^u(du \wedge \chi_1 - \omega_{12} \wedge \chi_2), \\
d(e^{-u}\omega_1) &= -e^{-u}(du \wedge \omega_1 + \omega_{12} \wedge \omega_2) &= & -e^u(du \wedge \chi_1 - \omega_{12} \wedge \chi_2); \\
d(e^{-u}\omega_2) &= & -d(e^u \chi_2) &= & -e^u(du \wedge \chi_2 - \omega_{21} \wedge \chi_1), \\
d(e^{-u}\omega_2) &= -e^{-u}(du \wedge \omega_2 + \omega_{21} \wedge \omega_1) &= & e^u(du \wedge \chi_2 - \omega_{21} \wedge \chi_1).
\end{aligned}
$$

Hence $d(e^{-u}\omega_i) = 0$ (as well as $d(e^u \chi_i) = 0$), so that there are (local) coordinates (x_1, x_2) with $dx_i = e^{-u}\omega_i = \pm e^u \chi_i$. Since

$$
I_f = \sum_i \omega_i^2 = e^{2u} \sum_i dx_i^2 \quad \text{and} \quad I_{\hat{f}} = e^{-2u} \sum_i dx_i^2,
$$

these are conformal coordinates for both surfaces, f as well as $\hat{f}$; and, because the $\frac{\partial}{\partial x_i}$'s are the eigendirections of A and $\hat{A}$, they are curvature line parameters.

We have proved that both envelopes ought to be "isothermic surfaces":

3.2.2 Definition. *A surface $f : M^2 \to S^3$ is called isothermic if, around each (nonumbilic) point, there exist conformal curvature line coordinates.*

3.2.3 Note that this definition for isothermic surfaces works globally. However, since isothermic surfaces need not be analytic, arbitrarily unpleasant configurations of umbilics can occur — thus we will always work with umbilic-free surface patches or, at least, with surface patches that carry regular curvature line parameters.

In order to get our hands on umbilics (and on the global geometry of isothermic surfaces), some "better" definition will be needed.

---

[11] By the determinant condition, only one, A or $\hat{A}$, can be a multiple of the identity.

3.2.4 Additionally, we want to show that the enveloped sphere congruences in case 1b (§3.1.2) cannot take values in a fixed sphere complex — this distinguishes the case at hand from case 1a geometrically.

Assume that $S : M^2 \to \mathcal{K}^\perp$ for some vector $\mathcal{K} \in \mathbb{R}^5_1$. Then $\mathcal{K} \perp S, dS$ and, consequently, $\mathcal{K} = \hat{\mu}f + \mu\hat{f}$ with some functions $\mu, \hat{\mu}$. Taking derivatives we obtain $0 = (\hat{\mu}\omega_1 + \mu\chi_1)S_1 + (\hat{\mu}\omega_2 + \mu\chi_2)S_2 + d\hat{\mu}f + d\mu\hat{f}$. Thus $d\mu = d\hat{\mu} = 0$ and

$$\begin{pmatrix} \hat{\mu} & \mu \\ \hat{\lambda} & -\lambda \end{pmatrix}\begin{pmatrix} \omega_1 \\ \chi_1 \end{pmatrix} = \begin{pmatrix} 0 \\ 0 \end{pmatrix} \quad \text{and} \quad \begin{pmatrix} \hat{\mu} & \mu \\ \hat{\lambda} & \lambda \end{pmatrix}\begin{pmatrix} \omega_2 \\ \chi_2 \end{pmatrix} = \begin{pmatrix} 0 \\ 0 \end{pmatrix}.$$

Consequently (λ and $\hat{\lambda}$ are positive) $\mu = \hat{\mu} \equiv 0$, so that $\mathcal{K} = 0$, and S cannot take values in a linear sphere complex.

Later, when reconsidering the Darboux transformation in the quaternionic setup, we will want to relax the requirement that the enveloped Ribaucour congruence not be immersed or not take values in a fixed sphere complex.[12] However, in our current situation, the following will be a good definition to work with:

3.2.5 Definition. *Two (isothermic) surfaces are said to form a regular Darboux pair if they envelope an immersed Ribaucour sphere congruence that does not take values in a (fixed) linear sphere complex and if their induced metrics are (weakly) conformally equivalent.*

One envelope is then said to be a Darboux transform of the other.

3.2.6 Note that by requiring our "Darboux sphere congruence" (that is, its envelopes form a Darboux pair) not to take values in a fixed linear complex, we rule out the sphere congruences occurring in case 1a. Thus, by the very definition, we obtain the converse of our characterization.

We will see later that any given isothermic surface allows (locally) a 4-parameter family (∞^4) of Darboux transforms, that is, extends in ∞^4 ways to a Darboux pair (cf., [106]).

3.2.7 Also note that we need not require the surfaces of a Darboux pair to be isothermic by definition: Because a regular Darboux pair envelopes a sphere congruence that falls into[13] case 1b of §3.1.2, it follows that both surfaces of a Darboux pair have to be isothermic where they are regular.

3.2.8 Case 1b. *The sphere congruence is Ribaucour and, around points where both envelopes are immersed, there exist (common) conformal curvature line coordinates for both envelopes; moreover, the sphere congruence*

---

[12] For example, we will want to consider two orthogonal nets on a 2-sphere as a (degenerate case of) a Darboux pair, in which case the sphere congruence is constant.

[13] "Ribaucour and same conformal structure" implies case 1, "not values in a fixed sphere complex" rules out, as discussed above, case 1a.

*does not take values in a fixed linear sphere complex: Its envelopes form a
regular Darboux pair of isothermic surfaces.*
The converse holds by the definition of a regular Darboux pair.

3.3 Aside: Transformations of isothermic surfaces

Isothermic surfaces have an exceptionally rich transformation theory related, for example, with their integrable systems description in terms of
"curved flats" (cf., [71]). Because most aspects of this transformation theory are dealt with much more smoothly in the quaternionic/Clifford algebra setup of Möbius geometry, we will only touch on the most important
topics in the present section: We will briefly discuss the curved flat description and the spectral transformation related to it [46]; then we will turn to
Christoffel's classical work [70] and discuss his transformation for isothermic
surfaces.

3.3.1 An investigation of the Gauss equation (1.2) shows that the "extended Gauss map" $p \mapsto \gamma(p) := \text{span}\{S(p), d_p S(T_p M)\}$ of the sphere congruence[14] S enveloped by a (regular) Darboux pair yields a flat vector space
bundle over M^2:

$$0 = \omega_1 \wedge \chi_2 + \chi_1 \wedge \omega_2 = \varrho_{12} - \psi_1 \wedge \psi_2 \quad \text{and}$$
$$0 = d\psi_1 + \omega_{12} \wedge \psi_2 = d\psi_2 - \omega_{12} \wedge \psi_1.$$

Note that flatness of the extended Gauss map is a frame-independent notion.
Because the normal 2-plane bundle $p \mapsto \text{span}\{f(p), \hat{f}(p)\}$ of S is flat,
too (since S is Ribaucour; cf., §3.1.8), the map $\gamma : M^2 \to \frac{O_1(5)}{O(3) \times O_1(2)}$ into
the (pseudo-Riemannian symmetric) space of point pairs in S^3 defines a
curved flat [46]. Remember that a curved flat $\gamma : M \to G/K$ in a reductive
homogeneous space (or semi-Riemannian symmetric space) is characterized
by the fact that the $\mathfrak{p}$-part $\Phi_\mathfrak{p}$ of the connection form $\Phi = \Phi_\mathfrak{k} + \Phi_\mathfrak{p}$ of any
lift $F : M \to G$ is an abelian 1-form, that is, $[\Phi_\mathfrak{p} \wedge \Phi_\mathfrak{p}] = 0$ (see §2.2.3).

3.3.2 Lemma. *The extended Gauss map*

$$M^2 \ni p \mapsto \gamma(p) := \{f(p), \hat{f}(p)\}^\perp \in O_1(5)/K, \quad \text{where} \quad K = O(3) \times O_1(2),$$

*of a Darboux sphere congruence S with envelopes f and $\hat{f}$ (that form a
Darboux pair of isothermic surfaces) is a curved flat in the (symmetric)
space $\frac{O_1(5)}{O(3) \times O_1(2)}$ of point pairs in the conformal 3-sphere.*

---

[14] Note that this extended Gauss map does not depend on a choice of homogeneous coordinates; it only depends on the honest sphere congruence $S : M^2 \to \mathbb{RP}^4_O$.

3.3.3 It is an entirely general fact that curved flats appear in associated, or spectral, families (cf., [112]): implanting a "spectral parameter" μ into the connection form Φ of a lift $F : M \to G$ of a curved flat $\gamma : M \to G/K$,

$$\Phi_\mu := \Phi_\mathfrak{k} + \mu\Phi_\mathfrak{p},$$

it is a direct consequence of the Maurer-Cartan equations together with the curved flat condition $[\Phi_\mathfrak{p} \wedge \Phi_\mathfrak{p}] = 0$ that the obtained connection forms Φ_μ satisfy the integrability condition for all μ, again[15]:

$$d\Phi_\mu + \tfrac{1}{2}[\Phi_\mu \wedge \Phi_\mu] \;=\; \left\{ \begin{array}{c} d\Phi_\mathfrak{k} + \tfrac{1}{2}[\Phi_\mathfrak{k} \wedge \Phi_\mathfrak{k}] + \tfrac{1}{2}\mu^2[\Phi_\mathfrak{p} \wedge \Phi_\mathfrak{p}] \\ + \, \mu\,(d\Phi_\mathfrak{p} + [\Phi_\mathfrak{k} \wedge \Phi_\mathfrak{p}]) \end{array} \right\} \;=\; 0.$$

Thus, integrating these connections forms Φ_μ, we obtain a family of lifts F_μ of curved flats γ_μ.[16] Moreover, the resulting curved flats γ_μ do not depend on the choice of lift for $\gamma = \gamma_1$: If $F \mapsto FH$, $H : M \to K$, is a gauge transformation of the original lift F of the curved flat γ, then

$$\Phi_\mu = \Phi_\mathfrak{k} + \mu\Phi_\mathfrak{p} \mapsto (\mathrm{Ad}_H\Phi_\mathfrak{k} + H^{-1}dH) + \mu\mathrm{Ad}_H\Phi_\mathfrak{p},$$

showing that $F_\mu \mapsto (FH)_\mu = F_\mu H$ are gauge transforms of the F_μ, and hence are projecting to the same curved flats γ_μ as the F_μ's.

3.3.4 By applying these observations to our situation of Darboux pairs of isothermic surfaces,[17] we obtain a 1-parameter family of Darboux pairs from a given Darboux pair: Introducing the spectral parameter μ into the connection form of an adapted frame for a Darboux pair of isothermic surfaces (normalized as in §3.2.1) yields a 1-parameter family

$$\Phi_\mu \;=\; \begin{pmatrix} 0 & \omega_{12} & -\psi_1 & \mu\omega_1 & \mu\chi_1 \\ -\omega_{12} & 0 & -\psi_2 & \mu\omega_2 & \mu\chi_2 \\ \psi_1 & \psi_2 & 0 & 0 & 0 \\ -\mu\chi_1 & -\mu\chi_2 & 0 & 0 & 0 \\ -\mu\omega_1 & -\mu\omega_2 & 0 & 0 & 0 \end{pmatrix} \qquad (3.1)$$

of connection forms of the same structure. In particular,

$$e^{-u}\omega_i = \pm e^u\chi_i = dx_i$$

---

[15] The integrability of the Φ_μ's is actually a characterization for curved flats, by this computation. Remember that we defined $[\Phi \wedge \Psi](x,y) := [\Phi(x), \Psi(y)] - [\Phi(y), \Psi(x)]$ (see Footnote 1.33).

[16] Note that the frames F_μ are unique up to left translation by a constant element in the group G; hence, the curved flats γ_μ are well defined up to their "position" in G/K.

[17] It can actually be shown that, under suitable regularity assumptions, any curved flat in the symmetric space of point pairs gives rise to a Darboux pair of isothermic surfaces [46]. However, for our purposes, it is sufficient to see that the curved flats γ_μ obtained by μ-deforming a Darboux pair of isothermic surfaces yields Darboux pairs, again.

provide simultaneous conformal curvature line parameters (x_1, x_2) for all the surfaces $f_\mu := F_\mu e_0$ and $\hat{f}_\mu := F_\mu e_\infty$, where F_μ, $dF_\mu = F_\mu \Phi_\mu$, denote lifts of the curved flats $\gamma_\mu = \{f_\mu, \hat{f}_\mu\}^\perp$.

These observations are worth being condensed to a lemma—definition:

3.3.5 Lemma and Definition. *Let $\Phi = \Phi_{\mathfrak{k}} + \Phi_{\mathfrak{p}}$ denote the connection form of a frame F of the curved flat $\gamma = \{f, \hat{f}\}^\perp$ corresponding to a Darboux pair of isothermic surfaces. Then, all connection forms $\Phi_\mu := \Phi_{\mathfrak{k}} + \mu \Phi_{\mathfrak{p}}$ are integrable to provide frames F_μ of curved flats corresponding to Darboux pairs $(f_\mu, \hat{f}_\mu)$.*

The family $(f_\mu, \hat{f}_\mu)_{\mu \in \mathbb{R}}$ will be called the spectral family of Darboux pairs, the transformation $(f, \hat{f}) \mapsto (f_\mu, \hat{f}_\mu)$ will be called spectral transformation.[18]

3.3.6 Note that, on one hand, the frames F_μ are uniquely determined by the Φ_μ's up to postcomposition by a constant element of $O_1(5)$, and, on the other hand, $\Phi = \Phi_1$ does not depend on the position of the Darboux pair in the conformal 3-sphere. Thus the Darboux pairs $(f_\mu, \hat{f}_\mu)$ are unique up to Möbius transformation and do not depend on the position of the original Darboux pair, that is, the spectral transformation is a transformation between Möbius equivalence classes of Darboux pairs.

We will see later that f_μ does not even depend on the Darboux pair, but only on the isothermic surface f. This will prove the classical notion of the "T-transformation," a slightly modified version of our spectral transformation (instead of μ we use $\lambda = \mu^2$ as a real parameter), to be a well-defined transformation for Möbius equivalence classes of isothermic surfaces. In this context the quaternionic model for Möbius geometry, as discussed in the following chapter, will turn out to be very useful.

3.3.7 Example. Writing the connection form (3.1) in terms of conformal curvature line parameters (x_1, x_2) and making an ansatz for the forms ψ_i,

$$\begin{aligned} \psi_1 &= (He^u + \hat{H}e^{-u})dx_1, & \omega_1 &= e^u dx_1, & \chi_1 &= e^{-u} dx_1, \\ \psi_2 &= (He^u - \hat{H}e^{-u})dx_2, & \omega_2 &= e^u dx_2, & \chi_2 &= -e^{-u} dx_2, \end{aligned} \quad (3.2)$$

with suitable functions H and $\hat{H}$, the Codazzi equations (1.3) show that the form $\omega_{12} = \frac{\partial u}{\partial x_2} dx_1 - \frac{\partial u}{\partial x_1} dx_2 = - \star du$, and $dH = 0$ if and only if $d\hat{H} = 0$.

Now we assume H (and, consequently, $\hat{H}$) to be constant; excluding the case $H = \hat{H} \equiv 0$, where the surfaces f and $\hat{f}$ take values in a fixed 2-sphere S, we may assume $\hat{H} \neq 0$. Then, by a (constant) gauge transformation

---

[18] Later a slightly modified version of this transformation will go by the name Calapso transformation, or T-transformation for short; it also provides a conformal deformation for isothermic surfaces, as will be discussed later.

$(f, \hat{f}) \mapsto (-\frac{1}{H}f, -\hat{H}\hat{f})$ and altering u appropriately, we achieve $\hat{H} \equiv -1$. Consequently,

$$\mathcal{K}_\mu := \mu^2(S_\mu + H(\tfrac{1}{\mu}f_\mu)) - (\mu\hat{f}_\mu)$$

is a constant vector, $d\mathcal{K}_\mu = 0$, and the surfaces

$$\tfrac{1}{\mu}f_\mu : M^2 \to Q^3_{(2H-\mu^2)\mu^2}$$

take values in the spaces of constant curvature $2H\mu^2 - \mu^4 = -|\mathcal{K}_\mu|^2$ determined by $\mathcal{K}_\mu$ since $\langle\tfrac{1}{\mu}f_\mu, \mathcal{K}_\mu\rangle \equiv -1$ (see §1.4.1). Determination of the tangent plane congruence $T_\mu = S_\mu + \mu^2 f_\mu$ of a surface $\tfrac{1}{\mu}f_\mu$, by $\langle\mathcal{K}_\mu, T_\mu\rangle \equiv 0$, allows one to compute the second fundamental form of $\tfrac{1}{\mu}f_\mu$ since T_μ can be interpreted as a unit normal field of $\tfrac{1}{\mu}f_\mu$ as a hypersurface of $Q^3_{(2H-\mu^2)\mu^2}$ — thus we have

$$I = e^{2u}(dx_1^2 + dx_2^2) \quad \text{and} \quad I\!I = (H - \mu^2)\, I - (dx_1^2 - dx_2^2),$$

showing that the surfaces $\tfrac{1}{\mu}f_\mu$ are all isometric and have constant mean curvature $H_\mu = H - \mu^2$. Hence we have identified the (spectral) transformation $\mu \mapsto \tfrac{1}{\mu}f_\mu$ as Lawson's correspondence. Note that the mean curvatures H_μ of $\tfrac{1}{\mu}f_\mu$ and the ambient space's curvatures $\kappa_\mu = (2H - \mu^2)\mu^2$ are related by a quadratic equation, $\kappa_\mu + H_\mu^2 \equiv H^2 = \kappa_{\max}$. Here $\kappa_{\max}$ denotes the maximum value of the ambient space's curvature κ_μ that is attained[19] for $\mu^2 = H$, where the surface $\tfrac{1}{\mu}f_\mu$ becomes minimal, $H_\mu = 0$.

3.3.8 Given a Darboux pair $(f, \hat{f})$, its spectral transform becomes singular as $\mu \to 0$, when they are computed using the form (3.1) of the connection forms for the family of curved flats: Clearly, $df_0 = 0$ as well as $d\hat{f}_0 = 0$. However, applying μ-dependent gauge transformations

$$(f_\mu, \hat{f}_\mu) \mapsto (\tfrac{1}{\mu}f_\mu, \mu\hat{f}_\mu) \quad \text{or} \quad (f_\mu, \hat{f}_\mu) \mapsto (\mu f_\mu, \tfrac{1}{\mu}\hat{f}_\mu) \qquad (3.3)$$

while sending $\mu \to 0$, only one surface degenerates, say $\hat{f}_\mu$, and can be interpreted as the point ∞ at infinity of the Q_0^3 determined by $\mathcal{K} := -\hat{f}_0$, so that f_0 takes values in that Euclidean space. Since $S_0 \perp \hat{f}_0$, the (Darboux) sphere congruence S_0 then becomes the tangent plane congruence of f_0. Thus, with the ansatz (3.2), we compute the first and second fundamental forms of f_0 (and of $\hat{f}_0$ by taking the reverse gauge in (3.3)) to find that

---

[19] In the present setup it actually makes sense to consider also $\mu^2 < 0$, that is, $\mu \in i\mathbb{R}$; this will be discussed in more detail later.

the fundamental forms of one surface are determined by those of the other surface:

$$\left.\begin{array}{l} I = e^{2u}(dx_1^2 + dx_2^2) \\ I\!I = H\,I + \hat{H}\,(dx_1^2 - dx_2^2) \end{array}\right\} \quad \leftrightarrow \quad \left\{\begin{array}{l} \hat{I} = e^{-2u}(dx_1^2 + dx_2^2) \\ \hat{I\!I} = \hat{H}\,\hat{I} + H\,(dx_1^2 - dx_2^2). \end{array}\right. \qquad (3.4)$$

Hence both surfaces are — in "different" Euclidean spaces — isothermic, with corresponding conformal curvature line parameters, reciprocal conformal factors, and with their mean curvature and (real) Hopf differentials exchanged.

3.3.9 Conversely, two such isothermic surfaces f_0 and $\hat{f}_0$ in Euclidean space can be used to generate a family of Darboux pairs of isothermic surfaces in the conformal 3-sphere via (3.2) and (3.1), with these surfaces as limiting cases as $\mu \to 0$, by defining Φ_μ in terms of their fundamental quantities and then integrating Φ_μ.

3.3.10 Another observation is the following: Since one of the surfaces, f_λ or $\hat{f}_\lambda$, shrinks to a point (at infinity), the sphere congruence S becomes the tangent plane congruence of the other — while the circles intersecting the spheres of the congruence orthogonally in the points of contact with its envelopes become more and more straight lines. Thus the two surfaces have more and more parallel tangent planes (cf., [152]). This observation may motivate the following problem of Christoffel[20] [70]:

3.3.11 Christoffel's problem. Let $f, f^* : M^2 \to \mathbb{R}^3$ be two surfaces with parallel Gauss maps, $n = \pm n^*$, such that their induced metrics are conformally equivalent, $\lambda^2 I = (\lambda^*)^2 I^*$; characterize the surfaces and their relation.[21]

3.3.12 As for our investigations of Blaschke's problem (see Section 3.1) we distinguish three cases that we formulate in terms of the Weingarten tensor fields A, A^* of f and f^*, respectively:

1. A, A^* and the identity I are linearly dependent, and the Gauss curvatures $K = \det A$ and $K^* = \det A^*$ of f and f^*
   a. have the same sign, $KK^* > 0$, or
   b. have opposite signs, $KK^* < 0$.
2. A, A^* and I are linearly independent.

3.3.13 Regularity assumption. We will consider n (as well as n^*) to be regular, that is, $KK^* \neq 0$. Moreover, we will assume that our surface pair (f, f^*) *globally* falls into one of the mentioned categories.

---

[20] Well, Christoffel himself was motivated by another observation concerning the relation between associated minimal surfaces, as we will discuss below.

[21] Variations and generalizations of this problem have been considered by various authors after Christoffel; see [249], [230], [215], [290], [291], [84], and [47].

3.3.14 Christoffel's problem is a Euclidean problem — the problem does not make any sense in Möbius geometry since the term "parallel" does not. As a purely Euclidean problem we will give the problem a Euclidean treatment, using the classical technology. Most of the analysis of this problem will be taken care of by the following lemma. To simplify the computation, we will assume the existence of curvature line parameters on one of the two surfaces, say f. In cases 1b and 2, this is not a restriction because it is implied by the assumptions; in case 1a, the general statement will follow from the "generic" situation by a continuity argument.

3.3.15 Lemma. *Let* $f : M^2 \to \mathbb{R}^3$ *be an immersion given in terms of curvature line coordinates,* $I = E\,dx_1^2 + G\,dx_2^2$ *and* $I\!I = L\,dx_1^2 + N\,dx_2^2$. *Then the integrability conditions for an immersion* $f^* : M^2 \to \mathbb{R}^3$ *with parallel tangent planes,* $\frac{\partial}{\partial x_1} f^* = \alpha \frac{\partial}{\partial x_1} f + \gamma \frac{\partial}{\partial x_2} f$ *and* $\frac{\partial}{\partial x_2} f^* = -\beta \frac{\partial}{\partial x_1} f + \delta \frac{\partial}{\partial x_2} f$, *read*

$$
\begin{aligned}
0 &= \tfrac{\partial}{\partial x_2}\alpha + \tfrac{\partial}{\partial x_1}\beta + \tfrac{1}{2E}[(\alpha - \delta)\tfrac{\partial}{\partial x_2}E + \beta\tfrac{\partial}{\partial x_1}E - \gamma\tfrac{\partial}{\partial x_1}G] \\
0 &= \tfrac{\partial}{\partial x_1}\delta - \tfrac{\partial}{\partial x_2}\gamma + \tfrac{1}{2G}[(\delta - \alpha)\tfrac{\partial}{\partial x_1}G + \beta\tfrac{\partial}{\partial x_2}E - \gamma\tfrac{\partial}{\partial x_2}G] \qquad (3.5) \\
0 &= \beta L + \gamma N,
\end{aligned}
$$

and its first and second fundamental forms (with respect to n*) are given by*

$$
\begin{aligned}
I^* &= (\alpha^2 E + \gamma^2 G)dx_1^2 - 2(\alpha\beta E - \gamma\delta G)dx_1 dx_2 + (\beta^2 E + \delta^2 G)dx_2^2, \\
I\!I^* &= \alpha L\,dx_1^2 - (\beta L - \gamma N)dx_1 dx_2 + \delta N\,dx_2^2.
\end{aligned} \qquad (3.6)
$$

3.3.16 *Proof.* By a straightforward computation using

$$
\begin{aligned}
\tfrac{\partial}{\partial x_1}\tfrac{\partial}{\partial x_1}f &= \tfrac{1}{2E}(\tfrac{\partial}{\partial x_1}E)\tfrac{\partial}{\partial x_1}f &- \tfrac{1}{2G}(\tfrac{\partial}{\partial x_2}E)\tfrac{\partial}{\partial x_2}f &+ L\,n, \\
\tfrac{\partial}{\partial x_1}\tfrac{\partial}{\partial x_2}f &= \tfrac{1}{2E}(\tfrac{\partial}{\partial x_2}E)\tfrac{\partial}{\partial x_1}f &+ \tfrac{1}{2G}(\tfrac{\partial}{\partial x_1}G)\tfrac{\partial}{\partial x_2}f, \\
\tfrac{\partial}{\partial x_2}\tfrac{\partial}{\partial x_2}f &= -\tfrac{1}{2E}(\tfrac{\partial}{\partial x_1}G)\tfrac{\partial}{\partial x_1}f &+ \tfrac{1}{2G}(\tfrac{\partial}{\partial x_2}G)\tfrac{\partial}{\partial x_2}f &+ N\,n.
\end{aligned}
$$

The details shall be left as an exercise for the reader. ◁

3.3.17 In the cases 1a and 1b of §3.3.12, the Weingarten tensor fields A and A^* are, just as in §3.1.2, simultaneously diagonalizable, that is, the curvature lines of f and f^* correspond where they are (uniquely) defined. Thus, in the above lemma in §3.3.15, we have $0 = \beta L + \gamma N$ from (3.5) as well as $0 = \beta L - \gamma N$ from (3.6), since the coordinate lines we use are curvature lines for f^*, too. As a consequence, $\beta = \gamma \equiv 0$ since we assumed n to be regular, $K \neq 0$.

Conformality of the induced metrics now implies that $\alpha^2 = \delta^2$, and from $K^* = \frac{1}{\alpha\delta}K$ we deduce $\delta = +\alpha \neq 0$ in case 1a and $\delta = -\alpha \neq 0$ in case 1b.

In case 1a, the integrability conditions (3.5) then read $d\alpha = 0$. Hence:

3.3.18 Case 1a. *The two surfaces f and f^* are similar: $f^* = \alpha f + c$ with constant $\alpha \in \mathbb{R} \setminus \{0\}$ and $c \in \mathbb{R}^3$. Conversely, any surface $f : M^2 \to \mathbb{R}^3$ and a scale translation of f form a pair of surfaces with parallel tangent planes and conformally equivalent metrics.*

3.3.19 In case 1b, $\delta = -\alpha \neq 0$ and $\beta = \gamma = 0$, as above, and (3.5) becomes

$$0 = \tfrac{1}{\alpha} \tfrac{\partial}{\partial x_2} \alpha + \tfrac{1}{E} \tfrac{\partial}{\partial x_2} E \quad \text{and} \quad 0 = \tfrac{1}{\alpha} \tfrac{\partial}{\partial x_1} \alpha + \tfrac{1}{G} \tfrac{\partial}{\partial x_1} G.$$

As a consequence, $0 = \tfrac{\partial}{\partial x_1} \tfrac{\partial}{\partial x_2} \ln E - \tfrac{\partial}{\partial x_2} \tfrac{\partial}{\partial x_1} \ln G = \tfrac{\partial}{\partial x_1} \tfrac{\partial}{\partial x_2} \ln \tfrac{E}{G}$, showing that $\tfrac{E}{G}$ is a fraction of two functions of one variable. Hence, by a suitable change $\tilde{x}_1 = \tilde{x}_1(x_1)$ and $\tilde{x}_2 = \tilde{x}_2(x_2)$ of curvature line coordinates, a conformal parametrization is obtained, $\tilde{E} = \tilde{G} =: e^{2u}$, identifying f (as well as f^*) as isothermic surfaces.

What we just discussed is nothing but Weingarten's criterion [299] for a surface to be isothermic[22]: Using Rothe's form from [247] of Weingarten's invariant 1-form, as discussed in §P.6.5, we find

$$\pm d\Omega = \tfrac{\partial}{\partial x_1} \tfrac{\partial}{\partial x_2} (\ln \tfrac{E}{G}) \, dx_1 \wedge dx_2,$$

so that f is an isothermic surface if and only if $d\Omega = 0$.

Moreover, $\alpha = e^{-2u}$ up to some multiplicative constant, so that I^* and $I\!\!I^*$ in (3.6) take the form of $\hat{I}$ and $\hat{I\!\!I}$ from (3.4).

3.3.20 Case 1b. *The two surfaces f and f^* are isothermic such that they allow common conformal curvature line coordinates with reciprocal conformal factors. In these coordinates their mean curvatures and Hopf differentials interchange roles, that is, they appear as limiting surfaces of a spectral family of Darboux pairs:*

$$\left. \begin{array}{rcl} I & = & e^{2u}(dx_1^2 + dx_2^2) \\ I\!\!I & = & H\,I + H^*(dx_1^2 - dx_2^2) \end{array} \right\} \quad \leftrightarrow \quad \left\{ \begin{array}{rcl} I^* & = & e^{-2u}(dx_1^2 + dx_2^2) \\ I\!\!I^* & = & H^*I^* + H\,(dx_1^2 - dx_2^2). \end{array} \right.$$

3.3.21 Such surface pairs (f, f^*) are usually called "dual isothermic surfaces," or "Christoffel pairs": We will use the geometric characterization from Christoffel's problem in §3.3.11 to define this terminology. To exclude the two cases 1a and 2 from §3.3.12, we explicitly exclude the trivial case 1a (as we did for Darboux pairs in §3.2.5), and we require the curvature lines of f and f^* to correspond, thusly excluding case 2:

3.3.22 Definition. *Two (isothermic) surfaces $f, f^* : M^2 \to \mathbb{R}^3$ that are not similar are said to form a Christoffel pair, or a pair of dual surfaces,*

---

[22] Compare also [65], [164], [244], and [246]; [53]; [233], [235], and [236].

if they have (pointwise) parallel tangent planes such that their curvature directions correspond, and if their induced metrics are conformally equivalent. One surface is then said to be a Christoffel transform of the other.

3.3.23 As for Darboux pairs (cf., §3.2.8), Christoffel pairs give, by definition, a surface pair falling into case 1b of our classification of §3.3.12.

However, we can do better: Suppose we are given an isothermic surface in terms of conformal curvature line coordinates, and let e^{2u} be the conformal factor. Then it is straightforward to see that the 1-form

$$df^* = e^{-2u}[\tfrac{\partial}{\partial x_1} f - \tfrac{\partial}{\partial x_2} f] \tag{3.7}$$

is closed and therefore (locally) integrates to give a Christoffel partner for f, from the lemma in §3.3.15. Thus:

3.3.24 Corollary. *Given an umbilic-free isothermic surface $f : M^2 \to \mathbb{R}^3$, there is (locally) a Christoffel transform $f^* : M^2 \supset U \to \mathbb{R}^3$ of f.*

3.3.25 Example. If $f : M^2 \to \mathbb{R}^3$ is a surface of constant mean curvature $H \neq 0$, then its parallel surface $f^* := f + \frac{1}{H} n$ is a surface of constant mean curvature $H^* = H$ when using $n^* := -n$ as its Gauss map (cf., §2.7.4). Moreover, by the Cayley-Hamilton identity, the third fundamental form of f is $I\!I\!I = 2H I\!I - K I$, so that $I^* = I - \frac{2}{H} I\!I + \frac{1}{H^2} I\!I\!I = \frac{H^2 - K}{H^2} I$ is conformally equivalent to I. Clearly the curvature lines of f and f^* coincide. Hence f and f^* form a Christoffel pair of isothermic surfaces.

On the other hand, f and f^* envelope a (Ribaucour) congruence of spheres, with constant radius $\frac{1}{2H}$. Consequently, they also form a Darboux pair — in fact, this can be shown to be a characterization of constant mean curvature surfaces in $\mathbb{R}^3$ (cf., [147]).

3.3.26 In the previous example we had $dn + H df = H^* df^*$, where $H^* = H$ was the (constant) mean curvature of f^*. This formula can be generalized for arbitrary isothermic surfaces:

3.3.27 Lemma. *If $f, f^* : M^2 \to \mathbb{R}^3$ form a Christoffel pair, then*

$$dn + H df = H^* df^*, \tag{3.8}$$

where H^* is the mean curvature of f^* with respect to $n^* := -n$.

3.3.28 *Proof.* Away from umbilics we may choose conformal curvature line parameters (x_1, x_2) for f; then $-dn = k_1 \tfrac{\partial}{\partial x_1} f \, dx_1 + k_2 \tfrac{\partial}{\partial x_2} f \, dx_2$. Consequently, by (3.7),

$$dn + H df = -\tfrac{k_1 - k_2}{2} [\tfrac{\partial}{\partial x_1} f \, dx_1 - \tfrac{\partial}{\partial x_2} f \, dx_2] = \lambda df^*$$

is a multiple of the differential of f^*, with a suitable function λ. By symmetry, $-dn + H^* df^* = \lambda^* df$. Combining the two, the claim follows since df and df^* are linearly independent. ◁

3.3.29 Example. Knowing that a minimal surface $f : M^2 \to \mathbb{R}^3$ is isothermic (3.8) shows that its Gauss map $n : M^2 \to S^2$ complements f as a Christoffel partner,[23] which is unique up to scaling and translation. However, given a map $n : M^2 \to S^2$, its Christoffel partner is unique (up to translation) only after fixing a "curvature line net" or, equivalently, a "Hopf differential." Without this additional information, the family of minimal surfaces mentioned in case 2 below provides as many Christoffel partners for n as there are holomorphic functions — as one expects, knowing about the Weierstrass representation.

Using a minimal surface together with its Gauss map as the limiting surfaces for a spectral family of Darboux pairs (cf., §3.3.8), we obtain a family of surfaces of constant mean curvature $H_\mu = -\mu^2$ in hyperbolic spaces $Q^3_{-\mu^4}$. This gives[24] the Umehara-Yamada perturbation of minimal surfaces into cmc-1 surfaces in hyperbolic space (cf., [153]) — we will study this fact in more detail later, after introducing the quaternionic model for Möbius geometry.

3.3.30 Now we turn to the remaining case 2 of §3.3.12: This is the case that actually caught Christoffel's interest, as the title of his paper [70] indicates.

In the case at hand, the Weingarten operators do not simultaneously diagonalize. On the other hand, $n^* = \pm n$, so that the third fundamental forms of f and f^* coincide, that is, $I(., A^2.) = I^*(., (A^*)^2.)$. Hence, by the conformality assumption in Christoffel's problem in §3.3.11, we conclude that $(\lambda^* A)^2 = (\lambda A^*)^2$. Then, by the same Cayley-Hamilton argument as in §3.1.2, both surfaces f and f^* have to be minimal, $H = 0$ and $H^* = 0$.

Minimal surfaces are isothermic. Therefore, in the formulas of §3.3.15, we can assume that $E = G$ and $L + N = 0$ — note that, by the regularity of n, we have automatically excluded umbilics. The third equation in (3.5) then yields $\gamma = \beta$ with $\beta \neq 0$, since we assumed that the curvature lines of f and f^* do not line up, and conformality gives $\alpha\beta = \gamma\delta \Leftrightarrow \alpha = \delta$, by (3.6).

Assuming that $\frac{\partial}{\partial x_1} f$, $\frac{\partial}{\partial x_2} f$, and n form a positively oriented basis in $\mathbb{R}^3$ the ansatz for df^* from §3.3.15 can now be rewritten as

$$df^* = \alpha df + \beta n \times df;$$

---

[23] This can also be argued directly: Conformality is checked by using the Cayley-Hamilton identity, $I\!I\!I = -KI$, and curvature lines are not an issue because, for a totally umbilic surface, any orthogonal net is a "curvature line net."

[24] We may also allow imaginary values of μ: This is equivalent to choosing $-f^*$ instead of f^* as a Christoffel transform to generate the spectral family of Darboux pairs, or to interchanging the roles of the curvature lines.

moreover, the integrability conditions (3.5) become the Cauchy-Riemann equations for $\alpha + i\beta$,

$$\tfrac{\partial}{\partial x_1}\alpha - \tfrac{\partial}{\partial x_2}\beta = 0 \quad \text{and} \quad \tfrac{\partial}{\partial x_2}\alpha + \tfrac{\partial}{\partial x_1}\beta = 0.$$

The converse follows by straightforward computation, using the fact that f is minimal. Thus we have:

3.3.31 Case 2.   *The two surfaces f and f^* are minimal surfaces whose differentials are holomorphic multiples of each other:*

$$df^* = (\alpha + \beta n\times)df, \quad \text{where} \quad \alpha + i\beta : M^2 \to \mathbb{C}$$

is a holomorphic function. Conversely, any two such minimal surfaces have parallel tangent planes and conformally equivalent metrics.

3.3.32 Note that the associated family of minimal surfaces occurs as a special case where $\alpha + i\beta = e^{i\vartheta}$ is a constant of unit length.

3.4 Dual pairs of conformally minimal surfaces

In the last case 2 of §3.1.2, the enveloped sphere congruence is not Ribaucour; if it was the Weingarten tensors A and $\hat{A}$ could be diagonalized simultaneously by the lemma in §3.1.6 and, consequently, A, $\hat{A}$, and id would be linearly dependent since we are in dimension $\dim M = 2$. However, the condition $\mathrm{tr}\, A = 0$ characterizes the sphere congruence as a special sphere congruence for its envelope f; we will start by discussing the corresponding relation between a surface and an enveloped sphere congruence in a more general setup:

3.4.1 Lemma and Definition.   *Given an immersion $f : M^{n-1} \to S^n$, there is exactly one sphere congruence $Z : M^{n-1} \to S_1^{n+1}$ such that the corresponding Weingarten tensor A_Z (of f with respect to Z as a normal field) has trace $\mathrm{tr}\, A_Z = 0$.*

   This sphere congruence is called the central sphere congruence of f; in the case of dimension $\dim M = 2$, it is also called the conformal Gauss map of f.

3.4.2 *Proof.* Choose any light cone representative $f : M^{n-1} \to L^{n+1}$ of the immersion and any sphere congruence $S : M^{n-1} \to S^{n+1}$ enveloped by f to obtain a strip (f, S). Any different choice ("admissible change") of the strip, $(f, S) \mapsto (e^u f, Z := S + a\, f)$, has the effect

$$\langle df, df\rangle \mapsto e^{2u}\langle df, df\rangle \quad \text{and} \quad \langle df, dS\rangle \mapsto e^u(\langle df, dS\rangle + a\langle df, df\rangle).$$

Hence the Weingarten tensor A_S changes by

$$A_S \mapsto A_Z = e^{-u}(A_S - a\,id).$$

This shows two things:

1. There is a unique sphere congruence $Z = S + \frac{\text{tr}A_S}{n-1} \cdot f$ such that $\text{tr}A_Z = 0$.
2. This sphere congruence does not depend on the choice of homogeneous coordinates for f, that is, the central sphere congruence is a well-defined object for a hypersurface in the conformal n-sphere S^n.   ◁

3.4.3 Sometimes, the central sphere $Z(p)$ of a hypersurface $f : M^{n-1} \to S^n$ at a point $f(p)$ is also called its "mean curvature sphere." To justify this notion we compute the central sphere in terms of a stereographic projection of f (cf., §1.4.4).

Thus, let $f : M^{n-1} \to Q^n_\kappa$ denote an immersion into a quadric of constant curvature $\kappa = -|\mathcal{K}|^2$ (cf., §1.4.1) and let $T : M^{n-1} \to \mathcal{K}^\perp$ denote its tangent plane congruence (cf., §1.4.11). Since $T(p) \in \{f(p), \mathcal{K}\}^\perp = T_{f(p)}Q^n_\kappa$ at every point $p \in M$, the tangent plane map T can be interpreted as a unit normal vector field of $f : M^{n-1} \to Q^n_\kappa$, and, consequently, A_T becomes the Weingarten tensor field of f as a hypersurface in the space Q^n_κ of constant curvature. Hence $Z = T + \frac{1}{n-1}\text{tr}A_T f$ consists of spheres that have the same mean curvature[25] $-\langle Z, \mathcal{K} \rangle = \frac{1}{n-1}\text{tr}A_T$ as the hypersurface f (cf., p. 60).

This observation is important enough to formulate it as a lemma:

3.4.4 Lemma. *If $f : M^{n-1} \to Q^n_\kappa$ is an immersion into a space Q^n_κ of constant curvature κ, then the central sphere congruence Z consists of those spheres $Z(p)$ that have the same mean curvature $-\langle Z(p), \mathcal{K} \rangle = H(p)$ as the hypersurface.*

3.4.5 As an immediate consequence we obtain a Möbius geometric characterization for minimal hypersurfaces or, more generally, constant mean curvature hypersurfaces in spaces of constant curvature:

3.4.6 Corollary. *A hypersurface $f : M^{n-1} \to S^n$ allows a minimal (constant mean curvature) stereographic projection $f : M^{n-1} \to Q^n_\kappa$ into a quadric of constant curvature if and only if its central sphere congruence takes values in a linear (affine: $\{S \mid \langle S, \mathcal{K} \rangle = const\}$) sphere complex $\mathcal{K}$.*

3.4.7 Next we want to motivate the notion of the "conformal Gauss map" in the 2-dimensional case. By the Cayley-Hamilton identity, the induced

---

[25] We chose Z and T to have "oriented contact" — by replacing Z by $-Z$ we pick up a sign that comes from differently orienting the hypersurface and the touching mean curvature sphere.

metric[26] of a sphere congruence $S : M^2 \to S_1^4$ enveloped by an immersion satisfies $I_S = I_f(., A_S^2.) = \mathrm{tr}A_S \cdot I_f(., A_S.) - \det A_S \cdot I_f$, where $f : M^2 \to L^4$ is any lift of the immersion.

At nonumbilic points, A_S and the identity are linearly independent; hence I_S and I_f are conformally equivalent if and only if $\mathrm{tr}A_S = 0$.

At umbilics, on the other hand, A_S is (for *any* S) a multiple of the identity, so that $\mathrm{tr}A_S = 0 \Leftrightarrow A_S = 0$, that is, I_S degenerates exactly when $\mathrm{tr}A_S = 0$.

Thus we obtain a characterization of the central sphere congruence as the "conformal Gauss map" of a surface in the conformal 3-sphere:

3.4.8 Lemma. *A surface* $f : M^2 \to S^3$ *and its central sphere congruence* $Z : M^2 \to \mathbb{RP}_O^4$ *induce (weakly) conformally equivalent metrics — the metric of* Z *degenerates exactly at the umbilics[27] of* f.

This is a characterization of the conformal Gauss map.

3.4.9 Example. If $f : M^2 \to Q_0^3 \cong \mathbb{R}^3$, then the induced metric of the central sphere congruence $Z = T + Hf$ becomes

$$I_Z = I_f(., A_Z^2.) = \tfrac{1}{4}(a_1 - a_2)^2\, I_f = (H^2 - K)\, I_f,$$

where a_i are the principal curvatures of f and $K = a_1 a_2$ its Gauss curvature. Thus I_Z turns out to be exactly the "conformal metric" from §P.6.4.

3.4.10 A special feature in dimension 2 is that an invertible endomorphism A has trace $\mathrm{tr}A = 0$ if and only if its inverse has $\mathrm{tr}A^{-1} = 0$. As a consequence, a regular sphere congruence S is the central sphere congruence of an immersion f if and only if the Weingarten tensor field $A_f = A_S^{-1}$ of S with respect to f as an (isotropic) normal field has trace $\mathrm{tr}A_f = 0$. Thus we can check a sphere congruence for being central by observing its second fundamental form (cf., §1.7.6): Since

$$\begin{aligned} I\!I &= -\langle dS, df\rangle\hat{f} - \langle dS, d\hat{f}\rangle f \\ &= I(., A.)\hat{f} + I(., \hat{A}.)f, \end{aligned}$$

where $A = A_f$ and $\hat{A} = A_{\hat{f}}$ denote the Weingarten tensors with respect to f and $\hat{f}$, respectively, we see that $\mathrm{tr}I\!I$ becomes isotropic if and only if S is the central sphere congruence for one of its envelopes, and S becomes

---

[26] The metric $I_S = \langle dS, dS\rangle$ does not depend on the lift $S : M^2 \to S_1^4$ of S (sign of S); so the induced metric I_S of a sphere congruence is a well-defined *global* quantity.

[27] This is where the mean curvature sphere becomes a curvature sphere; see §1.7.10.

minimal,[28] $\operatorname{tr} \mathit{I\!I} = 0$, if and only if it is the central sphere congruence for both of its envelopes (if they are regular).

3.4.11 Applying these observations to case 2 from §3.1.2, which we are currently investigating, tells us that the sphere congruence (that we assumed to be immersed) is minimal and is therefore the central sphere congruence of both of its envelopes, where those are immersed. Otherwise said, both envelopes are "conformally minimal":

3.4.12 Definition. *A surface $f : M^2 \to S^3$ is called conformally minimal if its central sphere congruence is, away from umbilics, a minimal surface.*

3.4.13 Case 2. *Around points where both envelopes f and $\hat{f}$ of the sphere congruence S are immersed, it is the central sphere congruence of both envelopes at the same time, $S = Z = \hat{Z}$. This is the case if and only if the sphere congruence $S : M^2 \to \mathbb{R}P_O^4$ is minimal.*

On the other hand, the central sphere congruence of any conformally minimal surface falls into case 2 of §3.1.2, as soon as it is regular; the second envelope of the sphere congruence is then also conformally minimal where it is regular.

3.4.14 This last statement follows from the next theorem, below. In our setup, defining conformal minimality via the minimality of the central sphere congruence, this theorem is a rather direct consequence of the definition; however, choosing a different approach, which we will address in the following section, it requires a quite involved proof: See [40].

3.4.15 Theorem and Definition. *The second envelope $\hat{f}$ of the central sphere congruence Z of a conformally minimal surface f is (where regular) conformally minimal itself; it is called the dual surface of f.*

3.4.16 *Proof.* In our context, this theorem follows from the symmetry that we built into the definition of conformal minimality and from the following lemma, which is merely a reformulation of the considerations in §3.4.10. ◁

3.4.17 Lemma. *Let $f : M^2 \to S^3$ be a surface such that the second envelope $\hat{f}$ of its central sphere congruence Z is immersed. Then f is conformally minimal if and only if Z is the central sphere congruence of $\hat{f}$.*

3.4.18 Example. Let us, for a moment, equip our Minkowski space $\mathbb{R}_1^5$ with an orthonormal basis $(e_0, e_1, \ldots, e_4)$, that is, $\langle e_i, e_j \rangle = \pm \delta_{ij}$ with -1 for $i = 0$ and $+1$ otherwise. Let

$$\mathbb{R}^2 / 2\pi \mathbb{Z}^2 \ni (u, v) \mapsto f(u, v) = (1, \tfrac{\cos u}{\sqrt{2}}, \tfrac{\sin u}{\sqrt{2}}, \tfrac{\cos v}{\sqrt{2}}, \tfrac{\sin v}{\sqrt{2}}) \in Q_1^3$$

---

[28] Remember that we equipped the space $\mathbb{R}P_O^4$ of (unoriented) hyperspheres with the metric that makes $S_1^4 \to \mathbb{R}P_O^4$ a (double) pseudo-Riemannian cover.

parametrize the square Clifford torus. Because f is a minimal surface, its tangent plane congruence

$$T(u,v) = (0, \tfrac{\cos u}{\sqrt 2}, \tfrac{\sin u}{\sqrt 2}, -\tfrac{\cos v}{\sqrt 2}, -\tfrac{\sin v}{\sqrt 2}) \in S_1^4$$

coincides with its central sphere congruence, $Z = T$; alternatively, one easily checks that the Weingarten tensor field $A_T = -\frac{\partial}{\partial u}du + \frac{\partial}{\partial v}dv$ has $\operatorname{tr}A_T = 0$. The second envelope of the central sphere congruence T becomes

$$\hat f(u,v) = (-\tfrac{1}{2}, \tfrac{\cos u}{2\sqrt 2}, \tfrac{\sin u}{2\sqrt 2}, \tfrac{\cos v}{2\sqrt 2}, \tfrac{\sin v}{2\sqrt 2}).$$

This is just a reparametrization of f, $\hat f(u,v) \simeq f(u+\pi, v+\pi)$. Because the Weingarten tensor field $\hat A_T = -\frac{1}{2}\frac{\partial}{\partial u}du + \frac{1}{2}\frac{\partial}{\partial v}dv$ of $\hat f$ with respect to T as a unit normal field also has trace $\operatorname{tr}\hat A_T = 0$, the square Clifford torus f is a conformally minimal surface whose dual is just a reparametrization — it is a "self-dual" conformally minimal surface.

3.4.19 At this point we wish to switch viewpoints, in the usual manner, and extend our definition of conformally minimal surfaces slightly: We want to relax the requirement on the central sphere congruence to be immersed.

For a map $\varphi : (M, g) \to (\hat M, \hat g)$ between (pseudo)Riemannian manifolds we define the Hessian by $\operatorname{hess}\varphi(x,y) := \nabla_x d\varphi(y) := \hat\nabla_x(d\varphi(y)) - d\varphi(\nabla_x y)$. In case φ is an isometric immersion, $I\!\!I = \operatorname{hess}\varphi$ and $\operatorname{tr}I\!\!I = \Delta\varphi$. A conformal change $g \to \tilde g = e^{2u}g$ of metric on the domain manifold yields (see §P.2.1)

$$\begin{aligned}
\tilde\Delta\varphi &= \sum_j \tfrac{1}{\tilde g(v_j, v_j)}(\hat\nabla_{v_j}d\varphi(v_j) - d\varphi(\nabla_{v_j}v_j + 2du(v_j)v_j - g(v_j, v_j)\operatorname{grad}u)) \\
&= e^{-2u}(\Delta\varphi + (n-2)d\varphi(\operatorname{grad}u)),
\end{aligned}$$

where v_i, $i = 1, \ldots, n$, denotes an orthonormal basis of $T_p M$ with respect to g. Thus, for $\dim M = 2$, $\tilde\Delta\varphi = 0$ if and only if $\Delta\varphi = 0$, that is, harmonicity of a map from a 2-dimensional domain does not depend on a metric but only on a conformal structure on the domain manifold.

In particular, the central sphere congruence $Z : M^2 \to \mathbb{RP}_O^4$ of an immersion $f : M^2 \to S^3$ is minimal if and only if it is harmonic with respect to the conformal structure induced by the immersion f. This condition on a surface f to be conformally minimal works across umbilics, too.[29] Note that, in our definition of conformal minimality, we did not impose any condition at umbilics, so that, in particular, sphere pieces are conformally minimal.

Summarizing we obtain the following characterization of conformally minimal surfaces that requires only the surface to be immersed, not its central sphere congruence:

3.4.20 Theorem. *A surface $f : M^2 \to S^3$ is conformally minimal if and only if its central sphere congruence $Z : M^2 \to \mathbb{RP}_O^4$ is harmonic.*

---

[29] By a continuity argument $\Delta Z = 0$ on the closure of the set of regular points of I_Z; on open neighborhoods, where $I_Z \equiv 0$, we have $dZ = -\chi f$; but, using an f-adapted frame here, the Codazzi equations for Z show that $\chi = 0$. Hence $I_Z \equiv 0$ implies $dZ \equiv 0$, so that $\Delta Z \equiv 0$.

3.5 Aside: Willmore surfaces

Motivated by the observation that the total Gauss curvature $\int_M K da$ of a compact surface $f : M^2 \to \mathbb{R}^3$ in Euclidean 3-space is a topological invariant, $\frac{1}{2\pi} \int_M K dA = \chi(M)$ by the Gauss-Bonnet theorem, T. Willmore introduced the total squared mean curvature $\tau(f) := \frac{1}{2\pi} \int_M H^2 dA$ and raised the question for relations between its infimum value and the topology of the underlying domain manifold (see [306]). He gave a complete analysis in the case $M^2 = S^2$ of genus 0. In the case where the underlying manifold is a 2-torus, $M^2 = T^2$, the problem of finding the infimum (minimum: [256]) value of $\tau(f)$ is still open — in fact, many people have since worked on the famous "Willmore conjecture," saying that $\tau(f) \geq \pi$ for a 2-torus.[30] At the end of this chapter we will touch on this conjecture in a special case.

3.5.1 Another point of view is to consider the "Willmore functional" τ as the "bending energy" of a (closed) surface, analogous to the bending energy for (planar) curves that is given by their total squared geodesic curvature: This was the approach that led Germain [126] in the beginning of the nineteenth century to consider the mean curvature of a surface as the "elastic force" of a thin plate (cf., [83]). Indeed, in modern literature on the elasticity of membranes (see for example [289] or [186]; cf., [176]), a weighted sum of the total squared mean, the total mean, and the total Gauss curvatures is considered as the "free energy" of a membrane. By neglecting the total mean curvature (by physical considerations) and observing that the total Gauss curvature is a topological invariant, one is led to consider the Willmore functional as the free energy of the membrane.

And just as the (free) elastic curves are the critical curves for the bending energy of rods (see [41] and [177]), "Willmore surfaces" are the (closed) critical surfaces for the total squared mean curvature. In fact, we will even discover a close relationship between elastic curves (in planes of constant curvature) and special Willmore surfaces [276] (cf., [218] or [178]).

In this section we will deal with the corresponding variational problem.

---

[30] Just to give a few references: After the Möbius symmetry of the problem was recognized (see [302] and [67]), Li and Yau obtained a partial solution by introducing an important new technology based on the notion of "conformal volume" (see [180]), and Bryant gave the corresponding variational problem a Möbius geometric treatment in [40] (see also [43]). Bryant's work is of particular interest to us because it establishes the relation with conformally minimal surfaces. An extension of the Willmore conjecture for different genuses has been formulated by R. Kusner [172] (cf., [171]), and generalizations to higher dimensions and codimensions have been discussed in [220], [169], [241], [107], [202], and [225] (see also [226]). Important examples were given in [219] (compare with our discussion of Willmore channel surfaces later), and (constrained) Willmore surfaces were treated using quaternions in [240]. For a survey see [221] or [307]. Very recently, a proof of the Willmore conjecture, using integrable systems techniques, has been claimed: See [251].

3.5.2 Our starting point is the following observation: If $f : M^2 \to \mathbb{R}^3$ is a *compact* surface in Euclidean 3-space, then its total Gaussian curvature is a topological invariant, so that, fixing the genus of the underlying manifold, our example in §3.4.9 shows that the total squared mean curvature $\tau(f)$ of the immersed surface f is given by the area of its conformal Gauss map, up to constants:

$$\tfrac{1}{2\pi} \int_{M^2} dA_Z \;\; = \;\; \tfrac{1}{2\pi} \int_{M^2}(H^2 - K)dA \;\; = \;\; \tau(f) - \chi(M^2). \qquad (3.9)$$

This observation has some interesting consequences: First, the total squared mean curvature $\tau(f)$ is a *conformal* invariant of the surface, that is, it is not effected by Möbius transformation of the surface (cf., [302] and [67]) or, more generally, by conformal changes of the ambient space's metric. Thus it makes sense to consider the Willmore functional for (compact) surfaces in the conformal 3-sphere.

3.5.3 Definition. *A compact[31] surface $f : M^2 \to S^3$ is called a Willmore surface, or elastic surface, if it is critical for the Willmore functional*

$$W(f) \;\; := \;\; \tfrac{1}{2\pi} \int_{M^2} dA_Z,$$

where dA_Z denotes the area element of the central sphere congruence (conformal Gauss map) $Z : M^2 \to \mathbb{R}P^4_O$ of f.

3.5.4 As a second consequence of the relation (3.9) between the total squared mean curvature and the Willmore functional we learn that compact conformally minimal surfaces are critical for the Willmore functional, that is, they are Willmore surfaces: By our previous definition, $f : M^2 \to S^3$ is conformally minimal if its central sphere congruence is a minimal surface, that is, the first variation of its area vanishes when varying through sphere congruences. Consequently, the first variation of its area also vanishes when restricting it to variations through central sphere congruences of surfaces into the conformal 3-sphere: Hence, the first variation of the Willmore functional (as a functional defined for surfaces) vanishes.

We want to show that the converse is true — also extending our argument to surfaces whose central sphere congruence is not immersed (cf., §3.4.20):

3.5.5 Theorem. *A compact surface $f : M^2 \to S^3$ is a Willmore surface if and only if its central sphere congruence $Z : M^2 \to \mathbb{R}P^4_O$ is harmonic. Otherwise said, f is Willmore if and only if it is conformally minimal.*

---

[31] One may replace the compactness assumption by requiring the integral to exist and considering only compactly supported variations. However, without the compactness assumption, the functionals W and τ are obviously no longer related by (3.9).

3.5.6 *Proof.* Compare [49] or [110]; the second assertion follows by §3.4.20.

To begin with, we compute the first variation of the Willmore functional. For this first note that the Willmore functional of an immersion f coincides with the harmonic map energy of the conformal Gauss map Z, since Z is conformal and the domain is 2-dimensional:

$$2dA_Z = \langle \star dZ \wedge dZ \rangle = \langle dZ \circ J \wedge dZ \rangle, \tag{3.10}$$

where we define[32)]

$$\langle \omega \wedge \eta \rangle(x,y) := \langle \omega(x), \eta(y) \rangle - \langle \omega(y), \eta(x) \rangle$$

for two $\mathbb{R}^5_1$-valued 1-forms $\omega, \eta : TM \to \mathbb{R}^5_1$, and $\star$ denotes the Hodge-$\star$ operator of the conformal structure induced by f, $\star\omega = \omega \circ J$ with the almost complex structure of f. Here the advantage of dealing with the harmonic map energy rather than with the area of the conformal Gauss map becomes apparent: Umbilics, where the induced metric of the conformal Gauss map degenerates (see §3.4.8), do not have to be treated separately (cf., [110]).

The equality (3.10) is straightforwardly verified by applying both 2-forms to a basis (x, Jx) of a tangent space, $x \in T_p M^2 \setminus \{0\}$:

$$2dA_Z(x, Jx) = 2|dZ(x)|^2 = |dZ(Jx)|^2 + |dZ(x)|^2 = \langle \star dZ \wedge dZ \rangle(x, Jx),$$

where the first equality holds since $dZ(x)$ and $dZ(Jx)$ form a positively oriented orthogonal basis of $d_p Z(T_p M^2) \subset T_{Z(p)} S^4_1$ — as long as the metric induced by Z is not degenerate at p; otherwise, both sides in the equality vanish — and the second equality holds since J is an isometry for any metric in the conformal class of the metric induced by f.

Now let $t \mapsto f_t$ be a variation of $f_0 = f : M^2 \to S^3$ and $t \mapsto Z_t$ the corresponding variation of central sphere congruences. We denote the respective variational vector fields by $\frac{\partial}{\partial t}|_{t=0} f =: f'$, Z', and so on. Note that the conformal structure induced by f_t also varies, so that variations J_t of the almost complex structure (or the associated Hodge-$\star$ operator, respectively) also have to be taken into account. However, $J^2 = -id$ implies $JJ' + J'J = 0$, so that

$$
\begin{aligned}
\langle dZ \circ J' \wedge dZ \rangle(x, Jx) &= \langle dZ(J'x), dZ(Jx) \rangle - \langle dZ(J'Jx), dZ(x) \rangle \\
&= -\langle dZ((JJ' + J'J)x), dZ(x) \rangle \\
&= 0,
\end{aligned}
$$

[32)] Because the scalar product is symmetric, the product $\langle \omega \wedge \eta \rangle$ is skew-symmetric in ω and η (compare Footnote 1.33); we also will need a Leibniz rule for the product of a function g and a 1-form ω:

$$d\langle g, \omega \rangle = \langle dg \wedge \omega \rangle + \langle g, d\omega \rangle.$$

where we have also used $\langle dZ(Jy), dZ(x) \rangle + \langle dZ(y), dZ(Jx) \rangle = 0$ since J is skew for any metric in the conformal class of the metric induced by f.

Moreover, it is an entirely general fact that J is skew for the product $\langle \omega \wedge \eta \rangle$. Namely, a test on the basis (x, Jx), $x \in T_p M \setminus \{0\}$, yields

$$
\begin{aligned}
(\langle \omega \circ J \wedge \eta \rangle + \langle \omega \wedge \eta \circ J \rangle)(x, Jx) &= \langle \omega(Jx), \eta(Jx) \rangle + \langle \omega(x), \eta(x) \rangle \\
&\quad - \langle \omega(x), \eta(x) \rangle - \langle \omega(Jx), \eta(Jx) \rangle \\
&= 0.
\end{aligned}
$$

In particular, we also have $\langle dZ' \circ J \wedge dZ \rangle + \langle dZ' \wedge dZ \circ J \rangle = 0$.

With these preparations we are now well equipped to compute the variation of the Willmore density $2dA_Z = \langle \star dZ \wedge dZ \rangle = \langle dZ \circ J \wedge dZ \rangle$:

$$
\begin{aligned}
2dA'_Z &= \langle dZ' \circ J + dZ \circ J' \wedge dZ \rangle + \langle dZ \circ J \wedge dZ' \rangle \\
&= -\langle dZ' \wedge dZ \circ J \rangle + \langle dZ \circ J \wedge dZ' \rangle \\
&= -2d\langle Z', \star dZ \rangle + 2\langle Z', d \star dZ \rangle.
\end{aligned}
$$

Integration by parts (Stokes' theorem) then gives us

3.5.7 Lemma (First Variation Formula). *Let $f_t : M^2 \to S^3$ denote a variation of a compact surface $f_0 = f$ and $Z_t : M^2 \to \mathbb{RP}^4_O$ the corresponding variation of conformal Gauss maps. Then*

$$
(W(f))' = \tfrac{1}{2\pi} \int_{M^2} \langle d \star dZ, Z' \rangle = \tfrac{1}{2\pi} \int_{M^2} \langle \Delta Z, Z' \rangle \, dA, \qquad (3.11)
$$

where $Z' = \frac{\partial}{\partial t}\big|_{t=0} Z_t$ denotes the variational vector field of the variation of conformal Gauss maps, and Δ and dA are the Laplacian and the area element of any metric in the conformal structure induced by f on M^2.

3.5.8 Note that (3.11) already implies that f is Willmore as soon as its central sphere congruence Z is harmonic. However, the formula does not prove the converse, since we only consider variations Z_t of Z through conformal Gauss maps of variations f_t of f; hence, Z' cannot be chosen to be an arbitrary section of $Z^\star T S^4_1$, and we cannot conclude the argument as usual.

Thus, to see the converse we analyze the right-hand side of (3.11) further. For that purpose we have to choose a metric representative of the conformal structure induced by f. We do that by assuming that f takes values in a metric $S^3 \cong Q^3_1$ (see Section 1.4): Consider $\mathbb{R}^5_1 = \mathbb{R} \times \mathbb{R}^4$ and let $f = (1, \mathfrak{f})$, where $\mathfrak{f} : M^2 \to S^3 \subset \mathbb{R}^4$, that is, $\mathcal{K} = (1, 0)$. The tangent plane congruence of f is then given by $T = (0, \mathfrak{n})$, where $\mathfrak{n} : M^2 \to S^3$ denotes a unit normal field of $\mathfrak{f}$ (see §1.4.11), and the central sphere congruence is given by the mean curvature spheres $Z = T + Hf$ of f, where H denotes the mean curvature of $\mathfrak{f}$ with respect to $\mathfrak{n}$ (see §3.4.4). Computing the Laplacians of $\mathfrak{f}$ and $\mathfrak{n}$ gives

$$\begin{aligned}
\Delta\mathfrak{f} &= -\mathrm{tr}_I I\,\mathfrak{f} &+& \mathrm{tr}_I I\hspace{-2pt}I\,\mathfrak{n} &=& -2\mathfrak{f} &+& 2H\,\mathfrak{n} \\
\Delta\mathfrak{n} &= \mathrm{tr}_I I\hspace{-2pt}I\,\mathfrak{f} &-& \mathrm{tr}_I I\hspace{-2pt}I\hspace{-2pt}I\,\mathfrak{n} &=& 2H\,\mathfrak{f} &-& 2(2H^2 - K)\,\mathfrak{n},
\end{aligned}$$

where $\mathrm{tr}_I I\hspace{-2pt}I\hspace{-2pt}I = 2H\,\mathrm{tr}_I I\hspace{-2pt}I - K\,\mathrm{tr}_I I$ by the Cayley-Hamilton identity, and K denotes the *extrinsic* Gauss curvature of $\mathfrak{f}$. As a consequence,

$$\begin{aligned}
\Delta Z &= \Delta H \mathfrak{f} - 2(H^2 - K)T \\
&= (\Delta H + 2H(H^2 - K))\,\mathfrak{f} - 2(H^2 - K)\,Z.
\end{aligned}$$

On the other hand, $Z' = H'\,\mathfrak{f} + H\,\mathfrak{f}' + T' \perp Z$, so that we obtain

$$\begin{aligned}
\langle \Delta Z, Z' \rangle &= (\Delta H + 2H(H^2 - K))\,(\langle T', \mathfrak{f} \rangle + H'\langle \mathfrak{f}, \mathfrak{f} \rangle + H\langle \mathfrak{f}', \mathfrak{f} \rangle) \\
&= (\Delta H + 2H(H^2 - K))\,(-\mathfrak{n} \cdot \mathfrak{f}' + 0 + 0).
\end{aligned}$$

Hence $(W(f))'$ vanishes for all variations f_t of f if and only if

$$\Delta H + 2H(H^2 - K) = 0, \tag{3.12}$$

that is, if and only if $Z : M^2 \to \mathbb{R}P_O^4 = S_1^4/\pm$ is harmonic, $\Delta Z \equiv 0 \bmod Z$.

This completes the proof of our theorem in §3.5.5. ◁

3.5.9 As a by-product we obtain the Euler-Lagrange equation (3.12) of the Willmore surfaces in terms of the mean and (extrinsic) Gauss curvatures of the surface in the metric 3-sphere. It is no surprise that this Euler-Lagrange equation has exactly the same form as in the Euclidean setup: By §3.4.4, the mean curvature of $f : M^2 \to Q_\kappa^3$ is given by $H = -\langle Z, \mathcal{K} \rangle$, where Z is (a lift of) the central sphere congruence of f. Thus

$$\begin{aligned}
\Delta H + 2H(H^2 - K) &= -\langle \Delta Z + 2(H^2 - K)Z, \mathcal{K} \rangle \\
&= -\langle \Delta Z - \langle \Delta Z, Z \rangle Z, \mathcal{K} \rangle
\end{aligned}$$

provides, in any space Q_κ^3, the "f-component" of the Laplacian of Z, that is, the Laplacian of $Z : M^2 \to \mathbb{R}P_O^4$ as a map into $\mathbb{R}P_O^4$. This seems worth stating a corollary (cf., [294]):

3.5.10 Corollary. *A (compact) surface* $f : M^2 \to Q_\kappa^3$ *in a space of constant curvature is Willmore if and only if*

$$\Delta H + 2H(H^2 - K) = 0,$$

where H and K denote the mean and (extrinsic) Gauss curvatures of f.

3.5.11 As an immediate consequence of the Euler-Lagrange equation for the Willmore surfaces, we learn that minimal surfaces in spaces of constant curvature are Willmore. In fact, we learn that these are the only surfaces of constant mean curvature in a space form that are Willmore besides spheres. In the following section, we will attack this phenomenon from a more general viewpoint.

3.6 Thomsen's theorem

To analyze the geometric configurations that occur as solutions of Blaschke's problem (see Section 3.1) we assumed that the cases that we were examining did not mix. However, an interesting phenomenon happens when assuming that a surface is conformally minimal and isothermic at the same time: We will see that assuming these two Möbius geometric properties for a surface causes symmetry-breaking, that is, it forces the surface to be characterized in terms of a (metric) subgeometry of Möbius geometry.

First we give a more geometric characterization of isothermic surfaces that does not refer to special coordinates (cf., §3.2.2):

3.6.1 Lemma. *An umbilic-free[33] surface $f : M^2 \to S^3$ is isothermic if and only if its central sphere congruence $Z : M^2 \to \mathbb{RP}_O^4$ is Ribaucour.*

3.6.2 Proof. Let F be a local S-adapted frame that diagonalizes the Weingarten tensor A_f of S with respect to f, such that

$$\omega_1 = e^{-u}\psi_1 \quad \text{and} \quad \omega_2 = -e^{-u}\psi_2$$

with a suitable function u; the Codazzi equations (1.3) then yield

$$
\begin{aligned}
0 &= d\psi_1 + e^u(d\omega_1 - \nu \wedge \omega_1) &= e^u(2d\omega_1 + (du - \nu) \wedge \omega_1), \\
0 &= -d\psi_2 + e^u(d\omega_2 - \nu \wedge \omega_2) &= e^u(2d\omega_2 + (du - \nu) \wedge \omega_2).
\end{aligned}
$$

If f is isothermic, we can choose f to be an isometric light cone lift, so that we may assume $d\omega_i = 0$. Then $\nu = du$, showing that S is Ribaucour.

Conversely, if S is Ribaucour, we may assume $\nu = 0$. Then $d(e^{\frac{u}{2}}\omega_i) = 0$, so that the forms $e^{\frac{u}{2}}\omega_i$ can be integrated to obtain conformal curvature line coordinates for the surface f that is, as a consequence, isothermic. ◁

3.6.3 This lemma provides a simple and Möbius geometric argument that surfaces of constant mean curvature in space forms are isothermic: Remember, from §3.4.6, that a surface $f : M^2 \to S^3$ has constant mean curvature in some quadric Q_κ^3 of constant curvature if and only if its central sphere congruence $Z : M^2 \to S_1^4$ takes values in an affine (or linear, if it is minimal) sphere complex, given by the same vector $\mathcal{K}$ that determines the quadric Q_κ^3. Thus the central sphere congruence of a surface of constant mean curvature in some quadric Q_κ^3 of constant curvature has flat normal bundle,[34] that is, it is Ribaucour.

---

[33] Remember from §3.4.8 that Z degenerates exactly at the umbilics of f.

[34] From §3.4.4, a normal field of Z is given by $\mathcal{K} + H Z$; since the mean curvature H is constant, this field is clearly parallel.

3.6.4 Example. In general, the central sphere congruence of an isothermic surface is Ribaucour, but it does *not* give rise to a Darboux transform of the surface (see §3.2.5). However, it does in the following special case.[35]

Let $f : M^2 \to Q^3_{-1}$ be an immersion into hyperbolic space with infinity boundary $\mathcal{K} \simeq S_\infty$; let $T : M^2 \to \mathcal{K}^\perp$ denote the tangent plane congruence of f; and let $F : M^2 \to O_1(5)$ be an f-adapted frame such that the Weingarten tensor A_T of f is diagonal (F is a "principal curvature frame"). Then $-\frac{1}{2}f - S_\infty$ is the second envelope of T, and the connection form of F takes the form (note that im $df = $ im dT)

$$
\Phi \;=\; \begin{pmatrix}
0 & \omega_{12} & -(H+Q)\omega_1 & \omega_1 & -\frac{1}{2}\omega_1 \\
-\omega_{12} & 0 & -(H-Q)\omega_2 & \omega_2 & -\frac{1}{2}\omega_2 \\
(H+Q)\omega_1 & (H-Q)\omega_2 & 0 & 0 & 0 \\
\frac{1}{2}\omega_1 & \frac{1}{2}\omega_2 & 0 & 0 & 0 \\
-\omega_1 & -\omega_2 & 0 & 0 & 0
\end{pmatrix},
$$

where H is the mean curvature of f and Q some function. Now let the mean curvature $H \equiv 1$ and consider $Z := T + f : M^2 \to S^4_1$; the second envelope of Z then becomes $\hat{f} = -(T + S_\infty + f) = -(Z + S_\infty)$. Since $\langle Z, S_\infty \rangle \equiv -1$, the spheres $Z(p)$ are horospheres in Q^3_{-1}, and $\hat{f}$ takes values in S_∞. $\hat{f}$ is commonly called the "hyperbolic Gauss map" of f (cf., [42]). The connection form of the transformed adapted frame,

$$
\Phi \;=\; \begin{pmatrix}
0 & \omega_{12} & -Q\omega_1 & \omega_1 & Q\omega_1 \\
-\omega_{12} & 0 & Q\omega_2 & \omega_2 & -Q\omega_2 \\
Q\omega_1 & -Q\omega_2 & 0 & 0 & 0 \\
-\omega_1 & -\omega_2 & 0 & 0 & 0 \\
-Q\omega_1 & Q\omega_2 & 0 & 0 & 0
\end{pmatrix},
$$

shows that f and $\hat{f}$ form a Darboux pair.

In fact, reversing our viewpoint, it can be shown that any Darboux transform of an immersion $\hat{f} : M^2 \to S_\infty$ into a sphere (with prescribed "curvature lines"; cf., §3.3.29) yields a constant mean curvature 1 surface $f : M^2 \to Q^3_{-1}$ in hyperbolic space with hyperbolic Gauss map $\hat{f}$, and, moreover, that constant mean curvature 1 surfaces in hyperbolic space are the only surfaces whose central sphere congruence gives rise to a nonconstant Darboux transformation (see [153]).

This will be discussed in more detail later, in the quaternionic setup.

3.6.5 Now assume that $f : M^2 \to S^3$ is isothermic and conformally minimal at the same time. Then, by the previous lemma, the central sphere

---

[35] Compare [281], [282], [253], and [234].

congruence Z of f is Ribaucour. Hence the Weingarten tensors A and $\hat{A}$ of Z with respect to f and $\hat{f}$, respectively, commute and, therefore, simultaneously diagonalize, that is, A, $\hat{A}$, and the identity id are linearly dependent.

On the other hand, f is conformally minimal, so that $\operatorname{tr} A = \operatorname{tr} \hat{A} = 0$. This implies that $\det A, \det \hat{A} < 0$ have the same sign. Hence (case 1a of §3.1.2) the central sphere congruence Z of f is its tangent plane congruence in a suitable quadric Q_κ^3, that is, f is minimal in Q_κ^3 (see §3.4.6).

3.6.6 We already discussed the converse in conjunction with the Euler-Lagrange equation for Willmore surfaces (see §3.5.10). However, it might be of interest to see a different, more direct proof of the statement.

Thus observe that the central sphere congruence $Z : M^2 \to \mathbb{R}P_O^4$ of a minimal surface $f : M^2 \to Q_\kappa^3$ is its tangent plane congruence. This is clearly Ribaucour, and it is minimal since its second fundamental form (note that $\hat{f} = \frac{\kappa}{2} f - \mathcal{K}$ is the second envelope of Z; see §1.7.8)

$$
\begin{aligned}
I\!I &= -\langle dZ, df \rangle (\tfrac{\kappa}{2} f - \mathcal{K}) &-& \langle dZ, d(\tfrac{\kappa}{2} f - \mathcal{K}) \rangle f \\
&= -\langle dZ, df \rangle (\kappa f - \mathcal{K}) &=& I_Z(., A.)(\kappa f - \mathcal{K})
\end{aligned}
$$

is trace-free. Consequently, f is isothermic as well as conformally minimal.

Summarizing, we obtain the following famous theorem by Thomsen [29]:

3.6.7 Theorem (Thomsen). *If $f : M^2 \to S^3$ is an umbilic-free isothermic, conformally minimal surface, then it allows a minimal stereographic projection $f : M^2 \to Q_\kappa^3$ into some quadric of constant curvature.*

Conversely, any minimal surface $f : M^2 \to Q_\kappa^3$ is isothermic and conformally minimal.

3.7 Isothermic and Willmore channel surfaces

As an example that nicely illustrates some of the developed theory, we conclude this chapter with some results on channel surfaces (cf., Section 1.8). In particular, we will obtain a classification of isothermic and Willmore channel surfaces (cf., [144]) presenting a unified proof for a classification result from a diploma thesis [276] written under the supervision of Konrad Voss (see also [295]). A similar classification result may be found in [205]. Finally, by computing the Willmore functional for (umbilic-free) channel surfaces, we make contact with a proof of the Willmore conjecture for channel tori [141].

In order to carry out our investigations, we will use adapted frames as well as the description from §1.8.3 of the channel surface in terms of the enveloped sphere curve, and switch between them as it seems appropriate. In order to make the technical details more transparent, we begin this section with a brief discussion of the setup.

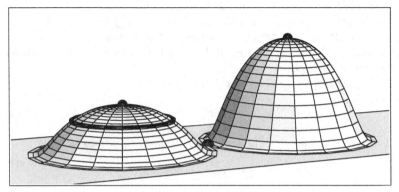

Fig. 3.1. Umbilics on a channel surface

3.7.1 Let $f : M^2 \to S^3$ be a branched channel surface (see §1.8.15), and let $S : M^2 \to S_1^4$ denote the enveloped 1-parameter family of spheres, that is, $\mathrm{rk}\, dS \le 1$.

Since we wish to work with principal curvature direction fields for f we *assume* their existence. Note that the C^∞-surface shown in Figure 3.1 is channel as a C^∞-composition of rotational surfaces and a sphere piece. However, at some of the umbilics (indicated by the three thick points) the curvature line net becomes singular; other umbilics (for example, those indicated by the thick line) are no problem, because the curvature line net extends smoothly through them.

According to our definition in §3.2.2, isothermic-ness is tested away from umbilics — hence nothing is to be investigated at umbilics anyway, so that we will exclude umbilics when discussing isothermic channel surfaces and the isothermic-ness of Willmore channel surfaces. For example, the surface shown in Figure 3.1 will turn out to be isothermic in the sense of §3.2.2. On the other hand, we do not wish to exclude umbilics on Willmore channel surfaces later, in order to make our discussions also apply to Willmore channel surfaces with lines of umbilics (cf., [11]).

3.7.2 Thus let $F : M^2 \to O_1(5)$ be an f-adapted principal curvature frame for the strip (f, S) (see §1.7.1). Then its connection form $\Phi = F^{-1}dF$ takes the form (cf., (1.1))

$$
\Phi = \begin{pmatrix}
0 & -c_1\omega_1 - c_2\omega_2 & -a\omega_1 & \omega_1 & b_{11}\omega_1 + b_{12}\omega_2 \\
c_1\omega_1 + c_2\omega_2 & 0 & 0 & \omega_2 & b_{21}\omega_1 + b_{22}\omega_2 \\
a\omega_1 & 0 & 0 & 0 & b\omega_1 \\
-b_{11}\omega_1 - b_{12}\omega_2 & -b_{21}\omega_1 - b_{22}\omega_2 & -b\omega_1 & 0 & 0 \\
-\omega_1 & -\omega_2 & 0 & 0 & 0
\end{pmatrix}
$$

$$\tag{3.13}$$

with suitable functions $a, b, b_{ij}, c_i : M^2 \to \mathbb{R}$ and forms $\omega_i : TM \to \mathbb{R}$ whose dual basis fields v_i provide the principal directions of f (see §1.7.7).

Note that $dS(v_2) = 0$, which characterizes f as a channel surface.

As usual, we analyze the integrability conditions (see §1.7.4) to extract further information about f and S.

The Codazzi equations (1.3) for f then read

$$d\omega_1 = c_1\,\omega_1 \wedge \omega_2 \quad \text{and} \quad d\omega_2 = c_2\,\omega_1 \wedge \omega_2 \qquad (3.14)$$

which determines the c_i's. The remaining Codazzi equations become

$$\partial_2 a - ac_1 = 0 \quad \text{and} \quad -ac_2 = b,$$

$$\begin{aligned}
-\partial_2 b_{11} + \partial_1 b_{12} + c_1(b_{11} - b_{22}) + c_2(b_{12} + b_{21}) &= 0, \\
-\partial_2 b_{21} + \partial_1 b_{22} + c_1(b_{12} + b_{21}) - c_2(b_{11} - b_{22}) &= 0,
\end{aligned} \qquad (3.15)$$

where ∂_i is the derivative in direction v_i, $\partial_i u = du(v_i)$ for $u \in C^\infty(M^2)$. The Gauss and Ricci equations, (1.2) and (1.4), become

$$\begin{aligned}
\partial_2 c_1 - \partial_1 c_2 - c_1^2 - c_2^2 &= b_{11} + b_{22}, \\
b_{12} = b_{21} \quad \text{and} \quad \partial_2 b - bc_1 &= ab_{12}.
\end{aligned} \qquad (3.16)$$

3.7.3 To make contact with our discussions from §1.8.5, let us assume that

$$d\omega_1 = c_1\omega_1 \wedge \omega_2 = 0 \quad \text{so that} \quad \omega_1 = dt$$

provides a parameter t for the sphere curve S; we denote the respective derivative by $\partial_1(.) = \frac{\partial}{\partial t}(.) =: (.)'$. Note that (3.15) then yields $\partial_2 a = 0$, so that $a = a(t)$, as it should be.

If f has no umbilics, that is, if the sphere curve S is spacelike ($a \neq 0$), then the representative $f : M^2 \to L^4$ of our channel surface can be scaled such that $a \equiv 1$; as a consequence, the first of the Codazzi equations (3.15) together with (3.14) yields $d\omega_1 = 0$. Hence, in the absence of umbilics, we can alway introduce the above parameter t for the sphere curve.

Now we construct a Frenet-type frame for the sphere curve by computing

$$\begin{aligned}
S' &= -a\,(S_1 - c_2 f) & &=: \; -aT \\
T' &= aS + [-c_2 S_1 + (c_2^2 + b_{22})f - \hat{f}] & &=: \; aS - N \quad (3.17) \\
N' &= -(2b_{22} + c_2^2)T + c_2 N + b_{21}S_2 - (\partial_2 b_{21})f.
\end{aligned}$$

Note that one verifies $\partial_2 T = \partial_2 N = 0$ from the integrability conditions, as it should be: From (3.15) and (3.16) we derive

$$0 = \partial_2(ac_2 + b) = a(\partial_2 c_2 + b_{12}), \qquad (3.18)$$

which yields $\partial_2 T = 0$; from (3.14) we have $[\partial_1, \partial_2] = -c_1\partial_1 - c_2\partial_2 = -c_2\partial_2$ since $c_1 = 0$, so that the Gauss equation (3.16) combined with (3.15) yields

$$c_2 b_{12} + \partial_2(c_2^2 + b_{22}) = -(c_2 + \partial_1)(\partial_2 c_2 + b_{21}) = 0,$$

which provides us with $\partial_2 N = 0$. In particular, the function

$$h := -\langle N, N \rangle = 2b_{22} + c_2^2 \qquad (3.19)$$

depends on t only: This remark will prove useful later.

3.7.4 Now assume that f is isothermic, that is, away from umbilics there exist conformal curvature line parameters (see §3.2.2). Thus we can choose a light cone representative $f : M^2 \to L^4$ with $d\omega_1 = d\omega_2 = 0$, namely,

$$\omega_1 = dt \quad \text{and} \quad \omega_2 = d\vartheta$$

are the derivatives of the conformal coordinate functions t and ϑ. Then, taking into account (3.14), the connection form (3.13) takes the form

$$\Phi = \begin{pmatrix} 0 & 0 & -a\,dt & dt & b_{11}dt + b_{12}d\vartheta \\ 0 & 0 & 0 & d\vartheta & b_{21}dt + b_{22}d\vartheta \\ a\,dt & 0 & 0 & 0 & b\,dt \\ -b_{11}dt - b_{12}d\vartheta & -b_{21}dt - b_{22}d\vartheta & -b\,dt & 0 & 0 \\ -dt & -d\vartheta & 0 & 0 & 0 \end{pmatrix},$$

and the integrability conditions (3.15) and (3.16) reduce to

$$\tfrac{\partial}{\partial\vartheta}a = 0, \quad 0 = b, \quad \tfrac{\partial}{\partial t}b_{12} = \tfrac{\partial}{\partial\vartheta}b_{11}, \quad \tfrac{\partial}{\partial t}b_{22} = \tfrac{\partial}{\partial\vartheta}b_{21};$$

$$0 = b_{11} + b_{22}, \quad b_{12} = b_{21}, \quad \text{and} \quad \tfrac{\partial}{\partial\vartheta}b = 0 = ab_{12}. \qquad (3.20)$$

In particular, $ab_{12} = 0$ so that, away from umbilics, $b_{21} = b_{12} = 0$.

Consequently, the sphere curve S is a planar curve by (3.17) and our discussions from §1.8.5 apply (cf., [293]):

3.7.5 Theorem (Vessiot). *An isothermic channel surface is, away from umbilics, (a piece of) a cylinder, a cone, or a surface of revolution in a suitably chosen Euclidean subgeometry of the conformal 3-sphere.*

3.7.6 By contemplating the compatibility conditions (3.20) further we learn that $-b_{22} = b_{11} =: \tfrac{\kappa}{2}$ and $d\kappa = 0$. So the connection form (3.13) reduces to

$$\Phi = \begin{pmatrix} 0 & 0 & -a\,dt & dt & \tfrac{\kappa}{2}dt \\ 0 & 0 & 0 & d\vartheta & -\tfrac{\kappa}{2}d\vartheta \\ a\,dt & 0 & 0 & 0 & 0 \\ -\tfrac{\kappa}{2}dt & \tfrac{\kappa}{2}d\vartheta & 0 & 0 & 0 \\ -dt & -d\vartheta & 0 & 0 & 0 \end{pmatrix}, \qquad (3.21)$$

with $a = a(t)$ and constant κ, and all integrability conditions are satisfied.

Hence we have $\frac{\partial}{\partial t}S_2 = 0$ and $\frac{\partial}{\partial t}(\frac{\kappa}{2}f - \hat{f}) = 0$: As a consequence, each "meridian curve" $\gamma_\vartheta = f(., \vartheta)$, $\vartheta = const$, takes values in the *metric* 2-plane of constant curvature $\kappa = -|\mathcal{K}|^2$ defined by

$$S^2 \simeq S_2(\vartheta) \quad \text{and} \quad \mathcal{K}(\vartheta) := (\tfrac{\kappa}{2}f - \hat{f})(\vartheta) :$$

For all t we have

$$\langle \gamma_\vartheta(t), S_2(\vartheta) \rangle = 0 \quad \text{and} \quad \langle \gamma_\vartheta(t), \mathcal{K}(\vartheta) \rangle = -1,$$

that is, γ_ϑ takes values in the quadric $Q^3_\kappa(\vartheta)$ of constant curvature defined by $\mathcal{K}(\vartheta)$ and in the conformal 2-sphere $S_2(\vartheta)$ at the same time — since $S_2(\vartheta)$ belongs to the sphere complex $\mathcal{K}(\vartheta)$, $S_2 \perp \mathcal{K}$, the 2-sphere $S_2(\vartheta)$ is in fact a hyperplane in $Q^3_\kappa(\vartheta)$, so that it defines a plane of constant curvature κ indeed (see §1.4.11).

The meridian curve γ_ϑ is parametrized by arc length, $|\gamma'_\vartheta|^2 = |\frac{\partial}{\partial t}f|^2 \equiv 1$, and has $S = S(t)$ as a (unit) normal field since $S \perp S_2(\vartheta), \mathcal{K}(\vartheta), \gamma_\vartheta, \gamma'_\vartheta$. Hence its curvature is given by $\langle \gamma'', S \rangle = a$.

By (3.17), the normal field of the sphere curve S is given by

$$N(t) = (\tfrac{\kappa}{2}f + \hat{f})(t) = \kappa\gamma_\vartheta(t) - \mathcal{K}(\vartheta);$$

note that $\frac{\partial}{\partial \vartheta}(\frac{\kappa}{2}f + \hat{f}) = 0$, so that $N = N(t)$. Thus, if $\kappa \neq 0$, we can identify the meridian curve $\gamma_\vartheta = \frac{1}{\kappa}\mathcal{K}(\vartheta) + \frac{1}{\kappa}N(t)$ with the normal field $\frac{1}{\kappa}N$ of the sphere curve, up to a proper choice of $0 \simeq \frac{1}{\kappa}\mathcal{K}(\vartheta)$ in the affine subspace

$$E^3_\vartheta = \{y = \tfrac{1}{\kappa}\mathcal{K}(\vartheta) + y^\perp \mid y^\perp \perp S_2(\vartheta), \mathcal{K}(\vartheta)\}$$

(compare with Section 1.4 for a similar setup). Note that $|\frac{1}{\kappa}N|^2 = \frac{1}{\kappa}$. By rewriting the Frenet frame formulas (3.17) with this in mind,[36]

$$S' = -aT, \quad T' = aS - \kappa(\tfrac{1}{\kappa}N), \quad \tfrac{1}{\kappa}N' = T$$

with $T = S_1$, we learn that $\frac{1}{\kappa}N$ is a curve with a unit normal field S and curvature $a = \langle \frac{1}{\kappa}N'', S \rangle$ in a standard space form of curvature κ in the (Euclidean or Minkowski, respectively) space E^3_ϑ.

Finally note that $\text{span}\{S_2, \mathcal{K}\}$ is a *fixed* 2-plane in $\mathbb{R}^5_1$: By (3.21), we have

$$dS_2 = \mathcal{K}\,d\vartheta \quad \text{and} \quad d\mathcal{K} = \kappa S_2\,d\vartheta.$$

---

[36] These formulas can also be read off directly from (3.21).

Setting, for abbreviation,

$$c_\kappa(\vartheta) := \begin{cases} \cos(\sqrt{-\kappa}\vartheta) \\ 1 \\ \cosh(\sqrt{\kappa}\vartheta) \end{cases} \quad \text{and} \quad s_\kappa(\vartheta) := \begin{cases} \frac{1}{\sqrt{-\kappa}}\sin(\sqrt{-\kappa}\vartheta) & \text{for} \quad \kappa < 0 \\ \vartheta & \text{for} \quad \kappa = 0 \\ \frac{1}{\sqrt{\kappa}}\sinh(\sqrt{\kappa}\vartheta) & \text{for} \quad \kappa > 0, \end{cases}$$

we therefore have

$$\begin{aligned} S_2(\vartheta) &= c_\kappa(\vartheta)\,S_2(0) + s_\kappa(\vartheta)\,\mathcal{K}(0) \quad \text{and} \\ \mathcal{K}(\vartheta) &= \kappa s_\kappa(\vartheta)\,S_2(0) + c_\kappa(\vartheta)\,\mathcal{K}(0). \end{aligned} \tag{3.22}$$

Consequently, we can express f (cf., §1.8.3) in terms of the parallel normal frame $(N, \mathcal{K}(0), S_2(0))$ of S:

$$\begin{aligned} f(t,\vartheta) &= \tfrac{1}{\kappa}N(t) + s_\kappa(\vartheta)\,S_2(0) + \tfrac{1}{\kappa}c_\kappa(\vartheta)\,\mathcal{K}(0) \\ &= \gamma_0(t) + s_\kappa(\vartheta)\,S_2(0) + \tfrac{1}{\kappa}(c_\kappa(\vartheta) - 1)\,\mathcal{K}(0). \end{aligned} \tag{3.23}$$

Observe that the first expression only makes sense for $\kappa \neq 0$, whereas the second expression, in terms of the "honest" meridian curve γ_0, also remains valid for $\kappa = 0$, when taking into account that $\frac{c_\kappa(\vartheta)-1}{\kappa} \to \tfrac{1}{2}\vartheta^2$ for $\kappa \to 0$. In this way we have described isothermic channel surfaces as equivariant surfaces in the conformal 3-sphere: They are obtained by letting a 1-parameter group $\vartheta \mapsto \mu_\vartheta$ of Möbius transformations act on a meridian curve γ_0, where the μ_ϑ's are given[37] by (3.22) on span$\{S_2, \mathcal{K}\}$ and by $\mu_\vartheta = id$ on $\{S_2, \mathcal{K}\}^\perp$.

Thus the Möbius transformations μ_ϑ leave the plane of the sphere curve S, span$\{S, T, N\} = \{S_2, \mathcal{K}\}^\perp$, alone. This remark will be useful later.

3.7.7 To prove the converse of our theorem in §3.7.5, we use the parametrization for f from §1.8.3. Since we have to check for the existence of conformal curvature line parameters *away from umbilics*, we may assume that the enveloped sphere curve $S : I \to S_1^4$ is parametrized by arc length, $|S'|^2 \equiv 1$. Let $N_0, N_1, N_2 : I \to \mathbb{R}_1^5$ be a parallel normal frame of the sphere curve S, and let $\kappa_i := \langle S'', N_i \rangle$ denote the corresponding curvatures. Then we have

$$S'' = -S - \kappa_0 N_0 + \kappa_1 N_1 + \kappa_2 N_2.$$

Assuming the sphere curve to be planar, we must have

$$S''' = (-1 + \kappa_0^2 - \kappa_1^2 - \kappa_2^2)S' - \kappa_0' N_0 + \kappa_1' N_1 + \kappa_2' N_2 \in \text{span}\{S, S', S''\},$$

---

[37] If $\kappa \neq 0$, this clearly defines the μ_ϑ. If, on the other hand, $\kappa = 0$, then $\mathcal{K} \equiv const$ is isotropic and is left alone by the μ_ϑ, as one sees from (3.22); moreover, the μ_ϑ's preserve any sphere that contains $\mathcal{K}$ and is orthogonal to (any) S_2: The μ_ϑ's are translations of the Euclidean quadric given by $\mathcal{K}$ in the direction orthogonal to the (parallel) planes S_2.

so that the curvature vector field $(\kappa_0, \kappa_1, \kappa_2)$ and its derivative have to be linearly dependent. As a consequence, $\lambda := (\kappa_0 + \kappa_1 \cos \vartheta + \kappa_2 \sin \vartheta)$ satisfies

$$\frac{\partial}{\partial t} \frac{\partial}{\partial \vartheta} (\ln \lambda) = \frac{(\kappa_1 \kappa_2' - \kappa_1' \kappa_2) - (\kappa_0 \kappa_1' - \kappa_0' \kappa_1) \sin \vartheta + (\kappa_0 \kappa_2' - \kappa_0' \kappa_2) \cos \vartheta}{\lambda^2} = 0,$$

that is, λ is a product of two functions of one variable, of t resp. ϑ, alone.

Now let $f(t, \vartheta) := N_0(t) + N_1(t) \cos \vartheta + N_2(t) \sin \vartheta$ be the parametrization of the channel surface as in §1.8.3. Then the induced metric of f and its second fundamental form with respect to S as a normal field are given by

$$\langle df, df \rangle = \lambda^2 dt^2 + d\vartheta^2 \quad \text{and} \quad -\langle df, dS \rangle = \lambda dt^2.$$

Hence (t, ϑ) are curvature line parameters, and, since λ is a product of two functions of one variable, we can choose new curvature line parameters

$$(\tilde{t}, \tilde{\vartheta}) = (\tilde{t}(t), \tilde{\vartheta}(\vartheta)) \quad \text{so that} \quad \langle df, df \rangle = \tilde{\lambda}^2 (d\tilde{t}^2 + d\tilde{\vartheta}^2).$$

As a consequence, f is isothermic.

3.7.8 Note that our theorem from §3.7.5 usually does not hold globally: By §3.7.7, the surface shown in Figure 3.1 is isothermic because it is assembled from two surfaces of revolution that are smoothly glued to a sphere (plane). However, this surface is not a surface of revolution (nor a cylinder or a cone) globally. In fact, it is not even locally of this form if we do not restrict it to umbilic-free parts: consider any neighborhood of the umbilic, where the two gluing circles touch.

Arbitrarily nasty surfaces that are "isothermic" in the sense of §3.2.2 can be constructed in this way — it shall be an interesting task to give a reasonable global definition of an "isothermic surface."

3.7.9 As a next step towards a classification of Willmore channel surfaces, we intend to show that Willmore channel surfaces are isothermic. For this purpose we restrict ourselves to umbilic-free pieces of the surface and go back to the frame formulas. To simplify computations, we assume that the light cone representative $f : M^2 \to L^4$ of our channel surface is chosen so that $a \equiv 1$, as discussed in §3.7.3. Then we have $c_1 = 0$ and $\omega_1 = dt$ with a parameter t of the sphere curve. The connection form (3.13) takes the form

$$\Phi = \begin{pmatrix} 0 & -c_2 \omega_2 & -a\omega_1 & \omega_1 & b_{11}\omega_1 + b_{12}\omega_2 \\ c_2 \omega_2 & 0 & 0 & \omega_2 & b_{21}\omega_1 + b_{22}\omega_2 \\ a\omega_1 & 0 & 0 & 0 & b\omega_1 \\ -b_{11}\omega_1 - b_{12}\omega_2 & -b_{21}\omega_1 - b_{22}\omega_2 & -b\omega_1 & 0 & 0 \\ -\omega_1 & -\omega_2 & 0 & 0 & 0 \end{pmatrix},$$

and the integrability conditions (3.15) and (3.16) become

$$-c_2 = b, \quad \partial_1 b_{12} = \partial_2 b_{11} - c_2(b_{12} + b_{21}), \quad \partial_2 b_{21} = \partial_1 b_{22} - c_2(b_{11} - b_{22});$$

$$\partial_1 c_2 + c_2^2 + (b_{11} + b_{22}) = 0, \quad b_{12} = b_{21}, \quad \partial_2 b = b_{12}.$$

We use the first Codazzi equation, (3.19), and the first Ricci equation to eliminate b, $b_{11} + b_{22}$, and b_{21} from these equations:

$$
\begin{aligned}
0 &= \partial_1 c_2 + b_{11} - b_{22} + h \quad \text{and} \quad \partial_2 c_2 = -b_{12}; \\
\partial_1 b_{12} &= \partial_2(b_{11} - b_{22}) - c_2 b_{12}, \\
\partial_2 b_{12} &= \tfrac{1}{2}h' - c_2(\partial_1 c_2 + b_{11} - b_{22}) = \tfrac{1}{2}h' + c_2 h.
\end{aligned}
\tag{3.24}
$$

Remember that $-|N|^2 = h = 2b_{22} + c_2^2$ is a function of t only.

The central sphere congruence of f is given by $Z = S + \tfrac{1}{2}f$. We compute the Laplacian of Z:

$$
\begin{aligned}
\Delta Z \, dA = d \star dZ &= -\tfrac{1}{2}[S + (2\partial_1 b + 2c_2 b - b_{11} + b_{22})f]\,\omega_1 \wedge \omega_2 \\
&= -\tfrac{1}{2}[Z - (2\partial_1 c_2 + 2c_2^2 + b_{11} - b_{22} + \tfrac{1}{2})f]\,\omega_1 \wedge \omega_2,
\end{aligned}
$$

where we used $\star\omega_1 = \omega_2$, $\star\omega_2 = -\omega_1$, and $d\omega_2 = c_2\omega_1 \wedge \omega_2$ besides the derivatives that are read off from the connection form. Hence, by §3.5.5, our channel surface f is Willmore if and only if

$$2(\partial_1 c_2 + c_2^2) + (b_{11} - b_{22}) + \tfrac{1}{2} = 0. \tag{3.25}$$

Using the Gauss equation from (3.24) to eliminate $\partial_1 c_2$, we deduce

$$b_{11} - b_{22} = \tfrac{1}{2} + 2c_2^2 - 2h,$$

and the remaining two equations give us a differential equation for b_{12}:

$$db_{12} = -5c_2 b_{12}\omega_1 + (\tfrac{1}{2}h' + c_2 h)\omega_2.$$

Eliminating, on the other hand, $b_{11} - b_{22}$, we get

$$\partial_1 c_2 + 2c_2^2 = h - \tfrac{1}{2}.$$

Thus, using (3.24) and the differential equation for b_{12},

$$
\begin{aligned}
0 = d^2 b_{12} &= [\tfrac{1}{2}h'' + 4c_2 h' + (\partial_1 c_2 + 6c_2^2)h - 5b_{12}^2]\,\omega_1 \wedge \omega_2 \\
&= [\tfrac{1}{2}h'' + 4c_2 h' + (h + 4c_2^2 - \tfrac{1}{2})h - 5b_{12}^2]\,\omega_1 \wedge \omega_2,
\end{aligned}
$$

so that

$$5b_{12}^2 \;=\; \tfrac{1}{2}h'' + 4c_2 h' + (h + 4c_2^2 - \tfrac{1}{2})h.$$

If $h \equiv 0$, then $b_{12} = 0$; if, on the other hand, $h \neq 0$, we differentiate this last equation twice with ∂_2 to find

$$0 \;=\; (\tfrac{1}{2}h' + c_2 h)\, b_{12}$$
$$0 \;=\; (\tfrac{1}{2}h' + c_2 h)^2 - h b_{12}^2.$$

Hence we must have $b_{12} = 0$ in any case. As a consequence, our sphere curve S is planar by (3.17) and f is isothermic by §3.7.7:

3.7.10 Theorem. *Any Willmore channel surface is isothermic.*

3.7.11 This result will help us to analyze Willmore channel surfaces further: We may use a flat lift $f : M^2 \to L^4$ for a given Willmore channel surface. Then, according to our considerations in §3.7.6, the connection form (3.13) takes the form

$$\Phi = \begin{pmatrix} 0 & 0 & -a\,dt & dt & \tfrac{\kappa}{2}dt \\ 0 & 0 & 0 & d\vartheta & -\tfrac{\kappa}{2}d\vartheta \\ a\,dt & 0 & 0 & 0 & 0 \\ -\tfrac{\kappa}{2}dt & \tfrac{\kappa}{2}d\vartheta & 0 & 0 & 0 \\ -dt & -d\vartheta & 0 & 0 & 0 \end{pmatrix}, \qquad (3.26)$$

where (t, ϑ) are conformal curvature line parameters (remember that we assumed the existence of curvature line parameters), with t denoting the parameter of the sphere curve, and $a = a(t)$ is the curvature of a meridian curve $\gamma_\vartheta = f(., \vartheta)$, $\vartheta = const$, that takes values in a fixed plane of constant curvature defined by $S_2(\vartheta)$ and $\mathcal{K}(\vartheta) = (\tfrac{\kappa}{2}f - \hat{f})(\vartheta)$. By recomputing the Laplacian of the central sphere congruence $Z = S + \tfrac{1}{2}af$ in this setup, we find

$$\Delta Z\, dA \;=\; d \star dZ \;=\; \tfrac{1}{2}[-a^2 S + (a'' + \kappa a)f]\, dt \wedge d\vartheta$$
$$= \tfrac{1}{2}[-a^2 Z + (a'' + \tfrac{1}{2}a^3 + \kappa a)f]\, dt \wedge d\vartheta,$$

where we used $\star dt = d\vartheta$ and $\star d\vartheta = -dt$ as for (3.25). Thus the Willmore condition becomes

$$0 = a'' + \tfrac{1}{2}a^3 + \kappa a, \qquad (3.27)$$

the condition on a meridian curve γ_ϑ to be a free elastic curve[38] in its 2-plane of constant curvature κ (see [177]). We have proved (cf., [41], [178]):

---

[38] Compare also the construction of Willmore Hopf tori in [219].

3.7.12 Theorem. *A Willmore channel surface is Möbius equivalent to a*

$$
\left.\begin{array}{r} cone \\ cylinder \\ surface\ of\ revolution \end{array}\right\} over\ a\ free\ elastic\ curve\ in\ a \left\{\begin{array}{l} \text{2-sphere,} \\ \text{Euclidean 2-plane,} \\ \text{hyperbolic 2-plane.} \end{array}\right.
$$

3.7.13 Free elastic curves in planes of constant curvature have been clas- sified by Langer and Singer in [177] (cf., [41]). By our above theorem, this gives rise to a complete classification of Willmore channel surfaces. Explicit formulas for Willmore cones, cylinders, and surfaces of revolution (which, by the theorem, comprise all Willmore channel surfaces) were given in [218], [276], and [205].

Note that we required the existence of curvature line parameters on our Willmore channel surface — however we, did not exclude umbilics. Indeed, there are Willmore channel surfaces with (lines of) umbilics that arise from the zeros of a: For examples see [218] or [11].

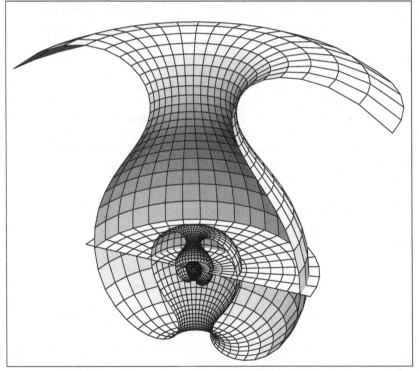

Fig. 3.2. A rotational Willmore channel surface (cf., [11])

3.7.14 Since any Willmore channel surface is isothermic, it must be minimal in some suitable space of constant curvature, by Thomsen's theorem in §3.6.7. Indeed, given a frame with connection form (3.26) for a Willmore channel surface f, the vector

$$\mathcal{M} := -a'S_1 + \tfrac{1}{2}a^2 S - a(\tfrac{\kappa}{2}f + \hat{f}) = \tfrac{1}{2}a^2 S - a'T - a\,N$$

turns out to be constant, $d\mathcal{M} = -(a'' + \tfrac{1}{2}a^3 + \kappa a)S_1 dt = 0$. Assuming that no umbilics are present, that is, $a \neq 0$, the surface $\tfrac{1}{a}f$ takes values in the corresponding space of constant curvature[39]

$$c = -\langle \mathcal{M}, \mathcal{M} \rangle = -(a'^2 + \tfrac{1}{4}a^4 + \kappa a^2). \tag{3.28}$$

Since the central sphere congruence consists of planes in the respective quadric of constant curvature, $Z \perp \mathcal{M}$, the surface $\tfrac{1}{a}f$ is minimal in this quadric by §3.4.6. From (3.28) we immediately deduce that, in the cases of a cone ($\kappa > 0$) and a cylinder ($\kappa = 0$), the ambient curvature of the minimal surface is $c < 0$, that is, the surface is minimal in some hyperbolic space. On the other hand, in the case of a surface of revolution ($\kappa < 0$), all types of ambient curvatures may occur — appropriate examples are the (minimal) Clifford torus in S^3, the catenoid in $\mathbb{R}^3$, and the minimal surface of revolution[40] in (two copies of) H^3 shown in Figure 3.2.

Moreover, by describing f as the orbit of the 1-parameter group $\vartheta \mapsto \mu_\vartheta$ of Möbius transformations from §3.7.6, we learn that these surfaces are equivariant minimal surfaces in the quadric of constant curvature given by $\mathcal{M}$. Namely, remember from §3.7.6 that the μ_ϑ's fix the points of the plane $\mathrm{span}\{S, T, N\}$ of the enveloped sphere curve S — in particular, the μ_ϑ leave $\mathcal{M} \in \mathrm{span}\{S, T, N\}$ invariant and hence are isometries of the quadric of constant curvature (cf., §1.4.13).

---

[39] Note that this equation is exactly the integrated form of (3.27).

[40] Given a suitable curvature function a that satisfies (3.28) with $\kappa = -1$, $c = -r^2$ ($r > 0$), the meridian curve γ can be parametrized, as a curve in the Poincaré half-plane (see §P.1.3), by

$$\gamma = \frac{e^\tau}{\sqrt{r^2 + a^2}}(r, a) \quad \text{where} \quad \tau = \int \frac{ra^2 \, dt}{2(r^2 + a^2)}$$

(cf., [41]); for Figure 3.2 we chose the Jacobi elliptic function (compare [177])

$$a(t) = \frac{2p\sqrt{1-p^2}}{\sqrt{2p^2-1}} \, \mathrm{sd}_p\left(\frac{1}{\sqrt{2p^2-1}} t\right) \quad \text{with} \quad p = \frac{4}{\sqrt{17}} > \frac{1}{\sqrt{2}} \quad \text{so that} \quad r = \frac{2p\sqrt{1-p^2}}{2p^2-1} = \frac{8}{15}.$$

A parametrization in terms of theta functions is given in [11]. At this point I would like to thank A. Bobenko for providing me with *Mathematica* code, which greatly facilitated the experiment with the computer graphics.

Note that, at the zeros $a = 0$ of the curvature function, $f \perp \mathcal{M}$: The lines of umbilics are the intersection lines of the surface with the common infinity plane of the two hyperbolic ambient spaces, as indicated in Figure 3.2 (cf., [11]).

Thus we have proved the following theorem (compare [8] and [218]):

3.7.15 Theorem. *Any Willmore channel surface is an equivariant minimal surface in a space of constant curvature.*

3.7.16 Finally, we want to compute the Willmore functional of a regular (umbilic-free) channel surface. Our previous results strongly suggest that the Willmore functional should have something to do with the bending energy for curves since free elastic curves are exactly the critical curves for the bending energy (see [177]). Thus let $S : I \to S_1^4$ be (a lift of) the 1-parameter family of spheres enveloped by the channel surface; excluding, for a moment, umbilics, we may assume the sphere curve S to be regular: Without loss of generality $|S'|^2 \equiv 1$. As in §3.7.7, we parametrize the channel surface by

$$f(t, \vartheta) \quad = \quad N_0(t) + N_1(t) \cos \vartheta + N_2(t) \sin \vartheta$$

where the N_i's are parallel normal fields of $S : I \to S_1^4$. If we require the surface f to be regular on all of $I \times S^1$, then the osculating planes of the sphere curve $S : I \to \mathbb{RP}_O^4$ all have to intersect the absolute quadric $S^3 \subset \mathbb{RP}^4$ transversally (see §1.8.3). For the curvatures $\kappa_i = \langle S'', N_i \rangle$ of S, this means that

$$\kappa_0^2 - \kappa_1^2 - \kappa_2^2 \quad =: \quad \kappa_g^2 \quad > \quad 0, \tag{3.29}$$

that is, $N := S + S'' = -\kappa_0 N_0 + \kappa_1 N_1 + \kappa_2 N_2$ is timelike. As in §3.7.7, the derivatives of f become

$$\begin{aligned} \tfrac{\partial}{\partial t} f &= -(\kappa_0 + \kappa_1 \cos \vartheta + \kappa_2 \sin \vartheta) S' &=: & -\lambda S', \\ \tfrac{\partial}{\partial \vartheta} f &= -N_1 \sin \vartheta + N_2 \cos \vartheta &=: & S_2, \end{aligned}$$

and the induced metric and second fundamental form with respect to S as a unit normal field are

$$\langle df, df \rangle = \lambda^2 dt^2 + d\vartheta^2 \quad \text{and} \quad -\langle df, dS \rangle = \lambda \, dt^2.$$

Thus, by §3.4.8, the central sphere congruence of f is given by $Z = S + \frac{1}{2\lambda} f$ since

$$\langle dZ, dZ \rangle = \tfrac{1}{4} dt^2 + \tfrac{1}{4\lambda^2} d\vartheta^2 = \tfrac{1}{4\lambda^2} (\lambda^2 dt^2 + d\vartheta^2) = \tfrac{1}{4\lambda^2} \langle df, df \rangle.$$

With this we compute the Willmore functional of f, that is, the area of the conformal Gauss map:

$$W(f) = \tfrac{1}{2\pi} \int_{T^2} dA_Z = \tfrac{1}{2\pi} \int_{t \in I} \int_{\vartheta=0}^{2\pi} \tfrac{1}{4|\lambda|} d\vartheta \, dt = \tfrac{1}{4} \int_S \tfrac{1}{\kappa_g},$$

where the last equality is an integration that is best looked up from some table of definite integrals — note that (3.29) ensures that λ does not change sign, so that the remaining integrand has no singularities. As a consequence, the Willmore functional of an umbilic-free channel surface is given in terms of the sphere curve only. In particular (cf., [141]),

3.7.17 Theorem. *The Willmore functional of an umbilic-free channel torus is the total inverse geodesic curvature of the enveloped sphere curve.*

3.7.18 When dropping the assumption on S to be parametrized by arc length, we have

$$
\begin{aligned}
S' &= -aT \\
T' &= aS - a(\kappa_0 N_0 - \kappa_1 N_1 - \kappa_2 N_2) \quad =: \quad aS - a\kappa_g N
\end{aligned}
\tag{3.30}
$$

with the speed $a = a(t)$ and the normalized Frenet normal field N of S, $|N|^2 \equiv -1$. Choosing a new parametrization of S such that $a\kappa_g \equiv 1$, the Willmore functional becomes

$$
W(f) = \tfrac{1}{4}\int_S \tfrac{1}{\kappa_g} = \tfrac{1}{4}\int_{t\in I} a^2 dt.
$$

In the case where the sphere curve is planar, that is, in the case of a surface of revolution where the normal field N of S can be identified with its meridian curve in a hyperbolic plane (see §3.7.6), the Willmore functional becomes the bending energy of the meridian curve [218] (see also [41] and [141]). As mentioned above, this is no surprise after knowing about the relation between Willmore surfaces of revolution and free elastic curves in a hyperbolic plane (see §3.7.12), that is, the critical curves for the total squared geodesic curvature.

3.7.19 We want to take our result from §3.7.17 a little further: Consider the enveloped sphere curve of our channel surface to be given in terms of a curve $m : \mathbb{R} \to S^3$ of centers and a function $r : \mathbb{R} \to (0, \pi)$ so that

$$
S = \tfrac{1}{\sin r}(\cos r, m) : \mathbb{R} \to S_1^4 \subset \mathbb{R} \times \mathbb{R}^4
$$

(cf., Figure 1.2). Now we compute the geodesic curvature of S. First,

$$
S' = \tfrac{1}{\sin^2 r}(-r', -r'm\cos r + m'\sin r) = -aT.
$$

Because $0 < a^2 = \tfrac{m'^2 - r'^2}{\sin^2 r}$, we see that the center curve m is regular, $m'^2 > 0$; so we may assume that it is parametrized by arc length, $m'^2 \equiv 1$. Next we compute the derivative of the unit tangent field of S:

$$
\begin{aligned}
T &= \tfrac{1}{\sqrt{1-r'^2}\,\sin r}(r', r'm\cos r - m'\sin r), \\
T' &= \tfrac{1}{\sqrt{1-r'^2}^3\,\sin^2 r}\{ r''\sin r - r'^2(1 - r'^2)\cos r, \\
&\qquad\qquad m\,(r''\cos r\sin r - r'^2(1 - r'^2)) \\
&\qquad + \; m'((1 - r'^2)\cos r - r''\sin r)\,r'\sin r \\
&\qquad - \; m''(1 - r'^2)\sin^2 r\}.
\end{aligned}
$$

Now remember that the Frenet normal N of S is timelike and the geodesic curvature κ_g of S does not change sign, so that it can be chosen to be positive. Hence we may use the equation $|T' - aS|^2 = -a^2\kappa_g^2$ (see (3.30)) to determine κ_g:

$$\kappa_g^2 = \tfrac{1}{(1-r'^2)^3}\{(r'' \sin r - (1 - r'^2)\cos r)^2 - (m''^2 - 1)(1 - r'^2)\sin^2 r\}.$$
(3.31)

As a consequence, the Willmore functional $W(f) = \int_S \frac{1}{\kappa_g}$ only depends on, besides the radius function r, the geodesic curvature $(m''^2 - 1)$ of the center curve m.

3.7.20 In particular, (3.31) shows that $W(f)$ decreases with the geodesic curvature of the center curve: This was the key observation leading to a proof of the Willmore conjecture for (umbilic-free) channel tori in [141]. Namely, replacing the center curve m by any curve of a (pointwise) lower geodesic curvature while keeping the radius function r will provide us with a channel surface with a lower Willmore functional than the original one. In the optimal case, where m parametrizes (part of) a great circle, we obtain a surface of revolution (the sphere curve becomes planar then) with a lower Willmore functional, actually, with the lowest possible Willmore functional for a given radius function r:

3.7.21 Theorem. *For any umbilic-free channel surface $f : I \times S^1 \to S^3$ there is a surface of revolution $\tilde{f} : I \times S^1 \to S^3$ with $W(\tilde{f}) \leq W(f)$.*

3.7.22 In the case of a torus this procedure can be used to reduce the Willmore conjecture to a statement about surfaces of revolution, that is, to a problem for curves in the hyperbolic plane by §3.7.18. However, the main problem is to keep the curve periodic: For this to be possible, we need the original center curve to have the same length 2π as a great circle in S^3. For a proof that this can be achieved (using an intermediate value argument) by a suitable Möbius transformation for any embedded channel torus,[41] as well as for a direct proof that the total squared geodesic curvature $\frac{1}{4}\int_\gamma k^2 ds \geq \pi$ for any curve in the hyperbolic plane, see [141]. An alternative proof for the second step, that is, the Willmore conjecture for surfaces of revolution, can be found in [178].

---

[41] Note that, in order to prove the Willmore conjecture, it is sufficient to consider embedded tori by a result of Li and Yau [180].

Chapter 4
A quaternionic model

In dimension $n = 2$, the (conformal) 2-sphere can be identified with the complex projective line, $\mathbb{C} \cup \{\infty\} \cong \mathbb{C}P^1 := \{v\mathbb{C} \mid v \in \mathbb{C}^2\}$, with the (orientation-preserving) Möbius transformations acting by fractional linear transformations,

$$\mathbb{C} \cup \{\infty\} \ni z \mapsto \mu(z) = \tfrac{az+b}{cz+d} \in \mathbb{C} \cup \{\infty\}.$$

In terms of homogeneous coordinates $v \in \mathbb{C}^2$ for points $z = v\mathbb{C} \in \mathbb{C}P^1$, this action will just be by linear transformations, $\mu(v\mathbb{C}) = (M_\mu v)\mathbb{C}$, where M_μ denotes a suitable endomorphism of $\mathbb{C}^2$ projecting to μ:

$$\mathbb{C}P^1 \ni \begin{pmatrix} z \\ 1 \end{pmatrix}\mathbb{C} \mapsto \begin{pmatrix} a & b \\ c & d \end{pmatrix}\begin{pmatrix} z \\ 1 \end{pmatrix}\mathbb{C} = \begin{pmatrix} az+b \\ cz+d \end{pmatrix}\mathbb{C} = \begin{pmatrix} \tfrac{az+b}{cz+d} \\ 1 \end{pmatrix}\mathbb{C} \in \mathbb{C}P^1.$$

Note that the result $(M_\mu v)\mathbb{C}$ does not depend on the choice of homogeneous coordinates v: If $\tilde{v}\mathbb{C} = v\mathbb{C}$, then $(M_\mu \tilde{v})\mathbb{C} = (M_\mu v)\mathbb{C}$ since M_μ is linear. Obviously, M_μ is not uniquely determined by μ, because any nonzero multiple of M_μ will provide a valid choice for M_μ. Thus the group of orientation-preserving Möbius transformations becomes $PGl(2, \mathbb{C})$. Using the determinant to normalize representatives, the special linear group $Sl(2, \mathbb{C})$ becomes a double cover of the group of orientation-preserving Möbius transformations.

Our mission in this chapter is to generalize this model of 2-dimensional Möbius geometry to dimensions 3 and 4, using quaternions. We will start by establishing some basic facts from quaternionic linear algebra. In particular, we will emphasize the differences from the usual linear algebra, which are caused by the lack of commutativity of the quaternions, and we will introduce a kind of determinant (cf., [10]) that allows us to define the "quaternionic special linear group." To establish the relation with the classical model of Möbius geometry — which will allow us to "translate" results from one model into the other — we will consider quaternionic Hermitian forms. However, this is not the only possible way to establish this relation: Another possibility is to give the relation by identifying (hyper)spheres with inversions (compare [304], where explicit formulas are given). We will present a variant of this approach in the 3-dimensional case, that is, for 2-spheres in S^4 (cf., [49]). These ideas are not new: The idea of using quaternions in Möbius geometry goes back (at least) to E. Study, who already discussed most of the (nondifferential) geometric aspects of the quaternionic model that we are going to discuss in this chapter in his series of papers [262].

The quaternionic model of Möbius geometry provides a powerful setup for the study of isothermic surfaces. In the second half of this chapter we will collect the technical details to reexamine the transformation theory of smooth and discrete isothermic surfaces: We will study the symmetric space of point pairs and establish properties of the cross-ratio as an important invariant of four points in conformal geometry. This will become important for our discussion of "discrete isothermic nets."

The contents of this chapter are organized as follows:

Section 4.1. In this section some viewpoints on the field of quaternions and various basic facts in quaternionic linear algebra are collected. In particular, the effects of the noncommutativity of the quaternions are discussed.

Section 4.2. The Study determinant for quaternionic 2×2 matrices and for endomorphisms of a 2-dimensional quaternionic vector space is introduced. Some basic properties of the Study determinant are derived. Definitions for the general and special linear groups of $\mathbb{H}^2$ are given.

Section 4.3. The notion of quaternionic Hermitian forms on a quaternionic vector space is introduced. It is shown that the space of quaternionic Hermitian forms on a 2-dimensional quaternionic vector space becomes a 6-dimensional Minkowski space in a natural way. Points and hyperspheres in the conformal 4-sphere $\mathbb{H}P^1 = \mathbb{H} \cup \{\infty\}$ are identified with null and spacelike directions in the space of quaternionic Hermitian forms on $\mathbb{H}^2$, respectively.

Section 4.4. Here we present how the quaternionic special linear group acts on the space of quaternionic Hermitian forms by isometries. Using this representation $Sl(2, \mathbb{H}) \to O_1(6)$, it is shown that $Sl(2, \mathbb{H})$ is the (double) universal cover of the group of orientation-preserving Möbius transformations.

Section 4.5. The space of point pairs in S^4 is shown to be a reductive homogeneous space. The notion of adapted frames of a point pair map is introduced, the effect of a gauge transformation on its connection form is established, and the Maurer-Cartan equations are analyzed. A "translation table" between quaternionic connection forms and Lorentzian connection forms is given.

Section 4.6. The space of single points, that is, the conformal 4-sphere $\mathbb{H}P^1$, is described as a (nonreductive) homogeneous space. The notion of adapted frames for a map into the conformal 4-sphere is introduced, and the effect of a gauge transformation on its connection form is examined.

Section 4.7. The enveloping conditions for a map into $\mathbb{H}P^1$ and a hypersphere congruence, thought of as a map into the space of quaternionic Hermitian forms, are established. Considering a congruence of 2-

spheres in S^4 as a congruence of elliptic hypersphere pencils, the respective enveloping conditions are discussed. A technical lemma concerning adapted frames of a point pair map that has a 2-sphere congruence attached is provided.

Section 4.8. A more effective treatment of 2-sphere congruences in terms of involutions is given: 2-spheres are identified with Möbius involutions, it is investigated when a 2-sphere is contained in a 3-sphere, and the enveloping conditions are formulated in this setup.

Section 4.9. The action of the Möbius group on the dual space $(\mathbb{H}P^1)^\perp$ of $\mathbb{H}P^1$ is discussed, and a notion of stereographic projection $\mathbb{H}P^1 \to \mathbb{H}$ is introduced. The (complex) cross-ratio of four points in the conformal 4-sphere is introduced, and its properties are investigated — in particular, the cross-ratio identities are derived. The "hexahedron lemma" is proven: This lemma will become important in the context of the Darboux transformation for isothermic surfaces as well as in the context of discrete net theory.

Remark. It may be possible to read this chapter independently; in order to follow various arguments, as well as to get an understanding of the geometric interpretations of the presented material, it will be of great help to be familiar with the classical model of Möbius (differential) geometry, as it is presented in Chapter 1.

4.1 Quaternionic linear algebra

The quaternions form a noncommutative field. This major difference from the reals and the complex numbers seems to make it necessary to establish some basic properties and to clarify some notations before we can start to discuss Möbius geometry. In particular, we want to introduce some quaternionic analog of the determinant for 2×2 matrices or endomorphisms of 2-dimensional quaternionic vector spaces: The Study determinant. However, before doing so in the following section, we will have to discuss some basic quaternionic linear algebra.

4.1.1 The quaternions $\mathbb{H}$ can be considered as a 4-dimensional real vector space, $\mathbb{R}^4$, or as a 2-dimensional complex vector space, $\mathbb{C}^2$, equipped with a suitable multiplication that makes $\mathbb{H}$ a (noncommutative) field. Thinking of $\mathbb{H}$ as a real vector space, it is common to introduce the multiplication in terms of a (real) basis $(1, i, j, k)$ of $\mathbb{H}$: 1 is the identity of the multiplication on $\mathbb{H}$, and the products of the other basis vectors are given by

$$i^2 = j^2 = k^2 = -1, \quad ij = -ji = k, \quad jk = -kj = i, \quad ki = -ik = j.$$

On the other hand, considering $\mathbb{H}$ as a 2-dimensional complex vector space, the multiplication is defined in terms of a basis $(1, j)$, where 1 denotes the identity as before and $j^2 = -1$. But now $zj = j\bar{z} \neq jz$, unless $z \in \mathbb{R}$ is real since $ij = -ji$, as above: Quaternionic multiplication is not commutative.

4.1.2 On $\mathbb{H}$ one introduces an $\mathbb{R}$-linear map $q \mapsto \bar{q}$, the "quaternionic conjugation," by $\bar{1} = 1$, $\bar{i} = -i$, $\bar{j} = -j$, and $\bar{k} = -k$. It is easily checked that quaternionic conjugation is an antiautomorphism, $\overline{pq} = \bar{q}\bar{p}$. Hence, in terms of the complex setup, $\overline{x + yj} = \bar{x} - j\bar{y}$

The real and imaginary parts of a quaternion $q \in \mathbb{H}$ are then defined by

$$2\mathrm{Re}\, q = q + \bar{q} \in \mathbb{R} \quad \text{and} \quad 2\mathrm{Im}\, q = q - \bar{q} \in \mathrm{span}\{i, j, k\}.$$

And, the norm of a quaternion (identifying $\mathbb{H}$ with the 4-dimensional *Euclidean* space $\mathbb{R}^4$) is given by $|q|^2 = q\bar{q} = (\mathrm{Re}\, q)^2 + |\mathrm{Im}\, q|^2$.

4.1.3 Another possibility is to realize quaternions as (special) complex matrices or endomorphisms: a common way is to write a quaternion in terms of the following matrices

$$1 \simeq \begin{pmatrix} 1 & 0 \\ 0 & 1 \end{pmatrix}, \quad i \simeq \begin{pmatrix} i & 0 \\ 0 & -i \end{pmatrix}, \quad j \simeq \begin{pmatrix} 0 & 1 \\ -1 & 0 \end{pmatrix}, \quad k \simeq \begin{pmatrix} 0 & i \\ i & 0 \end{pmatrix},$$

which are closely related to the Pauli matrices — quaternionic multiplication then becomes matrix multiplication. More invariantly, this identification reads

$$\mathbb{H} \cong \{Q \in End(\mathbb{C}^2) \,|\, \mathrm{tr}\, Q \in \mathbb{R}, Q + Q^* \in \mathbb{R} \cdot 1\},$$

where Q^* is the adjoint of Q, taken with respect to the canonical Hermitian inner product on $\mathbb{C}^2$. Note that, in this model, quaternionic conjugation is taking adjoints, $\bar{Q} = Q^*$.

4.1.4 For the purposes of surface geometry in 3-dimensional (Euclidean) space, it is often convenient to decompose the quaternions[1] into their real and imaginary parts, $\mathbb{H} \cong \mathbb{R} \oplus \mathbb{R}^3$, $q = r + v$. Then the multiplication reads

$$(r + v)(s + w) = (rs - v \cdot w) + sv + rw + v \times w,$$

where $v \cdot w$ is the canonical Euclidean scalar product on $\mathbb{R}^3$, and $v \times w$ denotes the cross-product on $\mathbb{R}^3$. In particular, the product of two imaginary quaternions (vectors in 3-space) gives $vw = -v \cdot w + v \times w$.

Note that $S^3 = \{q \in \mathbb{H} \,|\, |q| = 1\}$ acts by isometries on $\mathbb{R}^3$ via the adjoint action

$$S^3 \times \mathbb{R}^3 \ni (q, x) \mapsto qx\bar{q} \in \mathbb{R}^3,$$

---

[1] For the relation of the quaternions with the Clifford algebra $\mathcal{A}(\mathbb{R}^3)$ of Euclidean 3-space, see, for example, [76] or [134]. See also Section 7.1.

$|qx\bar{q}|^2 = |x|^2$ for $|q| = 1$. Moreover, its differential at the identity,

$$T_{id}S^3 = \mathbb{R}^3 \ni y \mapsto 2y \times . \in \mathfrak{o}(3),$$

is a Lie algebra isomorphism, so that S^3 is the universal cover of the identity component of $O(3)$. In particular, S^3 acts transitively on its Lie algebra $\mathbb{R}^3$.

4.1.5 For quaternion-valued 1-forms $\omega, \eta \in \Lambda^1(TM)$ on a manifold, one defines, as usual, the exterior product

$$\omega \wedge \eta(v, w) = \omega(v)\eta(w) - \omega(w)\eta(v)$$

and the exterior differentiation

$$d\omega(v, w) = \partial_v(\omega(w)) - \partial_w(\omega(v)) - \omega([v, w]),$$

where $v, w \in \Gamma(TM)$ denote some vector fields. Note that, by the lack of commutativity, $\omega \wedge \omega \neq 0$, in general. Also, it seems worth collecting a couple of formulas: Conjugation of a wedge product yields

$$\overline{\omega \wedge \eta} = -\bar{\eta} \wedge \bar{\omega},$$

and the Leibniz rule reads

$$d(g\omega) = dg \wedge \omega + g\,d\omega \quad \text{or} \quad d(\omega g) = d\omega\,g - \omega \wedge dg.$$

4.1.6 Vector spaces over the quaternions are defined in the usual way — with the exception that (because of the lack of commutativity) it has to be fixed whether scalar multiplication is from the left or from the right. We will usually assume that the quaternionic vector space we are working with is a *right vector space*. Then associativity reads $(v\lambda)\mu = v(\lambda\mu)$ for $v \in V$ and $\lambda, \mu \in \mathbb{H}$.

Linear (in)dependency of families of vectors is defined in the usual way, and a basis of a quaternionic vector space is a linearly independent family of vectors that span the space (if such a family exists). The common proof of Steinitz' basis exchange theorem goes through so that the quaternionic dimension of a (finitely generated) vector space is well defined.

Hence, any finitely generated quaternionic vector space V is isomorphic to $\mathbb{H}^n$, where $n = \dim V$.

4.1.7 Linear maps between quaternionic vector spaces will only be considered for both vector spaces being of the same type, left vector spaces or right vector spaces, respectively. In this case, the linearity of a map is defined in the usual way.

Given bases $(v_1, \ldots, v_n)$ and $(w_1, \ldots, w_m)$ in both domain V and target W of a linear map $L : V \to W$, L can be represented by a (quaternionic) matrix: If $Lv_j = \sum_i w_i l_{ij}$, then, for any vector $v = \sum_j v_j b_j$, the matrix $L \simeq (l_{ij})_{ij}$ acts on its "coordinates" by left multiplication, $Lv = \sum_{ij} w_i l_{ij} b_j$.

4.1.8 For endomorphisms we will consider the notion of "eigenvectors": A vector $v \in V \setminus \{0\}$ will be called an eigenvector of $L \in End(V)$ if $Lv = v\lambda$ for some $\lambda \in \mathbb{H}$.

However, the notion of "eigenvalue" does *not* make sense: If v is an eigenvector of L, $Lv = v\lambda$, then $L(va) = (Lv)a = v\lambda a = (va)(a^{-1}\lambda a)$, so that the multiple λ cannot be associated with the "eigendirection" $v\mathbb{H}$ but, in general, it depends on the choice of vector spanning the quaternionic line $v\mathbb{H}$.

4.1.9 The "dual space" $V^* := \{\nu \mid \nu : V \to \mathbb{H} \text{ linear}\}$ of a quaternionic vector space V can be equipped with a quaternionic linear structure itself[2]: If $\nu : V \to \mathbb{H}$ is linear, then $(\lambda\nu)(v) := \lambda(\nu(v))$ yields a well-defined scalar multiplication since $(\lambda\nu)(va) = (\lambda\nu)(v)a$.

Note that this multiplication makes the dual space V^* a quaternionic *left* vector space.[3]

4.2 The Study determinant

We are mainly interested in obtaining a model of Möbius geometry in dimension $n \leq 4$. Thus, in what follows, we will restrict ourselves to 2-dimensional quaternionic vector spaces: This is all we need. In dimension $n = 2$, the special linear group $Sl(2, \mathbb{C})$ provides a double cover of the group of orientation-preserving Möbius transformations. To obtain a quaternionic analog we seek a notion of a determinant. However, in the quaternionic case, the dimension of the Möbius group is $\dim M\ddot{o}b(4) = 15 = 16 - 1$, that is, our determinant should reduce the (real) dimension $16 = \dim_R End(\mathbb{H}^2)$ by 1 and should hence be real-valued.

Thus, we define the "Study determinant" (cf., [262], [10]) for quaternionic 2×2 matrices. For this, first remember that a quaternion can be realized as a complex 2×2 matrix of a certain form (see §4.1.3), so that a quaternionic 2×2 matrix can be written as a complex 4×4 matrix.

4.2.1 Definition.   *Given $A \in M(2 \times 2, \mathbb{H}) \widetilde{\subset} M(4 \times 4, \mathbb{C})$, its Study determinant $[A] := \det A^{\mathbb{C}}$ is the determinant of the corresponding complex 4×4*

---

[2] Note that, in general, it is difficult to supply the space of linear maps between quaternionic vector spaces with a quaternionic linear structure: With the target space, the space of linear maps should be a right vector space, but $(L\lambda)(va) = L(va)\lambda \neq L(v)\lambda a = (L\lambda)(v)a$ in general.

[3] We could consider V^* as a right vector space by means of $(\nu\lambda)(v) := \bar{\lambda}(\nu(v))$.

matrix.

4.2.2 As an immediate consequence, the Study determinant is multiplicative,

$$[AB] = [A] \cdot [B].$$

Consequently, the Study determinant yields a well-defined notion for quaternionic endomorphisms $L \in End(V)$, where V is a quaternionic vector space of dimension $\dim V = 2$: If $(\tilde{e}_1, \tilde{e}_2) = (e_1 b_{11} + e_2 b_{21}, e_1 b_{12} + e_2 b_{22})$ denotes a change of basis, $B = (b_{ij})_{ij}$, then the matrices M_L and $\tilde{M}_L$ associated with L transform by $\tilde{M}_L = B^{-1} M_L B$. Thus the following definition is independent of the involved choice of basis.

4.2.3 Definition. *Let $L \in End(V)$ be an endomorphism of a quaternionic vector space of dimension $\dim V = 2$. Then $[L] := [M_L]$, where M_L denotes the representing matrix of L with respect to any basis of V, is called the Study determinant of L.*

4.2.4 Considering, after a choice of basis[4] (e_1, e_2) for our 2-dimensional quaternionic vector space V, the Study determinant as a map

$$V \times V \to \mathbb{R}, \quad (v, w) \mapsto [v, w]$$

on pairs of vectors, it follows from the definition that $[v, w] = [w, v]$: Interchanging the columns (rows) of a quaternionic 2×2 matrix amounts to two column (row) interchangements of the corresponding complex 4×4 matrix.

A straightforward computation from the formula for $[A]$ of the following lemma shows that, for $\lambda \in \mathbb{H}$, $[v, w\lambda] = |\lambda|^2 [v, w]$ and $[v + w, w] = [v, w]$. Moreover, reformulating the corresponding statement in the lemma below, $0 \leq [v, w] \in \mathbb{R}$ and $0 = [v, w]$ if and only if v and w are linearly dependent:

4.2.5 Lemma. *For $A = \begin{pmatrix} a & b \\ c & d \end{pmatrix} \in M(2 \times 2, \mathbb{H})$, the Study determinant[5]*

$$[A] = |a|^2 |d|^2 + |b|^2 |c|^2 - \bar{a}b\bar{d}c - \bar{c}d\bar{b}a.$$

Moreover, $0 \leq [A] \in \mathbb{R}$ for any $A \in M(2 \times 2, \mathbb{H})$; and if $[A] \neq 0$, then

$$A^{-1} = \frac{1}{[A]} \begin{pmatrix} |d|^2 \bar{a} - \bar{c}d\bar{b} & |b|^2 \bar{c} - \bar{a}b\bar{d} \\ |c|^2 \bar{b} - \bar{d}c\bar{a} & |a|^2 \bar{d} - \bar{b}a\bar{c} \end{pmatrix} ;$$

---

[4] We identify a pair (v, w) of vectors with the endomorphism $(e_1, e_2) \mapsto (v, w)$ of V.

[5] Note that, even though the usual determinant for 2×2 matrices makes no sense in the quaternionic setting (caused by the lack of commutativity), it does make sense for self-adjoint matrices since, in that case, the order of multiplication is irrelevant (cf., [10]). Thus we may write the following formula as $[A] = \det(A^* A)$.

if, on the other hand, $[A] = 0$, then A is not invertible.

4.2.6 *Proof.* We first consider the special case where one of the entries of A vanishes; since the Study determinant is invariant under column or row interchangings, without loss of generality $c = 0$. Then $[A] = |a|^2 |d|^2$ by a straightforward computation with the complex 4×4 matrix. If, on the other hand, none of the entries of A vanishes, then we write

$$\begin{pmatrix} a & b \\ c & d \end{pmatrix} = \begin{pmatrix} a & 0 \\ c & 1 \end{pmatrix} \cdot \begin{pmatrix} 1 & a^{-1}b \\ 0 & d - ca^{-1}b \end{pmatrix},$$

so that $[A] = |a|^2 |d - ca^{-1}b|^2 \in \mathbb{R}$ with $[A] \geq 0$, and

$$[A] = |a|^2 |d|^2 + |b|^2 |c|^2 - 2\mathrm{Re}\,(\bar{d}c\bar{a}b),$$

which yields the above formula since the real part of the quaternionic product is symmetric, $\mathrm{Re}(ab) = \mathrm{Re}(ba)$.

The formula for the inverse in case $[A] \neq 0$ is verified by direct computation. If $[A] = 0$, we again distinguish two cases: If A has an entry $= 0$, without loss of generality $c = 0$, then $[A] = 0$ implies either $a = 0$ or $d = 0$, and, therefore, A has a kernel and cannot be bijective; if, on the other hand, all entries $\neq 0$, then $[A] = 0$ implies $d = c \cdot a^{-1}b$, so that the columns of A are linearly dependent since $b = a \cdot a^{-1}b$: Again, A has a kernel and is not invertible. ◁

4.2.7 Thus, from the previous lemma, the Study determinant can be used to characterize the invertible linear transformations, that is, the general linear group $Gl(2, \mathbb{H})$. Moreover, since the Study determinant is real-valued, the corresponding quaternionic special linear group $Sl(2, \mathbb{H})$ has the same real dimension $\dim_{\mathbb{R}} Sl(2, \mathbb{H}) = 15$ as the Möbius group $M\ddot{o}b(4)$ of the conformal 4-sphere: We will learn later how $Sl(2, \mathbb{H})$ does indeed act by Möbius transformations on S^4.

4.2.8 Definition. *The general and special linear groups are given by:*

$$\begin{aligned} Gl(2, \mathbb{H}) &:= \{L \in End(\mathbb{H}^2) \,|\, [L] \neq 0\}, \\ Sl(2, \mathbb{H}) &:= \{L \in End(\mathbb{H}^2) \,|\, [L] = 1\}. \end{aligned}$$

4.3 Quaternionic Hermitian forms

This is the central section of our presentation: We will link the classical model of the Möbius geometry of S^4 to the quaternionic setup. Here the key idea is to use quaternionic Hermitian forms on $\mathbb{H}^2$ as the homogeneous coordinates for hyperspheres and points (cf., [262]). The rest will then just be a matter of translating the notions we developed when presenting the classical model.

4.3.1 Definition. *Let V denote a quaternionic (right) vector space. Then a map $S : V \times V \to \mathbb{H}$ is called a (quaternionic) Hermitian form if, for all $v, w, \tilde{w} \in V$ and $\lambda \in \mathbb{H}$,*

(i) $S(v, w) = \overline{S(w, v)}$, *and*

(ii) $S(v, w + \tilde{w}) = S(v, w) + S(v, \tilde{w})$ *and* $S(v, w\lambda) = S(v, w)\lambda$.

$\mathfrak{H}(V) := \{ S : V \times V \to \mathbb{H} \mid S$ *quaternionic Hermitian*$\}$ *will denote the space of quaternionic Hermitian forms.*

4.3.2 As usual, the linear structure on the target space $\mathbb{H} \cong \mathbb{R}^4$ induces a (real) linear structure on the space $\mathfrak{H}(V)$ of quaternionic Hermitian forms via

$$(S + \tilde{S})(v, w) := S(v, w) + \tilde{S}(v, w) \quad \text{and} \quad (r \cdot S)(v, w) := r \cdot S(v, w)$$

for all $v, w \in V$, where $S, \tilde{S} \in \mathfrak{H}(V)$ and $r \in \mathbb{R}$.

Fixing a basis $(e_1, \dots, e_n)$ of V, a quaternionic Hermitian form is determined by its values $S(e_i, e_j)$, $i \leq j$, on the basis vectors, where $S(e_i, e_j) \in \mathbb{H}$ for $i < j$ and $S(e_i, e_i) \in \mathbb{R}$. Consequently, a quaternionic Hermitian form depends on $n + 4\frac{n(n-1)}{2}$ real variables:

4.3.3 Lemma. *The space $\mathfrak{H}(V)$ of Hermitian forms on an n-dimensional quaternionic vector space V is a $(2n - 1)n$-dimensional real vector space.*

4.3.4 Now let $V \cong \mathbb{H}^2$ be a 2-dimensional quaternionic vector space. By the previous lemma, $\mathfrak{H}(V)$ is a 6-dimensional real vector space, $\mathfrak{H}(V) \cong \mathbb{R}^6$; next we want to make it a Minkowski space, that is, we want to introduce a Lorentz inner product. For that purpose we choose a basis (e_1, e_2) of V; then we define a symmetric bilinear form by its quadratic form[6)]

$$\mathfrak{H}(V) \to \mathbb{R}, \quad S \mapsto \langle S, S \rangle := |S(e_1, e_2)|^2 - S(e_1, e_1)S(e_2, e_2). \tag{4.1}$$

Clearly this scalar product has Lorentz signature $(5, 1)$.

Choosing a different basis $(\tilde{e}_1, \tilde{e}_2) = (e_1 b_{11} + e_2 b_{21}, e_1 b_{12} + e_2 b_{22})$, we use the same method as in the lemma in §4.2.5 to determine the relation of the associated scalar products $\langle ., . \rangle$ and $\langle ., . \rangle$: We decompose the change of basis in two,

$$
\begin{aligned}
(e_1, e_2) &\mapsto (e_1 b_{11} + e_2 b_{21}, e_2) =: (\hat{e}_1, \hat{e}_2) \quad \text{and} \\
(\hat{e}_1, \hat{e}_2) &\mapsto (\hat{e}_1, \hat{e}_1[b_{11}^{-1} b_{12}] + \hat{e}_2[b_{22} - b_{21} b_{11}^{-1} b_{12}]),
\end{aligned}
$$

[6)] Note that this is exactly the same way in which the (real) vector space of complex Hermitian matrices $\begin{pmatrix} x_0 + x_3 & x_1 + ix_2 \\ x_1 - ix_2 & x_0 - x_3 \end{pmatrix}$ is made into a 4-dimensional Minkowski space (cf., [42]).

where we assume $b_{11} \neq 0$, without loss of generality, because we may interchange the roles of $\tilde{e}_1$ and $\tilde{e}_2$. Then, for any $S \in \mathfrak{H}(V)$, we compute

$$
\begin{aligned}
\langle S, S \rangle &= & |b_{11}|^2 \langle S, S \rangle & \quad \text{and} \\
\langle S, S \rangle &= |b_{22} - b_{21} b_{11}^{-1} b_{12}|^2 \langle S, S \rangle &= [B] \cdot \langle S, S \rangle.
\end{aligned}
\tag{4.2}
$$

4.3.5 Lemma. *The scalar product* (4.1) *makes the space* $\mathfrak{H}(V)$ *of quaternionic Hermitian forms on a 2-dimensional quaternionic vector space* V *a 6-dimensional Minkowski space* $\mathbb{R}_1^6$; *this scalar product is invariantly defined up to (positive) rescaling.*[7]

4.3.6 As a consequence of this lemma, the causal type is a well-defined notion for quaternionic Hermitian forms, and so are the notions of the "absolute quadric" $S^4 \subset P\mathfrak{H}(V) = \mathfrak{H}(V)/\mathbb{R} \cong \mathbb{R}P^5$ and of "polarity" in that $\mathbb{R}P^5$. Hence $\mathfrak{H}(V)$ is a suitable classical model space for 4-dimensional Möbius geometry.

We want to elaborate on this further: Next we wish to establish the relation between the absolute quadric $S^4 \subset \mathbb{R}P^5$ and the quaternionic projective line

$$
\mathbb{H}P^1 := \{ v\mathbb{H} \mid v \in \mathbb{H}^2 \setminus \{0\} \} :
\tag{4.3}
$$

4.3.7 Lemma. *The null directions* $\mathbb{R}S \subset \mathfrak{H}(\mathbb{H}^2)$ *are in 1-to-1 correspondence with the points of the quaternionic projective line* $\mathbb{H}P^1$: $\mathbb{H}P^1 \cong S^4$.

4.3.8 *Proof.* For a nonnull isotropic Hermitian form $S \in \mathfrak{H}(\mathbb{H}^2)$, we have $|S(e_1, e_2)|^2 = S(e_1, e_1)S(e_2, e_2)$ and at least one, $S(e_1, e_1)$ or $S(e_2, e_2)$, does not vanish; without loss of generality, $S(e_1, e_1) \neq 0$.

Now suppose that $v = e_1 \lambda_1 + e_2 \lambda_2 \in \mathbb{H}^2 \setminus \{0\}$ is an isotropic vector for S. In particular, $v \nparallel e_1$ since $S(e_1, e_1) \neq 0$, so that (e_1, v) is a basis. Since the Minkowski product $\langle ., . \rangle$ on $\mathfrak{H}(\mathbb{H}^2)$ only changes by a real multiple when changing the basis ($|\lambda_2|^2$ in our case), we have

$$
0 = S(e_1, e_1)S(v, v) = |S(e_1, v)|^2;
$$

hence $S(e_1, v) = 0$. Consequently, $S(e_1, e_1)\lambda_1 + S(e_1, e_2)\lambda_2 = 0$, so that[8]

$$
v = -e_1 S(e_1, e_2)\lambda + e_2 S(e_1, e_1)\lambda \quad \text{with} \quad \lambda \in \mathbb{H} \setminus \{0\},
$$

that is, S has exactly one null direction $(-e_1 S(e_1, e_2) + e_2 S(e_1, e_1))\mathbb{H} \subset \mathbb{H}^2$.

---

[7] Remember, from §1.1.2, that we had the same scaling ambiguity when equipping the space $\mathbb{R}^{n+2}$ of homogeneous coordinates for the classical model of Möbius geometry with a Minkowski product.

[8] Remember that $S(e_1, e_1) \in \mathbb{R}$ commutes with any quaternion.

On the other hand, given $v = e_1 v_1 + e_2 v_2 \in \mathbb{H}^2 \setminus \{0\}$,

$$S_v(e_1, e_1) := |v_2|^2, \quad S_v(e_1, e_2) := -v_1 \bar{v}_2, \quad \text{and} \quad S_v(e_2, e_2) := |v_1|^2 \quad (4.4)$$

define an isotropic Hermitian form S_v having $v\mathbb{H}$ as its null direction.   ◁

4.3.9 Having established a 1-to-1 correspondence between $\mathbb{H}P^1$ and the conformal 4-sphere S^4, we need to characterize hyperspheres in $\mathbb{H}P^1$ and the incidence of hyperspheres $S \in S_1^5 \subset \mathfrak{H}(\mathbb{H}^2)$ and points $v\mathbb{H} \in \mathbb{H}P^1$. There are two strategies to attack the problem: a direct way, using affine coordinates, or using the characterizations from the classical model. We will indicate both approaches. Fix a basis (e_1, e_2) of $\mathbb{H}^2$ so that we have a well-defined Minkowski scalar product on $\mathfrak{H}(\mathbb{H}^2)$, and fix a unit $S \in \mathfrak{H}(\mathbb{H}^2)$.

4.3.10 To determine the form of S, we distinguish two cases, depending on whether $S(e_1, e_1) = 0$ or $S(e_1, e_1) \neq 0$:

$$S = \begin{pmatrix} 0 & -n \\ -\bar{n} & 2d \end{pmatrix} \quad \text{or} \quad S = \frac{1}{r}\begin{pmatrix} 1 & -m \\ -\bar{m} & |m|^2 - r^2 \end{pmatrix}$$

with suitable $n, m \in \mathbb{H}$ and $d, r \in \mathbb{R}$, respectively. Then, in terms of the affine coordinates, $v = e_1 q + e_2$ is isotropic for S if and only if $q \in \mathbb{H}$ lies on the plane with unit normal n and distance d from the origin or it has distance r from the point $m \in \mathbb{H}$:

$$S(v, v) = -\bar{q}n - \bar{n}q + 2d \quad \text{or} \quad S(v, v) = \frac{1}{r}\{|q - m|^2 - r^2\}.$$

4.3.11 In the classical model, the incidence of a point and a sphere is encoded by orthogonality (polarity; cf., Figure T.1). With the Hermitian form S_v, defined in (4.4) above, a straightforward computation yields

$$\langle S, S_v \rangle = -\tfrac{1}{2} S(v, v),$$

so that $v\mathbb{H} \in S^4 \cong \mathbb{H}P^1$ lies on the hypersphere $\mathbb{R}S \in \mathbb{R}P_O^5 \widetilde{\subset} P\mathfrak{H}(\mathbb{H}^2)$ if and only if $S(v, v) = 0$. Note that this condition does not depend on the choices S and v of homogeneous coordinates.

In either way we obtain the following:

4.3.12 Lemma. *The spacelike lines $\mathbb{R}S \subset \mathfrak{H}(\mathbb{H}^2)$ are in 1-to-1 correspondence with the hyperspheres $S \subset S^4$ via isotropy:*

$$v\mathbb{H} \in S \subset \mathbb{H}P^1 \cong S^4 \quad \Leftrightarrow \quad S(v, v) = 0.$$

4.4 The Möbius group

The general linear group $Gl(2, \mathbb{H})$ canonically acts on the space $\mathfrak{H}(\mathbb{H}^2)$ of quaternionic Hermitian forms via

$$Gl(2, \mathbb{H}) \times \mathfrak{H}(\mathbb{H}^2) \to \mathfrak{H}(\mathbb{H}^2), \quad (A, S) \mapsto AS := S(A^{-1}., A^{-1}.). \quad (4.5)$$

Fixing a basis (e_1, e_2) of $\mathbb{H}^2$ and, therefore, a Lorentz scalar product $\langle ., . \rangle$ on $\mathfrak{H}(\mathbb{H}^2)$, we have

$$[A]\langle AS, AS \rangle = \langle S, S \rangle;$$

namely, we clearly have $\langle S, S \rangle = \langle AS, AS \rangle^{\sim}$ with the scalar product $\langle ., . \rangle^{\sim}$ taken with respect to the basis $(\tilde{e}_1, \tilde{e}_2) = (Ae_1, Ae_2)$ (see (4.1)). The transformation formula (4.2) then yields the result. In particular,

4.4.1 Lemma. *After fixing a basis (e_1, e_2) of $\mathbb{H}^2$, the special linear group $Sl(2, \mathbb{H})$ canonically acts by isometries on $\mathfrak{H}(\mathbb{H}^2) \cong \mathbb{R}^6_1$.*

Two endomorphisms $A, \tilde{A} \in Sl(2, \mathbb{H})$ yield the same isometry if and only if $\tilde{A} = \pm A$.

4.4.2 *Proof.* The first part is clear from the above considerations; it remains to prove the second assertion.

Clearly, $AS = (-A)S$; for the converse, it suffices to show that $AS = S$ for all $S \in \mathfrak{H}(\mathbb{H}^2)$ implies $A = \pm id$. Since the isotropic Hermitian forms $S \in \mathfrak{H}(\mathbb{H}^2)$ are in 1-to-1 correspondence with the points of the conformal 4-sphere $\mathbb{H}P^1$, we deduce that A^{-1} induces the identity on $\mathbb{H}P^1$, that is, every vector $v \in \mathbb{H}^2$ is an eigenvector of A^{-1}. In particular, if (e_1, e_2) denotes a basis of $\mathbb{H}^2$, we have $A^{-1}e_i = e_i\lambda_i$. Defining $S \in \mathfrak{H}(\mathbb{H}^2)$ by

- $S(e_i, e_i) = 1$ and $S(e_1, e_2) = 0$, we deduce that $|\lambda_i|^2 = 1$; and,
- $S(e_i, e_i) = 0$ and $S(e_1, e_2) = c$, we find $\lambda_1 c = c\lambda_2$ for arbitrary $c \in \mathbb{H}$.

As a consequence, $\lambda_1 = \lambda_2 = \pm 1$. ◁

4.4.3 Corollary. *The action of $Sl(2, \mathbb{H})$ on $\mathbb{H}P^1$ preserves hyperspheres: $Sl(2, \mathbb{H})$ acts on $\mathbb{H}P^1 \cong S^4$ by Möbius transformations.*

4.4.4 This corollary follows with §4.3.7 from the previous lemma when employing the classical model of Möbius geometry (cf., Figure T.1), or it follows directly with the description of hyperspheres in $\mathbb{H}P^1$ and the characterization of incidence given in §4.3.12.

4.4.5 Using some basis (e_1, e_2) of $\mathbb{H}^2$, we may introduce affine coordinates

$$\mathbb{H}P^1 = \mathbb{H} \cup \{\infty\} = \{(e_1 q + e_2)\mathbb{H} \mid q \in \mathbb{H}\} \cup \{e_1\mathbb{H}\}$$

to find that the action $(A, v\mathbb{H}) \mapsto (Av)\mathbb{H}$ of $Sl(2, \mathbb{H})$ — or, equally, of $Gl(2, \mathbb{H})$ — on $\mathbb{H}P^1$ is by fractional linear transformations:

$$A(e_1 q + e_2)\mathbb{H} = (e_1(a_{11}q + a_{12})(a_{21}q + a_{22})^{-1} + e_2)\mathbb{H}.$$

Thus we succeeded to describe certain Möbius transformations as fractional linear transformations, just as in the complex case. Note that *real* multiples cancel in the action, so that we indeed obtain an action of the 15-dimensional group $PGl(2, \mathbb{H}) = Gl(2, \mathbb{H})/\mathbb{R} = Sl(2, \mathbb{H})/ \pm 1$.

4.4.6 So far we have learned that any transformation $A \in Sl(2, \mathbb{H})$ projects to a Lorentz transformation of $\mathbb{R}_1^6 \cong \mathfrak{H}(\mathbb{H}^2)$ and to a Möbius transformation of the conformal 4-sphere $S^4 \cong \mathbb{H}P^1$, respectively; moreover, we know that $A, \tilde{A} \in Sl(2, \mathbb{H})$ project to the same Lorentz transformation iff $\tilde{A} = \pm A$.

We want to study the structure of this map $Sl(2, \mathbb{H}) \to O_1(6)$, given by (4.5), and $Sl(2, \mathbb{H}) \to M\ddot{o}b(4)$, respectively, further. Clearly, the map

$$r : Sl(2, \mathbb{H}) \to O_1(6), \quad r(A)S := S(A^{-1}., A^{-1}.) \qquad (4.6)$$

is a group homomorphism, that is, $r(AB) = r(A)r(B)$. Consequently, its differential at the identity,

$$\varrho := d_{id}r : \mathfrak{sl}(2, \mathbb{H}) \to \mathfrak{o}_1(6), \quad \varrho(X)S = -S(X.,.) - S(.,X.), \qquad (4.7)$$

is a Lie algebra homomorphism: It is $\mathbb{R}$-linear and $[\varrho(X), \varrho(Y)] = \varrho([X,Y])$.

Moreover, we can show that ϱ is injective: Suppose $\varrho(X) = 0$, that is, $\varrho(X)S = 0$ for all $S \in \mathfrak{H}(\mathbb{H}^2)$. Writing $Xe_j = \sum_i e_i x_{ij}$ in terms of some basis (e_1, e_2) of $\mathbb{H}^2$ (see §4.1.7) and denoting $S(e_i, e_j) =: s_{ij}$, we have

$$0 = -\varrho(X)S(e_i, e_j) = \bar{x}_{1i}s_{1j} + s_{i1}x_{1j} + \bar{x}_{2i}s_{2j} + s_{i2}x_{2j}, \qquad (4.8)$$

where the $s_{ii} \in \mathbb{R}$ and $\bar{s}_{21} = s_{12} \in \mathbb{H}$ can be arbitrarily chosen: choosing $s_{12} = 0$ in the $(i,j) = (1,2)$-term we learn that $x_{12} = x_{21} = 0$; $s_{12} = 0$ in the (i,i)-terms yields $x_{ii} \in \text{Im}\,\mathbb{H}$; and then $s_{ii} = 0$ in the $(1,2)$-term reads $s_{12}x_{22} = -\bar{x}_{11}s_{12} = x_{11}s_{12}$ for all $s_{12} \in \mathbb{H}$, that is, $x_{22} = x_{11} \in \text{Im}\,\mathbb{H}$ $(s_{12} = 1)$ commutes with all $s_{12} \in \text{Im}\,\mathbb{H}$ or, otherwise said, $\forall s_{12} \in \mathbb{R}^3 : x_{11} \times s_{12} = 0$ with the cross-product $\times$ on $\mathbb{R}^3$, so that $x_{ii} = 0$ also.
Hence $X = 0$ and ϱ is shown to be injective.

Surjectivity then follows from a dimension argument: Considering quaternionic 2×2 matrices as complex 4×4 matrices, $\mathfrak{gl}(2, \mathbb{H}) \tilde{\subset} \mathfrak{gl}(4, \mathbb{C})$, our definition of $Sl(2, \mathbb{H})$ in §4.2.8 tells us that[9]

$$\mathfrak{sl}(2, \mathbb{H}) = \mathfrak{gl}(2, \mathbb{H}) \cap \mathfrak{sl}(4, \mathbb{C}) = \{X \in \mathfrak{gl}(2, \mathbb{H}) \mid \text{Re}\,\text{tr}\,X = 0\}$$

---

[9] Note that $\text{Re}\,\text{tr}\,X$ makes sense since the real part of a quaternionic product is symmetric: Hence the real part of the trace of a quaternionic endomorphism is independent of the choice of basis used to represent the endomorphism as a matrix.

has real dimension $\dim \mathfrak{sl}(2, \mathbb{H}) = 15 = \dim \mathfrak{o}_1(6)$: Note that the *complex* trace $\mathrm{tr}_{\mathbb{C}} : \mathfrak{gl}(2, \mathbb{H}) \to \mathbb{R}$ is real-valued (see §4.1.3).

Thus we have proved the following:

4.4.7 Lemma. *The differential* (4.7) *of the above map* $Sl(2, \mathbb{H}) \to O_1(6)$, *given in* (4.6), *is a real Lie algebra isomorphism* $\varrho : \mathfrak{sl}(2, \mathbb{H}) \to \mathfrak{o}_1(6)$.

4.4.8 Hence, by the preceding lemma, the identity component of $Sl(2, \mathbb{H})$ is a double cover of the identity component of $O_1(6)$ (see [298]). Since the identity component of $O_1(6)$ is just the orientation-preserving Möbius transformations,[10] we learn that the identity component of $Sl(2, \mathbb{H})$ is a double cover of the orientation-preserving Möbius transformations.

Hence, in order to obtain the following theorem, it remains to show that $Sl(2, \mathbb{H})$ is simply connected, that is, we have to show that it is connected and any loop $F : S^1 \to Sl(2, \mathbb{H})$ is homotopic to a constant loop.

4.4.9 Theorem. $Sl(2, \mathbb{H})$ *is the universal cover of the identity component of* $O_1(6)$; *as such, it is the (double) universal cover of the group* $M\ddot{o}b^+(4)$ *of orientation-preserving Möbius transformations of* $S^4 \cong \mathbb{HP}^1$.

4.4.10 *Proof.* $Sl(2, \mathbb{H})$ is connected: Let $A \in Sl(2, \mathbb{H})$; we want to show that there is a path joining A to the identity $id \in Sl(2, \mathbb{H})$.

Choose a basis (e_1, e_2) of $\mathbb{H}^2$ and consider the two points $Ae_i \mathbb{H} \in \mathbb{HP}^1$. Since the space of point pairs $\mathfrak{P} := \{(p_1, p_2) \mid p_i \in \mathbb{HP}^1, p_1 \neq p_2\}$ is, with $\mathbb{HP}^1$, connected, we can join the pair $(Ae_1\mathbb{H}, Ae_2\mathbb{H})$ to $(e_1\mathbb{H}, e_2\mathbb{H})$; a lift of this path to $Sl(2, \mathbb{H})$ joins A to an endomorphism B with eigenvectors e_1 and e_2, that is, we have $Be_i = e_i b_i$. $B \in Sl(2, \mathbb{H})$ implies $|b_1|^2|b_2|^2 = 1$. Since S^3 is connected, we can now join B to some endomorphism C with real positive "eigenvalues," $Ce_1 = e_1 e^c$ and $Ce_2 = e_2 e^{-c}$. Finally, a continuous rescaling joins C to the identity.

4.4.11 $Sl(2, \mathbb{H})$ is simply connected: Now let $A : S^1 \to Sl(2, \mathbb{H})$ be a loop in $Sl(2, \mathbb{H})$; we want to show that A is null homotopic.

Here we use the same idea as above: Choosing a basis (e_1, e_2) of $\mathbb{H}^2$, we obtain a loop $(Ae_1\mathbb{H}, Ae_2\mathbb{H}) : S^1 \to \mathfrak{P}$ in the space of point pairs. Since $\mathfrak{P}$

---

[10] Recall that the Lorentz group $O_1(6)$ decomposes into four components, determined by the orientation and the future orientation: The identity component consists of those transformations that preserve both orientation and future orientation. From §1.3.17 we know that any orientation-preserving Möbius transformation μ can be obtained as the composition of an even number of inversions, that is, of an even number of reflections in Minkowski hyperplanes in $\mathbb{R}_1^6$. Hence this composition $A \in O_1(6)$ preserves future orientation and has determinant $\det A = +1$. Therefore, μ comes from an element of the identity component of $O_1(6)$. Note that, equally, μ is given by $-A$, which is orientation-preserving but does not preserve future orientation.

is, with $\mathbb{H}P^1$, simply connected,[11] this loop can be contracted to a constant loop. Hence A is homotopic to a loop $B : S^1 \to Sl(2, \mathbb{H})$ that fixes, without loss of generality, the points $e_i \mathbb{H} \in \mathbb{H}P^1$. Then, by the simple connectedness of S^3, B is homotopic to a loop C with $Ce_1 = e_1 e^u$ and $Ce_2 = e_2 e^{-u}$ with some function $u : S^1 \to \mathbb{R}$, as above. Finally, a uniform scaling $u \to 0$ yields the homotopy to a constant loop in $Sl(2, \mathbb{H})$. ◁

4.4.12 In both parts of the proof of the previous theorem we made heavy use of the space of point pairs $\mathfrak{P}$ in $S^4 \cong \mathbb{H}P^1$. Thus this seems to be a good time to investigate the geometry of this space a bit further — in particular, as it will reappear when we discuss Darboux pairs of isothermic surfaces.

4.5 The space of point pairs

When describing m-spheres in the classical model, we already discussed 0-spheres, or "point pairs," as a special case: They could be identified with hyperbolic sphere pencils, that is, with Minkowski 2-planes in the classical coordinate Minkowski space $\mathbb{R}_1^{n+2}$ (see Figure T.1). Thus, as a Grassmannian, the space of point pairs is a (pseudo-Riemannian) symmetric space, as we already discussed briefly in the context of our lemma in §3.3.2; in particular, it is a reductive homogeneous space.

4.5.1 In the quaternionic setup the space of point pairs

$$\mathfrak{P} = \{(\hat{p}, p) \in \mathbb{H}P^1 \times \mathbb{H}P^1 \,|\, p \neq \hat{p}\}$$

can be described as follows[12]: Fixing two base points $\infty, 0 \in \mathbb{H}P^1$, any pair of points $p, \hat{p} \in \mathbb{H}P^1$ can be obtained as the image of the two base points under a Möbius transformation $\mu \in M\ddot{o}b(4)$, $\hat{p} = \mu(\infty)$ and $p = \mu(0)$. Of course, this Möbius transformation is not unique: Precomposing[13] μ with any Möbius transformation that has ∞ and 0 as fixed points provides another choice $\tilde{\mu}$ of a suitable Möbius transformation.

4.5.2 In terms of homogeneous coordinates, fixing the two base points $\infty, 0$ amounts to fixing a basis (e_∞, e_0) of $\mathbb{H}^2$, where $0 = e_0 \mathbb{H}$ and $\infty = e_\infty \mathbb{H}$,

---

[11] Note that, for a point pair map $(f_1, f_2) : M \to \mathfrak{P}$, one requires $f_1(p) \neq f_2(p)$ for all $p \in M$; it does *not* have to satisfy $f_1(M) \cap f_2(M) = \emptyset$.

[12] This particularly simple description may, in fact, be a reason why the transformation theory of isothermic surfaces, as described in the following chapter, is treated with such ease in the quaternionic model.

[13] Note that this is the same as postcomposing with a Möbius transformation that has p and $\hat{p}$ as fixed points: The Möbius transformations with $\hat{p}, p$ as fixed points are obtained from those with $\infty, 0$ as fixed points by conjugation with μ.

up to (quaternionic) rescalings of e_∞ and e_0. Then a point pair $(\hat{v}\mathbb{H}, v\mathbb{H})$ is given as the image of a transformation $A \in Sl(2, \mathbb{H})$, that is, up to scaling $v = Ae_0$ and $\hat{v} = Ae_\infty$. Precomposition with a Möbius transformation that has fixed points ∞ and 0 now becomes a precomposition of A with some $H \in Sl(2, \mathbb{H})$ that has e_∞ and e_0 as eigenvectors. Note how this gauge freedom exactly compensates for the ambiguity in choosing homogeneous coordinates e_∞ and e_0 for the base points ∞ and 0. Hence,

4.5.3 Lemma. *The space of point pairs in $S^4 \cong \mathbb{H}P^1$ is a homogeneous space*

$$\mathfrak{P} = Sl(2, \mathbb{H})/K_{\infty,0},$$

where $K_{\infty,0} := \{H \in Sl(2, \mathbb{H}) \,|\, He_i = e_i\lambda_i, i = 0, \infty\}$ denotes the isotropy group of some chosen base point $(\infty, 0) = (e_\infty\mathbb{H}, e_0\mathbb{H}) \in \mathfrak{P}$.

4.5.4 Notation. Note that the matrix M_A of A in terms of the basis (e_∞, e_0) contains the coordinates of the point pair $(Ae_\infty\mathbb{H}, Ae_0\mathbb{H})$ as its columns. For this reason, we will often write $A = (Ae_\infty, Ae_0)$.

4.5.5 We may identify the isotropy group $K_{\infty,0}$ of our chosen base point as the subgroup $\{A \in Sl(2, \mathbb{H}) \,|\, AR = RA\}$ of special linear transformations that commute with the reflection $R : (e_\infty, e_0) \mapsto (-e_\infty, e_0)$ in the base point.

Differentiating, we learn that the isotropy algebra $\mathfrak{k}_{\infty,0}$ is the $+1$-eigenspace of the involution $\mathrm{Ad}_R : \mathfrak{sl}(2, \mathbb{H}) \to \mathfrak{sl}(2, \mathbb{H})$. The -1-eigenspace of Ad_R then provides a canonical complement $\mathfrak{p}_{\infty,0}$ of $\mathfrak{k}_{\infty,0} \subset \mathfrak{sl}(2, \mathbb{H})$ and hence a model for the tangent space $T_{(\infty,0)}\mathfrak{P} \cong \mathfrak{sl}(2, \mathbb{H})/\mathfrak{k}_{\infty,0} \cong \mathfrak{p}_{\infty,0}$ of $\mathfrak{P}$ at the base point $(\infty, 0)$. Left translation $\mathfrak{p}_{\infty,0} \mapsto A\mathfrak{p}_{\infty,0}$ by $A \in Sl(2, \mathbb{H})$ with $Ae_\infty\mathbb{H} = \hat{p}$ and $Ae_0\mathbb{H} = p$ shifts that model to any tangent space $T_{(\hat{p},p)}\mathfrak{P}$ at $(\hat{p}, p) \in \mathfrak{P}$. Note that $Sl(2, \mathbb{H})$ acts transitively on $\mathfrak{P}$ and that $\mathfrak{p}_{\infty,0}$ is invariant under left translation by the isotropy group.

4.5.6 Since $\mathfrak{p}_{\infty,0} := \{X \in \mathfrak{sl}(2, \mathbb{H}) \,|\, \mathrm{Ad}_R X = -X\}$ is defined as the -1-eigenspace of Ad_R, the decomposition

$$\mathfrak{sl}(2, \mathbb{H}) = \mathfrak{k}_{\infty,0} \oplus \mathfrak{p}_{\infty,0}$$

is a symmetric decomposition of the Lie algebra $\mathfrak{sl}(2, \mathbb{H})$, that is, we have

$$[\mathfrak{k}_{\infty,0}, \mathfrak{k}_{\infty,0}] \subset \mathfrak{k}_{\infty,0}, \quad [\mathfrak{k}_{\infty,0}, \mathfrak{p}_{\infty,0}] \subset \mathfrak{p}_{\infty,0}, \quad [\mathfrak{p}_{\infty,0}, \mathfrak{p}_{\infty,0}] \subset \mathfrak{k}_{\infty,0}, \qquad (4.9)$$

making the space $\mathfrak{P}$ of point pairs a reductive homogeneous space (cf., (2.6)).

4.5.7 We now turn to maps into the space of point pairs: Often we will use frames to describe their geometry. Thus, in the rest of this section, we want to discuss frames, their gauge transformations, the structure equations, and compatibility conditions. In order to be able to "translate" results and computations from the classical setup (see Figures T.1 and T.2), we will finally derive a formula that gives the relation between a quaternionic connection form and the "classical connection form" (1.1).

4.5.8 Definition. *Let* $(\hat{f}, f) : M \to \mathfrak{P}$ *be a point pair map and fix a base point* $(\infty, 0) \in \mathfrak{P}$. *Then a map* $F : M \to Sl(2, \mathbb{H})$ *is called an adapted frame for the point pair map if* $(\hat{f}, f) = (Fe_\infty\mathbb{H}, Fe_0\mathbb{H})$.

4.5.9 Now let $(\hat{f}, f) : M \to \mathfrak{P}$ with adapted frame $F : M \to Sl(2, \mathbb{H})$. Then we may choose homogeneous coordinates for both maps, also denoted by f and $\hat{f}$, so that the frame can be identified with the (homogeneous coordinates of the) point pair map, $F = (\hat{f}, f) : M \to Sl(2, \mathbb{H})$.

The structure equations $dF = F\Phi$ then read

$$\Phi = \begin{pmatrix} \hat{\varphi} & \psi \\ \hat{\psi} & \varphi \end{pmatrix} \quad \Leftrightarrow \quad \begin{cases} d\hat{f} = \hat{f}\hat{\varphi} + f\hat{\psi} \\ df = \hat{f}\psi + f\varphi, \end{cases} \tag{4.10}$$

where $\Phi = F^{-1}dF : TM \to \mathfrak{sl}(2, \mathbb{H})$ denotes the connection form of F. A gauge transformation $(\hat{f}, f) \mapsto (\hat{f}\hat{a}, fa)$, that is, $F \mapsto \tilde{F} = FH$ with $H : M \to K_{\infty,0}$, $He_0 = e_0a$ and $He_\infty = e_\infty\hat{a}$, has the effect

$$\begin{aligned} d(\hat{f}\hat{a}) &= \hat{f}(d\hat{a} + \hat{\varphi}\hat{a}) &+& \quad f\hat{\psi}\hat{a} \\ d(fa) &= \hat{f}\psi a &+& \quad f(da + \varphi a) \end{aligned}$$

or, in terms of connection forms,

$$\tilde{\Phi} = H^{-1}(\Phi H + dH) = \begin{pmatrix} \hat{a}^{-1}(d\hat{a} + \hat{\varphi}\hat{a}) & \hat{a}^{-1}\psi a \\ a^{-1}\hat{\psi}\hat{a} & a^{-1}(da + \varphi a) \end{pmatrix}. \tag{4.11}$$

4.5.10 The integrability conditions (Maurer-Cartan equation) for the connection form (4.10) read[14]

$$0 = d\Phi + \tfrac{1}{2}[\Phi \wedge \Phi] \quad \Leftrightarrow \quad 0 = d^2f = d^2\hat{f}.$$

According to the symmetric decomposition (4.9) of $\mathfrak{sl}(2, \mathbb{H})$, the Maurer-Cartan equation splits into four "scalar" equations: The $\mathfrak{k}_{\infty,0}$-part yields the Gauss-Ricci equations

$$\begin{aligned} 0 &= d\hat{\varphi} + \hat{\varphi} \wedge \hat{\varphi} + \psi \wedge \hat{\psi}, \\ 0 &= d\varphi + \varphi \wedge \varphi + \hat{\psi} \wedge \psi, \end{aligned} \tag{4.12}$$

---

[14] Remember (see Footnote 1.33) that we defined $[\Phi \wedge \Psi](x, y) := [\Phi(x), \Psi(y)] - [\Phi(y), \Psi(x)]$.

while the $\mathfrak{p}_{\infty,0}$-part gives the Codazzi equations

$$
\begin{aligned}
0 &= d\psi + \psi \wedge \varphi + \hat{\varphi} \wedge \psi, \\
0 &= d\hat{\psi} + \hat{\psi} \wedge \hat{\varphi} + \varphi \wedge \hat{\psi}.
\end{aligned}
\tag{4.13}
$$

4.5.11 Now let $S, \tilde{S} \in \mathfrak{H}(\mathbb{H}^2) \cong \mathbb{R}_1^6$ denote two fixed (constant) quaternionic Hermitian forms. Using the fact that $r : Gl(2,\mathbb{H}) \to O_1(6)$ (defined in (4.6)) is a Lie group homomorphism, we obtain $d_p(r{\circ}F) = r(F(p))\varrho(\Phi|_p)$. Together with the fact that r takes values in the isometry group of $\mathbb{R}_1^6$, we then deduce

$$
\langle d(FS), F\tilde{S} \rangle = \langle \varrho(\Phi)S, \tilde{S} \rangle,
$$

where, for brevity, we write FS for $(r \circ F)S$ (cf., (4.5)). To compute the scalar product, we evaluate S and $\tilde{S}$ on a basis (e_∞, e_0) of $\mathbb{H}^2$ (see (4.1)) and find

$$
\begin{aligned}
\langle \varrho(\Phi)S, \tilde{S} \rangle &= \mathrm{Re}(\hat{\varphi})\tilde{S}(e_0,e_0)S(e_\infty,e_\infty) && - \quad \mathrm{Re}(\tilde{S}(e_\infty,e_0)S(e_0,e_\infty)\hat{\varphi}) \\
&+ \mathrm{Re}(\varphi)S(e_0,e_0)\tilde{S}(e_\infty,e_\infty) && - \quad \mathrm{Re}(\tilde{S}(e_0,e_\infty)S(e_\infty,e_0)\varphi) \\
&+ \mathrm{Re}(S(e_0,e_\infty)\psi)\tilde{S}(e_\infty,e_\infty) && - \quad \mathrm{Re}(\tilde{S}(e_0,e_\infty)\psi)S(e_\infty,e_\infty) \\
&+ \mathrm{Re}(S(e_\infty,e_0)\hat{\psi})\tilde{S}(e_0,e_0) && - \quad \mathrm{Re}(\tilde{S}(e_\infty,e_0)\hat{\psi})S(e_0,e_0).
\end{aligned}
$$

4.5.12 With this we are prepared to establish the relation between a classical $\mathfrak{o}_1(6)$-connection form (1.1) and its quaternionic counterpart (4.10): Choose a basis $(S_0, S_1, S_i, S_j, S_k, S_\infty)$ of $\mathfrak{H}(\mathbb{H}^2)$ such that, for $c = 1, i, j, k$,

$$
\begin{aligned}
1 &= S_0(e_\infty, e_\infty); & 0 &= S_0(e_\infty, e_0), & 0 &= S_0(e_0, e_0), \\
0 &= S_c(e_\infty, e_\infty); & \bar{c} &= S_c(e_\infty, e_0), & 0 &= S_c(e_0, e_0), \\
0 &= S_\infty(e_\infty, e_\infty), & 0 &= S_\infty(e_\infty, e_0), & 1 &= S_\infty(e_0, e_0).
\end{aligned}
$$

In this way the S's define a pseudo-orthonormal basis — however, in contrast to the basis we used in (1.1), here we have $\langle S_0, S_\infty \rangle = -\frac{1}{2}$. By combining this with the formulas from the previous paragraph, we finally can express the components of the connection form (4.10) in terms of the $\mathbb{R}_1^6$-moving frame:

$$
\begin{aligned}
\varphi &= \langle d(FS_\infty), FS_0 \rangle & &+ \tfrac{\langle d(FS_1),FS_i\rangle i + \langle d(FS_1),FS_j\rangle j + \langle d(FS_1),FS_k\rangle k}{2} \\
& & &+ \tfrac{\langle d(FS_j),FS_k\rangle i + \langle d(FS_k),FS_j\rangle j + \langle d(FS_i),FS_j\rangle k}{2}, \\
\hat{\varphi} &= \langle d(FS_0), FS_\infty \rangle & &- \tfrac{\langle d(FS_1),FS_i\rangle i + \langle d(FS_1),FS_j\rangle j + \langle d(FS_1),FS_k\rangle k}{2} \\
& & &+ \tfrac{\langle d(FS_j),FS_k\rangle i + \langle d(FS_k),FS_i\rangle j + \langle d(FS_i),FS_j\rangle k}{2}, \\
\psi &= -\langle d(FS_0), FS_1 \rangle \\
& \quad + \langle d(FS_0), FS_i \rangle i & &+ \langle d(FS_0), FS_j \rangle j + \langle d(FS_0), FS_k \rangle k, \\
\hat{\psi} &= -\langle d(FS_\infty), FS_1 \rangle \\
& \quad - \langle d(FS_\infty), FS_i \rangle i & &- \langle d(FS_\infty), FS_j \rangle j - \langle d(FS_\infty), FS_k \rangle k.
\end{aligned}
$$

This "translation table" provides the basis for carrying over the geometric interpretations in §1.7.6 or §1.7.7 of the structure equations in the classical setup to the quaternionic setup — in situations where f or $\hat{f}$ envelope one (or more) of the sphere congruences FS_1, FS_i, FS_j, or FS_k given by the frame. For example, this will become of interest when we discuss the isothermic transformations of surfaces of constant mean curvature in space forms in the next chapter.

4.6 The space of single points

Just as point pair maps will canonically turn up in various contexts, we will be interested in single maps into the conformal 4-sphere in other situations: There will not always be canonical partners for the surfaces that we are going to consider. Just as the space of point pairs, the conformal 4-sphere can be considered as a *Möb*(4)-homogeneous or a $Sl(2, \mathbb{H})$-homogeneous space, respectively. Fixing, as before, a base point $0 \in S^4$, any point $p \in S^4$ can be described as the image of a Möbius transformation $\mu \in Möb(4)$ that is unique up to precomposition with a Möbius transformation that fixes the base point 0:

4.6.1 Lemma. *The conformal 4-sphere $S^4 \cong \mathbb{H}P^1$ is a homogeneous space*

$$S^4 = Sl(2, \mathbb{H})/K_0, \quad where \quad K_0 := \{H \in Sl(2, \mathbb{H}) \,|\, He_0 = e_0\lambda_0\}$$

denotes the isotropy group of some chosen base point $0 = e_0\mathbb{H} \in S^4$.

4.6.2 Note that, choosing a basis (e_0, e_∞) of $\mathbb{H}^2$ with $0 = e_0\mathbb{H}$ in order to obtain a matrix representation, this isotropy subgroup K_0 consists of upper triangular matrices. Any change of basis that preserves the condition on the first vector providing homogeneous coordinates for the chosen base point is given by an element of K_0, that is, by an upper triangular matrix in terms of the old basis.

4.6.3 Also note that $S^4 = Sl(2, \mathbb{H})/K_0$ is a nonreductive homogeneous space: This corresponds to the fact that we are considering the *conformal* 4-sphere S^4. In fact, $S^4 = Sl(2, \mathbb{H})/K_0$ is a homogeneous $Sl(2, \mathbb{H})$-space with parabolic isotropy subgroup K_0, so that the conformal S^4 appears as a symmetric R-space (see [52] for details). Thus, in contrast to the situation of the space of point pairs, there is no symmetric decomposition of the Lie algebra $\mathfrak{sl}(2, \mathbb{H})$ nor a canonical complement of the isotropy algebra

$$\mathfrak{k}_0 = \{X \in \mathfrak{sl}(2, \mathbb{H}) \,|\, X0 \subset 0\} \subset \mathfrak{sl}(2, \mathbb{H})$$

to be identified with the tangent space[15] $T_0 S^4 \cong \mathfrak{sl}(2, \mathbb{H})/\mathfrak{k}_0$. However, after choosing a second point $\infty \in S^4 \setminus \{0\}$, there is a canonical complement

$$\mathfrak{p}_\infty = \{ X \in \mathfrak{sl}(2, \mathbb{H}) \, | \, \ker X = \operatorname{im} X = \infty \}$$

associated with that choice, and we recover our symmetric decomposition (4.9) relative to the choice of base point $(\infty, 0)$ in the space $\mathfrak{P}$ of point pairs with[16]

$$\mathfrak{k}_0 = \mathfrak{k}_{\infty,0} \oplus \mathfrak{p}_0, \quad \text{where} \quad \mathfrak{p}_0 = \{ X \in \mathfrak{k}_0 \, | \, \ker X = \operatorname{im} X = 0 \}.$$

Note that both subspaces $\mathfrak{p}_0$ and $\mathfrak{p}_\infty$ are $\mathrm{ad}_{\mathfrak{k}_{\infty,0}}$-stable, $[\mathfrak{k}_{\infty,0}, \mathfrak{p}_i] \subset \mathfrak{p}_i$.

4.6.4 As before, in the case of the space of point pairs, a main purpose of the above view on the conformal 4-sphere S^4 as a homogeneous space is to justify our view on frames for a map $f : M \to S^4$ as lifts $F : M \to Sl(2, \mathbb{H})$ of the map into the symmetry group of the ambient geometry, and to clarify the form of the corresponding gauge transformations.

4.6.5 Definition. *Fixing a base point $0 \in S^4$, a map $F : M \to Sl(2, \mathbb{H})$ is called an (adapted) frame for a map $f : M \to \mathbb{HP}^1$ if $F \cdot 0 = f$.*

4.6.6 After fixing a second base point $\infty \in S^4$, an adapted frame F provides a second map $\hat{f} = F \cdot \infty$, thus providing a point pair map. In terms of homogeneous coordinates $(\hat{f}, f) = (F e_\infty, F e_0)$, we again have the structure equations (4.10). However, now a gauge transformation is of the form $(\hat{f}, f) \mapsto (fb + \hat{f}a, fa)$, that is, $F \mapsto \tilde{F} = FH$ with $H : M \to K_0$, $H e_0 = e_0 a$, but $H e_\infty = e_\infty \hat{a} + e_0 b$. When computing the effect on the connection form, we find

$$\tilde{\Phi} = \begin{pmatrix} \hat{a}^{-1}[d\hat{a} + \varphi \hat{a} + \psi b] & \hat{a}^{-1} \psi a \\ a^{-1}(db + \varphi b + \hat{\psi}\hat{a} - b\hat{a}^{-1}[d\hat{a} + \varphi\hat{a} + \psi b]) & a^{-1}(da + \varphi a - b\hat{a}^{-1}\psi a) \end{pmatrix} \tag{4.14}$$

instead of (4.11).

Note that, as before, $\psi \mapsto \hat{a}^{-1}\psi a$.

4.7 Envelopes

Remember from §1.6.3 that a (differentiable) map $f : M^m \to S^n$ is called an envelope of a (hyper)sphere congruence $S : M^m \to \mathbb{RP}_O^{n+1}$, that is, of a (smooth) m-parameter family of hyperspheres in S^n, if, for all $p \in M$,

$$f(p) \in S(p) \quad \text{and} \quad T_{f(p)} f(M) \subset T_{f(p)} S(p) \, .$$

---

[15] This may indeed be considered to be the reason why, when discussing the classical model, we never referred to the tangent space of the conformal 4-sphere but always dealt with tangent lines or planes instead (cf., §1.1.4).

[16] The nilpotent subalgebra $\mathfrak{p}_0 \subset \mathfrak{k}_0$ does not depend on the choice of the second point ∞.

In terms of homogeneous coordinates (local lifts) $f : M^m \to L^{n+1} \subset \mathbb{R}_1^{n+2}$ and $S : M^m \to S_1^{n+1}$, these two conditions were translated to

$$\langle f, S \rangle = 0 \quad \text{and} \quad \langle df, S \rangle \equiv 0$$

(see §1.6.3), that is, to the fact that the pair $(f, S) : M^m \to L^{n+1} \times S_1^{n+1}$ forms a strip (see §1.6.12). In this section it is our mission to give a quaternionic formulation of the enveloping conditions and to extend the notion of an envelope to 2-sphere congruences in the conformal 4-sphere $\mathbb{HP}^1$, so that we will be prepared to study the geometry of (2-dimensional) surfaces in S^4.

4.7.1 In §4.3.12 we discussed the incidence relation of a point and a hypersphere: The image points of a map $f : M \to \mathbb{HP}^1$ lie on the spheres of a congruence $S : M \to S_1^5 \subset \mathfrak{H}(\mathbb{H}^2)$ if and only if $S(f, f) \equiv 0$.

There are several ways to obtain a quaternionic version of the second condition: One possibility is to use the explicit incarnation (4.4) of the correspondence between points in $\mathbb{HP}^1$ and isotropic lines in $\mathfrak{H}(\mathbb{H}^2)$ (see §4.3.7) to rewrite the above scalar product as $\langle df, S \rangle = -S(f, df) - S(df, f)$. Another possibility is to argue with the classical geometers: The map f envelopes the sphere congruence S if each point $f(p)$ lies in the intersection of the sphere $S(p)$ and any infinitesimally neighboring sphere $(S + dS)(p)$. Hence the enveloping conditions read

$$0 = S(f, f), \quad 0 = (S + dS)(f, f) = dS(f, f) = -S(f, df) - S(df, f), \tag{4.15}$$

where the last equality is obtained by differentiating $S(f, f) \equiv 0$.

4.7.2 Here, to entirely convince the reader, another rather affine argument shall be given: Given a sphere $S = S(p)$, we may locally choose a basis (e_∞, e_0) of $\mathbb{H}^2$ such that $S(e_\infty, e_\infty) \neq 0$, that is, $\infty = e_\infty \mathbb{H} \notin S$. Then S is of the form

$$S(e_\infty, e_\infty) = \tfrac{1}{r}, \quad S(e_\infty, e_0) = -\tfrac{1}{r} m, \quad \text{and} \quad S(e_0, e_0) = \tfrac{1}{r}(|m|^2 - r^2),$$

where $m \in \mathbb{H}$ and $r \in \mathbb{R}$ denote the center and radius of the sphere S in the affine Euclidean space $\mathbb{R}^4 \cong \mathbb{H} \cong \{e_\infty q + e_0 \mid q \in \mathbb{H}\}$ (see §4.3.10). Writing f in affine coordinates, too, $f = e_\infty \mathfrak{f} + e_0$, the two conditions in (4.15) read

$$0 = S(f, f) = \tfrac{1}{r}(|\mathfrak{f} - m|^2 - r^2) \quad \text{and} \quad 0 = \operatorname{Re} S(f, df) = \tfrac{1}{r} d\mathfrak{f} \cdot (\mathfrak{f} - m).$$

In either way we obtain the following:

4.7.3 Lemma. *A smooth map* $f : M \to \mathbb{HP}^1$ *envelopes*[17) *a (hyper)sphere congruence* $S : M \to S_1^5 \subset \mathfrak{H}(\mathbb{H}^2)$ *if and only if*

$$S(f, f) = 0 \quad and \quad S(f, df) + S(df, f) \equiv 0.$$

4.7.4 If $F : M \to Sl(2, \mathbb{H})$ is an adapted frame for the map $f : M \to \mathbb{HP}^1$, $f = Fe_0\mathbb{H}$, and $S = F\tilde{S} = (r \circ F)\tilde{S}$ with $\tilde{S} : M \to S_1^5 \subset \mathfrak{H}(\mathbb{H}^2)$ (see (4.6)), then the enveloping conditions reduce to

$$0 = \tilde{S}(e_0, e_0) \quad \text{and} \quad 0 = -(\varrho(\Phi)\tilde{S})(e_0, e_0) = 2\mathrm{Re}[\tilde{S}(e_0, e_\infty) \cdot \psi], \quad (4.16)$$

where $\varrho = d_{id}r : \mathfrak{sl}(2, \mathbb{H}) \to \mathfrak{o}_1(6)$ is the Lie algebra isomorphism discussed in (4.7) and ψ is the component form in the matrix representation (4.10) of Φ with respect to a basis (e_∞, e_0) of $\mathbb{H}^2$. This is a direct consequence of the lemma in §4.7.3 and of (4.7),

$$(F\tilde{S})(f, df) + (F\tilde{S})(df, f) = -(\varrho(\Phi)\tilde{S})(e_0, e_0).$$

When the sphere congruence is $S = FS_c$, with $S_c \in \mathfrak{H}(\mathbb{H}^2)$ being one of the fixed basis spheres used in the "translation formulas" of §4.5.12,

$$S_c(e_0, e_0) = 0, \quad S_c(e_0, e_\infty) = c \in \{1, i, j, k\}, \quad \text{and} \quad S_c(e_\infty, e_\infty) = 0,$$

then the first enveloping condition is automatic, and the second condition in (4.16) becomes $\mathrm{Re}(c\psi) = 0$. Comparison with the translation formula in §4.5.12 recovers the classical condition from §1.6.3.

4.7.5 Clearly a similar formula holds as the enveloping condition for a second map $\hat{f}$ in the case where F frames a point pair map $(\hat{f}, f) : M \to \mathfrak{P}$, that is, $f = Fe_0\mathbb{H}$ and also $\hat{f} = Fe_\infty\mathbb{H}$.

For example, consider the sphere congruence $S = FS_1$. Then S is enveloped by $f = Fe_0\mathbb{H}$ as well as $\hat{f} = Fe_\infty\mathbb{H}$ if and only if

$$\mathrm{Re}\,\psi = \mathrm{Re}\,\hat{\psi} = 0. \quad (4.17)$$

4.7.6 Now suppose FS_1 is a *fixed* 3-sphere, $S^4 \supset S^3 \simeq FS_1 \in \mathfrak{H}(\mathbb{H}^2)$. Then $d(FS_1) = 0$, and we may assume[18) $FS_1 \equiv S_1$ without loss of generality, that is, $F : M \to M\ddot{o}b(3)$. Since $d(FS_1) = 0$, the formulas from §4.5.12 show that

$$\mathrm{Re}\,\psi = \mathrm{Re}\,\hat{\psi} \equiv 0 \quad \text{and} \quad \hat{\varphi} + \bar{\varphi} \equiv 0$$

[17) Note that these conditions do not depend on any choice of homogeneous coordinates.

[18) This can be achieved by possibly postcomposing F with a (constant) Möbius transformation $A \in Sl(2, \mathbb{H})$.

or, otherwise said, $\Phi : TM \to \mathfrak{möb}(3)$, where

$$\mathfrak{möb}(3) = \{X \in \mathfrak{sl}(2, \mathbb{H}) \mid \varrho(X)S_1 = 0\}.$$

Note that with this choice, $S^3 \simeq S_1$, we have $S^3 = \operatorname{Im} \mathbb{H} \cup \{\infty\} \subset \mathbb{H} \cup \{\infty\}$.

4.7.7 In the sequel we will often be interested in surfaces $f : M^2 \to S^3$. If $F : M^2 \to M\ddot{o}b(3) \subset Sl(2, \mathbb{H})$ denotes an adapted frame for a codimension 1 surface $f : M^2 \to S^3$, then f envelopes $S = FS_i$ if and only if

$$\psi : TM^2 \to \mathbb{C}j, \quad \text{and similarly} \quad \hat{\psi} : TM^2 \to \mathbb{C}j \qquad (4.18)$$

for $\hat{f}$, since by construction the points of $f = Fe_0\mathbb{H}$ (as well as $\hat{f}$) lie on the spheres $S = FS_i$ of the congruence under consideration, $S_i(e_0, e_0) = 0$ as well as $S_i(e_\infty, e_\infty)$.

Here we can consider the spheres $S = FS_i$ as 2-spheres in $S^3 \simeq S_1$ since we are restricting our attention to the geometry of the 3-dimensional ambient space S^3. Thus we identify each 3-sphere $FS_i(p)$ with the 2-sphere given by its intersection $S(p) \cap S^3$ with $S^3 \simeq S_1$, that is, with the elliptic sphere pencil given by $\operatorname{span}\{FS_1(p), FS_i(p)\}$ (see Figure T.1).

4.7.8 Dropping the assumption on FS_1 to be constant, this situation generalizes and, in this way, provides a notion of envelope for a congruence of 2-spheres in $S^4 = \mathbb{H}P^1$. Note that, in the above situation, $f = Fe_0\mathbb{H}$ (and $\hat{f} = Fe_\infty\mathbb{H}$) also envelope the 3-sphere $S^3 \simeq FS_1$, so that f (and $\hat{f}$) envelope all hypersphere congruences in the congruence of elliptic sphere pencils simultaneously.

Thus we consider a congruence of 2-spheres in S^4 as a map into the space of elliptic sphere pencils, that is, into the Grassmannian $G_+(6, 2)$ of spacelike 2-planes in $\mathfrak{H}(\mathbb{H}^2) \cong \mathbb{R}_1^6$ (cf., Figure T.1). In this way, it appears as a rank 2 vector bundle over M^2. Choosing local (orthogonal) basis fields for this vector bundle, $S, \tilde{S} : M^2 \supset U \to S_1^5$, the 2-spheres of the congruence are described as the (orthogonal) intersection of 3-spheres.

4.7.9 Definition. *A map $f : M^2 \to S^4 = \mathbb{H}P^1$ is said to envelope a congruence $S : M^2 \to G_+(6, 2)$ of 2-spheres (elliptic sphere pencils) if it envelopes every congruence of 3-spheres in that pencil congruence.*

4.7.10 Given two maps f and $\hat{f}$ so that $f(p), \hat{f}(p) \in S(p) \cap \tilde{S}(p)$ at each point $p \in M^2$, we may choose homogeneous coordinates $f, \hat{f} : M^2 \to \mathbb{H}^2$ for the two maps such that $S(f, \hat{f}) \equiv 1$ and $\tilde{S}(f, \hat{f}) \equiv i$, as soon as S and $\tilde{S}$ intersect orthogonally.

Namely, if $f, \hat{f}$ denote *any* homogeneous coordinates of the two maps, we may assume that $S(f, \hat{f}) \equiv 1$ by possibly replacing $\hat{f}$ by $\hat{f}(S(f, \hat{f}))^{-1}$. Then any relative rescaling of f and $\hat{f}$ that preserves $S(f, \hat{f}) = 1$ is of the form

$(f, \hat{f}) \mapsto (f\bar{a}, \hat{f}a^{-1})$ with some function a; for the other sphere congruence $\tilde{S}$, such a rescaling yields $\tilde{S}(f, \hat{f}) \mapsto a\tilde{S}(f, \hat{f})a^{-1}$. But, this is exactly how S^3 acts on its Lie algebra $\mathbb{R}^3$ (cf., §4.1.4): Since this (adjoint) action is transitive, we can achieve $\tilde{S}(f, \hat{f}) \equiv i$ by a suitable choice of a, $|a| \equiv 1$.

A choice of basis (e_∞, e_0) of $\mathbb{H}^2$ gives rise to a frame $F : (e_\infty, e_0) \mapsto (\hat{f}, f)$, where f and $\hat{f}$ denote the above lifts of the maps into the 4-sphere. Then we have $S = FS_1$ and $\tilde{S} = FS_i$. Moreover, since the action (4.5) of $Gl(2, \mathbb{H})$ on $\mathfrak{H}(\mathbb{H}^2)$ satisfies $[F]|FS|^2 = |S|^2$, we conclude that $[F] = \frac{|S_1|^2}{|S|^2} = 1$: The frame constructed in this way is $Sl(2, \mathbb{H})$-valued, $F : M^2 \to Sl(2, \mathbb{H})$.

We collect these rather technical results for later reference:

4.7.11 Lemma. *Given a congruence of 2-spheres $S : M^2 \to G_+(6, 2)$ and two maps $f, \hat{f} : M^2 \to \mathbb{HP}^1$, so that $f(p), \hat{f}(p) \in S(p)$ at each point $p \in M^2$, there is (locally) an adapted frame $F : M^2 \to Sl(2, \mathbb{H})$ for $f = Fe_0\mathbb{H}$ and $\hat{f} = Fe_\infty\mathbb{H}$ such that $S = \mathrm{span}\{FS_1, FS_i\}$.*

4.7.12 As a consequence, the condition on a surface $f : M^2 \to \mathbb{HP}^1$ to envelope a congruence $S : M^2 \to G_+(6, 2)$ of 2-spheres can (locally) always be written in the form (4.18), after an appropriate choice of frame. Note that we can always extend f to a map into the space of point pairs such that both maps have their points on the 2-spheres of a given congruence.

4.8 Involutions and 2-spheres

There is another way of describing 2-sphere congruences in S^4 and their enveloping conditions (cf., [49]). The key idea is to identify a sphere with its inversion, that is, with a Möbius involution that has this sphere as its fixed point set. This is a rather universal idea that can be used for any codimension spheres in S^n (cf., §1.3.21) or $S^4 = \mathbb{HP}^1$ (cf., [262]). However, in the quaternionic setup, $Sl(2, \mathbb{H})$ describes the *orientation-preserving* Möbius transformations, so that the corresponding involution can only be described by an element of $Sl(2, \mathbb{H})$ if the codimension of the sphere is even.[19]

This is completely analogous to the complex situation, where an involution can be identified with its two fixed points — which form a codimension 2 sphere in S^2. For example, the involution $z \mapsto \frac{1}{z}$ can be identified with its fixed point set $\{-1, 1\}$.

4.8.1 Lemma. *Any 2-sphere $S \subset S^4$ can be identified with the unique Möbius involution that has this 2-sphere as its fixed point set.*

---

[19] Otherwise, one has to deal with orientation-reversing Möbius transformations, for example, with quaternionic conjugation. In contrast to the complex picture, quaternionic conjugation is an antiautomorphism, so that the formalism becomes relatively unpleasant (cf., [262], [304]).

Any such involution comes, unique up to sign, from some

$$\mathcal{S} \in \mathfrak{S}(\mathbb{H}^2) := \{\mathcal{S} \in \text{End}(\mathbb{H}^2) \,|\, \mathcal{S}^2 = -id\} \subset Sl(2, \mathbb{H}),$$

so that we obtain a double cover $\mathfrak{S}(\mathbb{H}^2) \to G_+(6, 2)$ *of the space of 2-spheres.*

4.8.2 *Proof.* At this point we will give a straightforward proof using suitable affine coordinates. However, another interesting proof of this lemma will turn up when examining the relation of the description of a 2-sphere as the fixed point set of a Möbius involution versus its description by an elliptic hypersphere pencil (that is, a 2-dimensional linear subspace of quaternionic Hermitian forms) below.

First note that we have $[\mathcal{S}] = +\sqrt{[\mathcal{S}^2]} = 1$ for any $\mathcal{S} \in \mathfrak{S}(\mathbb{H}^2)$, so that, consequently, $\mathfrak{S}(\mathbb{H}^2) \subset Sl(2, \mathbb{H})$ and $\mathcal{S} \in \mathfrak{S}(\mathbb{H}^2)$ is a Möbius involution indeed.

We will need the fact that, given $\mathcal{S} \in \mathfrak{S}(\mathbb{H}^2)$, there is some $v\mathbb{H} \in \mathbb{H}P^1$ not in the fixed point set $\text{Fix}(\mathcal{S})$ of $\mathcal{S}$. Suppose this was not the case, that is, suppose *any* vector $v \in \mathbb{H}^2$ is an eigenvector of $\mathcal{S}$. Then, in particular, there is a basis (e_1, e_2) of eigenvectors, $\mathcal{S}e_i = e_i\lambda_i$. Also, for any $q \in \mathbb{H}$, there is a $\lambda \in \mathbb{H}$ such that $(e_1q + e_2)\lambda = \mathcal{S}(e_1q + e_2) = e_1\lambda_1q + e_2\lambda_2$. Eliminating λ we deduce $\lambda_1q = q\lambda_2$ for all $q \in \mathbb{H}$. Hence $\lambda_1 = \lambda_2 \in \mathbb{R}$. On the other hand, $0 = (id + \mathcal{S}^2)e_1 = e_1(1 + \lambda_1^2)$, which is not possible if $\lambda_1 \in \mathbb{R}$.

4.8.3 Now fix $\mathcal{S} \in \mathfrak{S}(\mathbb{H}^2)$ and choose some $v \in \mathbb{H}^2$ that is *not* an eigenvector of $\mathcal{S}$. Set $e_0 := v$ and $e_\infty := \mathcal{S}v$. Then (e_∞, e_0) provides a basis of $\mathbb{H}^2$, and every eigenvector of $\mathcal{S}$ is of the form $e_\infty q + e_0$ for some $q \in \mathbb{H} \setminus \{0\}$: $e_\infty = \mathcal{S}e_0$ cannot be an eigenvector of $\mathcal{S}$ because e_0 is not an eigenvector. Eliminating λ from the condition $-e_0q + e_\infty = \mathcal{S}(e_\infty q + e_0) = (e_\infty q + e_0)\lambda$ that $e_\infty q + e_0$ is an eigenvector, we find $q^2 = -1$, that is, $|q| = 1$ and $q \in \text{Im}\,\mathbb{H} = \mathbb{R}^3$. Hence the fixed point set of $\mathcal{S}$ is a 2-sphere in $\mathbb{H}P^1$.

To construct $\mathcal{S}$ from a given 2-sphere we just reverse this computation: Note that, given a 2-sphere in $\mathbb{H}P^1$, we can always choose[20] a basis (e_∞, e_0) of $\mathbb{H}^2$ so that this 2-sphere is $\{e_\infty q + e_0 \,|\, q^2 + 1 = 0\}$. Then we define $\mathcal{S}$ by

$$\mathcal{S}e_0 := e_\infty \quad \text{and} \quad \mathcal{S}e_\infty := -e_0. \tag{4.19}$$

Clearly, $\mathcal{S} \in \mathfrak{S}(\mathbb{H}^2)$ and the Möbius involution $\mathcal{S}$ has the given 2-sphere as its fixed point set by the above argument.

4.8.4 It remains to prove the uniqueness statement. For this suppose that $\mathcal{S} \in \mathfrak{S}(\mathbb{H}^2)$ has $\{e_\infty q + e_0 \,|\, q^2 + 1 = 0\}$ as its fixed point set, as in the

---

[20] Otherwise said, after fixing affine coordinates on $\mathbb{H}P^1 = \{e_\infty q + e_0 \,|\, q \in \mathbb{H}\} \cup \{e_\infty \mathbb{H}\}$ there is a Möbius transformation that maps the given 2-sphere to the unit sphere in $\text{Im}\,\mathbb{H} \subset \mathbb{H}P^1$.

previous paragraph. We wish to deduce that (4.19) holds up to sign. Thus, for each q with $q^2 = -1$, there is some λ_q with $\mathcal{S}(e_\infty q + e_0) = (e_\infty q + e_0)\lambda_q$; the fact that $\mathcal{S}^2 = -id$ implies that $\lambda_q^2 = -1$. By using the eigenvector condition for antipodal points q and $-q$, we obtain two equations that we solve for $\mathcal{S}e_i$:

$$
\begin{aligned}
2\mathcal{S}e_0 &= e_0 \left(\lambda_q + \lambda_{-q}\right) & + & \quad e_\infty q(\lambda_q - \lambda_{-q}), \\
-2\mathcal{S}e_\infty &= e_0 \left(\lambda_q - \lambda_{-q}\right)q & + & \quad e_\infty q(\lambda_q + \lambda_{-q})q.
\end{aligned}
\tag{4.20}
$$

Observe that neither the e_i's nor the $\mathcal{S}e_i$'s depend on q; hence, $\lambda_q + \lambda_{-q} =: \tilde{c}$ does not depend on q, and neither does $q\tilde{c}q$. Choosing $q \perp \tilde{c} \in \operatorname{Im} \mathbb{H}$, we deduce that $q\tilde{c}q = -q^2\tilde{c} = \tilde{c}$. But this equation has to hold for all q with $q^2 = -1$, since $q\tilde{c}q$ does not depend on q — which is only possible if $\tilde{c} = 0$. Hence $\lambda_{-q} = -\lambda_q$, and (4.20) reduces to

$$\mathcal{S}e_0 = e_\infty q\lambda_q \quad \text{and} \quad \mathcal{S}e_\infty = -e_0 \lambda_q q,$$

where $q\lambda_q =: c$ and $-\lambda_q q = qcq$ are constant. Note that this time $|c| = 1$. Choosing $q \perp c$, we find $qcq \equiv \bar{c}$. Hence $c \perp \operatorname{Im} \mathbb{H}$, so that $c = \pm 1$, and (4.20) indeed reduces to (4.19), up to sign. ◁

4.8.5 Our next goal is to understand the relation between the two descriptions of 2-spheres in $\mathbb{HP}^1$ that we have at hand. For this let $v \in \mathbb{H}^2$ be an eigenvector of some given $\mathcal{S} \in \mathfrak{G}(\mathbb{H}^2)$, $\mathcal{S}v = v\lambda$. Then

$$(|\lambda|^2 + 1) S(v, v) = S(\mathcal{S}v, \mathcal{S}v) - S(v, \mathcal{S}^2 v)$$

for any $S \in \mathfrak{H}(\mathbb{H}^2)$. Hence $\mathcal{S}$ is contained in S as soon as the endomorphism $\mathcal{S}$ is symmetric with respect to the quaternionic Hermitian form S.

To see the converse, assume that any eigenvector of $\mathcal{S}$ is isotropic for S. Choosing a basis (e_∞, e_0) as in (4.19), so that $\mathcal{S}(e_\infty q + e_0) = (e_\infty q + e_0) q$ for all q with $q^2 = -1$, we find

$$
\begin{aligned}
0 &= S(e_\infty q + e_0, e_\infty q + e_0) \\
&= \{S(e_0, e_0) + S(e_\infty, e_\infty)\} + \{S(e_0, e_\infty) q + \overline{S(e_0, e_\infty) q}\}.
\end{aligned}
$$

Consequently, since this holds for all q with $q^2 + 1 = 0$ so that both terms have to vanish separately,

$$S(e_\infty, \mathcal{S}e_0) - S(\mathcal{S}e_\infty, e_0) = S(e_\infty, e_\infty) + S(e_0, e_0) = 0 \quad \text{and}$$

$$
\left.
\begin{aligned}
S(e_0, \mathcal{S}e_0) - S(\mathcal{S}e_0, e_0) \\
S(e_\infty, \mathcal{S}e_\infty) - S(\mathcal{S}e_\infty, e_\infty)
\end{aligned}
\right\} = S(e_0, e_\infty) - S(e_\infty, e_0) = 0.
\tag{4.21}
$$

Hence $\mathcal{S}$ is symmetric with respect to S and we have established:

4.8.6 Lemma. *Let $\mathcal{S} \in \mathfrak{S}(\mathbb{H}^2)$ be a 2-sphere and $S \in S_1^5 \subset \mathfrak{H}(\mathbb{H}^2)$ a 3-sphere in S^4. Then $\mathcal{S}$ is contained in S if and only if the endomorphism $\mathcal{S}$ is symmetric with respect to the quaternionic Hermitian form S.*

4.8.7 In this situation, $\mathfrak{S}(\mathbb{H}^2) \ni \mathcal{S} \subset S \in S_1^5 \subset \mathfrak{H}(\mathbb{H}^2)$, we can easily construct a second 3-sphere $\mathcal{J}S$ that contains $\mathcal{S}$ and intersects the given 3-sphere $S \in S_1^5$ orthogonally. First observe that if $\mathcal{S}$ is symmetric with respect to S, then

$$\mathcal{J}S := S(.,\mathcal{S}.) \in \mathfrak{H}(\mathbb{H}^2) \tag{4.22}$$

is a quaternionic Hermitian form so that $\mathcal{S}$ is symmetric with respect to $\mathcal{J}S$, too: By using the symmetry[21] of $\mathcal{S}$ with respect to S, it is straightforward to check both,

$$\mathcal{J}S(w,v) = \overline{\mathcal{J}S(v,w)} \quad \text{and} \quad \mathcal{J}S(\mathcal{S}v,w) = \mathcal{J}S(v,\mathcal{S}w).$$

Note that $S \mapsto \mathcal{J}S$ defines a complex structure on the (real) vector space of quaternionic Hermitian forms that make $\mathcal{S}$ symmetric since $\mathcal{J}^2 S = -S$. In particular, S and $\mathcal{J}S$ are linearly independent (over the reals).

To verify the orthogonality claim we choose a basis that gives the scalar product on $\mathfrak{H}(\mathbb{H}^2)$ up to scale: Thus, let (e_∞, e_0) be a basis as in (4.19). Then

$$\begin{aligned}
\mathcal{J}S(e_0, e_0) &= S(e_0, e_\infty) &= -\mathcal{J}S(e_\infty, e_\infty), \\
S(e_\infty, e_\infty) &= \mathcal{J}S(e_0, e_\infty) &= -S(e_0, e_0).
\end{aligned}$$

In particular, $S(e_0, e_\infty), \mathcal{J}S(e_0, e_\infty) \in \mathbb{R}$ are real and, up to scale,

$$|S|^2 = (S(e_0, e_\infty))^2 + (\mathcal{J}S(e_0, e_\infty))^2 = |\mathcal{J}S|^2, \quad |S + \mathcal{J}S|^2 = |S|^2 + |\mathcal{J}S|^2.$$

Consequently, $\mathcal{J}S \perp S$ are orthogonal and have the same length — as we might have expected for a complex structure.

4.8.8 Later, for a motivation of the Vahlen matrix model, we will use this construction to identify the whole coordinate Minkowski space $\mathbb{R}_1^5$ of 3-dimensional Möbius geometry with certain quaternionic endomorphisms that are symmetric with respect to a fixed 3-sphere $S^3 \in \mathfrak{H}(\mathbb{H}^2)$.

At this point it may serve the interested reader as an exercise to show that the Minkowski space $\mathbb{R}_1^5 \cong (S^3)^\perp \subset \mathfrak{H}(\mathbb{H}^2)$ can be identified with the (real) vector space of endomorphisms of $\mathbb{H}^2$ that are symmetric with respect to S^3 and square to a (real) multiple of the identity.

4.8.9 By the lemma in §4.8.1 and geometry, we already know that there is a 2-dimensional spacelike subspace of quaternionic Hermitian forms (an

---

[21] In fact, the symmetry of $\mathcal{S}$ is *equivalent* to $\mathcal{J}S$ being quaternionic Hermitian.

elliptic sphere pencil) that make a given 2-sphere $S \in \mathfrak{S}(\mathbb{H}^2)$ symmetric. However, it is now a simple *algebraic* matter to prove this statement.

Namely, if (e_∞, e_0) denotes a basis as before, $e_\infty = Se_0$, the computation in (4.21) showed that S is symmetric with respect to $S \in \mathfrak{H}(\mathbb{H}^2)$ if and only if

$$S(e_0, e_0) + S(e_\infty, e_\infty) = 0 \quad \text{and} \quad S(e_0, e_\infty) \in \mathbb{R}.$$

Hence, given S, we may define S by

$$S(e_0, e_0) = S(e_\infty, e_\infty) := 0 \quad \text{and} \quad S(e_0, e_\infty) := 1$$

to obtain an elliptic sphere pencil $\mathrm{span}\{S, \mathcal{J}S\}$, where $\mathcal{J}S$ is given by (4.22), that consists of 3-spheres that make S symmetric.

4.8.10 To complete this alternative proof we should show that any elliptic sphere pencil gives rise to some $S \in \mathfrak{S}(\mathbb{H}^2)$. To do so, let $S_1, S_2 \in S_1^5$ be two orthogonal hyperspheres that span the sphere pencil and choose a basis (e_∞, e_0) of $\mathbb{H}^2$ such that $S_i(e_0, e_0) = S_i(e_\infty, e_\infty) = 0$ for $i = 1, 2$, that is, in contrast to our choices above, $e_0\mathbb{H}$ and $e_\infty\mathbb{H}$ are now points on the 2-sphere under consideration. By suitably normalizing the basis vectors, we may assume $S_1(e_0, e_\infty) = 1$. The orthogonality of S_1 and S_2 then implies that $S_2(e_0, e_\infty) =: \lambda \in \mathrm{Im}\,\mathbb{H}$, and $|S_1|^2 = |S_2|^2$ gives $|\lambda|^2 = 1$, so that $\lambda^2 = -1$. With the ansatz $Se_j = \sum_i e_i s_{ij}$, $i, j \in \{0, \infty\}$, the conditions that S is symmetric with respect to S_1 and S_2 compute to

$$
\begin{array}{rclcrcl}
s_{\infty 0} & = & \bar{s}_{\infty 0} & \text{and} & \lambda s_{\infty 0} & = & -\bar{s}_{\infty 0}\lambda, \\
s_{\infty \infty} & = & \bar{s}_{00} & \text{and} & \lambda s_{\infty \infty} & = & \bar{s}_{00}\lambda, \\
s_{0\infty} & = & \bar{s}_{0\infty} & \text{and} & -\lambda s_{0\infty} & = & \bar{s}_{0\infty}\lambda,
\end{array}
$$

showing that S should be given by

$$Se_0 = \pm e_0\lambda \quad \text{and} \quad Se_\infty = \mp e_\infty\lambda$$

since $S^2 = -id$.

Thus, to an elliptic sphere pencil given by $S_1, S_2 \in S_1^5$, there is an endomorphism $S \in \mathfrak{S}(\mathbb{H}^2)$, unique up to sign, that is symmetric for all quaternionic Hermitian forms of the pencil. This completes the alternative proof of our lemma in §4.8.1.

4.8.11 Now let $\mathcal{K} \in \mathbb{R}_1^6 \setminus \{0\}$ be a sphere complex — remember that such a sphere complex defines a quadric Q_κ^4 of constant curvature $\kappa = -|\mathcal{K}|^2$ (see Section 1.4). Further, let $S \in \mathfrak{S}(\mathbb{H}^2)$ be a 2-sphere. We want to investigate

$$\mathcal{K}_S := \tfrac{1}{2}\{\mathcal{K}(., S.) + \mathcal{K}(S., .)\}.$$

Note that $\mathcal{K}_\mathcal{S}$ is quaternionic Hermitian.

Let $S \in S_1^5$ be a 3-sphere that contains $\mathcal{S}$ and choose a basis (e_∞, e_0) that consists of eigenvectors of S, $Se_i = e_i\lambda_i$ for $i = 0, \infty$. We normalize the basis vectors[22] so that $S(e_0, e_\infty) = 1$. As a consequence, $\lambda_\infty = -\lambda_0 =: \lambda$, since S is symmetric with respect to S, and

$$S(e_0, e_0) = 0, \quad S(e_0, e_\infty) = 1, \quad S(e_\infty, e_\infty) = 0,$$
$$JS(e_0, e_0) = 0, \quad JS(e_0, e_\infty) = \lambda, \quad JS(e_\infty, e_\infty) = 0.$$

With this basis we find

$$
\begin{aligned}
2\mathcal{K}_\mathcal{S}(e_0, e_\infty) &= \mathcal{K}(e_0, e_\infty)\lambda + \lambda\mathcal{K}(e_0, e_\infty) \\
&= \{\mathcal{K}(e_0, e_\infty) + \mathcal{K}(e_\infty, e_0)\}\lambda + \{\lambda\mathcal{K}(e_0, e_\infty) - \mathcal{K}(e_\infty, e_0)\lambda\} \\
&= 2\{\langle \mathcal{K}, S \rangle \, JS + \langle \mathcal{K}, JS \rangle \, S\}(e_0, e_\infty).
\end{aligned}
$$

Since, moreover, $\mathcal{K}_\mathcal{S}(e_i, e_i) = 0$ for $i = 0, \infty$, we have equality of the two quaternionic Hermitian forms,

$$\mathcal{K}_\mathcal{S} = \langle \mathcal{K}, S \rangle \, JS + \langle \mathcal{K}, JS \rangle \, S. \qquad (4.23)$$

Thus $|\mathcal{K}_\mathcal{S}|^2 \geq 0$ and $|\mathcal{K}_\mathcal{S}|$ yields the curvature of the 2-sphere $\mathcal{S}$ in Q_κ^4: By choosing, for example, S to be a totally geodesic 3-sphere in the elliptic sphere pencil[23] given by $\mathcal{S}$, we have $|\mathcal{K}_\mathcal{S}| = |\langle \mathcal{K}, JS \rangle|$, which is, up to sign (orientation), the mean curvature of the 3-sphere JS that intersects the hyperplane S in $\mathcal{S}$ orthogonally (cf., p. 60).

In particular, $\mathcal{S} \subset Q_\kappa^4$ is totally geodesic if and only if $\mathcal{K}_\mathcal{S} = 0$:

4.8.12 Lemma. *Let $\mathcal{K} \in \mathfrak{H}(\mathbb{H}^2) \cong \mathbb{R}_1^6$ be a linear sphere complex in the conformal 4-sphere S^4, and let $\mathcal{S} \in \mathfrak{S}(\mathbb{H}^2)$ be a 2-sphere. Then $\mathcal{S}$ is a 2-plane in the quadric Q_κ^4 of constant curvature defined by $\mathcal{K}$ if and only if*

$$\mathcal{K}_\mathcal{S} = \tfrac{1}{2}\{\mathcal{K}(., \mathcal{S}.) + \mathcal{K}(\mathcal{S}., .)\} = 0,$$

that is, if and only if $\mathcal{S}$ is skew with respect to $\mathcal{K}$.

4.8.13 Note that (4.23) also yields another criterion for the incidence of a point $p \in S^4$ and a 2-sphere $\mathcal{S} \subset S^4$: If the point p is given as an (isotropic) quaternionic Hermitian form, $p \simeq S_p$ (cf., §4.3.7), then $p \in \mathcal{S}$ if and only if $\mathcal{S}$ is skew with respect to S_p.

---

[22] Note that this normalization ensures that (e_∞, e_0) is a "correct" basis to compute the scalar products of quaternionic Hermitian forms: $|S|^2 = |S(e_0, e_\infty)|^2 - S(e_0, e_0)S(e_\infty, e_\infty) = 1 - 0$, so that $S \in S_1^5$ is correctly detected with the basis at hand, (cf., §4.3.5).

[23] Any elliptic sphere pencil intersects the space $\mathcal{K}^\perp$ of hyperplanes in Q_κ^4 (cf., §1.4.11).

4.8.14 We want to conclude this section by formulating the enveloping condition in the setup of this section. The incidence condition is already clear from the above discussions: The points of a map $f : M^2 \to I\!H P^1$ lie on the corresponding 2-spheres of a congruence $\mathcal{S} : M^2 \to \mathfrak{S}(I\!H^2)$ if and only if $f(p)$ is a fixed point of the Möbius transformation $\mathcal{S}(p)$. In terms of homogeneous coordinates, also denoted by f, this means that $\mathcal{S}f = f\lambda$ with a function λ that satisfies $\lambda^2 = -1$ since $\mathcal{S}^2 = -id$.

By differentiating the incidence condition $\mathcal{S}f = f\lambda$, we obtain

$$d\mathcal{S}\, f \;=\; df\,\lambda - \mathcal{S}df + f d\lambda. \tag{4.24}$$

Now remember from §4.7.9 that f envelopes $\mathcal{S}$ if and only if it envelopes any of the congruences of hyperspheres that contain the 2-spheres $\mathcal{S}$. Thus choose (locally) a 3-sphere congruence $S : M^2 \to \mathfrak{H}(I\!H^2)$ so that $\mathcal{S}$ is symmetric with respect to S: Then f envelopes $\mathcal{S}$ if and only if it envelopes S as well as $\mathcal{J}S$, that is, according to §4.7.3, if and only if

$$\operatorname{Re} S(df, f) = 0 \quad \text{and} \quad \operatorname{Re} \mathcal{J}S(df, f) = \operatorname{Re}(S(df, f)\lambda) = 0.$$

These two equations are equivalent to $0 = S(df, f)\lambda + \lambda S(df, f)$. Combination with (4.24) then yields $0 = S(d\mathcal{S}\, f, f)$ as the enveloping condition. Since S is spacelike, we have $\{v \in I\!H^2 \mid S(v, f) = 0\} = \operatorname{span}\{f\}$, and we obtain the following:

4.8.15 Lemma. *A map $f : M^2 \to I\!H P^1$ envelopes a 2-sphere congruence[24] given by $\mathcal{S} : M^2 \to \mathfrak{S}(I\!H^2)$ if and only if*

$$\mathcal{S}f \parallel f \quad \text{and} \quad d\mathcal{S} \cdot f \parallel f.$$

4.9 The cross-ratio

An important invariant of four[25] points $p_i = v_i \, \mathbb{C} \in \mathbb{C}P^1$, $i = 1, \dots, 4$, is the cross-ratio. When we write $v_i = \left(\begin{smallmatrix} z_i \\ 1 \end{smallmatrix}\right)$, the cross-ratio[26] is given by

$$[p_1; p_2; p_3; p_4] = \tfrac{z_1 - z_2}{z_2 - z_3} \tfrac{z_3 - z_4}{z_4 - z_1} = \tfrac{\det(v_1, v_2)}{\det(v_2, v_3)} \tfrac{\det(v_3, v_4)}{\det(v_4, v_1)}.$$

---

[24] Note that $\mathfrak{S}(I\!H^2)$ is a double cover of the space of 2-spheres (cf., §4.8.1). Thus a 2-sphere congruence may not lift to a map into $\mathfrak{S}(I\!H^2)$. A simple example is provided by the 1-parameter family of curvature spheres enveloped by a channel Klein bottle in S^3. However, the enveloping condition is a local matter, so that we can formulate it in this way without loss of generality.

[25] Given any two sets z_1, z_2, z_3 and w_1, w_2, w_3 of three points in S^2, there is a Möbius transformation μ mapping one set to the other, $\mu(z_i) = w_i$; for example, $z \mapsto \frac{z - z_1}{z - z_3} \frac{z_2 - z_3}{z_2 - z_1}$ will map any given three points z_1, z_2, z_3 to $0, 1, \infty$. As a consequence, there can be no Möbius invariant for three points because any configurations of three points are equivalent in Möbius geometry.

[26] This definition of the cross-ratio differs from the classical one by the order of the points.

Note that the right-hand side term is independent of the choice of homogeneous coordinates; moreover, it shows that the cross-ratio is invariant under Möbius transformations: If $\mu \in Sl(2, \mathbb{C})$ denotes a Möbius transformation, then the cross-ratio $[\mu(p_1); \mu(p_2); \mu(p_3); \mu(p_4)] = [p_1; p_2; p_3; p_4]$.

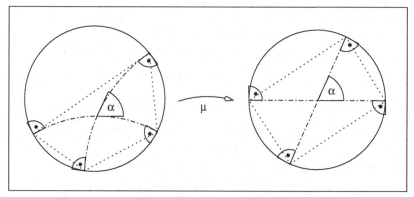

Fig. 4.1. Rectangles in Möbius geometry

Using a Möbius transformation μ to map the three points p_2, p_3, p_4 to $0, 1, \infty$, the cross-ratio of the four points computes to $\mu(p_1)$. Consequently, the four points are concircular (lie on a circle) if and only if their cross-ratio $[p_1; p_2; p_3; p_4] = \mu(p_1)$ is real, and it equals -1 if and only if the two circles intersecting their circle orthogonally in p_1, p_3, and p_2, p_4, respectively, intersect orthogonally (cf., Figure 4.1) — in that case, the two point pairs p_1, p_3 and p_2, p_4 are said to separate harmonically. These notions will become crucial when we discuss discrete isothermic surfaces in the next chapter.

In the sequel we will reformulate these facts in the quaternionic setting.

4.9.1 Given four points $p_i = v_i \mathbb{H} \in \mathbb{H}P^1$, $i = 1, 2, 3, 4$, in the conformal 4-sphere, we may just use the middle term of the above formula to define the quaternionic cross-ratio (cf., [30]): Writing, as above, $v_i = \binom{\mathfrak{p}_i}{1}$ with suitable quaternions $\mathfrak{p}_i \in \mathbb{H}$,

$$cr_{[p_1; p_2; p_3; p_4]} := (\mathfrak{p}_1 - \mathfrak{p}_2)(\mathfrak{p}_2 - \mathfrak{p}_3)^{-1}(\mathfrak{p}_3 - \mathfrak{p}_4)(\mathfrak{p}_4 - \mathfrak{p}_1)^{-1} \qquad (4.25)$$

provides a generalization for the cross-ratio of four points in $\mathbb{C}P^1$. However, to discuss the invariance of $cr_{[p_1; p_2; p_3; p_4]}$ under Möbius transformations, it would be convenient to express this quantity more invariantly, as above.[27] This will be our next goal.

---

[27] Note that we do not have a notion of a "quaternionic determinant": The Study determinant is real valued, so that we would loose (crucial) information by using it.

4.9.2 Lemma. *Given a point $p = v\mathbb{H} \in \mathbb{HP}^1$, there is a unique line of linear forms*[28]

$$p^\perp := \mathbb{H}\nu \in (\mathbb{HP}^1)^\perp := \{\mathbb{H}\nu \,|\, \nu \in (\mathbb{H}^2)^\star \setminus \{0\}\}$$

so that $p^\perp p = 0$.

The map $p \mapsto p^\perp$ *provides an identification* $\mathbb{HP}^1 \leftrightarrow (\mathbb{HP}^1)^\perp$, *and the induced action of the Möbius group* $M\ddot{o}b^+(4)$ *on* $(\mathbb{HP}^1)^\perp$ *is given by*

$$Sl(2, \mathbb{H}) \times (\mathbb{H}^2)^\star \ni (\mu, \nu) \mapsto \nu \circ \mu^{-1} \in (\mathbb{H}^2)^\star.$$

4.9.3 *Proof.* Let $p = v\mathbb{H} \in \mathbb{HP}^1$, $0 \neq v \in \mathbb{H}^2$, be given; let $w \in \mathbb{H}^2$ so that (v, w) is a basis of $\mathbb{H}^2$; and define $\nu \in (\mathbb{H}^2)^\star$ by $\nu(v) := 0$ and $\nu(w) := 1$. Clearly, $\mathbb{H}\nu \in (\mathbb{HP}^1)^\perp$ and $\mathbb{H}\nu(v\mathbb{H}) = 0$. If $\tilde{\nu} \in (\mathbb{H}^2)^\star \setminus \{0\}$ with $\tilde{\nu}(v) = 0$, then $\tilde{\nu} = \tilde{\nu}(w)\nu$, that is, $\mathbb{H}\tilde{\nu} = \mathbb{H}\nu$.

It is clear that the above action of $Sl(2, \mathbb{H})$ on $(\mathbb{H}^2)^\star$ descends to an action of the Möbius group on $(\mathbb{HP}^1)^\perp$ that is compatible with the above identification $\mathbb{HP}^1 \leftrightarrow (\mathbb{HP}^1)^\perp$. ◁

4.9.4 Note that, by using the notions from the lemma in §4.9.2, we may define a "stereographic projection" $\mathbb{HP}^1 \to \mathbb{H}$ (cf., §1.4.4): By choosing homogeneous coordinates $v_0, v_\infty \in \mathbb{H}^2$ for two points $0, \infty \in \mathbb{HP}^1$, we define uniquely two unique linear forms $\nu_0, \nu_\infty \in (\mathbb{H}^2)^\star$ by[29]

$$\nu_0 v_0 = 0, \quad \nu_0 v_\infty = 1 \quad \text{and} \quad \nu_\infty v_0 = 1, \quad \nu_\infty v_\infty = 0,$$

the "pseudodual basis" of the basis (v_∞, v_0) of $\mathbb{H}^2$. Note that the defining equations also can be written as $v_\infty \nu_0 + v_0 \nu_\infty = id_{\mathbb{H}^2}$. Clearly, the map

$$\mathbb{H}^2 \setminus \{\infty\} \ni v \mapsto (\nu_0 v)(\nu_\infty v)^{-1} \in \mathbb{H}$$

descends to a map $\mathbb{HP}^1 \setminus \{\infty\} \to \mathbb{H}$ with $\mathbb{HP}^1 \ni (v_0 + v_\infty \mathfrak{p})\mathbb{H} \mapsto \mathfrak{p} \in \mathbb{H}$. Equivalently, $\mathfrak{p} = -(\nu v_\infty)^{-1}(\nu v_0)$, where $\mathbb{H}\nu = (v\mathbb{H})^\perp$ as in §4.9.2:

$$(\nu v_\infty)^{-1}(\nu v_0) + (\nu_0 v)(\nu_\infty v)^{-1} = 0$$
$$\Leftrightarrow (\nu v_0)(\nu_\infty v) + (\nu v_\infty)(\nu_0 v) = \nu v = 0.$$

A change of basis (v_∞, v_0), while keeping $\infty \in \mathbb{HP}^1$ fixed, results in

$$\left.\begin{array}{rcl} \tilde{v}_0 &=& v_0 a_0 + v_\infty b \\ \tilde{v}_\infty &=& v_\infty a_\infty \end{array}\right\} \quad \leftrightarrow \quad \left\{\begin{array}{rcl} \tilde{\nu}_0 &=& a_\infty^{-1}\nu_0 - a_\infty^{-1}ba_0^{-1}\nu_\infty \\ \tilde{\nu}_\infty &=& a_0^{-1}\nu_\infty \end{array}\right.$$

---

[28] Remember (see §4.1.9) that we consider the dual space $(\mathbb{H}^2)^\star$ of $\mathbb{H}^2$ to be a *left* vector space.

[29] For short, we write $\nu v := \nu(v)$ for the natural contraction of $v \in \mathbb{H}^2$ with $\nu \in (\mathbb{H}^2)^\star$.

(cf., Section 4.6). Then

$$(\tilde{\nu}_0 v)(\tilde{\nu}_\infty v)^{-1} = a_\infty^{-1}(\nu_0 v)(\nu_\infty v)^{-1} a_0 - a_\infty^{-1} b,$$

that is, our projection $\mathbb{HP}^1 \to \mathbb{H}$ changes by a similarity.

4.9.5 Definition and Lemma. *Let $v_\infty, v_0 \in \mathbb{H}^2$ be a choice of homogeneous coordinates for a pair of points $\infty, 0 \in \mathbb{HP}^1$. Then the map*

$$\sigma : \mathbb{HP}^1 \setminus \{\infty\} \to \mathbb{H}, \quad p = v\mathbb{H} \mapsto \mathfrak{p} := (\nu_0 v)(\nu_\infty v)^{-1} = -(\nu\nu_\infty)^{-1}(\nu\nu_0),$$

where (ν_∞, ν_0) is the pseudodual basis of (v_∞, v_0), $v_\infty\nu_0 + v_0\nu_\infty = id_{\mathbb{H}^2}$, will be called a stereographic projection. The point $\infty \in \mathbb{HP}^1$ determines the stereographic projection up to postcomposition by a similarity.

4.9.6 Before coming back to our main topic in this section, the cross-ratio, we want to briefly examine this stereographic projection in the 3-dimensional case. Thus, let $S \in \mathfrak{H}(\mathbb{H}^2)$ be a quaternionic Hermitian form that determines a 3-sphere in $\mathbb{HP}^1$, and let $0, \infty$ be points in that 3-sphere so that $S(v_0, v_0) = S(v_\infty, v_\infty) = 0$ (see §4.3.12). Then $S(v_0, v_\infty) = 1$ is a canonical relative normalization for v_0 and v_∞ that causes S to have length $|S| = 1$ when using the basis (v_∞, v_0) to determine the scalar product (4.1) on $\mathfrak{H}(\mathbb{H}^2)$. This relative scaling of v_0 and v_∞ also causes the pseudodual basis of (v_∞, v_0) to be given by

$$\nu_\infty = S(v_\infty, .) \quad \text{and} \quad \nu_0 = S(v_0, .).$$

If $v = v_0(\nu_\infty v) + v_\infty(\nu_0 v) = (v_0 + v_\infty\mathfrak{p})(\nu_\infty v)$ are homogeneous coordinates for a point p of our 3-sphere S, $p \neq \infty$, then

$$S(v, v) = \overline{(\nu_\infty v)}(\mathfrak{p} + \bar{\mathfrak{p}})(\nu_\infty v) = 0 \quad \Leftrightarrow \quad \mathfrak{p} + \bar{\mathfrak{p}} = 0,$$

that is, the stereographic projection takes values in $\mathbb{R}^3 \cong \operatorname{Im} \mathbb{H}$.

4.9.7 To come back to our quaternionic cross-ratio (4.25), we first express the difference of the stereographic projections of two points $p_1 = v_1\mathbb{H}$ and $p_2 = v_2\mathbb{H}$ by using our newly developed technology:

$$\mathfrak{p}_2 - \mathfrak{p}_1 = (\nu_1 v_\infty)^{-1}(\nu_1 v_0) + (\nu_0 v_2)(\nu_\infty v_2)^{-1} = (\nu_1 v_\infty)^{-1}(\nu_1 v_2)(\nu_\infty v_2)^{-1}, \tag{4.26}$$

since $v_0\nu_\infty + v_\infty\nu_0 = id_{\mathbb{H}^2}$. Substituting (4.26) into (4.25) yields

$$cr_{[p_1;p_2;p_3;p_4]} = (\nu_1 v_\infty)^{-1}(\nu_1 v_2)(\nu_3 v_2)^{-1}(\nu_3 v_4)(\nu_1 v_4)^{-1}(\nu_1 v_\infty).$$

This quantity is clearly independent of the choice of homogeneous coordinates v_i and ν_i for the points $p_i \simeq p_i^\perp$, and it is invariant under the action of the Möbius group (see §4.9.2). A different choice $\tilde{v}_\infty = v_\infty a_\infty$ of homogeneous coordinates for ∞, however, causes a conjugation of $cr_{[p_1;p_2;p_3;p_4]}$:

$$v_\infty \to \tilde{v}_\infty$$
$$\Rightarrow \quad cr_{[p_1;p_2;p_3;p_4]} \to (\nu_1\tilde{v}_\infty)^{-1}(\nu_1 v_\infty)cr_{[p_1;p_2;p_3;p_4]}(\nu_1 v_\infty)^{-1}(\nu_1\tilde{v}_\infty).$$

Thus the quantity $cr_{[p_1;p_2;p_3;p_4]}$ itself is not Möbius invariant, and there seems to be no way to assign Möbius geometric meaning to (the direction of) its imaginary part.[30] However, its real part as well as the length of its imaginary part are Möbius invariant, so that we obtain

$$\operatorname{Re} cr_{[p_1;p_2;p_3;p_4]} + i\left|\operatorname{Im} cr_{[p_1;p_2;p_3;p_4]}\right|$$

as the "correct" invariant of four points.

Note that, in case $p_i \in \mathbb{CP}^1 \subset \mathbb{HP}^1$, this just becomes the usual complex cross-ratio that we discussed at the beginning of this section — up to complex conjugation. Since, on the other hand, four points always lie on a 2-sphere that can be identified with $\mathbb{CP}^1 \subset \mathbb{HP}^1$ via a Möbius transformation, our quaternionic cross-ratio is just the usual complex cross-ratio of the four points when considering this 2-sphere as a Riemann sphere. The ambiguity of orientation in this identification is reflected by the sign ambiguity in the imaginary part — which we brutally resolve by always choosing the imaginary part to be positive.

4.9.8 Lemma. *Let $p_1 \in \mathbb{HP}^1 \cong (\mathbb{HP}^1)^\perp$, $i = 1, \ldots, 4$, be four points and let $v_2, v_4 \in \mathbb{H}^2$ and $\nu_1, \nu_3 \in (\mathbb{H}^2)^\star$ denote homogeneous coordinates for p_i. Then*

$$\operatorname{Re}(\nu_1 v_2)(\nu_3 v_2)^{-1}(\nu_3 v_4)(\nu_1 v_4)^{-1} + i\left|\operatorname{Im}(\nu_1 v_2)(\nu_3 v_2)^{-1}(\nu_3 v_4)(\nu_1 v_4)^{-1}\right|$$

is, up to complex conjugation, the cross-ratio of the four points when interpreting a 2-sphere that contains the four points as a Riemann sphere.

4.9.9 This complex cross-ratio will become the key tool in our research of discrete isothermic nets in the following chapter. Because we have identified it as the usual complex cross-ratio, the two geometric properties that we discussed in the introduction to this section can be directly reformulated in the quaternionic setting:

(i) Four points $p_1, \ldots, p_4 \in \mathbb{HP}^1$ are concircular, that is, they form a "conformal rectangle" (see Figure 4.1), if and only if their cross-ratio is real, $[p_1;p_2;p_3;p_4] \in \mathbb{R}$.

---

[30] This is in contrast to a Clifford algebra version of the cross-ratio that we will discuss in Chapter 6: There, the direction of the "imaginary part" of a quantity that provides the cross-ratio can be interpreted as the (unique if it does not vanish) 2-sphere that contains the four points.

(ii) Four concircular points $p_1, \ldots, p_4$ form a "conformal square,[31]" that is, the pairs p_1, p_3 and p_2, p_4 separate harmonically on their circle, if and only if $[p_1; p_2; p_3; p_4] = -1$.

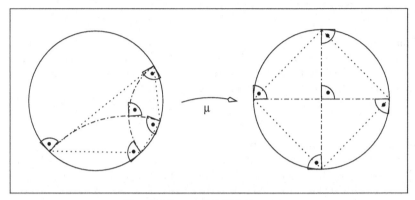

Fig. 4.2. Squares in Möbius geometry

To justify the notions "conformal rectangle" and "conformal square," respectively, first note that it is clear that (a Möbius transform of) a rectangle (in some Euclidean plane) has concircular vertices; to see that any quadrilateral with concircular vertices is Möbius equivalent to a rectangle (in a suitably chosen Euclidean plane), consider the two circles that intersect the vertex circle in opposite vertices (see Figure 4.1) and take one of their intersection points as the point at infinity. The other intersection point then becomes the center of the vertex circle, and the quadrilateral becomes a rectangle in the corresponding Euclidean plane. Since these two "diagonal" circles of a conformal rectangle are the honest (Euclidean) diagonals of the rectangle once the point at infinity is chosen in the described way, it becomes obvious that a conformal rectangle is a conformal square if and only if these diagonal circles intersect orthogonally.

In this sense we define

4.9.10 Definition. *A quadrilateral $(p_1; p_2; p_3; p_4)$ in the conformal sphere S^n is called*

(i) *a conformal rectangle if $[p_1; p_2; p_3; p_4] \in \mathbb{R}$, and*

(ii) *a conformal square if $[p_1; p_2; p_3; p_4] = -1$.*

4.9.11 We also obtain the usual permutation rules for the cross-ratio, with

---

[31] As in the conformal rectangle case, one may use the cross-ratio condition as a definition. Geometrically, a conformal square can be characterized as a conformal rectangle whose "conformal diagonals" intersect orthogonally, so that it becomes a "proper" (Euclidean) square after a suitable choice of ∞.

a slight alteration that is caused by our convention for the sign of the imaginary part of the cross-ratio:

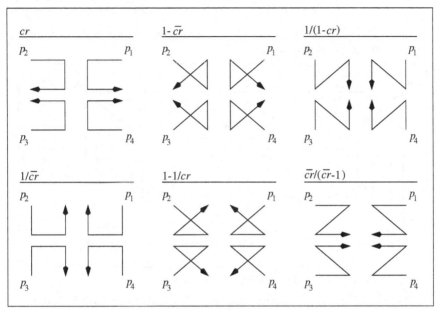

Fig. 4.3. Identities of the complex cross-ratio

4.9.12 For four points $p_1, \ldots, p_4 \in \mathbb{C}P^1$ in the complex plane, one of the points can be reconstructed from the other three and a prescribed cross-ratio — in fact, it is a Möbius transformation of z_4 that provides the missing point:

$$[p_1; p_2; p_3; p_4] = \frac{z_1 - z_2}{z_2 - z_3} \frac{z_3 - z_4}{z_4 - z_1} \quad \Leftrightarrow \quad z_1 = \frac{([p_1; p_2; p_3; p_4] - \frac{z_2}{z_2 - z_3}) z_4 + \frac{z_2}{z_2 - z_3} z_3}{-\frac{1}{z_2 - z_3} z_4 + ([p_1; p_2; p_3; p_4] + \frac{1}{z_2 - z_3} z_3)}.$$

For points in the conformal 4-sphere, this is obviously not true, since the 2-sphere of the four points cannot be determined from the cross-ratio and three points. However, if the cross-ratio is real, so that the fourth point must lie on the circle determined by the given points, then the fourth point can be reconstructed by a similar computation: By writing $p_i = \binom{\mathfrak{p}_i}{1} \mathbb{H}$, that is, the $\mathfrak{p}_i$'s are stereographic projections of the p_i, $i = 2, 3, 4$, we obtain

$$p_1 = \left(\begin{matrix} ([p_1; p_2; p_3; p_4] - \mathfrak{p}_2(\mathfrak{p}_2 - \mathfrak{p}_3)^{-1})\mathfrak{p}_4 + \mathfrak{p}_2(\mathfrak{p}_2 - \mathfrak{p}_3)^{-1}\mathfrak{p}_3) \\ -(\mathfrak{p}_2 - \mathfrak{p}_3)^{-1}\mathfrak{p}_4 + ([p_1; p_2; p_3; p_4] + (\mathfrak{p}_2 - \mathfrak{p}_3)^{-1}\mathfrak{p}_3) \end{matrix} \right) \mathbb{H}.$$

$$(4.27)$$

That is, $p_1 = \mu_{[p_1; p_2; p_3; p_4], \mathfrak{p}_2, \mathfrak{p}_3}(p_4)$ is a Möbius transform of p_4. Note that, given three points p_2, p_3, p_4 on a circle, the cross-ratio *parametrizes* the circle,[32] that is,

$$[\tilde{p}_1; p_2; p_3; p_4] \neq [p_1; p_2; p_3; p_4] \quad \text{for} \quad \tilde{p}_1 \neq p_1,$$

so that the point p_1 in (4.27) does not depend on the choice of stereographic projection used to determine p_1.

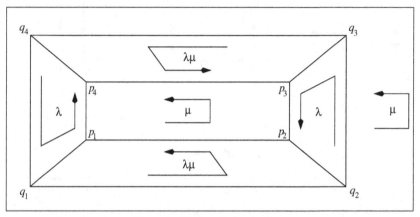

Fig. 4.4. The hexahedron lemma

We can use (4.27) to prove a version of Miguel's theorem [12] that will prove useful later, when we are considering the Darboux transformation for discrete isothermic surfaces:

4.9.13 Hexahedron Lemma. *Let p_1, p_2, p_3 and $q_1 \in \mathbb{HP}^1$ be four non-concircular points, and let $\lambda, \mu \in \mathbb{R}\backslash\{0, 1\}$, $\lambda\mu \neq 1$. Then there exist unique points p_4 and q_2, q_3, q_4 so that*

$$
\begin{aligned}
{[p_1; p_2; p_3; p_4]} &= [q_1; q_2; q_3; q_4] &&= \mu, \\
{[p_1; p_4; q_4; q_1]} &= [p_3; p_2; q_2; q_3] &&= \lambda, \\
{[q_1; q_2; p_2; p_1]} &= [q_3; q_4; p_4; p_3] &&= \lambda\mu.
\end{aligned}
$$

Moreover, all eight points lie on a 2-sphere in $\mathbb{HP}^1$.

4.9.14 *Proof.* Four non-concircular points define a unique 2-sphere that we consider as a Riemann sphere. Using (4.27), we then construct:

$$
\begin{aligned}
p_4 \quad \text{from} \quad [p_4; p_1; p_2; p_3] &= \tfrac{1}{\mu}, \\
q_2 \quad \text{from} \quad [q_2; q_1; p_1; p_2] &= \lambda\mu, \\
q_3 \quad \text{from} \quad [q_3; q_2; p_2; p_3] &= \lambda, \\
q_4 \quad \text{from} \quad [q_4; q_3; p_3; p_4] &= \lambda\mu.
\end{aligned}
$$

---

[32] This is most easily seen by using a Möbius transformation to map the three points to $0, 1, \infty$ so that the circle is the real line; then $[t; 0; 1; \infty] = t$.

Since each new point lies on a circle that is determined from three points on our Riemann sphere, all points lie on that 2-sphere. It remains to show that

$$[q_4; q_1; p_1; p_4] = \lambda \quad \text{and} \quad [q_1; q_2; q_3; q_4] = \mu :$$

Considering the p_i's and q_i's as complex numbers, this can be done by a lengthy but straightforward computation that can well be left to a computer algebra system (cf., [149]). ◁

4.9.15 As usual, some important identifications and relations developed throughout this chapter are summarized in a table (see Figure T.3).

Chapter 5
Application: Smooth and discrete isothermic surfaces

We already touched on isothermic surfaces and their transformation theory in Chapter 3 (see, in particular, Section 3.3). Here we will elaborate on the transformation theory in more detail, using the just-presented quaternionic model for Möbius differential geometry that seems particularly well adapted to the topic.[1] To make the current presentation relatively independent from our previous developments, we will start over from the beginning — however, it may be helpful to compare the viewpoint from Section 3.3 with the one elaborated here. In particular, we will extend our definitions and facts from Section 3.3 to higher codimension, enabling the codimension 2 feature provided by the quaternionic model.

We already gave some details of the history of isothermic surfaces in Chapter 3. Here we should add that L. Bianchi already gave very comprehensive expositions of the transformation theory of isothermic surfaces, mainly presented in his papers [21] and [20]. There, he does not only introduce the "T-transformation," which was independently introduced by P. Calapso [53] and [55], but he also gives an overview of all known transformations, including their interrelations in terms of "permutability theorems." In fact, there are authors who consider Bianchi's research on transformations and permutability theorems as his most important contribution to geometry (cf., [163]). On the other hand, it was Calapso's contribution to consider this T-transformation (or "transformation C_m," as he denoted it) in a more Möbius geometric context, in terms of conformal invariants of a surface.[2] We should also mention G. Tzitzéica's contribution to the theory of isothermic surfaces [284]: He already discussed isothermic surfaces in any dimension, in terms of the classical model. We will only touch on his very interesting viewpoint in the preliminary section of the present chapter.

Remember how we came about the transformations of isothermic surfaces in Section 3.3: The Christoffel transformation appeared in the context of Christoffel's problem in §3.3.11, similar to the Darboux transformation that

---

[1] Here the notion of a "retraction form," which I owe to discussions with Fran Burstall, deserves particular mentioning: It not only provides a truly invariant approach to the transformations, but also draws a formal relation with the theory of surfaces of constant mean curvature 1 in hyperbolic space — "horospherical surfaces."

[2] Therefore, we will denote it as "Calapso's transformation" and honor Bianchi by attributing his name to the permutability theorems. At times, we will also call this transformation the "spectral transformation," because of its relation to integrable system theory [46], or "conformal deformation," since it is a second-order deformation for isothermic surfaces (cf., [62] and [203]).

provided a solution to Blaschke's problem (see Section 3.1). The pairing of isothermic surfaces via the Darboux transformation then led us in §3.3.5, via a certain type of curved flats (see §3.3.2), to a third transformation, the spectral transformation. Here we will emphasize more the transformation viewpoint, carefully analyzing the geometric background of each transformation — in this way we will obtain a fourth transformation, the "Goursat transformation."

* *Christoffel's transformation.* This is a transformation that depends on the Euclidean geometry of the ambient space of an isothermic surface — even though isothermic surfaces naturally "live" in conformal space (the conformal n-sphere), their pairing via the Christoffel transformation only makes sense in a Euclidean subgeometry.

* *Goursat's transformation.* This transformation, generalizing the classical Goursat transformation for minimal surfaces [129], owes its existence to this interplay of the Euclidean and conformal geometries when Christoffel transforming isothermic surfaces: It is obtained by intertwining Christoffel and Möbius transformations.

* *Darboux's transformation.* In contrast to the previous two transformations this is a transformation for isothermic surfaces in the conformal sphere, that is, in Möbius geometry: Geometrically, it can be characterized by the fact that the conformal structures induced by the envelopes of a Ribaucour sphere congruence are equal (see §3.2.5). Thus the Darboux transformation provides a notion of "Darboux pairs" of isothermic surfaces in Möbius geometry, that is, the pair consisting of an isothermic surface and any of its Darboux transforms can be considered as a submanifold in the space of point pairs[3] (see §4.5.3).

* *Calapso's transformation.* Here we have a truly Möbius differential geometric transformation: Any Calapso transform of an isothermic surface is only well defined up to Möbius transformation, and any Möbius transform of the original isothermic surface provides the same Calapso transforms; that is, this transformation is a transformation between Möbius equivalence classes of isothermic surfaces rather than a transformation for individual surfaces.

After discussing each transformation in turn, we will discuss their interrelations in terms of Bianchi's "permutability theorems": Most of the presented theorems were, in some form, already known to Bianchi (see [21] and [20]). These permutability theorems have interesting applications, for example, in the theory of surfaces of constant mean curvature in space forms.

---

[3] Note the difference to the notion of "Christoffel pairs" (cf., §3.3.22), where the partners of a pair "live" in different Euclidean spaces (see §3.3.8).

Surfaces of constant mean curvature are isothermic (cf., §3.6.3) — in the case of a Euclidean ambient space, we will provide a proof from the "isothermic transformations viewpoint" (cf., §3.3.25 and §3.3.29). Also remember, from §3.4.6, that constant mean curvature surfaces can be characterized by the fact that their conformal Gauss map (central sphere congruence) has values in an "affine sphere complex," so that surfaces of constant mean curvature in space forms (together with their Möbius transforms) form a well-defined class of surfaces in Möbius geometry. Their isothermic transformation theory is rather interesting and yields various well-known transformations (as, for example, Lawson's correspondence (cf., §3.3.7) and the Bianchi-Bäcklund transformation for surfaces of nonzero constant mean curvature in Euclidean space (cf., [147])) as specializations of isothermic transformations. However, we will treat surfaces of constant mean curvature only in the form of examples.[4] The reason is that they are special "special isothermic surfaces," that were extensively studied by the classical geometers[5] — which seems to provide a more unified treatment of the isothermic theory of constant mean curvature surfaces. Nowadays, this theory of special isothermic surfaces still awaits to be fully rediscovered.

As another application of one of Bianchi's permutability theorems — the permutability theorem for different Darboux transformations, which is probably his most famous permutability theorem in the theory of isothermic transformations — one may view the theory of discrete isothermic nets: These can be obtained by repeated application of this permutability theorem. It turns out that, for the discrete isothermic nets constructed in this way, there is a transformation theory as rich as the transformation theory in the smooth case.

Discretizations for the Christoffel and the Darboux transformations have been given in [30] and [149], respectively, using a discrete version of Christoffel's original formula and of the Riccati-type partial differential equation that was derived in [147]: For both ansatzes it was crucial to have a Euclidean description of the transformation (even though, for the discrete analog of the Darboux transformation, a Möbius invariant definition was given — as it should be). As mentioned above, the Calapso transformation (or

---

[4] For more information on constant mean curvature surfaces in the context of isothermic surfaces and their transformations, see, for example, [153] and [206].

[5] The notion of a special isothermic surface is related to deformations of quadrics — this in fact may have been the way in which Darboux came about his transformation (cf., [89]). A comprehensive treatment can also be found in Bianchi's work (see [21] and [20]) and in [106]; for a modern account, see [206]. Special classes of special isothermic surfaces occur in [280] and [281], in particular, "horospherical surfaces" (see below); a modern description of a construction of horospherical surfaces by Bianchi can be found in [188]. Compare also [135].

The reader shall also be pointed to work concerning isothermic Weingarten surfaces, which applies to surfaces of constant mean curvature as well: See [305], [211], [227], and [228].

"T-transformation") is a transformation for Möbius equivalence classes of isothermic surfaces, not for surfaces with a fixed position in space (in contrast to the Darboux transformation, which incorporates the position of the surface); thus, to be consistent, any description should make use of the Möbius group. In order to introduce a discrete analog for the Calapso transformation, it therefore seems necessary to elaborate on a suitable setup. Here a quaternionic approach (or, in higher dimensions, a Clifford algebra approach, which will be discussed later: See [47]) seems preferable over the classical formalism because no normalization problems occur. With this setup we then obtain discrete analogs for *all* of the isothermic transformations as well as their interrelations in terms of permutability theorems.

In order to illustrate one way to introduce a certain type of discrete analogs for surfaces of constant mean curvature via their isothermic transformations, we finally discuss "horospherical nets," as discrete analogs of surfaces of constant mean curvature 1 in hyperbolic space, as an example. This case is particularly interesting because there are two different ways to characterize these surfaces in terms of isothermic transformations — the equivalence will then turn out to be provided by a permutability theorem.

The contents of this chapter are organized as follows:

Section 5.1. In this section, preliminary material is presented for later reference: The notion of isothermic surface is generalized to arbitrary codimension, a characterization of curvature line parameters of a surface in the conformal n-sphere is given, and a complex formulation for conformal parameters is discussed. The notion of the conformal Hopf differential and the Calapso potential are introduced.

Section 5.2. A notion of the Christoffel transformation for isothermic surfaces in Euclidean 4-space $\mathbb{H}$ is introduced. Its geometric properties are discussed, and an algebraic condition, which will become important later, is given. A formula for the differential of the Christoffel transform of an isothermic surface in Euclidean 3-space is derived and is applied to show that surfaces of constant mean curvature are isothermic. It is shown that the Hopf differential of an isothermic surface in $\mathbb{R}^3$ is a real multiple of a canonical holomorphic quadratic differential. The Christoffel transforms of minimal surfaces and of an ellipsoid are discussed as examples.

Section 5.3. A generalization of Goursat's transformation for minimal surfaces is given. A formula for the differentials of essential Goursat transforms of an isothermic surface is derived, and the classical (local) Weierstrass representation for minimal surfaces is shown to occur as a special case.

A map that encodes all possible Christoffel transforms of an isothermic surface in the conformal 4-sphere is introduced, and the notion of a retraction form of an isothermic surface is defined. As an example, the Goursat transforms of the catenoid are determined.

Section 5.4. After a general discussion on the relation between Riccati equations and linear systems of differential equations, the Darboux transformation is introduced analytically. Using the Christoffel transformation, it is shown that Darboux transforms of isothermic surfaces are isothermic, and the retraction form of a Darboux transform is determined. The Darboux transformation for surfaces of constant mean curvature is discussed.

A special framing for a Darboux pair of isothermic surfaces is introduced (which will also be important in the context of curved flats and of a permutability theorem) and is used to show that Darboux pairs in the present setup provide solutions to Blaschke's problem, that is, form conformal Ribaucour pairs. The Darboux transforms of Clifford tori are discussed.

Section 5.5. The Calapso transformation (or T-transformation) is introduced. It is shown that Sym's formula can be applied to obtain the general Christoffel transform of an isothermic surface in the conformal 4-sphere. The retraction forms for the Calapso transforms of an isothermic surface are determined, and it is shown that the Calapso transformation acts like a 1-parameter group on a given isothermic surface. It is proved that the Calapso transformation provides a second-order deformation for isothermic surfaces; in particular, the effect of a Calapso transformation on the central sphere congruence of an isothermic surface is investigated.

The relation between the Calapso transformation and the associated family of curved flats in the space of point pairs, that is, Darboux pairs of isothermic surfaces, is discussed, and it is shown how the Calapso transformation can be derived using any adapted frame for an isothermic surface. The Calapso transformations of Dupin cyclides are computed as an example. It is shown that Lawson's correspondence of constant mean curvature surfaces arises from the Calapso transformation — in particular, the Calapso transformation for minimal surfaces in Euclidean 3-space is discussed, and the notion of horospherical surfaces is introduced.

Section 5.6. Various permutability theorems for the presented transformations of isothermic surfaces are derived: Most important may be the permutability theorem for different Darboux transformations and the major permutability scheme intertwining the Christoffel, Darboux, and Calapso transformations. As an example, it is shown that any horospherical surface can be obtained as a Darboux transformation of its hyperbolic Gauss map. The Enneper cousins are discussed explicitly.

Section 5.7. Discrete principal and isothermic nets are defined, and the discrete retraction form of an isothermic net is introduced. On these grounds, discrete analogs are provided for all the isothermic transformations: the Christoffel, Goursat, Darboux, and Calapso transformations. Discrete versions of the permutability theorems are proven. Two possible ways of defining discrete horospherical nets via the transformations of isothermic surfaces are shown to be equivalent, as an example.

Remark. The presentation of the material in this chapter relies on the quaternionic formalism presented in Chapter 4. Some knowledge of the results concerning transformations of isothermic surfaces in Chapter 3 will be helpful but should not be necessary. Related material can be found in the last two sections of Chapter 8.

5.1 Isothermic surfaces revisited

In this preliminary section we discuss some introductory material for later reference, before starting to consider isothermic surfaces and their transformations in terms of quaternions: Besides extending the definition of an isothermic surface to arbitrary codimension — at least to codimension 2 is clearly desirable when considering surfaces in $I\!\!HP^1 \cong S^4$ — we will discuss the use of complex (conformal) parameters. We will touch on the notion of a "conformal Hopf differential" for surfaces in the conformal 3-sphere (cf., [48]) and introduce the "Calapso potential" (cf., [204]): the quantity that arises in the Calapso equation. Note that, in the context of higher codimension surfaces, we will also very briefly touch on some ideas that G. Tzitzéica elaborates in his remarkable book [284].

We start by extending our definition in §3.2.2 of an isothermic surface:

5.1.1 Definition. *An immersion $f : M^2 \to S^n$ into the conformal n-sphere is called isothermic if, around each (nonumbilic) point, there exist conformal curvature line coordinates.*

5.1.2 Obviously, a necessary condition for an immersion $f : M^2 \to S^n$ to be isothermic is the existence of curvature lines. Fixing, for example, a flat metric on the conformal n-sphere, we learned from the Ricci equation (cf., Section P.6) that the normal bundle of f has to be flat (remember that the normal curvature of a submanifold is conformally invariant; see §P.6.1) for f to carry well-defined curvature directions (cf., §P.6.7). Since we are considering *surfaces*, that is, 2-dimensional submanifolds of S^n, so that any codimension 1 distribution on the submanifold is integrable, the existence of well-defined curvature *directions* is also sufficient for the existence of curvature line *coordinates*.

Thus isothermic surfaces in S^n appear as a special case of "2-orthogonal systems," a notion that will be discussed in Chapter 8.

5.1.3 Considering $f : M^2 \to \mathbb{R}^n$ as an immersion into Euclidean space, curvature line coordinates $(x, y) : M^2 \to \mathbb{R}^2$ can (just as in $\mathbb{R}^3$) be characterized as orthogonal conjugate parameters, that is, by

$$f_x \cdot f_y = 0 \quad \text{and} \quad f_{xy} = \nabla_{\frac{\partial}{\partial x}} \tfrac{\partial}{\partial y} f \in \operatorname{span}\{f_x, f_y\},$$

where "$\cdot$" denotes the Euclidean scalar product of $\mathbb{R}^n$ and "∇" the (flat) Levi-Civita connection: The first condition says that (x, y) diagonalize the first fundamental form, and the second condition says that they diagonalize the (normal bundle—valued) second fundamental form (cf., Section P.6).

5.1.4 Identifying, as in §1.4.3, the Euclidean space $\mathbb{R}^n$ with the quadric Q_0^n of constant sectional curvature $\kappa = 0$ via

$$\mathbb{R}^n \ni p \quad \leftrightarrow \quad (\tfrac{1}{2}(1 + p \cdot p), p, \tfrac{1}{2}(1 - p \cdot p)) \in Q_0^n,$$

where the components of the vector in $Q_0^n \in \mathbb{R}_1^{n+2}$ are the components in terms of an orthonormal basis $(e_0, e_1, \ldots, e_n, e_{n+1})$ with $\langle e_0, e_0 \rangle = -1$, we obtain an interesting reformulation of the conditions on (x, y) to be curvature line coordinates: By computing the second mixed partial derivatives, we find

$$(\tfrac{1+p \cdot p}{2}, p, \tfrac{1-p \cdot p}{2})_{xy} = (p \cdot p_{xy}, p_{xy}, -p \cdot p_{xy}) + (p_x \cdot p_y, 0, -p_x \cdot p_y),$$

so that

$$(\tfrac{1+p \cdot p}{2}, p, \tfrac{1-p \cdot p}{2})_{xy} \in \operatorname{span}\{(\tfrac{1+p \cdot p}{2}, p, \tfrac{1-p \cdot p}{2})_x, (\tfrac{1+p \cdot p}{2}, p, \tfrac{1-p \cdot p}{2})_y\}$$

if and only if $p_{xy} \in \operatorname{span}\{p_x, p_y\}$ and $p_x \perp p_y$.

Thus $(x, y) : M^2 \to \mathbb{R}^2$ are curvature line parameters for f as an immersion into the flat quadric $Q_0^n \subset \mathbb{R}_1^{n+2}$ if and only if (x, y) are conjugate parameters for f considered as an immersion into $\mathbb{R}_1^{n+2}$.

Since the notion of curvature lines is conformally invariant — that is, it is invariant under rescalings of a light cone representative of an immersion into the conformal n-sphere — and the notion of conjugate parameters is projectively invariant — that is, the fact that f_{xy} takes values in the tangent planes of f (given by $\operatorname{span}\{f, f_x, f_y\}$ in projective geometry) is invariant under rescalings of the homogeneous coordinates of an immersion into projective $(n+1)$-space — we obtain the following lemma, which is the basis for G. Tzitzéica's investigations[6] of isothermic surfaces in [284]:

---

[6] In his very interesting book [284], Tzitzéica develops a theory of (conjugate) nets in quadrics and their transformations. This generalizes, for example, the transformation theory of isothermic surfaces in any codimension.

5.1.5 Lemma. *Let* $f : M^2 \to S^n \subset \mathbb{R}P^{n+1}$ *be an immersion into the conformal n-sphere. Then* $(x, y) : M^2 \to \mathbb{R}^2$ *are curvature line parameters for* f *if and only if they are conjugate for* $f : M^2 \to \mathbb{R}P^{n+1}$ *as an immersion into projective* $(n + 1)$*-space.*

5.1.6 To encode conformality of the (curvature line) parameters of a surface, it will sometimes be convenient to replace the real parameters (x, y) by a complex parameter[7] $z = x + iy : M^2 \to \mathbb{C}$. By introducing, as an analog of the Gaussian basis fields, the vector fields

$$\tfrac{\partial}{\partial z} = \tfrac{1}{2}\left(\tfrac{\partial}{\partial x} - i\tfrac{\partial}{\partial y}\right) \quad \text{and} \quad \tfrac{\partial}{\partial \bar z} = \tfrac{1}{2}\left(\tfrac{\partial}{\partial x} + i\tfrac{\partial}{\partial y}\right)$$

on the complexified tangent bundle $T^{\mathbb{C}}M$, and extending the Minkowski product $\langle .,. \rangle$ of our Minkowski coordinate space $\mathbb{R}_1^{n+2}$ bilinearly to $\mathbb{C}^{n+2}$, we find that (x, y) are conformal coordinates for an immersion $f : M^2 \to S^n$ into the conformal n-sphere if and only if $z = x + iy$ (or $\bar z = x - iy$, respectively) are isotropic coordinates for f:

$$\langle f_z, f_z \rangle = \langle \tfrac{\partial}{\partial z}f, \tfrac{\partial}{\partial z}f \rangle = \tfrac{1}{4}(\langle f_x, f_x \rangle - \langle f_y, f_y \rangle - 2i\,\langle f_x, f_y \rangle).$$

5.1.7 Lemma. $(x, y) : M^2 \to \mathbb{R}^2$ *are conformal parameters for an immersion* $f : M^2 \to S^n$ *into the conformal n-sphere if and only if* $z = x + iy$ *are null coordinates, that is, if and only if* f_z *is isotropic.*

5.1.8 For later reference we also compute the second derivative, which tells us when $z = x + iy$ provides curvature line parameters for $f : M^2 \to S^n$:

$$f_{zz} = \tfrac{1}{4}(f_{xx} - f_{yy}) - \tfrac{1}{2}i\,f_{xy}.$$

Consequently, $z = x + iy$ are curvature line parameters for f, that is, conjugate parameters for f viewed as an immersion into projective space $\mathbb{R}P^{n+1}$, if and only if the imaginary part $\operatorname{Im} f_{zz}$ is tangential. Or, otherwise said, any "Hopf differential"

$$Q_S = \langle f_{zz}, S \rangle \, dz^2, \tag{5.1}$$

where $S : M^2 \to S_1^{n+1}$ is any unit normal field of f (that is, a hypersphere congruence that is enveloped by f; cf., §1.6.3), is a *real* multiple of the holomorphic differential dz^2. Note that, in the case where $f : M^2 \to Q_0^3$ takes values in a 3-dimensional flat quadric and S is its tangent plane map (cf., §1.4.11), Q_S is *the* Hopf differential of f.

---

[7] In the context of constant mean curvature surfaces, it will turn out really wonderful to confuse the complex structure coming from some complex coordinates with the quaternionic structure of the ambient space.

5.1.9 In the codimension 1 case, this "Hopf differential" can be normalized to provide a Möbius invariant of the surface (see [48]): First note that

$$\langle f_{zz}, f \rangle = -\langle f_z, f_z \rangle \equiv 0$$

for any conformal coordinate chart $z = x + iy$ (see §5.1.7). As a consequence, $Q := Q_S$ does not depend on the choice of enveloped sphere congruence S: Remember that any sphere congruence enveloped by f is given by $S + af$ with some function a (cf., Figure T.1).

However, rescaling f rescales Q with the same factor,

$$f \to e^u f \quad \Rightarrow \quad Q = Q_S \to e^u Q.$$

If we fix the scaling of f relative to our choice of conformal coordinates so that the induced metric $\langle df, df \rangle = |dz|^2$, we obtain a conformally invariant quantity[8] by suitably adjusting the scaling of Q:

5.1.10 Lemma and Definition. *Let $z : M^2 \to \mathbb{C}$ be a conformal coordinate chart for an immersion $f : M^2 \to S^3$ into the conformal 3-sphere and choose a light cone lift $f : M^2 \to L^4$ so that $|df|^2 = |dz|^2$. Then the conformally invariant quantity*

$$Q_{con} := \tfrac{1}{|dz|} Q = \tfrac{1}{|dz|} \langle f_{zz}, S \rangle dz^2$$

is called the conformal Hopf differential *of f.*

5.1.11 Clearly, Q_{con} encodes information about the curvature directions of f because it is a rescaling of the ordinary Hopf differential. The remaining information encoded by Q_{con} is given by its norm. To compute $|Q_{con}|^2$ we may choose *any* enveloped sphere congruence — to simplify the calculation we choose $S = Z$ to be the central sphere congruence (cf., §3.4.1) so that

$$\langle f_x, Z_x \rangle + \langle f_y, Z_y \rangle = 0.$$

Then

$$|Q_{con}|^2 = |\langle f_z, Z_z \rangle|^2 |dz|^2 = \tfrac{1}{4}(\langle f_x, Z_x \rangle^2 + \langle f_x, Z_y \rangle^2) I_f,$$

where $I_f = \langle df, df \rangle$ denotes the (flat) metric induced by f.

On the other hand, since $(\frac{\partial}{\partial x}, \frac{\partial}{\partial y})$ is an orthonormal basis with respect to I_f, the matrix representations of the second fundamental form and the Weingarten tensor of f with respect to Z as a unit normal field coincide:

$$A_Z = - \begin{pmatrix} \langle f_x, Z_x \rangle & \langle f_x, Z_y \rangle \\ \langle f_y, Z_x \rangle & \langle f_y, Z_y \rangle \end{pmatrix}.$$

---

[8] Compare [5].

Therefore, the metric I_Z of the central sphere congruence Z becomes

$$I_Z = I_f(., A_Z^2.) = -\det A_Z \cdot I_f = (\langle f_x, Z_x \rangle^2 + \langle f_x, Z_y \rangle^2) I_f,$$

where the second equality follows with Cayley-Hamilton's formula since Z is the central sphere congruence of f, so that $\operatorname{tr} A_Z = 0$.

5.1.12 Lemma and Definition. *The norm of the conformal Hopf differential Q_{con} of an immersion $f : M^2 \to S^3$ into the conformal 3-sphere provides the induced metric of its conformal Gauss map Z: $|Q_{con}|^2 = \frac{1}{4} I_Z$.*

The factor relating the induced metric of the conformal Gauss map and the flat metric induced by f is called the Calapso potential.[9]

5.2 Christoffel's transformation

Now that we have discussed these preliminaries, we turn to the quaternionic setup that we developed in the previous chapter, considering isothermic surfaces of codimension 2 or 1, respectively. Thus, because we intend to discuss the Christoffel transformation, which is a transformation for isothermic surfaces in Euclidean space, let $f : M^2 \to \mathbb{H} \cong \mathbb{R}^4$ be an isothermic surface[10] and let (x, y) denote (local) conformal curvature line parameters. That is,

$$\left.\begin{array}{r} |f_x|^2 = |f_y|^2 \\ f_x \cdot f_y = 0 \end{array}\right\} \quad \Leftrightarrow \quad f_x f_y^{-1} + f_y f_x^{-1} = 0 \tag{5.2}$$

and $f_{xy} \in \operatorname{span}\{f_x, f_y\}$. Now define[11]

$$\psi^* := f_x^{-1} dx - f_y^{-1} dy.$$

Using that $f_{xy} = a f_x + b f_y$ with suitable functions $a, b : M^2 \to \mathbb{R}$ and (5.2) it is straightforward to verify that $d\psi^* = 0$. Consequently, there exists (locally) a map $f^* : M^2 \supset U \to \mathbb{H}$ with $df^* = \psi^*$.

Moreover, (x, y) are clearly conformal parameters for f^* — note that (5.2) is symmetric in f_x, f_y and f_x^{-1}, f_y^{-1} — and $f_{xy}^* = -a f_x^{-1} + b f_y^{-1}$. Consequently, (x, y) are also conformal curvature line parameters for f^*, which is therefore isothermic:

---

[9] This terminology was, to my knowledge, coined by E. Musso (cf., [204]), referring to the fact that this factor is the quantity arising in the Calapso equation (cf., [46]), which is the governing equation in the theory of isothermic surfaces.

[10] In the sequel we will notationally distinguish maps $f : M^2 \to \mathbb{H}$ into the Euclidean field (!) $\mathbb{H}$ from the corresponding maps $f : M^2 \to \mathbb{H}P^1$ into the conformal 4-sphere since, at certain points, we will have to consider both simultaneously.

[11] Here, $(.)^{-1}$ denotes the multiplicative inverse in the field $\mathbb{H}$.

5.2.1 Theorem and Definition. *Let* $\mathfrak{f} : M^2 \to \mathbb{R}^4 \cong \mathbb{H}$ *be isothermic with conformal curvature line parameters* $(x, y) : M^2 \supset U \to \mathbb{R}^2$. *Then*

$$\psi^* := \mathfrak{f}_x^{-1} dx - \mathfrak{f}_y^{-1} dy \qquad (5.3)$$

is locally integrable, $\psi^* = d\mathfrak{f}^*$, *and*[12] $\mathfrak{f}^* : U \to \mathbb{H}$ *is isothermic, too.*
$\mathfrak{f}^* : U \to \mathbb{H}$ *is called a Christoffel transform of* $\mathfrak{f} : U \to \mathbb{H}$.

5.2.2 Obviously, this notion of a Christoffel transformation generalizes our notion of a Christoffel transformation from §3.3.22: If $\mathfrak{f} : M^2 \to \mathbb{R}^3 \cong \operatorname{Im} \mathbb{H}$, then $\mathfrak{f}$ and $\mathfrak{f}^*$ have parallel tangent planes and induce conformally equivalent metrics. Consequently, the pair $(\mathfrak{f}, \mathfrak{f}^*)$ provides a solution to Christoffel's problem from §3.3.11 (compare [70]).

However, in general, $\mathfrak{f}$ and $\mathfrak{f}^*$ do *not* have parallel tangent planes, but $\mathfrak{f}$ and $\overline{\mathfrak{f}^*}$ — the quaternionic conjugate of $\mathfrak{f}^*$ — do. Therefore, the above Christoffel transformation does *not* provide a solution to Christoffel's problem in higher codimensions (or if we position the ambient $\mathbb{R}^3$ badly inside the quaternions). Nevertheless, in the quaternionic context, it seems to be "correct" to define the Christoffel transformation as we did, as will become apparent later.

Reconsidering, in Chapter 8, the Christoffel transformation in terms of the Vahlen matrix approach to Möbius geometry that we are going to discuss in Chapter 7, we will obtain a generalization of the Christoffel transformation to arbitrary codimension isothermic surfaces that does not have this "defect" (cf., [47] and [215]).

5.2.3 After fixing conformal curvature line parameters for an isothermic surface $\mathfrak{f}$, a Christoffel transform $\mathfrak{f}^*$ of $\mathfrak{f}$ is unique up to translation. A different choice of conformal curvature line parameters, $(\tilde{x}, \tilde{y}) = (\lambda x, \lambda y)$ with a constant $\lambda \in \mathbb{R} \setminus \{0\}$, yields a scaling (negative if the roles of x and y are interchanged) of the Christoffel transform: The Christoffel transformation is a well-defined notion for surfaces in the geometry of similarities on $\mathbb{H} \cong \mathbb{R}^4$. In a short, we will learn about a more invariant way than choosing coordinates to fix the scaling of the Christoffel transform — at least for surfaces $\mathfrak{f} : M^2 \to \operatorname{Im} \mathbb{H}$ in Euclidean 3-space — making it a well-defined notion in Euclidean geometry.

Thus, after fixing the scaling for the Christoffel transformation (by fixing a choice of conformal curvature line parameters, for example) the following is obtained as a trivial consequence of the definition:

5.2.4 Corollary. *The Christoffel transformation is involutive,* $\mathfrak{f}^{**} \simeq \mathfrak{f}$.

---

[12] For simplicity of notation, we assume that ψ^* can be integrated on all of U. This is, for example, the case when the parameter domain U is simply connected.

5.2.5 Using (5.2), it is a direct consequence of the definition (5.3) that

$$d\mathfrak{f}^* \wedge d\mathfrak{f} = (\mathfrak{f}_x^{-1}dx - \mathfrak{f}_y^{-1}dy) \wedge (\mathfrak{f}_x dx + \mathfrak{f}_y dy) = (\mathfrak{f}_x^{-1}\mathfrak{f}_y + \mathfrak{f}_y^{-1}\mathfrak{f}_x)dx \wedge dy = 0,$$

and the other way around. With this computation we have proved the "if" part of the following lemma, emphasizing the symmetry of the relation between an isothermic surface and its Christoffel transform:

5.2.6 Lemma. *Two immersions* $\mathfrak{f}, \mathfrak{f}^* : M^2 \to \mathbb{H}$ *satisfy*

$$d\mathfrak{f} \wedge d\mathfrak{f}^* = d\mathfrak{f}^* \wedge d\mathfrak{f} = 0 \tag{5.4}$$

if and only if both $\mathfrak{f}$ *and* $\mathfrak{f}^*$ *are isothermic and one is a Christoffel transform of the other: They form a "Christoffel pair," or "C-pair."*

5.2.7 It remains to prove the "only if" part; that is, given (5.4), we seek conformal curvature line parameters for $\mathfrak{f}$ so that $\mathfrak{f}^*$ satisfies (5.3).
Proof. To start with we introduce conformal coordinates[13] (x, y) for $\mathfrak{f}$. Then we have $\mathfrak{f}_x \bar{\mathfrak{f}}_y + \mathfrak{f}_y \bar{\mathfrak{f}}_x = 0$, that is, $\mathfrak{f}_y \bar{\mathfrak{f}}_x \in \operatorname{Im} \mathbb{H}$ and $|\mathfrak{f}_x|^2 = |\mathfrak{f}_y|^2 =: E$. Now (5.4) reads

$$\mathfrak{f}_x \mathfrak{f}_y^* = \mathfrak{f}_y \mathfrak{f}_x^* \quad \text{and} \quad \mathfrak{f}_x^* \mathfrak{f}_y = \mathfrak{f}_y^* \mathfrak{f}_x.$$

With these two equations we derive

$$(\mathfrak{f}_x \mathfrak{f}_x^*)(\mathfrak{f}_y \bar{\mathfrak{f}}_x) = E \mathfrak{f}_x \mathfrak{f}_y^* = E \mathfrak{f}_y \mathfrak{f}_x^* = (\mathfrak{f}_y \bar{\mathfrak{f}}_x)(\mathfrak{f}_x \mathfrak{f}_x^*).$$

Hence $\mathfrak{f}_x \mathfrak{f}_x^* \in \operatorname{span}\{1, \mathfrak{f}_y \bar{\mathfrak{f}}_x\}$, and we may write $\mathfrak{f}_x^* = \alpha \mathfrak{f}_x^{-1} - \beta \mathfrak{f}_y^{-1}$ with suitable functions $\alpha, \beta : M \to \mathbb{R}$ since $\bar{\mathfrak{f}}_y = E\mathfrak{f}_y^{-1}$. Subsequently, $\mathfrak{f}_y^* = -\beta \mathfrak{f}_x^{-1} - \alpha \mathfrak{f}_y^{-1}$.

The integrability condition $\mathfrak{f}_{xy}^* = \mathfrak{f}_{yx}^*$ for $\mathfrak{f}^*$ then reduces, after decomposition of the second derivatives of $\mathfrak{f}$ into tangential and normal parts, to

$$\alpha_x = \beta_y, \quad \alpha_y = -\beta_x \quad \text{and} \quad 0 = 2\alpha \mathfrak{f}_{xy}^{\perp} + \beta(\mathfrak{f}_{xx}^{\perp} - \mathfrak{f}_{yy}^{\perp}). \tag{5.5}$$

In particular, $\alpha + i\beta$ is holomorphic. With a conformal change of parameters,

$$\left. \begin{aligned} du &= a\,dx - b\,dy \\ dv &= b\,dx + a\,dy \end{aligned} \right\} \quad \Leftrightarrow \quad \left\{ \begin{aligned} \frac{\partial}{\partial u} &= \frac{1}{a^2+b^2}\left(a\frac{\partial}{\partial x} - b\frac{\partial}{\partial y}\right) \\ \frac{\partial}{\partial v} &= \frac{1}{a^2+b^2}\left(b\frac{\partial}{\partial x} + a\frac{\partial}{\partial y}\right), \end{aligned} \right.$$

where[14] $(a + ib)^2 = (\alpha + i\beta)$, we obtain

$$\begin{aligned} \mathfrak{f}_{uv}^{\perp} &= \tfrac{1}{(a^2+b^2)^2}(ab(\mathfrak{f}_{xx} - \mathfrak{f}_{yy}) + (a^2 - b^2)\mathfrak{f}_{xy}) \\ &= \tfrac{1}{2(\alpha^2+\beta^2)}(2\alpha\mathfrak{f}_{xy} + \beta(\mathfrak{f}_{xx} - \mathfrak{f}_{yy})) \\ &= 0, \end{aligned}$$

---

[13] In the codimension 1 case of a surface in $\mathbb{R}^3 \cong \operatorname{Im} \mathbb{H}$, it would as well be possible to start from curvature line parameters and then prove that they can be altered to become conformal (cf., §3.3.15). However, in the codimension 2 case, we do not know a priori that curvature line parameters do exist.

[14] Note that, away from the branch points of $\mathfrak{f}^*$, a square root can be taken without problems.

so that (u, v) are identified as (conformal) curvature line parameters for $\mathfrak{f}$; consequently, $\mathfrak{f} : M^2 \to \mathbb{H}$ is isothermic. Moreover,

$$\mathfrak{f}_u \mathfrak{f}_u^* = \tfrac{1}{(a^2+b^2)^2}(a\mathfrak{f}_x - b\mathfrak{f}_y)(a\,[\alpha\mathfrak{f}_x^{-1} - \beta\mathfrak{f}_y^{-1}] + b\,[\beta\mathfrak{f}_x^{-1} + \alpha\mathfrak{f}_y^{-1}]) = 1,$$
$$\mathfrak{f}_v \mathfrak{f}_v^* = \tfrac{1}{(a^2+b^2)^2}(b\mathfrak{f}_x + a\mathfrak{f}_y)(b\,[\alpha\mathfrak{f}_x^{-1} - \beta\mathfrak{f}_y^{-1}] - a\,[\beta\mathfrak{f}_x^{-1} + \alpha\mathfrak{f}_y^{-1}]) = -1,$$

which identifies $\mathfrak{f}^*$ as a Christoffel transform of $\mathfrak{f}$.   ◁

5.2.8 If $\mathfrak{f} : M^2 \to \mathbb{R}^3 \cong \operatorname{Im} \mathbb{H}$ is an isothermic immersion into Euclidean 3-space, we already derived (in the absence of umbilics) a coordinate-free formula for the Christoffel transform $\mathfrak{f}^*$ of $\mathfrak{f}$ in §3.3.27: Denote by the Gauss map of $\mathfrak{f}$ by $\mathfrak{n} : M^2 \to S^2 \subset \mathbb{R}^3$; then

$$d\mathfrak{f}^* = \varrho(d\mathfrak{n} + H\,d\mathfrak{f}),$$

where H is the mean curvature of $\mathfrak{f}$ (with respect to $\mathfrak{n}$) and $\varrho : M^2 \supset U \to \mathbb{R}$ is a suitable function. On the other hand, it is easy to check that we always have

$$d\mathfrak{f} \wedge (d\mathfrak{n} + H\,d\mathfrak{f}) = 0 : \tag{5.6}$$

Away from umbilics, we may introduce curvature line parameters (x, y) so that $\mathfrak{n}_x = -k_1\mathfrak{f}_x$ and $\mathfrak{n}_y = -k_2\mathfrak{f}_y$. Then

$$d\mathfrak{f} \wedge (d\mathfrak{n} + H\,d\mathfrak{f}) = \tfrac{k_1-k_2}{2}(\mathfrak{f}_x\mathfrak{f}_y + \mathfrak{f}_y\mathfrak{f}_x)\,dx \wedge dy = 0,$$

since curvature lines intersect orthogonally. And, at umbilics, there is nothing to check since $d\mathfrak{n} + H\,d\mathfrak{f} = 0$, anyway.

As a consequence we obtain the following extension of §3.3.27:

5.2.9 Corollary. *A surface* $\mathfrak{f} : M^2 \to \mathbb{R}^3 \cong \operatorname{Im} \mathbb{H}$ *is isothermic if and only if there is (locally) an integrating factor* $\varrho : M^2 \to \mathbb{R}$ *for* $d\mathfrak{n} + H\,d\mathfrak{f}$,

$$d[\varrho(d\mathfrak{n} + H\,d\mathfrak{f})] = 0.$$

5.2.10 Example. This corollary provides a new argument for the fact that minimal surfaces or, more generally, surfaces of constant mean curvature in Euclidean 3-space are isothermic (cf., §3.3.25 and §3.3.29): we always have[15]

$$d\mathfrak{f} \wedge (d\mathfrak{n} + H\,d\mathfrak{f}) = 0.$$

And, if $\mathfrak{f} : M^2 \to \mathbb{R}^3$ has constant mean curvature, then $d(d\mathfrak{n} + H\,d\mathfrak{f}) = 0$, so that the integrating factor ϱ can be chosen constant.

---

[15] The two equations (5.4) reduce, in the case of two immersions $\mathfrak{f}, \mathfrak{f}^* : M^2 \to \mathbb{R}^3 \cong \operatorname{Im} \mathbb{H}$, to one equation since one is the quaternionic conjugate of the other (cf., §4.1.5).

Thus, *a minimal surface is isothermic with its Gauss map as a Christoffel transform, and a surface of constant mean curvature is isothermic with its parallel (constant mean curvature) surface* $\mathfrak{f}^* = \mathfrak{f} + \frac{1}{H}\mathfrak{n}$ *as a Christoffel transform.*

In fact, if the factor ϱ can be chosen constant for an isothermic surface, that is, if $d(d\mathfrak{n} + H\,d\mathfrak{f}) = 0$, then the surface has to have constant mean curvature: If (x, y) denote *any* coordinates on M^2, then

$$0 = d(d\mathfrak{n} + H\,d\mathfrak{f})(\tfrac{\partial}{\partial x}, \tfrac{\partial}{\partial y}) = dH \wedge d\mathfrak{f}(\tfrac{\partial}{\partial x}, \tfrac{\partial}{\partial y}) = H_x \mathfrak{f}_y - H_y \mathfrak{f}_x$$

shows that $dH = 0$ since $\mathfrak{f}$ is an immersion, that is, $\mathfrak{f}_x, \mathfrak{f}_y \in \operatorname{Im} \mathbb{H}$ are linearly independent.

Thus we have the converse of the above statement (cf., [147]): *if an isothermic surface has its Gauss map as a Christoffel transform, then it is minimal; and if it has a suitable parallel surface* $\mathfrak{f}^* = \mathfrak{f} + r\mathfrak{n}$ *as a Christoffel transform, then it has constant mean curvature* $H \equiv \frac{1}{r}$.

5.2.11 Example. Another, rather trivial but interesting example is the following: If $\mathfrak{g}, \mathfrak{h} : M^2 \to \mathbb{C}$ are two holomorphic (or, meromorphic) functions on a Riemann surface, then

$$\mathfrak{f} := \mathfrak{h}j \quad \text{and} \quad \mathfrak{f}^* := -j\mathfrak{g}$$

form a Christoffel pair of two totally umbilic isothermic surfaces; namely, using complex coordinates $z : M^2 \supset U \to \mathbb{C}$, we have

$$d\mathfrak{f} \wedge d\mathfrak{f}^* = \mathfrak{h}' dz \wedge \mathfrak{g}' dz = 0.$$

Now suppose we have[16] $\mathfrak{h} = \int \frac{1}{\mathfrak{g}'} dz$ and write $z = x + iy$. Since $\mathfrak{g}$ is holomorphic, $\mathfrak{g}' = \mathfrak{g}_x = -i\mathfrak{g}_y$, and similarly for $\mathfrak{h}$. Hence

$$
\begin{aligned}
(\mathfrak{h}j)_x &= \mathfrak{h}'j &=& \quad \tfrac{1}{\mathfrak{g}'}(-j)^{-1} &=& \quad (-j\mathfrak{g}')^{-1} &=& \quad (-j\mathfrak{g})_x^{-1}, \\
(\mathfrak{h}j)_y &= i\mathfrak{h}'j &=& \ -\tfrac{1}{i\mathfrak{g}'}(-j)^{-1} &=& \ -(-ji\mathfrak{g}')^{-1} &=& \ -(-j\mathfrak{g})_y^{-1}:
\end{aligned}
$$

We have identified the proper "conformal curvature line parameters" (x, y) that make $\mathfrak{f}$ and $\mathfrak{f}^*$ Christoffel transforms of each other. The characteristic property that makes these coordinates the "correct" coordinates for the Christoffel transform is that

$$(d\mathfrak{f} \cdot d\mathfrak{f}^*)(\tfrac{\partial}{\partial x}) = dz^2(\tfrac{\partial}{\partial x}) = 1 \quad \text{and} \quad (d\mathfrak{f} \cdot d\mathfrak{f}^*)(\tfrac{\partial}{\partial y}) = dz^2(\tfrac{\partial}{\partial y}) = -1,$$

---

[16] Intertwining this Christoffel transformation for holomorphic maps with Möbius transformations (of the complex plane) yields a process to construct series of holomorphic maps. Using those as Gauss maps of minimal surfaces via the Weierstrass representation (compare our discussion of the Weierstrass representation below) provides an iteration process for minimal surfaces that has been studied in [192].

where we consider $d\mathfrak{f} \cdot d\mathfrak{f}^* = dz^2$ as a (holomorphic) quadratic differential.

In the general setup of two holomorphic functions $\mathfrak{h}$ and $\mathfrak{g}$ it remains true that the quantity $q := d\mathfrak{f} \cdot d\mathfrak{f}^*$ is a holomorphic quadratic differential, so that (locally, away from the zeros of q) there is a complex coordinate change $z = z(w)$ that restores the above situation:

$$q = d\mathfrak{f} \cdot d\mathfrak{f}^* = \mathfrak{h}'\mathfrak{g}'dz^2 = dw^2.$$

When writing $w = u + iv$ in terms of its real and imaginary parts, we obtain, as before,

$$1 = dw^2(\tfrac{\partial}{\partial u}) = q(\tfrac{\partial}{\partial u}) = \mathfrak{f}_u \cdot \mathfrak{f}_u^* \quad \text{and} \quad -1 = dw^2(\tfrac{\partial}{\partial v}) = q(\tfrac{\partial}{\partial v}) = \mathfrak{f}_v \cdot \mathfrak{f}_v^*,$$

characterizing (u, v) as "conformal curvature line coordinates" for $(\mathfrak{f}, \mathfrak{f}^*)$ as a Christoffel pair. Our next goal is to see that this is a general behavior of Christoffel pairs of isothermic surfaces in Euclidean 3-space (cf., [156]):

5.2.12 Lemma. *Let* $\mathfrak{f}, \mathfrak{f}^* : M^2 \to \mathbb{R}^3 \cong \operatorname{Im} \mathbb{H}$ *be a Christoffel pair of isothermic surfaces, and let* $\mathfrak{n} : M^2 \to S^2 = \{\mathfrak{x} \in \operatorname{Im} \mathbb{H} \,|\, \mathfrak{x}^2 = -1\}$ *denote the Gauss map of* $\mathfrak{f}$. *Then, after suitably identifying the (trivial) normal bundle* NM *of* $\mathfrak{f} : M^2 \to \mathbb{H}$ *with the trivial* $\mathbb{C}$-*bundle* $M \times \mathbb{C}$ *over* M, $N_p M = \operatorname{span}\{1, \mathfrak{n}(p)\} \cong \mathbb{C}$, *the map*

$$q := d\mathfrak{f} \cdot d\mathfrak{f}^* : TM \to NM$$

is a holomorphic quadratic differential. Moreover, the Hopf differential Q *of* $\mathfrak{f}$ *is a real multiple of* q, $Q = \varrho \cdot q$ *with* $\varrho : M^2 \to \mathbb{R}$.

5.2.13 *Proof.* Clearly, the normal bundle $p \mapsto \operatorname{span}\{1, \mathfrak{n}(p)\}$ of $\mathfrak{f} : M^2 \to \mathbb{H}$ becomes, equipped with the quaternionic multiplication, a trivial complex line bundle over M^2 since $\mathfrak{n}^2 \equiv -1$, so that each $\mathfrak{n}(p)$ can be identified with the (complex) imaginary unit.

Now suppose (x, y) are any conformal coordinates for $\mathfrak{f}$. Then, if we choose the "correct" orientation for $\mathfrak{n}$,

$$-\mathfrak{f}_x \mathfrak{f}_y^{-1} = \tfrac{1}{|\mathfrak{f}_y|^2} \mathfrak{f}_x \mathfrak{f}_y = \tfrac{1}{|\mathfrak{f}_y|^2} \mathfrak{f}_x \times \mathfrak{f}_y = \mathfrak{n} = \mathfrak{f}_y \mathfrak{f}_x^{-1},$$

since $|\mathfrak{f}_x| = |\mathfrak{f}_y|$ and $\mathfrak{f}_x \perp \mathfrak{f}_y$. With the conformal curvature line coordinates (x, y) that make the Christoffel formula (5.3) work we find

$$d\mathfrak{f} \cdot d\mathfrak{f}^* = (\mathfrak{f}_x dx + \mathfrak{f}_y dy) \cdot (\mathfrak{f}_x^{-1} dx - \mathfrak{f}_y^{-1} dy) = (dx^2 - dy^2) + 2\mathfrak{n}\, dx dy \simeq dz^2,$$

with the identifications $\mathfrak{n} \simeq i$ and $z = x + iy$.

Finally, by using the same conformal curvature line parameters (x, y), we compute the Hopf differential:

$$Q = \langle \mathfrak{f}_{zz}, \mathfrak{n} \rangle \, dz^2 = \tfrac{1}{4}(\langle (\mathfrak{f}_{xx} - \mathfrak{f}_{yy}), \mathfrak{n} \rangle - 2i\langle \mathfrak{f}_{xy}, \mathfrak{n} \rangle) \, dz^2 \simeq \tfrac{1}{4}\langle (\mathfrak{f}_{xx} - \mathfrak{f}_{yy}), \mathfrak{n} \rangle \cdot q,$$

since (x, y) are curvature line coordinates so that $\langle \mathfrak{f}_{xy}, \mathfrak{n} \rangle = 0$.   ◁

5.2.14 Note that, since $\mathfrak{f}, \mathfrak{f}^* : M^2 \to \mathrm{Im}\, \mathbb{H}$, we have

$$\bar{q} = \overline{d\mathfrak{f} \cdot d\mathfrak{f}^*} = d\mathfrak{f}^* \cdot d\mathfrak{f} = q^*.$$

However, if we choose $\mathfrak{n}^* = -\mathfrak{n}$ to identify the normal bundle of $\mathfrak{f}^*$ as a trivial complex line bundle, $\mathbb{C} \cong \mathrm{span}\{1, \mathfrak{n}^*\}$, then we have $q^* = q$. Otherwise said, $\mathfrak{f}$ and $\mathfrak{f}^*$ have parallel tangent planes but opposite orientations, as is also clear from Christoffel's formula (5.3).

5.2.15 Thus, for any isothermic surface $\mathfrak{f} : M^2 \to \mathbb{R}^3$, there exists (locally, away from umbilics) a decomposition of the Hopf differential $Q = \varrho \cdot q$ into a real function $\varrho : M^2 \supset U \to \mathbb{R}$ and a holomorphic differential $q : TU \to \mathbb{C}$.

In fact, this is a characterization of isothermic surfaces: To see the converse, assume that the Hopf differential of a surface $\mathfrak{f} : M^2 \to \mathbb{R}^3$ decomposes on some neighborhood $U \subset M^2$ in this way, $Q = \varrho \cdot q$. Away from umbilics we have $Q \neq 0$, so that also $q \neq 0$, and we can locally choose conformal coordinates (x, y) such that $q = dz^2$, where $z = x + iy$. Then, because Q is a real multiple of dz^2, $\mathrm{Im}\, Q(\tfrac{\partial}{\partial x}) = \mathrm{Im}\, Q(\tfrac{\partial}{\partial y}) = 0$, saying that (x, y) are (conformal) curvature line parameters.

5.2.16 Lemma. *An immersion $\mathfrak{f} : M^2 \to \mathbb{R}^3$ is isothermic if and only if, around each nonumbilic point $p \in M^2$, its Hopf differential $Q = \varrho q$ locally decomposes into a real function $\varrho : U \to \mathbb{R}$ and a holomorphic quadratic differential $q : TU \to \mathbb{C}$.*

5.2.17 As another consequence we can now fix the scaling of the Christoffel transformation of an isothermic surface in a more invariant way than by choosing certain conformal curvature line parameters: We can choose a factorization of its Hopf differential, $Q = \varrho q$. By holomorphicity, q is uniquely determined by Q, up to constant scaling — the same freedom that one has for the choice of conformal curvature line parameters.

This observation provides an option[17] extending the notion of the Christoffel transformation to points where the curvature line net of an isothermic surface becomes singular,[18] as well as to totally umbilic surface patches

---

[17] Another option would be to use the invariant formula (5.4). But this turns out to be less satisfactory when allowing totally umbilic surface pieces.

[18] Note that such points are necessarily umbilics. However, not all umbilics cause problems: For example, a line of umbilics on a surface of revolution is no problem at all because the curvature line net extends through these points without singularities (see also Figure 3.1).

where isothermic "curvature line parameters" are rather ambiguous. In particular, with the second issue in mind — this will be particularly interesting in the context of minimal surfaces (cf., §5.2.10) — we introduce the notion of a "polarized surface":

5.2.18 Definition. *A Riemann surface M^2 equipped with a holomorphic quadratic differential q will be called a polarized surface.*

An immersion $\mathfrak{f} : (M^2, q) \to \mathbb{R}^3$ of a polarized surface will be called isothermic if its Hopf differential $Q = \varrho q$ is a real multiple of q, $\varrho : M^2 \to \mathbb{R}$.

5.2.19 Example.   We first reconsider our example from §5.2.11 of a Christoffel pair of two holomorphic functions: Prescribing a holomorphic quadratic differential $q = a\,dz^2$, the Christoffel transform $\mathfrak{f}^*$ of $\mathfrak{f} = \mathfrak{h}j$, with $\mathfrak{h}$ a holomorphic function, can now be computed by

$$\mathfrak{f}^* = \int (\mathfrak{h}'j)^{-1} a\,dz = -j \int \tfrac{a}{\mathfrak{h}'}dz$$

since $d\mathfrak{f} = \mathfrak{h}'dz\,j$. Note that the zeros of a, the "true" umbilics of $\mathfrak{f}$ where the "curvature line net" may become singular, are the branch points of the Christoffel transform $\mathfrak{f}^*$.

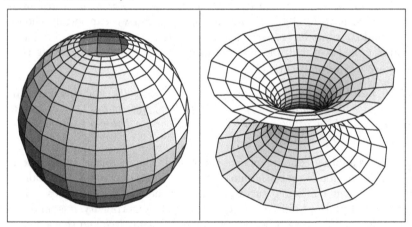

Fig. 5.1. The sphere and the catenoid as a Christoffel pair

5.2.20 Example. As a next example we want to consider surfaces of revolution. We already saw in §3.7.7 that any surface of revolution in Euclidean 3-space is isothermic. In fact, a surface

$$\mathfrak{f} : I \times S^1 \to \mathbb{R}^3, \quad (t, \vartheta) \mapsto (r(t)\cos\vartheta, r(t)\sin\vartheta, h(t))$$

of revolution is conformally parametrized (by curvature lines) if and only if its meridian curve $\gamma = (r, h) : I \to \mathbb{R}_+ \times \mathbb{R}$ is parametrized by hyperbolic arc length, that is, $r^2 = r'^2 + h'^2$: We consider $\mathbb{R}_+ \times \mathbb{R} \cong H^2$ as the Poincaré half-plane (see §P.1.3). Then the induced metric of $\mathfrak{f}$ is $r^2(dt^2 + d\vartheta^2)$, and the Christoffel transform $\mathfrak{f}^*$ of $\mathfrak{f}$ is obtained by integrating

$$
\begin{aligned}
d_{(t,\vartheta)}\mathfrak{f}^* &= \tfrac{1}{r^2(t)}(-\mathfrak{f}_t(t,\vartheta)dt + \mathfrak{f}_\vartheta(t,\vartheta)d\vartheta) \\
&= d(\tfrac{1}{r(t)}\cos\vartheta, \tfrac{1}{r(t)}\sin\vartheta, -\int \tfrac{h'(t)dt}{r^2(t)}).
\end{aligned}
$$

Hence $\mathfrak{f}^*$ is again a surface of revolution with the same axis of rotation as the original surface, up to translation. With $z = t + i\vartheta$ we compute the holomorphic differential

$$
q = d\mathfrak{f} \cdot d\mathfrak{f}^* = (dt^2 - d\vartheta^2) + \tfrac{2}{r}(-h'\cos\vartheta, -h'\sin\vartheta, r')\, dt\, d\vartheta \simeq dz^2,
$$

where we identify $i \simeq \tfrac{1}{r}(-h'\cos\vartheta, -h'\sin\vartheta, r') = \mathfrak{n}$, and the Hopf differential of $\mathfrak{f}$ is given by

$$
Q = \tfrac{r'h'' - (r+r'')h'}{4r}dz^2 = \tfrac{1}{4}r\kappa\, dz^2,
$$

where κ is the (hyperbolic) curvature of the meridian curve $\gamma : I \to H^2$ (cf., §3.7.6).

   Thus, for a surface of revolution, we have a globally defined holomorphic differential without zeros and the (conformal) curvature line net has no problems — the umbilics (which always appear as umbilic lines) are the zeros of the real factor in the decomposition of the Hopf differential: They are the points where the hyperbolic curvature of the meridian curve vanishes. And, with the holomorphic differential, the Christoffel transform of a surface of revolution is globally defined, up to translation periods in the case of a torus of revolution, that is, in case of a closed meridian curve.[19]

   A simple example is provided by revolving a hyperbolic geodesic that yields the catenoid as its Christoffel transform (see Figure 5.1):

$$
\begin{aligned}
\mathfrak{f}(t,\vartheta) &= (\tfrac{1}{\cosh t}\cos\vartheta, \tfrac{1}{\cosh t}\sin\vartheta, \tanh t), \\
\mathfrak{f}^*(t,\vartheta) &= (\cosh t \cos\vartheta, \cosh t \sin\vartheta, -t).
\end{aligned}
$$

However, building up an isothermic surface by attaching surfaces of revolution smoothly to a sphere as in Figure 3.1, the situation changes: If there is no global axis of revolution, there will be no globally defined holomorphic differential. And, at umbilics where two surfaces of revolution come together (see Figure 3.1), there will not even locally be a holomorphic differential.

---

[19] In that case, by Stokes' theorem, $-\int_\gamma \tfrac{dh}{r^2} = \int_G \tfrac{2}{r^3}dr \wedge dh$, where G denotes the domain enclosed by γ so that the Christoffel transform certainly does not close for embedded tori of revolution.

5.2.21 Example. As an example of a rather different flavor we finally want to discuss the ellipsoid: The Christoffel transforms of quadrics were already determined by K. Reinbek in [238]. However, we also intend to discuss the global existence of the Christoffel transform and the holomorphic differential.

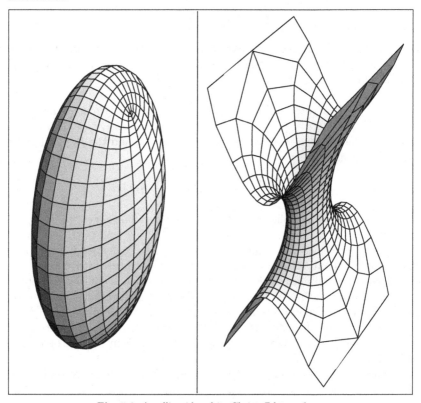

Fig. 5.2. An ellipsoid and its Christoffel transform

For this purpose we will start by giving a global (branched) curvature line parametrization of the ellipsoid[20)] using Jacobi elliptic functions:

$$\mathfrak{f}(x, y) = (ap\,\mathrm{cd}_p x\,\mathrm{nd}_q y,\, b\,pq\,\mathrm{sd}_p x\,\mathrm{sd}_q y,\, cq\,\mathrm{nd}_p x\,\mathrm{cd}_q y).$$

Here sd_p, cd_p, and nd_p denote the Jacobi elliptic functions of pole type d and module p, and q is the comodule, $p^2 + q^2 = 1$. It is straightforward

---

[20)] Using an ansatz that was suggested by our colleague, B. Springborn, this parametrization was mainly elaborated by my officemate, E. Tjaden, who also developed most of the code to verify the statements in this paragraph by symbolic algebra.

to verify that $\mathfrak{f}$ takes values on the ellipsoid with semiaxes $a < b < c$ by making use of the Pythagorean rules

$$1 + p^2\,\mathrm{sd}_p^2 \;=\; \mathrm{nd}_p^2 \;=\; \mathrm{sd}_p^2 + \mathrm{cd}_p^2.$$

The derivatives of these functions are given by

$$\mathrm{sd}_p' = \mathrm{cd}_p\,\mathrm{nd}_p, \quad \mathrm{cd}_p' = -q^2\,\mathrm{nd}_p\,\mathrm{sd}_p, \quad \text{and} \quad \mathrm{nd}_p' = p^2\,\mathrm{sd}_p\,\mathrm{cd}_p.$$

Then, one computes $\det(\mathfrak{f}_x, \mathfrak{f}_y, \mathfrak{f}_{xy}) = 0$, showing that (x,y) are conjugate parameters — which are orthogonal, $\mathfrak{f}_x \cdot \mathfrak{f}_y = 0$, as soon as

$$p = \sqrt{\tfrac{b^2 - a^2}{c^2 - a^2}}.$$

Thus, with the module p related to the semiaxes in this way, (x,y) are curvature line parameters of the ellipsoid. Computing the induced metric

$$
\begin{aligned}
|\mathfrak{f}_x|^2 &= (q^2\,\mathrm{nd}_p^2 x + p^2\,\mathrm{nd}_q^2 y - 1)\,(c^2 + (b^2 - c^2)\,\mathrm{nd}_p^2 x), \\
|\mathfrak{f}_y|^2 &= (q^2\,\mathrm{nd}_p^2 x + p^2\,\mathrm{nd}_q^2 y - 1)\,(a^2 + (b^2 - a^2)\,\mathrm{nd}_q^2 y),
\end{aligned}
\tag{5.7}
$$

we see that the curvature line parameters (x,y) can be replaced by *conformal* curvature line parameters $(u,v) = (u(x), v(y))$ by integrating

$$du = \sqrt{c^2 + (b^2 - c^2)\,\mathrm{nd}_p^2 x}\,dx, \quad dv = \sqrt{a^2 + (b^2 - a^2)\,\mathrm{nd}_q^2 y}\,dy. \tag{5.8}$$

Note that $1 \le \mathrm{nd}_p^2 \le \frac{1}{q^2}$, so that $c^2 + (b^2 - c^2)\,\mathrm{nd}_p^2 \ge a^2$ and the above square root is guaranteed to be real. In terms of these new parameters (u,v) the induced metric reads

$$|\mathfrak{f}_u|^2 du^2 + |\mathfrak{f}_v|^2 dv^2 = (q^2\,\mathrm{nd}_p^2 x(u) + p^2\,\mathrm{nd}_q^2 y(v) - 1)\,(du^2 + dv^2), \tag{5.9}$$

as we deduce from (5.7). Note that the metric degenerates exactly at the points where $\mathrm{nd}_p x = \mathrm{nd}_q y = 1$: These are the points that give the umbilics.

Rewriting (5.3) in terms of the parameters (x,y),

$$
\begin{aligned}
d_{(x,y)}\mathfrak{f}^* &= (c^2 + (b^2 - c^2)\,\mathrm{nd}_p^2 x)\mathfrak{f}_x^{-1}(x,y)\,dx \\
&\quad - (a^2 + (b^2 - a^2)\,\mathrm{nd}_q^2 y)\mathfrak{f}_y^{-1}(x,y)\,dy,
\end{aligned}
$$

an expression is obtained that can be integrated explicitly (cf., [238]): Up to translation, the Christoffel transform of $\mathfrak{f}$ is given by

$$\mathfrak{f}^*(x,y) = \left(-\tfrac{a}{p}\,\mathrm{artanh}\,\tfrac{\mathrm{cd}_p x}{\mathrm{nd}_q y},\; -\tfrac{b}{pq}\,\mathrm{arctan}\,\tfrac{\mathrm{sd}_p x}{\mathrm{sd}_q y},\; \tfrac{c}{q}\,\mathrm{artanh}\,\tfrac{\mathrm{cd}_q y}{\mathrm{nd}_p x}\right).$$

The Christoffel transform has four ends that correspond to the umbilics of the ellipsoid (see Figure 5.2) and it has a translation period in the direction of the middle semiaxis — that semiaxis that is orthogonal to the symmetry plane of the ellipsoid that contains its four umbilics.[21]

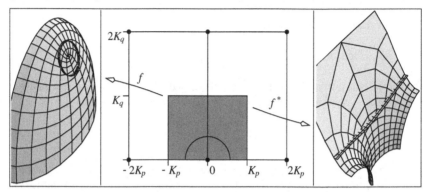

Fig. 5.3. A loop around an umbilic of the ellipsoid

Finally note that, in terms of conformal curvature line parameters (u, v), the holomorphic differential $q = dw^2$, where $w = u + iv$. Rewriting q in terms of the coordinates (x, y) using (5.8), we obtain a well-defined holomorphic[22] quadratic differential on the rectangular torus $\mathbb{R}^2/(4K_p\mathbb{Z} \times 4K_q\mathbb{Z})$ that is the natural domain of our parametrization.[23] This torus is a (branched) double cover of the ellipsoid, with $\mathfrak{f}$ being the covering map. Since any holomorphic quadratic differential on a Riemann surface of genus 0 has to vanish, our quadratic differential cannot descend (globally) to a holomorphic quadratic differential on the ellipsoid. However, because $d\mathfrak{f}^*$ is well defined on the ellipsoid punctured at its four umbilics, q has to descend to a holomorphic quadratic differential on the punctured ellipsoid.

To understand that q extends meromorphically to all of the ellipsoid,[24] we replace the curvature line parameters by (local) conformal coordinates. By symmetry it suffices to consider one umbilic, for example $\mathfrak{f}(0, 0)$. Now we first note that $\frac{du}{dx}$ and $\frac{dv}{dy}$ are even functions by (5.8) so that, if we arrange to have $u(0) = 0$ and $v(0) = 0$, then $x = x(u)$ and $y = y(v)$ are, with u and v, odd functions. Consequently, $\mathfrak{f}(-w) = \mathfrak{f}(w)$ since $\mathfrak{f}(-x, -y) = \mathfrak{f}(x, y)$,

---

[21] Note the similarity of this Christoffel transform with a Scherk tower (cf., [159]).

[22] That is, holomorphic with respect to the "correct" complex structure that is induced by $\mathfrak{f}$.

[23] The Jacobi elliptic functions with module p have period $4K_p$, where K_p denotes the complete elliptic integral of the first kind. As a consequence, our map $\mathfrak{f}$ is doubly periodic.

[24] We could also simply argue that $d\mathfrak{f}$ and $d\mathfrak{f}^*$ only contain meromorphic data, so that q, obtained from those by algebraic operations, must also be meromorphic. However, the following computation seems to be quite intuitive, showing also that q has simple poles at the umbilics.

and $z = w^2$ defines proper conformal[25] coordinates for the ellipsoid in a neighborhood of the umbilic. In terms of these new coordinates, the differential $q = dw^2 = \frac{1}{4z}dz^2$ has a simple pole — in particular, we see that it extends meromorphically to the entire ellipsoid, as proposed (cf., [158]).

5.3 Goursat's transformation

It is well known that (conformally parametrized: compare our discussions on conformality in §5.1.6) minimal surface patches $\mathfrak{f} : \mathbb{R}^2 \cong \mathbb{C} \supset U \to \mathbb{R}^3$ are exactly the real parts of holomorphic null curves[26] $\mathfrak{h} : U \to \mathbb{C}^3$: This is the classical (local) Weierstrass representation for minimal surfaces. Clearly, given an orthogonal transformation $A \in O(3, \mathbb{C})$, the map $\tilde{\mathfrak{h}} = A\mathfrak{h}$ is another holomorphic null curve giving rise to a new minimal surface $\tilde{\mathfrak{f}} = \operatorname{Re}\tilde{\mathfrak{h}}$, a "Goursat transform" of the original surface (cf., [129] and [167]; see [213] or [212]). However, any real orthogonal transformation $A \in O(3, \mathbb{R})$ only provides a motion of the original surface, so that, generically, there is a 3-parameter family of "essential" Goursat transformations for a given surface.

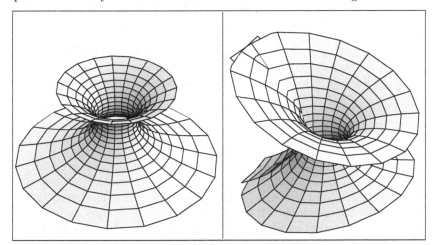

Fig. 5.4. Goursat transformations of the catenoid with respect to different axes

In [129] E. Goursat argues that every complex rotation can be decomposed into real rotations and an imaginary rotation (a rotation with an imaginary angle) about a prescribed axis, thus arriving at his formulas that

---

[25] Because $\operatorname{nd}_p^2(x) = 1 + p^2x^2 + o(x^2)$, the metric (5.9) vanishes quadratically (the lowest possible order) at $(x, y) = (0, 0)$. Since $|dz| = 2|w| \, |dw|$, the induced metric will appear nondegenerate in terms of z.

[26] $\mathfrak{h} : U \to \mathbb{C}^3$ is called a holomorphic null curve if $\mathfrak{h}_{\bar{z}} = 0$ and $|\mathfrak{h}_z|^2 = 0$. Note that we use the bilinearly extended scalar product on $\mathbb{C}^3$ and that $\mathfrak{h}_z = 2(\operatorname{Re}\mathfrak{h})_z$ by holomorphicity.

are commonly known as the Goursat transformation (cf., [184] or [101]). Note that the Goursat transformation depends on the direction of the (real) axis of the imaginary rotation relative to the surface: For example, choosing the axis of the imaginary rotation to be the axis of rotation of the catenoid just reparametrizes the catenoid whereas choosing it orthogonal to the axis of revolution gives a surface that is no longer a surface of revolution (see Figure 5.4). That is, Goursat parametrizes the 3-parameter family of essential Goursat transformations by the axis and the angle of the imaginary rotation.

It is a well-known fact (cf., [213] or [196]), which can be proved by linear algebra (see [258]), that the Hopf differential is invariant under Goursat transformations while the Gauss map is changed by a Möbius transformation. And conversely,[27] changing the Gauss map of a minimal surface by a Möbius transformation while keeping the Hopf differential gives rise to a Goursat transformation:

5.3.1 Lemma. *Two minimal surfaces* $\mathfrak{f}, \tilde{\mathfrak{f}} : M^2 \to \mathbb{R}^3$ *are Goursat transforms of each other if and only if their Hopf differentials are equal and their Gauss maps are Möbius equivalent.*

5.3.2 This, together with the fact that any minimal surface and its Gauss map form a Christoffel pair, is the key observation that justifies our terminology of a Goursat transformation for isothermic surfaces (cf., [148]):

5.3.3 Definition. *Given an isothermic surface* $\mathfrak{f} : M^2 \to \mathbb{H}$ *and a Möbius transformation* $\mu : \mathbb{H} \cup \{\infty\} \to \mathbb{H} \cup \{\infty\}$, *the surface* $\tilde{\mathfrak{f}} = (\mu \circ \mathfrak{f}^*)^*$ *will be called a Goursat transform of* $\mathfrak{f}$.

5.3.4 Note how this definition uses the interplay of the two symmetry groups of our notions: The notion of an "isothermic surface" is conformally invariant, while the notion of the "Christoffel transformation" is only invariant under similarities of the ambient space. Accordingly, we will see below that the Goursat transformation becomes inessential when using similarities in the construction, whereas an example will convince the reader that, generically, the Goursat transformations that correspond to "essential" Möbius transformations will indeed provide essential transforms of the original surface.

5.3.5 In a certain sense the Goursat transformations provide an action of the Möbius group on isothermic surfaces. Namely, let us, for a moment, assume that the Christoffel transformation is an honest involution, that is,

---

[27] Note that the Gauss map and the Hopf differential can serve as Weierstrass data for a minimal surface.

$\mathfrak{f}^{**} = \mathfrak{f}$ for any isothermic surface $\mathfrak{f}$ (cf., §5.2.4). Then

$$(\tilde{\mu} \circ ((\mu \circ \mathfrak{f}^*)^*)^*)^* = ((\tilde{\mu} \circ \mu) \circ \mathfrak{f}^*)^*$$

for Möbius transformations $\mu, \tilde{\mu} \in \mathit{M\ddot{o}b}(\mathbb{H} \cup \{\infty\})$. Of course, the Christoffel transformation is only well defined up to translation (if we somehow fix the scaling ambiguity), which spoils the above computation to a certain extent.

Nevertheless, it shows that, if $\tilde{\mathfrak{f}} = (\mu \circ \mathfrak{f}^*)^*$ is a Goursat transform of $\mathfrak{f}$, then $\mathfrak{f}$ is also a Goursat transform of $\tilde{\mathfrak{f}}$ — but we may have to adjust the inverse of the Möbius transformation μ by a translation to obtain a Möbius transformation $\tilde{\mu}$ that induces the "inverse" Goursat transformation $\mathfrak{f} = (\tilde{\mu} \circ \tilde{\mathfrak{f}}^*)^*$.

5.3.6 Writing the Möbius transformation μ as a fractional linear transformation, $\mu(q) = (a_{11}q + a_{12})(a_{21}q + a_{22})^{-1}$, we may use (5.4) to determine the (differential of the) corresponding Goursat transform $\tilde{\mathfrak{f}}$ of $\mathfrak{f}$ more explicitly: $\mathfrak{f}$ and $\tilde{\mathfrak{f}}$ are Goursat transforms of each other iff their Christoffel transforms satisfy

$$\tilde{\mathfrak{f}}^* = (a_{11}\mathfrak{f}^* + a_{12})(a_{21}\mathfrak{f}^* + a_{22})^{-1}. \tag{5.10}$$

As for the computation that led to our quaternionic formula of the Study determinant (see §4.2.5), we distinguish two cases:

5.3.7 If μ is a similarity, that is, if $a_{21} = 0$, then (5.10) yields

$$d\tilde{\mathfrak{f}}^* = a_{11}d\mathfrak{f}^* a_{22}^{-1}.$$

Note that, in this case, $a_{11} \neq 0$ as well as $a_{22} \neq 0$, since the Study determinant[28] of μ can be chosen to be $[\mu] = |a_{11}| \, |a_{22}| = 1$. Clearly, $\tilde{\psi} := a_{22}d\mathfrak{f}a_{11}^{-1}$ satisfies (5.4) and is closed, so that we may integrate $\tilde{\psi}$ to obtain a Christoffel transform of $\tilde{\mathfrak{f}}^*$; that is,

$$\tilde{\mathfrak{f}} = a_{22}\mathfrak{f}a_{11}^{-1}$$

provides the[29] corresponding Goursat transform of $\mathfrak{f}$. In particular, we have

5.3.8 Lemma. *If $\mu : \mathbb{H} \cup \{\infty\} \to \mathbb{H} \cup \{\infty\}$ is a similarity, $\mu(\infty) = \infty$, then any Goursat transform $\tilde{\mathfrak{f}} = (\mu \circ \mathfrak{f}^*)^*$ of an isothermic surface $\mathfrak{f} : M^2 \to \mathbb{H}$ is similar to the original surface.*

---

[28] Remember that the Study determinant is a well-defined notion for endomorphisms of $\mathbb{H}^2$.

[29] Remember that, away from umbilics, the Christoffel transformation is unique up to scaling and translation. At umbilics, the statement can easily be verified using conformal "curvature line coordinates" that we required to existx (see §5.2.1).

5.3.9 If, on the other hand, μ is an "essential" Möbius transformation, that is, not a similarity, then we may decompose it into a similarity and a "simple" essential Möbius transformation to obtain

$$
\begin{aligned}
\tilde{\mathfrak{f}}^* &= (a_{11}\mathfrak{f}^* + a_{12})(a_{21}\mathfrak{f}^* + a_{22})^{-1} \\
&= ((a_{12} - a_{11}a_{21}^{-1}a_{22})[\mathfrak{f}^* + a_{21}^{-1}a_{22}]^{-1} + a_{11})a_{21}^{-1}.
\end{aligned}
$$

By the "group structure" of the Goursat transformation (cf., §5.3.5), and since similarities yield similar surfaces, we may restrict ourselves to simple Möbius transformations of the form $q \mapsto (q - \mathfrak{m})^{-1}$, $\mathfrak{m} \in \mathbb{H}$. Then we compute

$$
d\tilde{\mathfrak{f}}^* = -(\mathfrak{f}^* - \mathfrak{m})^{-1}d\mathfrak{f}^*(\mathfrak{f}^* - \mathfrak{m})^{-1}.
$$

Employing (5.4) once more, we find that the ansatz $\tilde{\psi} := -(\mathfrak{f}^* - \mathfrak{m})d\mathfrak{f}\,(\mathfrak{f}^* - \mathfrak{m})$ provides the differential of the Goursat transform since $d\tilde{\psi} = 0$, by (5.4) again:

5.3.10 Lemma. *If $q \mapsto \mu(q) = (q - \mathfrak{m})^{-1}$ is an essential Möbius transformation, then the differential of the corresponding Goursat transform of an isothermic surface $\mathfrak{f} : M^2 \to \mathbb{H}$ is given by*

$$
d\tilde{\mathfrak{f}} = -(\mathfrak{f}^* - \mathfrak{m})d\mathfrak{f}\,(\mathfrak{f}^* - \mathfrak{m}). \tag{5.11}
$$

5.3.11 Note that we chose the sign (and scaling) of $d\tilde{\mathfrak{f}}$ in (5.11) such that, in case $\mathfrak{f}, \mathfrak{f}^* : M^2 \to \operatorname{Im} \mathbb{H}$ take values in Euclidean 3-space, the holomorphic differential $q = d\mathfrak{f} \cdot d\mathfrak{f}^*$ is not changed by the Goursat transformation (cf., §5.2.12). Namely, with (5.11) we find

$$
\tilde{q} = d\tilde{\mathfrak{f}} \cdot d\tilde{\mathfrak{f}}^* = (\mathfrak{f}^* - \mathfrak{m})q(\mathfrak{f}^* - \mathfrak{m})^{-1}.
$$

If $\mathfrak{n} : M^2 \to S^2$ denotes the Gauss map of $\mathfrak{f}$, then

$$
\tilde{\mathfrak{n}} = (\mathfrak{f}^* - \mathfrak{m})\mathfrak{n}(\mathfrak{f}^* - \mathfrak{m})^{-1} = (\mathfrak{f}^* - \mathfrak{m})^{-1}\mathfrak{n}(\mathfrak{f}^* - \mathfrak{m}),
$$

where the second equality follows since $(\mathfrak{f}^* - \mathfrak{m})^2 \in \mathbb{R}$ is real, so that $\tilde{\mathfrak{n}}$ is easily seen to be the Gauss map of $\tilde{\mathfrak{f}}$:

$$
\tilde{\mathfrak{n}} \cdot d\tilde{\mathfrak{f}} + d\tilde{\mathfrak{f}} \cdot \tilde{\mathfrak{n}} = (\mathfrak{f}^* - \mathfrak{m})(\mathfrak{n} \cdot d\mathfrak{f} + d\mathfrak{f} \cdot \mathfrak{n})(\mathfrak{f}^* - \mathfrak{m}) = 0.
$$

Hence, after identifying $\operatorname{span}\{1, \mathfrak{n}\} \cong M \times \mathbb{C} \cong \operatorname{span}\{1, \tilde{\mathfrak{n}}\}$, the two holomorphic differentials are identified, $q \simeq \tilde{q}$.

5.3.12 Example. The following example may convince the reader that the Goursat transformation for isothermic surfaces can indeed essentially transform a surface: Consider a Christoffel pair $\mathfrak{f} = \mathfrak{h}j$ and $\mathfrak{f}^* = -j\mathfrak{g}$ of holomorphic functions, as in §5.2.11, and let μ denote the inverse stereographic projection

$$q \mapsto \mu(q) = -i - 2(i + q)^{-1} \tag{5.12}$$

(cf., §P.1.3). It is a straightforward computation to see that (5.12) really does provide an inverse stereographic projection: With $q = {}_3j \in \mathbb{C}j$, we have

$$
\begin{aligned}
-i - 2(i + {}_3j)^{-1} &= i(i - {}_3j)(i + {}_3j)^{-1} \\
&= -\tfrac{i}{1 + |{}_3|^2}(i - {}_3j)(i + {}_3j) \\
&= \tfrac{1}{1 + |{}_3|^2}\left((1 - |{}_3|^2)\,i + 2{}_3j\right).
\end{aligned}
$$

Since the Christoffel transformation is only well defined up to translation, we may neglect the translation by $-i$ in (5.12), so that we can apply our formula (5.11) to obtain the differential of the Goursat transform $\tilde{\mathfrak{f}}$ of $\mathfrak{f}$:

$$
\begin{aligned}
d\tilde{\mathfrak{f}} &= \tfrac{1}{2}(i - j\mathfrak{g})d\mathfrak{h}j(i - j\mathfrak{g}) \\
&= i\mathrm{Re}\,(\mathfrak{g}d\mathfrak{h}) + \tfrac{1}{2}j\mathrm{Re}\,((1 - \mathfrak{g}^2)d\mathfrak{h}) - \tfrac{1}{2}k\mathrm{Re}\,(i(1 + \mathfrak{g}^2)d\mathfrak{h}).
\end{aligned} \tag{5.13}
$$

Thus the Goursat transform of the holomorphic function $\mathfrak{f} = \mathfrak{h}j$, with Christoffel partner $\mathfrak{f}^* = -j\mathfrak{g}$, yields a minimal surface $\tilde{\mathfrak{f}}$ with Weierstrass data $\mathfrak{g}, \mathfrak{h}$. In fact, (5.13) suggests that the Weierstrass representation for minimal surfaces in Euclidean 3-space can be considered as a special case of the Goursat transformation.

5.3.13 To define the Goursat transformation for isothermic surfaces in §5.3.3 we used the interplay of Möbius geometry as the natural ambient geometry of isothermic surfaces and Euclidean geometry (or, more accurately, the geometry of similarities of the flat ambient space) as the natural ambient geometry for the Christoffel transformation. Now we want to take a different viewpoint, studying the family of all possible Christoffel transforms of an isothermic surfaces in conformal geometry — this will lead to a "master Christoffel transform" that provides all possible Christoffel transforms by certain projections.

First we study the effect of different stereographic projections (see §4.9.5) of an isothermic surface $f : M^2 \to \mathbb{HP}^1$ on their Christoffel transforms:

5.3.14 Lemma. *Let* $\mathfrak{f} = (\nu_0 \circ f)(\nu_\infty \circ f)^{-1}$ *and* $\tilde{\mathfrak{f}} = (\tilde{\nu}_0 \circ f)(\tilde{\nu}_\infty \circ f)^{-1}$ *be two stereographic projections of an isothermic surface* $f : M^2 \to \mathbb{HP}^1$. *Then their Christoffel transforms* $\mathfrak{f}^*$ *and* $\tilde{\mathfrak{f}}^*$ *are Goursat transforms of each other, which are similar if the point* $\infty = \mathbb{H}\nu_\infty \in (\mathbb{HP}^1)^\perp$ *is the same for both projections.*

5.3.15 *Proof.* Writing[30] $\tilde{\nu}_\infty = a_{\infty\infty}\nu_\infty + a_{\infty 0}\nu_0$ and $\tilde{\nu}_0 = a_{0\infty}\nu_\infty + a_{00}\nu_0$,

---

[30] Note that (ν_∞, ν_0) and $(\tilde{\nu}_\infty, \tilde{\nu}_0)$ are bases of $(\mathbb{H}^2)^*$, since the points $\mathbb{H}\nu_0 = 0 \neq \infty = \mathbb{H}\nu_\infty$ and similarly for $\tilde{0}$ and $\tilde{\infty}$.

we obtain $\tilde{\mathfrak{f}}$ as a Möbius transform of $\mathfrak{f}$:

$$\begin{aligned}
\tilde{\mathfrak{f}} &= (a_{00}\nu_0 \circ f + a_{0\infty}\nu_\infty \circ f)(a_{\infty 0}\nu_0 \circ f + a_{\infty\infty}\nu_\infty \circ f)^{-1} \\
&= (a_{00}\mathfrak{f} + a_{0\infty})(a_{\infty 0}\mathfrak{f} + a_{\infty\infty})^{-1}.
\end{aligned}$$

Therefore, the Christoffel transforms $\mathfrak{f}^*$ of $\mathfrak{f}$ and $\tilde{\mathfrak{f}}^*$ of $\tilde{\mathfrak{f}}$ are Goursat transforms of each other. The above Möbius transformation is a similarity if and only if $a_{\infty 0} = 0$, that is, if and only if $\tilde{\infty} = \infty$ (see §4.9.5). In that case, the Goursat transformation is a similarity by §5.3.8.   ◁

5.3.16 We are now going to derive the "master" Christoffel transform: For this purpose we first restrict ourselves to "simple" essential Möbius transformations of the form $q \mapsto (q - \mathfrak{m})^{-1}$, as in §5.3.10.

Thus, let $f : M^2 \to \mathbb{H}P^1$ be an isothermic surface and choose homogeneous coordinates $v_0, v_\infty \in \mathbb{H}^2$ for two points $0, \infty \in \mathbb{H}P^1$. This gives us a stereographic projection $\mathfrak{f} = (\nu_0 \circ f)(\nu_\infty \circ f)^{-1} : M^2 \to \mathbb{H}$ of f, where $\nu_\infty, \nu_0 \in (\mathbb{H}^2)^*$ form the pseudodual basis of (v_∞, v_0), $v_0\nu_\infty + v_\infty\nu_0 = id_{\mathbb{H}^2}$. In order to simplify the formulas we work with homogeneous coordinates $f : M^2 \to \mathbb{H}^2$ of our isothermic surface such that $\nu_\infty \circ f \equiv 1$: Then $f = v_\infty\mathfrak{f} + v_0$.

Now we replace our first choice of homogeneous coordinates v_0 and v_∞ by

$$\left.\begin{aligned}
\tilde{v}_0 &= & v_\infty \\
\tilde{v}_\infty &= v_0 &+ v_\infty\mathfrak{m}
\end{aligned}\right\} \quad \leftrightarrow \quad \left\{\begin{aligned}
\tilde{\nu}_0 &= & \nu_\infty \\
\tilde{\nu}_\infty &= \nu_0 &- \mathfrak{m}\nu_\infty.
\end{aligned}\right.$$

Indeed, stereographically projecting f with this new basis yields

$$\tilde{\mathfrak{f}} = (\tilde{\nu}_0 \circ (v_0 + v_\infty\mathfrak{f}))(\tilde{\nu}_\infty \circ (v_0 + v_\infty\mathfrak{f}))^{-1} = (\mathfrak{f} - \mathfrak{m})^{-1}.$$

In particular, we have $\tilde{\nu}_\infty \circ f = \mathfrak{f} - \mathfrak{m}$, and $f^\perp\tilde{v}_\infty = -(\mathfrak{f} - \mathfrak{m})$ if we choose homogeneous coordinates $f^\perp = \nu_0 - \mathfrak{f}\nu_\infty$ for our isothermic immersion as a map into $(\mathbb{H}P^1)^\perp$. Hence, by §5.3.10, the differential of $\tilde{\mathfrak{f}}^*$ can be written as

$$d\tilde{\mathfrak{f}}^* = \tilde{\nu}_\infty \tau \tilde{v}_\infty \quad \text{where} \quad \tau := f \, d\mathfrak{f}^* f^\perp : TM \to \mathfrak{gl}(2, \mathbb{H}).$$

Because any Möbius transformation can be obtained by postcomposing an essential transformation $q \mapsto (q - \mathfrak{m})^{-1}$ by a similarity — that, by §4.9.5, yields only rescalings $\tilde{v}_\infty \to \tilde{v}_\infty a_\infty$ and $\tilde{\nu}_\infty \to a_0^{-1}\tilde{\nu}_\infty$ of v_∞ and ν_∞ — one can indeed get (the differentials of) *all* Goursat transforms of $\mathfrak{f}^*$, that is, *all possible* Christoffel transforms of our isothermic surface $f : M^2 \to \mathbb{H}P^1$ after stereographic projection, from τ.

Moreover, note that $d\tau = 0$, so that τ can be integrated to obtain a map $\mathfrak{F}^*$ that "contains" all possible Christoffel transformations of f, up to translation (the constant of integration for $\mathfrak{F}^*$):

5.3.17 Lemma. *Let $f : M^2 \to \mathbb{HP}^1$ be an isothermic surface, and let f^* denote a Christoffel transform of a stereographic projection f of f, obtained by a choice $v_0, v_\infty \in \mathbb{H}^2$ or $\nu_0, \nu_\infty \in (\mathbb{H}^2)^*$, respectively, of homogeneous coordinates for two points $0, \infty \in \mathbb{HP}^1 \cong (\mathbb{HP}^1)^\perp$. Then there exists a map*

$$\mathfrak{F}^* : M^2 \supset U \to \mathfrak{gl}(2, \mathbb{H}), \quad d\mathfrak{F}^* = \tau := (v_0 + v_\infty f) df^* (\nu_0 - f\nu_\infty)$$

that provides (up to translation) the Christoffel transform $\tilde{f}^*$ of any stereographic projection $\tilde{f} = (\tilde{\nu}_0 \circ f)(\tilde{\nu}_\infty \circ f)^{-1}$ of f via $\tilde{f}^* = \tilde{\nu}_\infty \mathfrak{F}^* \tilde{v}_\infty$.

5.3.18 So far we have used Euclidean data to define the form τ that integrates to give us this "master Christoffel transform" of an isothermic surface. Even though it is clear (by the property of being the universal Christoffel transform) that $\mathfrak{F}^*$ is indeed an object that belongs to f as an immersion into the conformal $\mathbb{HP}^1$ and does not depend on any choices, it might be good to convince ourselves by giving an invariant description:

5.3.19 Definition and Lemma. *Let $f : M^2 \to \mathbb{HP}^1$ be an immersion. Then a 1-form $\tau : TM \to \mathfrak{gl}(2, \mathbb{H})$ with $\ker \tau = \operatorname{im} \tau = f$ will be called a retraction form of f.*

Any isothermic immersion $f : M^2 \to \mathbb{HP}^1$ has a closed retraction form

$$\tau : TM \to \mathfrak{gl}(2, \mathbb{H}), \quad \tau = (v_0 + v_\infty f) df^* (\nu_0 - f\nu_\infty) \qquad (5.14)$$

where $v_0, v_\infty \in \mathbb{H}^2$ and $\nu_0, \nu_\infty \in (\mathbb{H}^2)^*$ are homogeneous coordinates of two points $0, \infty \in \mathbb{HP}^1 \cong (\mathbb{HP}^1)^\perp$ and f^* is a Christoffel transform of the corresponding stereographic projection $f = (\nu_0 f)(\nu_\infty f)^{-1}$ of f.

Conversely, if an immersion $f : M^2 \to \mathbb{HP}^1$ possesses a closed retraction form, then f is isothermic and τ is of the form (5.14).

5.3.20 *Proof.* We have already seen a proof of the first claim in §5.3.16.

To verify the second statement we choose v_∞, v_0 and ν_∞, ν_0 as in §5.3.16 and let f denote the corresponding stereographic projection, so that we have $f = (v_0 + v_\infty f)\mathbb{H}$ and $f^\perp = \mathbb{H}(\nu_0 - f\nu_\infty)$. The retraction form τ of f then must be of the form

$$\tau = (v_0 + v_\infty f)\omega(\nu_0 - f\nu_\infty)$$

with some 1-form $\omega : TM \to \mathbb{H}$. The closeness of τ, $0 = d\tau$, then reads

$$0 = v_\infty df \wedge \omega (\nu_0 - f\nu_\infty) + (v_0 + v_\infty f) d\omega (\nu_0 - f\nu_\infty) + (v_0 + v_\infty f)\omega \wedge df\nu_\infty.$$

Extracting "components" yields

$$\begin{aligned} 0 &= (\nu_0 - f\nu_\infty) d\tau v_\infty &= df \wedge \omega, \\ 0 &= \nu_\infty d\tau (v_0 + v_\infty f) &= \omega \wedge df, \quad \text{and} \\ 0 &= \nu_\infty d\tau v_\infty &= d\omega, \end{aligned}$$

so that $\mathfrak{f}$ must be isothermic and $\omega = d\mathfrak{f}^*$ is the differential of a Christoffel transform of $\mathfrak{f}$, by §5.2.6. ◁

5.3.21 Example. As an example we reconsider the case of a totally umbilic isothermic surface: We consider $f = (v_0 - v_\infty j\mathfrak{g})H$ with some holomorphic function $\mathfrak{g} : \mathbb{C} \supset U \to \mathbb{C} \cup \{\infty\}$. A Christoffel transform $\mathfrak{f}^* = \mathfrak{h}j$ of $\mathfrak{f} = -j\mathfrak{g}$ is then only determined after some choice — of a "curvature line net," of a holomorphic quadratic differential $q^* = d\mathfrak{h}\, d\mathfrak{g}$, or we may just prescribe the Christoffel transform $\mathfrak{f}^*$ itself, because any holomorphic function $\mathfrak{h}$ will provide a valid Christoffel transform $\mathfrak{f}^*$ of $\mathfrak{f}$ (cf., §5.2.11).

Instead of carrying our bases (v_∞, v_0) and (ν_∞, ν_0) through all computations, let us use matrix notations by choosing

$$v_0 = \begin{pmatrix} 0 \\ 1 \end{pmatrix}, \quad v_\infty = \begin{pmatrix} 1 \\ 0 \end{pmatrix} \quad \leftrightarrow \quad \begin{cases} \nu_0 &= (1,0) \\ \nu_\infty &= (0,1). \end{cases}$$

Then, with the above choice of $\mathfrak{h}j$ as a Christoffel transform of the stereographic projection $-j\mathfrak{g}$ of $f = \begin{pmatrix} -j\mathfrak{g} \\ 1 \end{pmatrix}$, we obtain[31]

$$\tau = \begin{pmatrix} \mathfrak{f}d\mathfrak{f}^* & -\mathfrak{f}d\mathfrak{f}^*\mathfrak{f} \\ d\mathfrak{f}^* & -d\mathfrak{f}^*\mathfrak{f} \end{pmatrix} = \begin{pmatrix} -j & 0 \\ 0 & 1 \end{pmatrix} \begin{pmatrix} \mathfrak{g}d\mathfrak{h} & -\mathfrak{g}^2 d\mathfrak{h} \\ d\mathfrak{h} & -\mathfrak{g}d\mathfrak{h} \end{pmatrix} \begin{pmatrix} j & 0 \\ 0 & 1 \end{pmatrix}.$$

Using $\tilde{v}_\infty = \begin{pmatrix} -i \\ 1 \end{pmatrix} \simeq (1,i) = \tilde{\nu}_\infty$ to project $\mathfrak{F}^*$, we recover, up to some factor,[32] our Weierstrass formula from §5.3.12:

$$(1,i)\tau \begin{pmatrix} -i \\ 1 \end{pmatrix} = (1,i) \begin{pmatrix} -j\mathfrak{g} & 0 \\ 0 & 1 \end{pmatrix} \begin{pmatrix} d\mathfrak{h}j & d\mathfrak{h}j \\ d\mathfrak{h}j & d\mathfrak{h}j \end{pmatrix} \begin{pmatrix} 1 & 0 \\ 0 & j\mathfrak{g} \end{pmatrix} \begin{pmatrix} -i \\ 1 \end{pmatrix}$$

$$= -(i - j\mathfrak{g})d\mathfrak{h}j(i - j\mathfrak{g}).$$

Note that, since $\tau = J^{-1}\tau^{\mathbb{C}}J$ is purely complex after a suitable change of basis, complex analysis[33] can be used to integrate τ in the present setup, so that the "recipe" to produce minimal surfaces described in this example really qualifies as a "Weierstrass representation" — for a minimal surface along with all its Goursat transforms.

To have an explicit example, let $\mathfrak{g}(z) = \tanh \frac{z}{2}$ and $\mathfrak{h}(z) = z + \sinh z$. Then our quadratic differential $q = d\mathfrak{g}\, d\mathfrak{h} = dz^2$, so that our "curvature

---

[31] We will see later that the striking similarity with the formulas of M. Umehara and K. Yamada in [285] is not by accident.

[32] This factor is caused by the stretching that was used in §5.3.12 to have the considered Möbius transformation restrict to the inverse stereographic projection.

[33] Also note how nicely the complex structure of the domain mixes with the quaternionic structure of the ambient space.

line parameters" become $x = \operatorname{Re} z$ and $y = \operatorname{Im} z$. Computing τ and then integrating we obtain

$$
\tau(z) = \begin{pmatrix} -j & 0 \\ 0 & 1 \end{pmatrix} \begin{pmatrix} \sinh z \, dz & dz - \cosh z \, dz \\ dz + \cosh z \, dz & -\sinh z \, dz \end{pmatrix} \begin{pmatrix} j & 0 \\ 0 & 1 \end{pmatrix} \quad \text{and}
$$

$$
\mathfrak{F}^*(z) = \begin{pmatrix} -j & 0 \\ 0 & 1 \end{pmatrix} \begin{pmatrix} \cosh z & z - \sinh z \\ z + \sinh z & -\cosh z \end{pmatrix} \begin{pmatrix} j & 0 \\ 0 & 1 \end{pmatrix}.
$$

Now we choose $\tilde{v}_\infty = \begin{pmatrix} -\lambda i \\ 1 \end{pmatrix}$ and $\tilde{v}_\infty = (1, \lambda i)$ to extract a Christoffel transform: With $z = x + iy$ we find

$$
\tilde{f}^*(x,y) = \tilde{v}_\infty \mathfrak{F}^*(z) \tilde{v}_\infty = -2\lambda \left\{ \begin{array}{l} i \cdot \cosh x \cos y \\ + \; j \cdot (\frac{1+\lambda^2}{2\lambda} x - \frac{1-\lambda^2}{2\lambda} \sinh x \cos y) \\ + \; k \cdot (\frac{1+\lambda^2}{2\lambda} \cosh x \sin y - \frac{1-\lambda^2}{2\lambda} y) \end{array} \right\}.
$$

For $\lambda = 1$, this is the catenoid with the j-axis as its axis of rotation; for other λ's, $\tilde{f}^*$ is a Goursat transform (of course!) of this catenoid that can be obtained by an imaginary rotation of the holomorphic null curve of the catenoid about the i-axis (cf., Figure 5.1): substituting $\lambda = e^\alpha$ we have

$$
\tilde{f}^*(x,y) = -2\lambda \operatorname{Re}\left[\begin{pmatrix} 1 & 0 & 0 \\ 0 & \cos i\alpha & \sin i\alpha \\ 0 & -\sin i\alpha & \cos i\alpha \end{pmatrix} \begin{pmatrix} \cosh z \\ z \\ -i \sinh z \end{pmatrix} \right].
$$

5.4 Darboux's transformation

We already learned about the Darboux transformation for isothermic surfaces in §3.2.5, in the context of Blaschke's problem (see Section 3.1). In this section we will take a completely different viewpoint, partially based on our developments in [147] and on discussions with F. Burstall and U. Pinkall (cf., [52]), and make contact with our earlier developments only at the end of this section. In fact, a notion of "Darboux transformation" can be introduced in the much more general setting of maps into an associative algebra, via a Riccati equation similar to the one below (see [49]).

So far, the transformation theory for isothermic surfaces that we developed in the present chapter has not been very different from the classical theory; we just "quaternionified" the theory that we touched on in Section 3.3. The first real achievement using quaternions might have been the retraction form from §5.3.19 that provided the "master Christoffel transform" (see §5.3.17). We will stress the use of this retraction form further — we will see that it can be viewed as *the* key object in the transformation theory of isothermic surfaces: In this way we will obtain an extremely powerful tool that will lead us quickly to the heart of the transformation theory.

As a preparation we will first recall a well-known relation between solutions of Riccati (-type partial) differential equations and linear systems:

5.4.1 Lemma. *Let* $\omega : TM \to \mathfrak{gl}(2, \mathbb{H})$ *be some* $\mathfrak{gl}(2, \mathbb{H})$*-valued 1-form.*

If $g : M^2 \to \mathbb{H}^2 \setminus \{0\}$ *satisfies the linear system* $dg + \omega g = 0$*, then any stereographic projection* $\mathfrak{g} = (\nu_0 g)(\nu_\infty g)^{-1} : M^2 \to \mathbb{H}$ *of* $g\mathbb{H} : M^2 \to \mathbb{HP}^1$ *satisfies the Riccati-type partial differential equation*

$$d\mathfrak{g} = \mathfrak{g}\,\omega_{\infty\infty}\,\mathfrak{g} + \mathfrak{g}\,\omega_{\infty 0} - \omega_{0\infty}\mathfrak{g} - \omega_{00}, \qquad (5.15)$$

where the $\omega_{ij} = \nu_i \omega \nu_j$ *denote the "components" of* ω *with respect to the basis* (ν_∞, ν_0) *used for the stereographic projection.*

Conversely, if[34] $d\omega + \frac{1}{2}[\omega \wedge \omega] = 0$ *and* $\mathfrak{g} : M^2 \to \mathbb{H}$ *satisfies* (5.15)*, then there is (locally) a function* $a : M^2 \supset U \to \mathbb{H}$ *such that* $g = (\nu_0 + \nu_\infty \mathfrak{g})a$ *satisfies the linear system* $dg + \omega g = 0$*.*

5.4.2 *Proof.* To verify the two statements we fix a basis as in the above example,

$$v_0 = \begin{pmatrix} 0 \\ 1 \end{pmatrix}, \quad v_\infty = \begin{pmatrix} 1 \\ 0 \end{pmatrix} \quad \leftrightarrow \quad \begin{cases} \nu_0 &= (1, 0) \\ \nu_\infty &= (0, 1), \end{cases}$$

and use the respective matrix representation,[35] so that

$$\omega = \begin{pmatrix} \omega_{0\infty} & \omega_{00} \\ \omega_{\infty\infty} & \omega_{\infty 0} \end{pmatrix} \quad \text{and} \quad g = \begin{pmatrix} \mathfrak{g} \\ 1 \end{pmatrix} a$$

with some function $a : M^2 \to \mathbb{H}$. Now the linear system $dg + \omega g = 0$ reads

$$
\begin{aligned}
0 &= (d\mathfrak{g} + \omega_{0\infty}\mathfrak{g} + \omega_{00})\,a + \mathfrak{g}\,da, \\
0 &= da + (\omega_{\infty\infty}\mathfrak{g} + \omega_{\infty 0})\,a.
\end{aligned}
$$

Inserting the second equation into the first yields equivalence of the first equation with the Riccati equation since $a \neq 0$. As a consequence, we have the first assertion. For the second assertion it suffices to verify the complete integrability of the second equation:

$$
\begin{aligned}
0 &= d[(\omega_{\infty\infty}\mathfrak{g} + \omega_{\infty 0})\,a] \\
&= [d(\omega_{\infty\infty}\mathfrak{g} + \omega_{\infty 0}) + (\omega_{\infty\infty}\mathfrak{g} + \omega_{\infty 0}) \wedge (\omega_{\infty\infty}\mathfrak{g} + \omega_{\infty 0})]\,a.
\end{aligned}
$$

---

[34] Note that the flatness condition on ω becomes obsolete in the case of an ordinary Riccati equation. Remember that, in terms of matrices, $[\omega \wedge \omega] = 2\omega \wedge \omega$, where the right term means matrix multiplication with $\wedge$ on the components (cf., §1.7.4).

[35] We could just as well carry our abstract bases (v_∞, v_0) and (ν_∞, ν_0) around. However, it might be profitable to get acquainted with the interplay of our pseudodual bases and their matrix representations; the abstract proof shall be left to the reader as an exercise.

This is easily obtained using our flatness assumption $d\omega + \frac{1}{2}[\omega \wedge \omega] = 0$ for ω and the Riccati equation for $\mathfrak{g}$. ◁

5.4.3 Note that this lemma can be reformulated in a projective setting: If g satisfies our linear system $dg + \omega g = 0$, then, in particular, $dg + \omega g \parallel g$, which is a projectively invariant statement. And, in fact, this projective version is enough to derive the Riccati equation — as one might expect, because the stereographic projection $\mathfrak{g}$ depends only on the map $g : M^2 \to \mathbb{HP}^1$: Writing $g = \left(\begin{smallmatrix} \mathfrak{g} \\ 1 \end{smallmatrix} \right)$ we find

$$ d \begin{pmatrix} \mathfrak{g} \\ 1 \end{pmatrix} + \begin{pmatrix} \omega_{0\infty} & \omega_{00} \\ \omega_{\infty\infty} & \omega_{\infty 0} \end{pmatrix} \begin{pmatrix} \mathfrak{g} \\ 1 \end{pmatrix} = \begin{pmatrix} \mathfrak{g} \\ 1 \end{pmatrix} \eta $$

if and only if $\eta = (\omega_{\infty\infty}\mathfrak{g} + \omega_{\infty 0})$ and $d\mathfrak{g} + \omega_{0\infty}\mathfrak{g} + \omega_{00} = \mathfrak{g}\eta$. This last equation is nothing but our Riccati equation when substituting η.

Thus the Maurer-Cartan equation for ω is only needed when seeking a parallel lift of $g\mathbb{H} : M^2 \to \mathbb{HP}^1$:

5.4.4 Corollary. *A map* $g : M^2 \to \mathbb{HP}^1$ *satisfies* $dg + \omega g \parallel g$ *if and only if any stereographic projection* $\mathfrak{g}$ *satisfies a Riccati-type partial differential equation* (5.15),

$$ d\mathfrak{g} = \mathfrak{g}\,\omega_{\infty\infty}\mathfrak{g} + \mathfrak{g}\,\omega_{\infty 0} - \omega_{0\infty}\mathfrak{g} - \omega_{00}, $$

with the components $\omega_{ij} = \nu_i \omega \nu_j$ *of* ω *taken with respect to the basis* (v_∞, v_0) *that is used for the stereographic projection.*

5.4.5 However, the Maurer-Cartan equation $0 = d\omega + \frac{1}{2}[\omega \wedge \omega]$ for ω guarantees the integrability of the Riccati equation: When taking exterior derivatives in (5.15) we obtain

$$
\begin{aligned}
d^2\mathfrak{g} &= d[\mathfrak{g}\,\omega_{\infty\infty}\mathfrak{g} + \mathfrak{g}\,\omega_{\infty 0} - \omega_{0\infty}\mathfrak{g} - \omega_{00}] \\
&= \mathfrak{g}(d\omega_{\infty\infty} + \omega_{\infty\infty} \wedge \omega_{0\infty} + \omega_{\infty 0} \wedge \omega_{\infty\infty})\mathfrak{g} \\
&\quad + \mathfrak{g}(d\omega_{\infty 0} + \omega_{\infty\infty} \wedge \omega_{00} + \omega_{\infty 0} \wedge \omega_{\infty 0}) \\
&\quad - (d\omega_{0\infty} + \omega_{0\infty} \wedge \omega_{0\infty} + \omega_{00} \wedge \omega_{\infty\infty})\mathfrak{g} \\
&\quad - (d\omega_{00} + \omega_{0\infty} \wedge \omega_{00} + \omega_{00} \wedge \omega_{\infty 0}).
\end{aligned}
$$

Hence the Riccati equation can be considered as a — locally solvable — initial value problem if ω satisfies the Maurer-Cartan equation:

5.4.6 Lemma. *The Riccati equation* $d\mathfrak{g} = \mathfrak{g}\,\omega_{\infty\infty}\mathfrak{g} + \mathfrak{g}\,\omega_{\infty 0} - \omega_{0\infty}\mathfrak{g} - \omega_{00}$ *is completely integrable as soon as* $d\omega + \frac{1}{2}[\omega \wedge \omega] = 0$.

5.4.7 Now we want to apply these general facts about linear systems and Riccati equations in the case where ω is a closed retraction form τ of an isothermic surface $f : M^2 \to \mathbb{HP}^1$ (see §5.3.19). Clearly τ satisfies the Maurer-Cartan equation: Since τ is a retraction form, we have $\tau \wedge \tau = 0$ and $d\tau = 0$ by assumption. Hence the linear system $d\hat{f} + \tau \hat{f} = 0$ is completely integrable,[36] as is the corresponding Riccati-type differential equation

$$d\hat{f} = \hat{f} \, df^* \hat{f} - \hat{f} \, df^* f - f \, df^* \hat{f} + f \, df^* f.$$

Note that, away from umbilics, all closed retraction forms are (constant, nonzero) real multiples of each other, just as the Christoffel transformation is unique up to real scaling (and translation). This suggests that we should consider the 1-parameter family

$$(\tau_\lambda)_{\lambda \in \mathbb{R} \setminus \{0\}}, \quad \tau_\lambda := \lambda \tau$$

of closed retraction forms, where $\tau = \tau_1$ is some "initial" retraction form — obtained from a choice of conformal curvature line parameters or a polarization $q = df \cdot df^*$ in the codimension 1 case (cf., §5.2.18).

In the sequel we will always assume the scaling of the retraction form τ_1 and "the" Christoffel transform f^* of a stereographic projection f of f to be adjusted,

$$df^* = \nu_\infty \tau_1 v_\infty. \tag{5.16}$$

Thus an isothermic surface in $\mathbb{HP}^1$ has (locally) a $(4 + 1)$-parameter family of Darboux transforms in the following sense:

5.4.8 Definition. *Let $\tau_\lambda : TM \to \mathfrak{gl}(2, \mathbb{H})$ be a closed retraction form of an isothermic surface $f : M^2 \to \mathbb{HP}^1$. Any solution $\hat{f} : M^2 \to \mathbb{HP}^1$ of the linear system $d\hat{f} + \tau_\lambda \hat{f} \parallel \hat{f}$ is called a Darboux transform of f.*

5.4.9 To see that this definition generalizes our previous definition of the Darboux transformation in §3.2.5, we derive the geometric properties of $\hat{f}$.

First, to see that $\hat{f}$ is isothermic, we employ the corresponding Riccati equation

$$d\hat{f} = \lambda(\hat{f} - f) df^* (\hat{f} - f) \tag{5.17}$$

that the stereographic projections $\hat{f}$ of $\hat{f}$ and f of f satisfy: If we now define

$$\tilde{f} := f^* + \tfrac{1}{\lambda}(\hat{f} - f)^{-1}, \tag{5.18}$$

it is straightforward to check that $\tilde{f} = \hat{f}^*$ is a Christoffel transform of $\hat{f}$. Namely, using (5.17) to substitute $d\hat{f}$ in $d\tilde{f}$, we find

$$d\tilde{f} = \tfrac{1}{\lambda}(\hat{f} - f)^{-1} df (\hat{f} - f)^{-1},$$

---

[36] That is, "completely integrable" in the sense that no additional integrability condition has to be fulfilled.

so that $d\tilde{\mathfrak{f}} \wedge d\hat{\mathfrak{f}} = d\hat{\mathfrak{f}} \wedge d\tilde{\mathfrak{f}} = 0$, which characterizes $\tilde{\mathfrak{f}}$ as a Christoffel transform of $\hat{\mathfrak{f}}$, by §5.2.6.

5.4.10 Lemma. *Any Darboux transform $\hat{f}$ of an isothermic surface f is isothermic.*

If $\mathfrak{f}$ and $\hat{\mathfrak{f}}$ are stereographic projections of f and $\hat{f}$, and the Christoffel transform $\mathfrak{f}^*$ of $\mathfrak{f}$ is scaled according to (5.16), *then the Christoffel transform $\hat{\mathfrak{f}}^*$ of $\hat{\mathfrak{f}}$ can be scaled and positioned so that*

$$1 = \lambda(\hat{\mathfrak{f}}^* - \mathfrak{f}^*)(\hat{\mathfrak{f}} - \mathfrak{f}),$$

where λ is the parameter of the Darboux transformation.

5.4.11 As a consequence of this formula for the Christoffel transform $\hat{\mathfrak{f}}^*$ of $\hat{\mathfrak{f}}$, we can now determine "the" closed retraction form (5.14) of $\hat{f}$: Using matrix representation with respect to the basis (v_∞, v_0) that we used for the stereographic projection,

$$\hat{\tau} = \frac{1}{\lambda} \begin{pmatrix} \hat{\mathfrak{f}}(\hat{\mathfrak{f}} - \mathfrak{f})^{-1} d\mathfrak{f}(\hat{\mathfrak{f}} - \mathfrak{f})^{-1} & -\hat{\mathfrak{f}}(\hat{\mathfrak{f}} - \mathfrak{f})^{-1} d\mathfrak{f}(\hat{\mathfrak{f}} - \mathfrak{f})^{-1}\hat{\mathfrak{f}} \\ (\hat{\mathfrak{f}} - \mathfrak{f})^{-1} d\mathfrak{f}(\hat{\mathfrak{f}} - \mathfrak{f})^{-1} & -(\hat{\mathfrak{f}} - \mathfrak{f})^{-1} d\mathfrak{f}(\hat{\mathfrak{f}} - \mathfrak{f})^{-1}\hat{\mathfrak{f}} \end{pmatrix}. \qquad (5.19)$$

Namely, $\hat{\tau}$ is a retraction form of $\hat{f}$ since $\operatorname{im}\hat{\tau} = \ker\hat{\tau} = \begin{pmatrix} \hat{\mathfrak{f}} \\ 1 \end{pmatrix} \mathbb{H} = \hat{f}$, and $\hat{\tau}$ is closed, $d\hat{\tau} = 0$, by construction.

Knowing $\hat{\tau}$, it is now a straightforward computation to check that the original isothermic surface f is a Darboux transform of its Darboux transform $\hat{f}$:

$$(d + \lambda\hat{\tau}) \begin{pmatrix} \mathfrak{f} \\ 1 \end{pmatrix} = \begin{pmatrix} \hat{\mathfrak{f}} - \mathfrak{f} \\ 0 \end{pmatrix} (\hat{\mathfrak{f}} - \mathfrak{f})^{-1} d\mathfrak{f} - \begin{pmatrix} \hat{\mathfrak{f}} \\ 1 \end{pmatrix} (\hat{\mathfrak{f}} - \mathfrak{f})^{-1} d\mathfrak{f} = \begin{pmatrix} \mathfrak{f} \\ 1 \end{pmatrix} (\mathfrak{f} - \hat{\mathfrak{f}}) d\mathfrak{f},$$

that is, $df + \lambda\hat{\tau}f \parallel f$.

5.4.12 Lemma. *The Darboux transformation is involutive, $\hat{\hat{f}} = f$. That is, if $\hat{f}$ is a Darboux transform of an isothermic surface f, then f is a Darboux transform of $\hat{f}$.*

Moreover, there is a canonical scaling $\hat{\tau} = \hat{\tau}_1$ for the closed retraction form of $\hat{f}$, relative to the scaling of the closed retraction form τ of f, so that

$$(d + \lambda\tau)\hat{f} \parallel \hat{f} \quad \Leftrightarrow \quad (d + \lambda\hat{\tau})f \parallel f.$$

The surfaces f and $\hat{f}$ form a "Darboux pair," or a "D_λ-pair."

5.4.13 Another, simple consequence of the Riccati equation is that a Darboux transform $\hat{f} : M^2 \to \mathbb{HP}^1$ of an isothermic surface $f : M^2 \to \mathbb{HP}^1$

induces the same conformal structure on M^2 as f does. Computing the metric induced by the stereographic projection $\hat{f}$ yields

$$|d\hat{f}|^2 = \lambda^2 |\hat{f} - f|^4 |df^*|^2.$$

Now the claim follows since the Christoffel transform f^* of f induces a conformally equivalent metric (cf., §5.2.1f).

This is the first step in making contact with the classical notion of the Darboux transformation from §3.2.5. It remains for us to show that f and $\hat{f}$ envelope a Ribaucour 2-sphere congruence.

In the codimension 1 case $f, \hat{f} : M^2 \to \mathrm{Im}\,\mathbb{H}$, this can also be seen from the Riccati equation:

5.4.14 Namely, in that case we compute the Gauss maps $n, \hat{n} : M^2 \to S^2$ of f and $\hat{f}$ using conformal curvature line parameters (x, y) of f:

$$
\begin{aligned}
n &= -f_x f_y^{-1} = -f_x^{-1} f_y \quad \text{and} \\
\hat{n} &= -\hat{f}_x \hat{f}_y^{-1} = (\hat{f} - f) f_x^{-1} f_y (\hat{f} - f)^{-1},
\end{aligned}
\tag{5.20}
$$

where we used the Riccati equation and Christoffel's formula (5.3).

Now we are seeking the radius and center functions, r and m, for the enveloped sphere congruence, that is, we want to have

$$f + r\,n = m = \hat{f} + r\,\hat{n}.$$

Thus, by eliminating m and substituting $\hat{n}$, we obtain

$$-|\hat{f} - f|^2 = (\hat{f} - f)^2 = r\left((\hat{f} - f)\,n + n\,(\hat{f} - f)\right) = -2\langle(\hat{f} - f), n\rangle\, r. \tag{5.21}$$

Unless $(\hat{f} - f) \perp n$, in which case $\hat{n} = -(\hat{f} - f)n(\hat{f} - f)^{-1} = n$ by (5.20), so that f and $\hat{f}$ share a tangent plane, (5.21) can be solved for r.

To see that this sphere congruence is Ribaucour, we compute the holomorphic differential

$$\hat{q} = d\hat{f} \cdot d\hat{f}^* = (\hat{f} - f)\, df^* df\, (\hat{f} - f)^{-1} = (\hat{f} - f)\, \bar{q}\, (\hat{f} - f)^{-1} \simeq q$$

since $\hat{n} = -(\hat{f} - f)n(\hat{f} - f)^{-1}$. Thus, if (x, y) are conformal curvature line parameters for f, so that $q = dx^2 - dy^2$, then they are for $\hat{f}$.

5.4.15 Example. This seems to be a good point to pause for a moment and to reconsider, as an example, surfaces of constant mean curvature. Remember that, in §5.2.10, we determined the Christoffel transforms of minimal and constant mean curvature surfaces in Euclidean space. Here we want to argue that minimal and constant mean curvature surfaces in Euclidean space have many Darboux transforms with the same mean curvature. In order to determine the mean curvature of a Darboux transform[37] $\hat{\mathfrak{f}}$ of an isothermic surface in Euclidean 3-space, we employ (5.6):

$$\begin{aligned} \hat{H}\,d\hat{\mathfrak{f}} \wedge d\hat{\mathfrak{f}} &= -d\hat{\mathfrak{n}} \wedge d\hat{\mathfrak{f}} \\ &= (\hat{\mathfrak{f}} - \mathfrak{f})(d\mathfrak{n} \wedge d\mathfrak{f}^* + d\mathfrak{f}^*((\hat{\mathfrak{f}} - \mathfrak{f})\mathfrak{n} + \mathfrak{n}(\hat{\mathfrak{f}} - \mathfrak{f})) \wedge d\mathfrak{f}^*)(\hat{\mathfrak{f}} - \mathfrak{f}) \\ &= \tfrac{1}{(\hat{\mathfrak{f}}-\mathfrak{f})^2}(\tfrac{1}{\lambda}H^* - 2\langle(\hat{\mathfrak{f}} - \mathfrak{f}), \mathfrak{n}\rangle)\,d\hat{\mathfrak{f}} \wedge d\hat{\mathfrak{f}}, \end{aligned}$$

where we used, besides the Riccati equation (5.17), that $\mathfrak{n}^* = -\mathfrak{n}$ is orthogonal to $d\mathfrak{f}^*$, that $(\hat{\mathfrak{f}} - \mathfrak{f})^2 \in \mathbb{R}$, and the characterization §5.2.6 for the Christoffel transformation. Thus a Darboux transform $\hat{\mathfrak{f}}$ has the same (constant) mean curvature $\hat{H} = H$ as $\mathfrak{f}$ if and only if the function

$$h := \tfrac{1}{\lambda}H^* - H\,(\hat{\mathfrak{f}} - \mathfrak{f})^2 + (\hat{\mathfrak{f}} - \mathfrak{f})\mathfrak{n} + \mathfrak{n}(\hat{\mathfrak{f}} - \mathfrak{f})$$

vanishes identically. Taking the derivative of h, we learn that h satisfies the differential equation

$$dh = \lambda((\hat{\mathfrak{f}} - \mathfrak{f})d\mathfrak{f}^* + d\mathfrak{f}^*(\hat{\mathfrak{f}} - \mathfrak{f}))\,h.$$

Here we used $H^*d\mathfrak{f}^* = d\mathfrak{n} + H d\mathfrak{f}$ to eliminate $d\mathfrak{n}$; this is a direct consequence of §5.2.9, since we are dealing with constant mean curvature surfaces.[38] Consequently, h must vanish identically as soon as it vanishes at one point: Adjusting an initial value $\hat{\mathfrak{f}}(p_0)$ so that $h(p_0) = 0$, we obtain a Darboux transform $\hat{\mathfrak{f}}$ of constant mean curvature $\hat{H} = H$.

To interpret the condition $h = 0$ geometrically, we have to distinguish two cases: In the case of a minimal surface, $H = 0$, the condition

$$h = 0 \quad \Leftrightarrow \quad \langle(\hat{\mathfrak{f}} - \mathfrak{f}), \mathfrak{n}\rangle = \tfrac{1}{2\lambda}H^*$$

just says that the Darboux transform $\hat{\mathfrak{f}}$ has a constant normal distance from $\mathfrak{f}$, that is, the points of $\hat{\mathfrak{f}}$ have a constant distance from the tangent planes of $\mathfrak{f}$ at the corresponding points; if, on the other hand, $H \neq 0$, then we may rewrite the condition

$$h = 0 \quad \Leftrightarrow \quad |H(\hat{\mathfrak{f}} - \mathfrak{f}) - \mathfrak{n}|^2 = 1 - \tfrac{1}{\lambda}HH^*,$$

---

[37] Considering the Riccati equation (5.17) as an initial value problem for the Darboux transform of an isothermic surface $\mathfrak{f}$ in $\mathbb{R}^3 \cong \operatorname{Im} \mathbb{H}$ (with Christoffel partner $\mathfrak{f}^*$ also in $\operatorname{Im} \mathbb{H}$), a solution will take values in $\operatorname{Im} \mathbb{H}$ as soon as we choose the initial value in $\operatorname{Im} \mathbb{H}$, since we then have $\operatorname{Re}(\hat{\mathfrak{f}} - \mathfrak{f})d\mathfrak{f}^*(\hat{\mathfrak{f}} - \mathfrak{f}) \equiv 0$. We will discuss the conformal version of this fact in greater detail below.

[38] In fact, if the scaling of the Christoffel transform $\mathfrak{f}^*$ is not fixed by some other considerations, we may just choose $\mathfrak{f}^* = \mathfrak{n} + H\mathfrak{f}$, in which case $H^* = 1$.

showing that $\hat{\mathfrak{f}}$ has constant distance $\frac{1}{H^2}(1 - \frac{1}{\lambda}HH^*)$ from the parallel constant mean curvature surface $\mathfrak{f} + \frac{1}{H}\mathfrak{n}$ of $\mathfrak{f}$. Note that this parallel constant mean curvature surface is, besides being a Christoffel transform of $\mathfrak{f}$, a Darboux transform of $\mathfrak{f}$ itself,[39] with $\lambda = HH^*$:

$$d(\mathfrak{f} + \tfrac{1}{H}\mathfrak{n}) = (HH^*)(\tfrac{1}{H}\mathfrak{n}) \, d\mathfrak{f}^*(\tfrac{1}{H}\mathfrak{n}).$$

To summarize,[40] *a surface of constant mean curvature in Euclidean 3-space has a 3-parameter family of Darboux transforms that have the same constant mean curvature as the original surface.*

We will now give a proof for the conformal version of our above assertion that, choosing a "correct" initial point, the Darboux transform of an isothermic surface in 3-space will also take values in 3-space:

5.4.16 Lemma. *If $f : M^2 \to S^3 \subset \mathbb{HP}^1$ takes values in a 3-sphere, then there is a $(1{+}3)$-parameter family of Darboux transforms $\hat{f} : M^2 \supset U \to S^3$ that take values in the same S^3.*

5.4.17 *Proof.* Let $S \in \mathfrak{H}(\mathbb{H}^2)$ denote the quaternionic Hermitian form that defines the 3-sphere $S^3 \subset \mathbb{HP}^1$; let v_∞, v_0 be homogeneous coordinates of two points $\infty, 0 \in S^3$, $S(v_0, v_\infty) = 1$; and let $\nu_i = S(v_i, .)$, $i = 0, \infty$, as in §4.9.6. Then S^3 stereographically projects to $\operatorname{Im} \mathbb{H}$.

In particular, $\mathfrak{f} : M^2 \to \operatorname{Im} \mathbb{H}$ and $d\mathfrak{f}^* : TM \to \operatorname{Im} \mathbb{H}$, so that τ is skew with respect to S: With (5.14) we recover the "components" of τ,

$$
\begin{array}{llll}
S(v_0, \tau v_\infty) = & \nu_0 \tau v_\infty = & \mathfrak{f} d\mathfrak{f}^*, & S(v_0, \tau v_0) = & \nu_0 \tau v_0 = & -\mathfrak{f} d\mathfrak{f}^* \mathfrak{f}, \\
S(v_\infty, \tau v_\infty) = & \nu_\infty \tau v_\infty = & d\mathfrak{f}^*, & S(v_\infty, \tau v_0) = & \nu_\infty \tau v_0 = & -d\mathfrak{f}^* \mathfrak{f},
\end{array}
$$

so that $S(v_i, \tau v_j) + S(\tau v_i, v_j) = 0$ since S is Hermitian.

Because $d\tau + \frac{1}{2}[\tau \wedge \tau] = 0$, we can choose homogeneous coordinates for any Darboux transform $\hat{f}$ so that $(d + \lambda\tau)\hat{f} = 0$ by §5.4.1. Then

$$d\{S(\hat{f}, \hat{f})\} = -\lambda[S(\tau\hat{f}, \hat{f}) + S(\hat{f}, \tau\hat{f})] = 0,$$

so that $S(\hat{f}, \hat{f})$ is constant. That is, if we choose an initial point of $\hat{f}$ on S^3, then $\hat{f}$ takes values in S^3. ◁

---

[39] This fact, that there is a surface that is a Christoffel and Darboux transform of $\mathfrak{f}$ at the same time, can in fact be used to characterize surfaces of constant mean curvature among isothermic surfaces in 3-space (see [147]). This was one crucial step in obtaining a definition for a discrete analog for constant mean curvature surfaces in Euclidean space (see [149]).

[40] Compare [16].

5.4.18 Let us examine the consequences of choosing homogeneous coordinates $f, \hat{f} : M^2 \to \mathbb{H}^2$ for a Darboux pair of isothermic surfaces such that

$$(d + \lambda \tau)\hat{f} = 0 \quad \text{and} \quad (d + \lambda \hat{\tau})f = 0$$

a bit further: Since $\operatorname{im}\tau = f\mathbb{H}$ and $\operatorname{im}\hat{\tau} = \hat{f}\mathbb{H}$, we obtain

$$df = -\lambda \hat{\tau} f = \hat{f} \cdot \psi \quad \text{and} \quad d\hat{f} = -\lambda \tau \hat{f} = -f \cdot (\lambda \psi^*)$$

with suitable forms $\psi, \psi^* : TM \to \mathbb{H}$. Analyzing the integrability conditions then gives

$$\begin{aligned} 0 &= d^2 f &= -f\,\lambda\psi^* \wedge \psi &-& \hat{f}\,d\psi^* \\ 0 &= d^2 \hat{f} &= -\hat{f}\,\lambda\psi \wedge \psi^* &+& f\,\lambda d\psi, \end{aligned}$$

showing that (locally) $\psi = d\mathfrak{g}$ and $\psi^* = d\mathfrak{g}^*$ with a Christoffel pair of (isothermic) surfaces $\mathfrak{g}$ and $\mathfrak{g}^*$:

5.4.19 Lemma. *If $f, \hat{f} : M^2 \to \mathbb{H}P^1$ form a Darboux pair, then there exists a choice of homogeneous coordinates $f, \hat{f} : M^2 \to \mathbb{H}^2$ so that*

$$df = \hat{f} \cdot d\mathfrak{g} \quad \text{and} \quad d\hat{f} = f \cdot \lambda d\mathfrak{g}^*$$

with some Christoffel pair $\mathfrak{g}, \mathfrak{g}^* : M^2 \supset U \to \mathbb{H}$ of isothermic surfaces.

5.4.20 This lemma provides the basis[41] for finally proving that the surfaces of a Darboux pair envelope a congruence of 2-spheres, which we want to define (at least locally,[42] on U) as an endomorphism field:

Let $\mathfrak{n}_1, \mathfrak{n}_2 : U \to \mathbb{H}$ denote two orthogonal unit normal fields of the surface $\mathfrak{g}$ from the above lemma. Note that, if $\mathfrak{n}_i$ are normal fields for $\mathfrak{g}$, then $\bar{\mathfrak{n}}_i$ are normal fields for $\mathfrak{g}^*$, since $\mathfrak{g}^*$ and $\bar{\mathfrak{g}}$ have parallel tangent planes (see §5.2.2). Then define an endomorphism field $\mathcal{S}$ of $\mathbb{H}^2$ by

$$\mathcal{S}f = f\,\bar{\mathfrak{n}}_2\mathfrak{n}_1 =: f\,a \quad \text{and} \quad \mathcal{S}\hat{f} = \hat{f}\,(-\mathfrak{n}_1\bar{\mathfrak{n}}_2) =: \hat{f}\,\hat{a}.$$

Because the $\mathfrak{n}_i$ are orthogonal, $\mathfrak{n}_1\bar{\mathfrak{n}}_2 + \mathfrak{n}_2\bar{\mathfrak{n}}_1 = 0$, and have length $|\mathfrak{n}_i|^2 = 1$, we derive that $a^2 = \hat{a}^2 = -1$, so that $\mathcal{S} : U \to \mathfrak{S}(\mathbb{H}^2)$ defines a congruence of 2-spheres (see §4.8.1).

---

[41] In fact, this lemma also provides the relation of the Darboux transformation with curved flats in the space of point pairs (cf., §3.3.2). We will come back to this point in the following section.

[42] Note that the spheres of a congruence enveloped by two surfaces are uniquely determined by the envelopes. Therefore, there is a global 2-sphere congruence, even if we can only locally give an analytic expression.

Since f and $\hat{f}$ give, by construction, (pointwise) eigendirections of $\mathcal{S}$, the points of f and $\hat{f}$ lie on the 2-spheres $\mathcal{S}$. Moreover,

$$
\begin{aligned}
d\mathcal{S}f &= d(\mathcal{S}f) - \mathcal{S}\,df &= f\,da + \hat{f}\,(d\mathfrak{g}\,a - \hat{a}\,d\mathfrak{g}) &= f\,da, \\
d\mathcal{S}\hat{f} &= d(\mathcal{S}\hat{f}) - \mathcal{S}\,d\hat{f} &= \lambda f\,(d\mathfrak{g}^*\hat{a} - a\,d\mathfrak{g}^*) + \hat{f}\,d\hat{a} &= \hat{f}\,d\hat{a}
\end{aligned}
$$

since $-\hat{a}\,d\mathfrak{g} = d\mathfrak{g}\,\bar{a}$ and $d\mathfrak{g}^*\hat{a} = -\bar{a}\,d\mathfrak{g}^*$. Consequently, f and $\hat{f}$ do indeed envelope $\mathcal{S}$ by §4.8.15.

5.4.21 Lemma. *The surfaces $f, \hat{f} : M^2 \to \mathbb{HP}^1$ of a Darboux pair of isothermic surfaces envelope a 2-sphere congruence.*

5.4.22 Finally, we should show that this 2-sphere congruence is Ribaucour, that is, the curvature directions on both surfaces do correspond (cf., §3.1.6). We already know that the two surfaces $f, \hat{f} : M^2 \to \mathbb{HP}^1$ of a Darboux pair induce the same conformal structure on their domain M^2 (see §5.4.13). Thus, if (x, y) denote conformal curvature line parameters for f, then these are conformal coordinates for $\hat{f}$ and it just remains to show that they are curvature line parameters — or conjugate parameters for any stereographic projection $\hat{\mathfrak{f}}$ of $\hat{f}$.

Now remember that we saw earlier, in (5.5), that the integrability of the dual surfaces, given in terms of conformal coordinates, is closely related to the fact that the coordinates are conjugate. In fact, if we choose conformal curvature line coordinates (x, y) for $\mathfrak{f}$ so that $\mathfrak{f}_x^* = \mathfrak{f}_x^{-1}$ and $\mathfrak{f}_y^* = -\mathfrak{f}_y^{-1}$, then we learn from (5.18) that

$$
\hat{\mathfrak{f}}_x^* = \hat{\mathfrak{f}}_x^{-1} \quad \text{and} \quad \hat{\mathfrak{f}}_y^* = -\hat{\mathfrak{f}}_y^{-1}.
$$

The integrability condition (5.5) then just reads $\hat{\mathfrak{f}}_{xy}^\perp = 0$, so that (x, y) must be conjugate parameters.

Thus we have finally collected all the geometric properties that characterize the Darboux transformation classically, as described in §3.2.5:

5.4.23 Corollary. *The surfaces $f, \hat{f} : M^2 \to \mathbb{HP}^1$ of a Darboux pair of isothermic surfaces induce the same conformal structure on M^2 and envelope a Ribaucour congruence of 2-spheres.*

In particular, if $f, \hat{f} : M^2 \to S^3$ take values in a 3-sphere, then they generically[43] form a Darboux pair of isothermic surfaces in the classical sense.

---

[43] The following examples will convince the reader that the enveloped sphere congruence will usually not take values in a fixed linear sphere complex. However, it seems desirable to slightly relax, or reformulate, this requirement anyway as the example of a "Darboux pair of holomorphic maps" below suggests. Note that, for our investigations of Blaschke's problem from Section 3.1, we assumed the enveloped sphere congruence to be regular (cf., §3.2.1).

5.4.24 Example. As a first example we resume our considerations in §5.3.21 (cf., §5.2.11). Thus let $\mathfrak{g}, \mathfrak{h} : \mathbb{C} \supset U \to \mathbb{C}$ denote two holomorphic functions and consider $f = \left(\begin{smallmatrix} -j\mathfrak{g} \\ 1 \end{smallmatrix} \right) \mathbb{H}$. We already discussed that

$$\tau = \begin{pmatrix} -j & 0 \\ 0 & 1 \end{pmatrix} \begin{pmatrix} \mathfrak{g}d\mathfrak{h} & -\mathfrak{g}^2 d\mathfrak{h} \\ d\mathfrak{h} & -\mathfrak{g}d\mathfrak{h} \end{pmatrix} \begin{pmatrix} j & 0 \\ 0 & 1 \end{pmatrix} = J^{-1}\tau^{\mathbb{C}} J$$

provides a retraction form for f, "the" retraction form related to $\mathfrak{h}j$ as a Christoffel transform (cf., §5.3.19). As in §5.3.21, we can use complex integration to determine a fundamental system $\hat{F}_\lambda : U \to Sl(2, \mathbb{H})$ of solutions for the Darboux linear system $(d + \tau_\lambda)\hat{F}_\lambda = 0$ (see §5.4.8): $\hat{F}_\lambda = J^{-1}\hat{F}^{\mathbb{C}}_\lambda$ with $\hat{F}^{\mathbb{C}}_\lambda : U \to Sl(2, \mathbb{C})$. Then any Darboux transform $\hat{f} = \hat{F}_\lambda c$, where $c \in \mathbb{H}P^1$ is constant. If c is chosen complex, $c \in \mathbb{C}P^1 \subset \mathbb{H}P^1$, then the Darboux transform $\hat{f} = \left(\begin{smallmatrix} -j\hat{\mathfrak{g}} \\ 1 \end{smallmatrix} \right) \mathbb{H}$ is totally umbilic; if, however, we choose $c \notin \mathbb{C}P^1$, then the Darboux transform $\hat{f}$ will not take values in $\mathbb{C}P^1$ and the surface will not be totally umbilic, in general. In fact, we will see later that these "essential" Darboux transforms of a holomorphic function are constant mean curvature 1 surfaces in some hyperbolic space (cf., [153]).

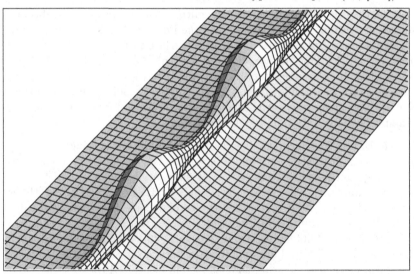

Fig. 5.5. A Darboux transform of $f(z) = -jz$

Let us consider a simple example, $\mathfrak{g}(z) = z = \mathfrak{h}(z)$. The computations done in this example will provide a motivation for those in our next example. Therefore, we will elaborate on these computations quite comprehensively. First note that the (complex) retraction form takes the form

$$\tau^{\mathbb{C}}_\lambda = \lambda \begin{pmatrix} z & -z^2 \\ 1 & -z \end{pmatrix} dz.$$

With the ansatz $\binom{a+zb}{b}$ for a solution, the Darboux system simplifies:

$$0 = \begin{pmatrix} a + zb \\ b \end{pmatrix}' + \tau_\lambda^{\mathbb{C}} \begin{pmatrix} a + zb \\ b \end{pmatrix} = \begin{pmatrix} 1 & z \\ 0 & 1 \end{pmatrix} \left(\begin{pmatrix} a \\ b \end{pmatrix}' + \begin{pmatrix} 0 & 1 \\ \lambda & 0 \end{pmatrix} \begin{pmatrix} a \\ b \end{pmatrix} \right).$$

To solve this system we derive the second-order linear differential equation,

$$a'' = -b' = \lambda a,$$

which yields $a(z) = \cosh\sqrt{\lambda}z$ and $a(z) = -\frac{1}{\sqrt{\lambda}}\sinh\sqrt{\lambda}z$ as two linearly independent solutions. With $b = -a'$ we obtain the fundamental system of solutions

$$\hat{F}_\lambda^{\mathbb{C}}(z) = \begin{pmatrix} \cosh\sqrt{\lambda}z - z\sqrt{\lambda}\sinh\sqrt{\lambda}z & -\frac{1}{\sqrt{\lambda}}\sinh\sqrt{\lambda}z + z\cosh\sqrt{\lambda}z \\ -\sqrt{\lambda}\sinh\sqrt{\lambda}z & \cosh\sqrt{\lambda}z \end{pmatrix}.$$

By choosing a complex initial point $\binom{c}{1}$, $c \in \mathbb{C}$, for the Darboux transformation, we obtain another holomorphic function as the Darboux transform of $-jz$: For example, choosing $c = 0$, we find $\hat{f}(z) = -j(z - \frac{1}{\sqrt{\lambda}}\tanh\sqrt{\lambda}z)$.

If, on the other hand, we choose the initial point not in $\mathbb{CP}^1$, we obtain a surface of constant mean curvature 1 in the Poincaré half-space model with $\mathbb{C}j$ as its infinity plane (see §P.1.3): The surface in Figure 5.5 is obtained from $\hat{F}_{1.33}\binom{ji}{1}$. This surface shows a "typical" behavior of the Darboux transformation: Periodic "bubbles" are added to the original surface along one curvature line, while the rest of the surface is only slightly changed, that is, it is asymptotic to the original surface. We will see in the following example that this intuition may be misleading.

5.4.25 Example. Surprisingly, the above method for solving Darboux's linear system in the holomorphic case can also be employed in certain cases of truly quaternionic systems of partial differential equations. In the following example we will recover tori found by H. Bernstein in her thesis [13] (cf., [45] and [207]). Note that the employed method works the same for cylinders of revolution in Euclidean 3-space: In that case, the constant mean curvature "bubbletons" from [259] (cf., [39]) can be recovered when choosing "correct" initial data for the transformation, so that the Darboux transform has constant mean curvature again (see §5.4.15).

Thus, let $p, q \in (0, 1)$ such that $p^2 + q^2 = 1$, and consider the corresponding Clifford torus in S^3 along with its Christoffel transform in $\mathbb{R}^4 = \mathbb{H}$ as the ambient space[44]:

$$\mathfrak{f}(x, y) = pe^{\frac{ix}{p}} + qe^{\frac{iy}{q}}j \quad \text{and} \quad \mathfrak{f}^*(x, y) = pe^{-\frac{ix}{p}} + qje^{-\frac{iy}{q}}.$$

---

[44] Note that, in contrast to the codimension 1 case, the Clifford tori in $S^3 \subset \mathbb{R}^4$ have closed Christoffel transforms (see §5.2.20).

It is easily checked that $\mathfrak{f}$ and $\mathfrak{f}^*$ indeed form a Christoffel pair by computing

$$\left.\begin{array}{ll} \mathfrak{f}_x(x,y) = ie^{\frac{ix}{p}}, & \mathfrak{f}_x^*(x,y) = -ie^{-\frac{ix}{p}} \\ \mathfrak{f}_y(x,y) = ie^{\frac{iy}{q}}j, & \mathfrak{f}_y^*(x,y) = -jie^{-\frac{iy}{q}} \end{array}\right\} \Rightarrow \left\{\begin{array}{l} \mathfrak{f}_x\mathfrak{f}_x^* = 1 \\ \mathfrak{f}_y\mathfrak{f}_y^* = -1. \end{array}\right.$$

Using the matrix description for our retraction form (cf., (5.14)), the very same ansatz as in the preceding example provides a substantial simplification of the Darboux system: With

$$\hat{f} = \begin{pmatrix} 1 & \mathfrak{f} \\ 0 & 1 \end{pmatrix}\begin{pmatrix} a \\ b \end{pmatrix} \quad \text{and} \quad \tau_\lambda = \lambda\begin{pmatrix} \mathfrak{f}d\mathfrak{f}^* & -\mathfrak{f}d\mathfrak{f}^*\mathfrak{f} \\ d\mathfrak{f}^* & -d\mathfrak{f}^*\mathfrak{f} \end{pmatrix}$$

Darboux's linear system becomes

$$0 = \begin{pmatrix} 1 & -\mathfrak{f} \\ 0 & 1 \end{pmatrix}(d\hat{f} + \tau_\lambda\hat{f}) = d\begin{pmatrix} a \\ b \end{pmatrix} + \begin{pmatrix} 0 & d\mathfrak{f} \\ \lambda d\mathfrak{f}^* & 0 \end{pmatrix}\begin{pmatrix} a \\ b \end{pmatrix}$$

or, writing the system in partial derivatives,

$$\begin{array}{ll} a_x = -ie^{\frac{ix}{p}}b, & b_x = \lambda ie^{-\frac{ix}{p}}a, \\ a_y = -ie^{\frac{iy}{q}}jb, & b_y = \lambda jie^{-\frac{iy}{q}}a. \end{array} \tag{5.22}$$

Just as in the above example, we may now take derivatives and eliminate b to obtain three second-order partial differential equations:

$$0 = a_{xx} - \tfrac{i}{p}a_x - \lambda a, \quad 0 = a_{yy} - \tfrac{i}{q}a_y + \lambda a, \quad 0 = a_{xy} + \lambda e^{\frac{ix}{p} + \frac{iy}{q}}ja. \tag{5.23}$$

Note that two of these equations become complex equations, so that the first two equations determine the function a as a complex function, up to multiplication by (constant) quaternions from the right — these three equations are quaternionic linear: A function a that solves the first two equations must be a linear combination of the four functions $e^{\alpha^\pm x + \beta^\pm y}$, where $\alpha^\pm$ and $\beta^\pm$ denote the roots of the two second-order equations of one variable,

$$\alpha^\pm := \tfrac{i}{2p}(1 \pm \sqrt{1 - 4p^2\lambda}) \quad \text{and} \quad \beta^\pm := \tfrac{i}{2q}(1 \pm \sqrt{1 + 4q^2\lambda}).$$

If we are interested in doubly periodic Darboux transforms, we may restrict ourselves to the cases where $\alpha^\pm, \beta^\pm \in i\mathbb{R}$, that is, $\lambda \in (-\frac{1}{4q^2}, \frac{1}{4p^2})$. With this assumption we have $je^{\alpha^\pm x} = e^{-\alpha^\pm x}j$ and $e^{\frac{ix}{p}}je^{\alpha^\pm x} = e^{\alpha^\mp x}j$, and similarly for $e^{\beta^\pm y}$. Then the third equation in (5.23) provides a relation between solutions with opposite sign configurations — hence there are two solutions to all three equations (5.23):

$$\begin{array}{ll} a_1(x,y) & = \alpha^-\beta^- e^{\alpha^+ x + \beta^+ y} - \lambda e^{\alpha^- x + \beta^- y}j, \\ a_2(x,y) & = \alpha^-\beta^+ e^{\alpha^+ x + \beta^- y} - \lambda e^{\alpha^- x + \beta^+ y}j. \end{array}$$

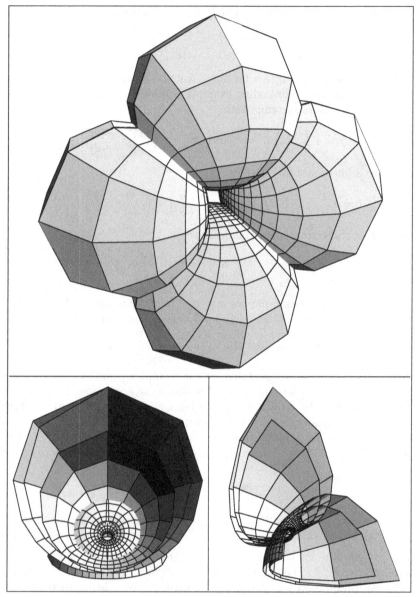

Fig. 5.6. A Darboux transform of a Clifford torus

Finally, going back to (5.22), we determine the corresponding second components for our solutions: Take $b_i = ie^{\frac{-ix}{p}}(a_i)_x$ and divide by $-\lambda = \alpha^+\alpha^-$ to obtain

$$\begin{pmatrix} a^{\pm} \\ b^{\pm} \end{pmatrix} = \beta^{\mp}e^{\beta^{\pm}y}\begin{pmatrix} \frac{1}{\alpha^+}e^{\alpha^+x} \\ ie^{-\alpha^-x} \end{pmatrix} + e^{\beta^{\mp}y}\begin{pmatrix} e^{\alpha^-x} \\ i\alpha^-e^{-\alpha^+x} \end{pmatrix}j.$$

A fundamental system of solutions is then provided by

$$\hat{F}_\lambda = \begin{pmatrix} 1 & \mathfrak{f} \\ 0 & 1 \end{pmatrix}\begin{pmatrix} a^+ & a^- \\ b^+ & b^- \end{pmatrix},$$

and a Darboux transform is, in affine coordinates, obtained by

$$\hat{\mathfrak{f}} = \mathfrak{f} + (a^+c^+ + a^-c^-)(b^+c^+ + b^-c^-)^{-1},$$

where $c^{\pm} \in \mathbb{H}$ are arbitrary constants. If we happen to choose these constants so that $\hat{\mathfrak{f}}(x_0, y_0) \in S^3$ for some point, then the Darboux transform $\hat{\mathfrak{f}}$ takes values in S^3 (see §5.4.16).

Note that $\hat{\mathfrak{f}}$ becomes (doubly) periodic under certain rationality conditions:

$$\left.\begin{array}{l} \sqrt{1 - 4\lambda p^2} = \frac{m_p}{n_p} \\ \sqrt{1 + 4\lambda q^2} = \frac{m_q}{n_q} \end{array}\right\} \quad\Rightarrow\quad \left\{\begin{array}{l} \hat{F}_\lambda(x + 2\pi p \cdot n_p, y) = \pm\hat{F}_\lambda(x, y) \\ \hat{F}_\lambda(x, y + 2\pi p \cdot n_q) = \pm\hat{F}_\lambda(x, y), \end{array}\right. \tag{5.24}$$

so that $\hat{\mathfrak{f}}(x + 2\pi p \cdot n_p, y + 2\pi p \cdot n_q) = \hat{\mathfrak{f}}(x, y)$. Note that this computation only provides the maximum periods — if, for example, m_p and n_p are both odd, then the x-period in (5.24) halves. To consider an explicit example, let

$$\sqrt{1 - 4\lambda p^2} = \frac{2}{3} \quad\text{and}\quad \sqrt{1 + 4\lambda q^2} = \frac{4}{3}.$$

Then $4\lambda = (\frac{4}{3})^2 - (\frac{2}{3})^2 = \frac{4}{3}$, $p^2 = \frac{5}{12}$, and $q^2 = \frac{7}{12}$, and any corresponding Darboux transform has periods $3\pi\sqrt{\frac{5}{3}}$ and $3\pi\sqrt{\frac{7}{3}}$ in the x- and y-directions, respectively. Thus $\hat{\mathfrak{f}}$ will be a 9-fold cover of the original Clifford torus $\mathfrak{f}$ with periods $\pi\sqrt{\frac{5}{3}}$ and $\pi\sqrt{\frac{7}{3}}$ (see Figure 5.6).

5.5 Calapso's transformation

Finally we turn to the T-transformation[45] for isothermic surfaces that was independently discovered by L. Bianchi [21], [20] and P. Calapso [53], [55]

---

[45] Bianchi called this transformation the "T-transformation." To give credit to P. Calapso's contributions to the theory of isothermic surfaces, we will use the name "Calapso transformation," but we will use "T-transformation" for abbreviation in order to avoid conflict with the "C"- (Christoffel) transformation.

(see also [59]). This transformation plays a key role in the theory of isothermic surfaces, as later developments will show: The existence of Calapso's transformation is responsible for the fact that isothermic surfaces are (the only second-order) nonrigid surfaces in Möbius geometry, as was pointed out by É. Cartan [62] and E. Musso [203]; moreover, this transformation gives the "spectral transformation" that we briefly touched on in §3.3.5 and that naturally comes with the integrable systems description of isothermic surfaces in terms of "curved flats" [112] (see [46]).

5.5.1 First remember that any isothermic surface $f : M^2 \to \mathbb{HP}^1$ has a 1-parameter family $(\tau_\lambda)_{\lambda \in \mathbb{R}}$ of closed retraction forms $\tau_\lambda = \lambda \tau_1$. We already discussed in §5.4.7, in the context of the Darboux transformation, that each τ_λ satisfies the Maurer-Cartan equation: Because τ_λ is *assumed* to be closed, the Maurer-Cartan equation reduces to $[\tau_\lambda \wedge \tau_\lambda] = 0$, which holds since τ_λ is a retraction form, $\operatorname{im} \tau_\lambda = \ker \tau_\lambda = f$. Hence each $\tau_\lambda : TM \to \mathfrak{sl}(2, \mathbb{H})$ can, at least locally, be integrated to provide a map $T_\lambda : M \supset U \to Sl(2, \mathbb{H})$.

Thus we will obtain Calapso's transformation by integrating τ_λ into the group instead of the algebra,[46] as we did in §5.3.17 for the "master" Christoffel transformation $\mathfrak{F}^*$ of f:

5.5.2 Definition. *Let $f : M^2 \to \mathbb{HP}^1$ be an isothermic surface with closed retraction forms $\tau_\lambda : TM^2 \to \mathfrak{sl}(2, \mathbb{H})$. A solution[47] $T_\lambda : M^2 \to Sl(2, \mathbb{H})$ of*

$$dT_\lambda = T_\lambda \tau_\lambda \tag{5.25}$$

is called a Calapso transformation, or T-transformation, for f.
The surfaces $f_\lambda := T_\lambda f$ are called the T_λ-transforms of f.

5.5.3 Note that the differential equation that has to be solved to obtain the Calapso transformation T_λ of an isothermic surface f is equivalent to the system that has to be solved to obtain its Darboux transformations (see §5.4.8):

$$dT_\lambda = T_\lambda \tau_\lambda \quad \Leftrightarrow \quad d(T_\lambda^{-1}) + \tau_\lambda T_\lambda^{-1} = 0.$$

Thus, given a Calapso transformation T_λ for an isothermic surface f, the Darboux transformations of f are given by $\hat{f} = T_\lambda^{-1} c$, where $c \in \mathbb{HP}^1$ is some (constant) point; and conversely, if we already know a fundamental system $(\hat{f}_1, \hat{f}_2)$ for Darboux's linear system $d\hat{f} + \tau_\lambda \hat{f} = 0$ for some λ, then the Calapso transformation T_λ of f can be derived by solely algebraic activities.

---

[46] The retraction form (5.14) of an isothermic surface always takes values in $\mathfrak{sl}(2, \mathbb{H})$, since the real part of a quaternionic product is symmetric so that $\operatorname{Re}(f df^*) = \operatorname{Re}(df^* f)$.

[47] As for the Darboux transformation, we assume global existence of a solution: A T-transform for an isothermic surface may consequently exist only after restriction to a simply connected domain.

5.5.4 A second formal observation is that the "master Christoffel trans-form" $\mathfrak{F}^*$ from §5.3.17 can be obtained from the family $(T_\lambda)_{\lambda \in \mathbb{R}}$ of T-transforms associated with an isothermic surface f by Sym's formula:

$$\mathfrak{F}^* = (T_\lambda^{-1} \tfrac{\partial}{\partial \lambda} T_\lambda)_{\lambda=0}.$$

Namely, $d(T_\lambda^{-1} \tfrac{\partial}{\partial \lambda} T_\lambda) = \tfrac{\partial}{\partial \lambda} \tau_\lambda + [T_\lambda^{-1} \tfrac{\partial}{\partial \lambda} T_\lambda, \tau_\lambda] = \tau_1 + [T_\lambda^{-1} \tfrac{\partial}{\partial \lambda} T_\lambda, \tau_\lambda]$; since $\tau_0 = 0$, the second term vanishes for $\lambda = 0$, and we have proved the claim.

5.5.5 Obviously, the T_λ-transforms f_λ of an isothermic surface f are only well defined up to Möbius transformation: The differential equation (5.25) determines T_λ uniquely up to postcomposition with a (constant) Möbius transformation.

On the other hand, any Möbius transform $\mu(f)$ of f yields $\mu \circ \tau_\lambda \circ \mu^{-1}$ for the retraction forms τ_λ of f. For T_λ this is precomposition by μ^{-1}, so that the T_λ-transforms $f_\lambda = (T_\lambda \circ \mu^{-1})(\mu f)$ of f remain unchanged.

In this sense:

5.5.6 Lemma. *The Calapso transformation is a well defined transformation for Möbius equivalence classes of isothermic surfaces.*

5.5.7 In this lemma we already anticipated the fact that the T_λ-transforms f_λ of an isothermic surface are isothermic — to prove this fact we will determine closed retraction forms for f_λ: Then it follows from §5.3.19 that the f_λ's are isothermic. Thus, define $\tau^\lambda := T_\lambda \tau_1 T_\lambda^{-1}$. Clearly, τ^λ is a retraction form for f_λ since

$$\operatorname{im} \tau^\lambda = T_\lambda \operatorname{im} \tau_1 = f_\lambda \quad \text{and} \quad \ker \tau^\lambda = T_\lambda \ker \tau_1 = f_\lambda.$$

Moreover, $d\tau^\lambda = 0$ since $d\tau_1 = 0$ and $[\tau_1 \wedge \tau_1] = 0$. Hence we have proved

5.5.8 Lemma. *The T_λ-transforms $f_\lambda = T_\lambda f$ of an isothermic surface f are isothermic with retraction forms*

$$\tau^\lambda = T_\lambda \tau_1 T_\lambda^{-1}. \tag{5.26}$$

5.5.9 From (5.26) we can now deduce the behavior of iterated Calapso transformations: Let $f_\lambda = T_\lambda f$ be the T_λ-transform of an isothermic surface f, and let $f_\mu^\lambda = T_\mu^\lambda f_\lambda$ denote the T_μ-transform of f_λ. Then $f_\mu^\lambda = T_\mu^\lambda T_\lambda f$. With (5.26) we compute

$$d(T_\mu^\lambda T_\lambda) = T_\mu^\lambda \, \mu \tau^\lambda \, T_\lambda + T_\mu^\lambda T_\lambda \, \lambda \tau_1 = (T_\mu^\lambda T_\lambda) \, (\lambda + \mu) \tau_1 = (T_\mu^\lambda T_\lambda) \tau_{\lambda+\mu}.$$

Therefore, $T_\mu^\lambda T_\lambda = T_{\lambda+\mu}$, up to postcomposition with a Möbius transformation.

In this sense, the T-transformations T_λ of a given isothermic surface f form a 1-parameter group — when fixing the scaling of the closed retraction forms of f_λ as in (5.26).

In particular, $T^\lambda_{-\lambda} = T_\lambda^{-1}$ yields the "inverse" T-transformation, that is, $f^\lambda_{-\lambda}$ is Möbius equivalent to the original surface f.

5.5.10 Next we want to verify that Calapso's transformation indeed provides a "second-order deformation" for isothermic surfaces; that is, at any point $p_0 \in M^2$, the surfaces $f_\lambda = T_\lambda f$ and $T_\lambda(p_0) \cdot f$ have second-order contact. To give a low technology proof of this statement we write both immersions in affine coordinates and show that their second-order Taylor expansions coincide. Thus, let

$$f = v_0 + v_\infty \mathfrak{f} \quad \text{and} \quad f_\lambda = T_\lambda f = (v_0 + v_\infty \mathfrak{f}_\lambda)a$$

with a suitable function $a : M^2 \to \mathbb{H}$. To simplify computations, we assume that $T_\lambda(p_0) = id$, so that $a(p_0) = 1$, where $p_0 \in M$ is the point where we intend to check second-order contact. Clearly, $\mathfrak{f}_\lambda(p_0) = \mathfrak{f}(p_0)$, so that the surfaces have zeroth-order contact. Denoting by ∂_i, $i = 1, 2$, some coordinate vector fields on M we find

$$v_\infty(\partial_j \mathfrak{f}_\lambda)\, a + (v_0 + v_\infty \mathfrak{f}_\lambda)\partial_j a = \partial_j T_\lambda f = T_\lambda(\tau_\lambda(\partial_j)f + \partial_j f) = T_\lambda v_\infty \partial_j \mathfrak{f}.$$

Comparing components at p_0 yields $\partial_j a(p_0) = 0$ and $\partial_j \mathfrak{f}_\lambda(p_0) = \partial_j \mathfrak{f}(p_0)$, that is, f and f_λ have first-order contact at p_0. Taking, on the other hand, one more derivative, we obtain

$$
\begin{aligned}
v_\infty(\partial_i \partial_j \mathfrak{f}_\lambda a &+ \partial_i \mathfrak{f}_\lambda \partial_j a + \partial_j \mathfrak{f}_\lambda \partial_i a) + (v_0 + v_\infty \mathfrak{f}_\lambda)\partial_i \partial_j a \\
&= T_\lambda(\tau_\lambda(\partial_i)v_\infty \partial_j \mathfrak{f} + v_\infty \partial_i \partial_j \mathfrak{f}) \\
&= T_\lambda((v_0 + v_\infty \mathfrak{f})\partial_i \mathfrak{f}^* \partial_j \mathfrak{f} + v_\infty \partial_i \partial_j \mathfrak{f}),
\end{aligned}
$$

where we used (5.14), $\tau_\lambda = \lambda(v_0 + v_\infty \mathfrak{f})\, d\mathfrak{f}^*(v_0 - \mathfrak{f}v_\infty)$, with the Christoffel transform $\mathfrak{f}^*$ of $\mathfrak{f}$. Since $\partial_j a(p_0) = 0$, evaluation at p_0 and comparing coefficients now gives $\partial_i \partial_j \mathfrak{f}_\lambda(p_0) = \partial_i \partial_j \mathfrak{f}(p_0)$, as desired: We have second-order contact.

5.5.11 Theorem. *The T-transforms f_λ of an isothermic surface f are second-order deformations of f; that is, for any given point $p_0 \in M$, the surfaces $f_\lambda = T_\lambda f$ and $T_\lambda(p_0) \cdot f$ have second-order contact.*

5.5.12 As a consequence, Calapso's transformation preserves all first- and second-order invariants of an isothermic surface.[48]

---

[48] There is an interesting relation with deformable surfaces in projective geometry and their corresponding transformation; see [56] and [58].

This observation is the key to the converse of our theorem, as it is proved by É. Cartan [62], E. Musso [203], or, more recently, by F. Burstall, F. Pedit and U. Pinkall [48]: In all cases, the authors prove that a surface in the conformal 3-sphere is uniquely determined by certain second-order invariants, unless it is isothermic. In [62] and [203], these invariants are given as functions in the connection form of a (Möbius invariantly) adapted frame; geometrically, these invariants comprise information about the curvature directions of the surface and the induced metric of its central sphere congruence (conformal Gauss map: Remember that the central sphere congruence is the unique enveloped sphere congruence with induced metric in the conformal class induced by the surface; see §3.4.8); in particular, the Calapso potential that we defined in §5.1.12 turns up (cf., [204], see also [246]). In [48] these invariants are replaced by *one* object, the conformal Hopf differential of the surface: Remember from §5.1.10 that this object can be obtained by suitably normalizing the Hopf differential, as the name indicates.

Here we skip this theorem and just note, for later reference, that:

5.5.13 Corollary. *The Calapso transformation preserves all first- and second-order invariants of an isothermic surface. In particular, it preserves the induced conformal structure and the curvature lines.*

5.5.14 Another way to describe first-order contact is by using an enveloped sphere congruence. Thus, let $\mathcal{S} : M^2 \to \mathfrak{S}(\mathbb{H}^2)$ denote a 2-sphere congruence that is enveloped by our isothermic surface $f : M^2 \to \mathbb{HP}^1$, that is, by §4.8.15 we have

$$\mathcal{S} \cdot f \parallel f \quad \text{and} \quad d\mathcal{S} \cdot f \parallel f.$$

Of course, at a given point $p_0 \in M$, the sphere $(T_\lambda \mathcal{S} T_\lambda^{-1})(p_0)$ will then touch the surface $T_\lambda(p_0)f$ at $(T_\lambda f)(p_0)$ — on the other hand, it is easy to show that the transformed sphere congruence

$$\mathcal{S}_\lambda := T_\lambda \mathcal{S} T_\lambda^{-1} : M^2 \to \mathfrak{S}(\mathbb{H}^2)$$

will envelope the T_λ-transform $f_\lambda = T_\lambda f$ of f: First note that $\mathcal{S}_\lambda^2 = -id$, so that $\mathcal{S}_\lambda$ indeed takes values in $\mathfrak{S}(\mathbb{H}^2)$ and therefore defines a sphere congruence; further, because f is an eigendirection of $\mathcal{S}$ and is annihilated by τ_λ, we compute

$$\mathcal{S}_\lambda f_\lambda = T_\lambda \mathcal{S} f \parallel f_\lambda \quad \text{and} \quad d\mathcal{S}_\lambda f_\lambda = T_\lambda(d\mathcal{S} + [\tau_\lambda, \mathcal{S}])f = T_\lambda d\mathcal{S} f \parallel f_\lambda,$$

which proves the claim.

5.5.15 In the codimension 1 case we can do even better: We will show directly that the T_λ-transformed principal and central sphere congruences

S_λ are the principal and central sphere congruences, respectively, for the Calapso transforms f_λ of f.

To work with affine coordinates we choose a basis (v_∞, v_0) of $I\!H^2$ so that the ambient (hyper)sphere $S^3 \in \mathfrak{H}(I\!H^2)$ is given by $S^3(v_i, v_i) = 0$, $i = 0, \infty$, and $S^3(v_0, v_\infty) = 1$, as usual. Then our isothermic surface $f = v_0 + v_\infty \mathfrak{f}$ with an Im $I\!H$-valued immersion $\mathfrak{f}$. From (5.14), the retraction forms

$$\tau_\lambda = \lambda(v_0 + v_\infty \mathfrak{f})d\mathfrak{f}^*(v_0 - \mathfrak{f}v_\infty),$$

where $\mathfrak{f}^*$ denotes the Christoffel transform of $\mathfrak{f}$. Since τ_λ is skew-symmetric with respect to S^3, we find that

$$d(T_\lambda S^3) = -S^3(\tau_\lambda T_\lambda^{-1}., T_\lambda^{-1}.) - S^3(T_\lambda^{-1}., \tau_\lambda T_\lambda^{-1}.) = 0,$$

and, by an appropriate choice of initial condition, we can achieve $T_\lambda S^3 \equiv S^3$, that is, $T_\lambda : M^2 \to M\ddot{o}b(S^3)$. Hence

$$S^3(f_\lambda, f_\lambda) = (T_\lambda^{-1} S^3)(f, f) \equiv 0,$$

showing that $f_\lambda = T_\lambda f$ also takes values in S^3 (cf., §4.3.12).

5.5.16 Lemma. *If an isothermic surface $f : M^2 \to S^3$ takes values in a 3-sphere, then, up to a Möbius transformation, each $T_\lambda : M^2 \to M\ddot{o}b(S^3)$ and the T_λ-transforms $f_\lambda = T_\lambda f$ of f also take values in S^3.*

5.5.17 Now let $\mathfrak{n} : M^2 \to S^2$ denote a unit normal field of our codimension 1 isothermic surface. By using the representation §4.3.10 of (hyper)spheres in terms of affine coordinates, where we write the center $\mathfrak{m} = \mathfrak{f} + \frac{1}{r}\mathfrak{n}$ and replace the radius $r = \frac{1}{a}$ by the inverse of the curvature a of the sphere, we have, for any 2-sphere congruence S enveloped by f,

$$JS^3 = \begin{pmatrix} a & -(\mathfrak{n} + a\,\mathfrak{f}) \\ \mathfrak{n} + a\,\mathfrak{f} & -(\mathfrak{fn} + \mathfrak{nf} + a\,\mathfrak{f}^2) \end{pmatrix} \leftrightarrow S = \begin{pmatrix} \mathfrak{n} + a\,\mathfrak{f} & -(\mathfrak{fn} + \mathfrak{nf} + a\,\mathfrak{f}^2) \\ a & -(\mathfrak{n} + a\,\mathfrak{f}) \end{pmatrix}$$

from (4.22). On the other hand, by writing our retraction form in matrix form,

$$\tau_\lambda = \lambda \begin{pmatrix} \mathfrak{f}d f^* & -\mathfrak{f}d f^* \mathfrak{f} \\ d f^* & -d f^* \mathfrak{f} \end{pmatrix},$$

we obtain

$$[\tau_\lambda, S] = \lambda \begin{pmatrix} \mathfrak{f}(d f^* \mathfrak{n} + \mathfrak{n}d f^*) & -\mathfrak{f}(d f^* \mathfrak{n} + \mathfrak{n}d f^*)\mathfrak{f} \\ (d f^* \mathfrak{n} + \mathfrak{n}d f^*) & -(d f^* \mathfrak{n} + \mathfrak{n}d f^*)\mathfrak{f} \end{pmatrix} = 0.$$

Hence $dS_\lambda = T_\lambda dS T_\lambda^{-1}$. As a consequence, employing $T_\lambda S^3 \equiv S^3$,

$$d(J_\lambda S^3) = S^3(., dS_\lambda.) = (T_\lambda S^3)(., dS_\lambda.) = S^3(T_\lambda^{-1}., dS T_\lambda^{-1}.) = T_\lambda d(JS^3).$$

Since the action (4.5) of $Möb(S^3) \subset Sl(2, \mathbb{H})$ on $\mathfrak{H}(\mathbb{H}^2)$ is by isometries with respect to the Minkowski product (4.1) on $\mathfrak{H}(\mathbb{H}^2)$, the metrics induced by S and S_λ coincide,

$$|d(\mathcal{J}_\lambda S^3)|^2 = |d(\mathcal{J}S^3)|^2 = -(dn + a\, df)^2.$$

From §1.7.10f we know that a sphere congruence is a curvature sphere congruence of f if and only if its induced metric degenerates — because S and S_λ induce the same metrics we learn that the Calapso transformation preserves the curvature sphere congruences of f.

Equally, we know from §3.4.8 that the central sphere congruence (conformal Gauss map) of f can be characterized by the fact that its induced metric is in the conformal structure induced by f — again, this is preserved by Calapso's transformation.

Thus, besides an alternative proof for the fact that Calapso's transformation preserves the conformal class as well as the curvature lines of an isothermic surface as stated in §5.5.13, we obtain

5.5.18 Lemma. *Let $f : M^2 \to S^3$ be a codimension 1 isothermic surface with conformal Gauss map $\mathcal{Z} : M^2 \to \mathfrak{S}(\mathbb{H}^2)$. Then $\mathcal{Z}_\lambda = T_\lambda \mathcal{Z} T_\lambda^{-1}$ are the conformal Gauss maps of its T-transforms $f_\lambda = T_\lambda f$, and all $\mathcal{Z}_\lambda$'s induce the same metric.*

5.5.19 In order to derive the relation of Calapso's transformation with the spectral transformation for Darboux pairs of isothermic surfaces from §3.3.5, we first observe the effect of a change

$$T_\lambda \to F_\lambda := T_\lambda F_0, \quad \text{where} \quad F_0 := \begin{pmatrix} 1 & f \\ 0 & 1 \end{pmatrix} \tag{5.27}$$

denotes the Euclidean frame of an isothermic surface f with respect to some choice of basis (v_0, v_∞) of $\mathbb{H}^2$, on the differential equation characterizing the Calapso transformation:

$$dF_\lambda = d(T_\lambda F_0^{-1}) = F_\lambda(F_0^{-1} \tau_\lambda F_0 + F_0^{-1} dF_0) =: F_\lambda \Phi_\lambda.$$

Writing the retraction form τ_λ, according to (5.14), in matrix form, we find

$$\Phi_\lambda = \begin{pmatrix} 0 & df \\ \lambda df^* & 0 \end{pmatrix} \tag{5.28}$$

with the Christoffel transform f^* of f. Note that, for positive $\lambda = \mu^2$, this connection form is gauge equivalent to the quaternionic version of the connection form (3.1) that defined the spectral transformation in §3.3.5:

$$\begin{pmatrix} 0 & df \\ \mu^2 df^* & 0 \end{pmatrix} = \begin{pmatrix} 1/\sqrt{\mu} & 0 \\ 0 & \sqrt{\mu} \end{pmatrix} \begin{pmatrix} 0 & \mu df \\ \mu df^* & 0 \end{pmatrix} \begin{pmatrix} \sqrt{\mu} & 0 \\ 0 & 1/\sqrt{\mu} \end{pmatrix}.$$

Otherwise said, if we let $\mathfrak{P} = Sl(2, I\!H)/K_{\infty,0}$ denote the homogeneous space of point pairs as in §4.5.3, then $F_\lambda : M^2 \to Sl(2, I\!H)$ defines an adapted frame for a curved flat $\gamma_\mu : M^2 \to \mathfrak{P}$. Namely, $\Phi_\lambda : TM \to \mathfrak{p}_{\infty,0}$ takes values in the tangent space $\mathfrak{p}_{\infty,0} = T_{(\infty,0)}\mathfrak{P}$ of the homogeneous space, so that its isotropy part $\Phi_{\mathfrak{k}}$ vanishes identically and it clearly satisfies the curved flat conditions from §2.2.3:

$$d\Phi_\lambda + \tfrac{1}{2}[\Phi_\lambda \wedge \Phi_\lambda] = 0 \quad \text{and} \quad [\Phi_{\mathfrak{p}} \wedge \Phi_{\mathfrak{p}}] = 0.$$

Thus, as in §3.3.1, we obtain a curved flat into the space of point pairs. If, on the other hand, $\lambda = -\mu^2$, then we may compensate for the sign by interchanging the roles of the curvature directions in (5.3), that is, by replacing our conformal curvature line coordinates $z = x + iy$ by $iz = -y + ix$ if we want to keep the orientation. In this way the Christoffel transform $\mathfrak{f}^*$ is replaced by $-\mathfrak{f}^*$. Then the same argument as before applies to see that we obtain a curved flat $\gamma_{i\mu} : M^2 \to \mathfrak{P}$ into the space of point pairs.[49]

In the current setup it is now straightforward to derive similar properties for these curved flats as in the classical case (cf., §3.3.5):

5.5.20 Theorem. *Let $f : M^2 \to I\!HP^1$ be an isothermic surface with Calapso transformations $T_\lambda : M^2 \to Sl(2, I\!H)$, and let $F_\lambda := T_\lambda F_0$, where F_0 is the Euclidean frame of f with respect to some basis (v_∞, v_0) of $I\!H^2$, as in (5.27). Then the F_λ frame a family of curved flats $\gamma_{\sqrt{\lambda}} : M^2 \to \mathfrak{P}$ into the space of point pairs. For each F_λ, its two legs*

$$f_\lambda := F_\lambda v_0 I\!H \quad \text{and} \quad \hat{f}_\lambda := F_\lambda v_\infty I\!H$$

are the T_λ-transforms of $f = v_0 + v_\infty \mathfrak{f}$ and $v_0 + v_\infty \mathfrak{f}^*$, respectively, and each $(\hat{f}_\lambda, f_\lambda)$ forms a $D_{-\lambda}$-pair.

5.5.21 Proof. We have just seen that $F_\lambda : M^2 \to Sl(2, I\!H)$ are frames for a family of curved flats $\gamma_{\sqrt{\lambda}} : M^2 \to \mathfrak{P}$.

To verify that its two legs are T_λ-transforms of f and $\lambda \mathfrak{f}^*$, respectively, first note that, by construction, $f_\lambda = T_\lambda F_0 v_0 = T_\lambda f$ are the T_λ-transforms of f. On the other hand, $\frac{1}{\lambda}\mathfrak{f} = (\lambda \mathfrak{f}^*)^*$ is the (properly scaled) Christoffel transform of $\lambda \mathfrak{f}^*$ since the Christoffel transformation is involutive by §5.2.4, so that, by the symmetry of (5.28) in $\mathfrak{f} = \lambda(\frac{1}{\lambda}\mathfrak{f})$ and $\lambda \mathfrak{f}^*$, $\hat{f}_\lambda$ is the T_λ-transform of $v_0 + v_\infty(\lambda \mathfrak{f}^*) \simeq v_0 + v_\infty \mathfrak{f}^*$: Remember that the T_λ-transform is the same for Möbius equivalent isothermic surfaces (see §5.5.6).

---

[49] In fact, replacing $\Phi_\lambda = \Phi_{\mathfrak{k}} + \mu\Phi_{\mathfrak{p}}$ by $\Phi_\lambda = \Phi_{\mathfrak{k}} + i\mu\Phi_{\mathfrak{p}}$ places the curved flat in the dual symmetric space $\mathfrak{P}^*$ of the space $\mathfrak{P}$ of point pairs, which is associated to the Cartan decomposition $\mathfrak{g}^* = \mathfrak{k} \oplus i\mathfrak{p} \subset \mathfrak{g}^{\mathbb{C}}$ (see [140]). But the space of point pairs is self-dual: Multiplication by i preserves the condition of being a Minkowski 2-plane in $\mathbb{C}^5$ equipped with the bilinearly extended Minkowski product (compare [47]). I am grateful to F. Burstall for pointing out this observation to me.

To verify the statement about the Darboux pair property of the two legs of our curved flat, we employ the retraction form of f_λ as it is given in §5.5.8: With this

$$d(F_\lambda v_\infty) + \tau^\lambda_{-\lambda}(F_\lambda v_\infty) = dT_\lambda v_\infty - \lambda T_\lambda \tau_1 v_\infty = 0,$$

since the Euclidean frame has v_∞ as an eigenvector, $F_0 v_\infty = v_\infty$.   ◁

5.5.22 Another interesting way to prove that $\hat{f}_\lambda$ is the Calapso transform of $v_0 + v_\infty f^*$ is by determining T^*_λ. Define

$$T^*_\lambda := T_\lambda \cdot (v_\infty \nu_\infty + \lambda(v_0 + v_\infty f)(\nu_0 - f^* \nu_\infty));$$

using (5.14), it is straightforward to verify that T^*_λ integrates τ^*_λ:

$$dT^*_\lambda - T^*_\lambda \tau^*_\lambda$$
$$= T_\lambda \{\tau_\lambda v_\infty \nu_\infty + \lambda v_\infty df (\nu_0 - f^* \nu_\infty) - \lambda(v_0 + v_\infty f) df^* \nu_\infty - v_\infty \nu_\infty \tau^*_\lambda\} = 0.$$

Then $T^*_\lambda (v_0 + v_\infty f^*) = T_\lambda v_\infty = F_\lambda v_\infty = \hat{f}_\lambda$, which proves the claim.

5.5.23 By using, from §5.3.19, that the existence of a closed retraction form for a surface $f : M^2 \to \mathbb{HP}^1$ ensures that f is isothermic, one can derive the converse of the theorem in §5.5.20:

Any curved flat $\gamma : M^2 \to \mathfrak{P}$ into the space of point pairs is a Darboux pair of isothermic surfaces, and the corresponding associated family provides Calapso transformations.

The proof shall be left to the reader as an exercise.[50] Just note that, using a *parallel* frame F for the curved flat and the characterization of Christoffel pairs from §5.2.6, allows us to extract the derivatives of the "limiting" Christoffel pair (cf., §3.3.8) from its connection form, as in (5.28) (compare §5.4.19).

5.5.24 The form of the connection form (5.28) of the curved flat frame F_λ indicates how to determine the Calapso transformation from *any* adapted frame of an isothermic surface: Suppose $F : M^2 \to Sl(2, \mathbb{H})$ denotes an adapted frame for an isothermic surface $f : M^2 \to \mathbb{HP}^1$, that is, $f = Fv_0 \mathbb{H}$ with some basis (v_∞, v_0) of $\mathbb{H}^2$, and let

$$\Phi = \begin{pmatrix} \hat{\varphi} & \psi \\ \hat{\psi} & \varphi \end{pmatrix}$$

denote its connection form (cf., (4.10)). If F was a Euclidean frame for f, that is, $\varphi = \hat{\varphi} = \hat{\psi} = 0$ and $\psi = df$ with $f = (v_0 + v_\infty f)\mathbb{H}$, then we could

---

[50] Compare [148] for an utterly complicated but direct proof.

just apply Christoffel's formula (5.3) to define the connection forms Φ_λ as in (5.28).

On the other hand, any adapted frame for f can be obtained from such a Euclidean frame by the kind of gauge transformation $F \to \tilde{F}$ that we discussed in Section 4.6. We computed the effect of such a gauge transformation on the connection form in (4.14):

$$\Phi \quad \to \quad \tilde{\Phi} = \begin{pmatrix} \dots & \hat{a}^{-1}\psi a \\ a^{-1}\hat{\psi}\hat{a} + \dots & \dots \end{pmatrix}.$$

The key observation is that $\hat{\psi}$ only appears once and changes in the "opposite" way than ψ does. Hence the (algebraic) activity of replacing $\hat{\psi}$ by $\hat{\psi} + \lambda\psi^*$, where

$$\psi^* := (\psi(\tfrac{\partial}{\partial x}))^{-1}dx - (\psi(\tfrac{\partial}{\partial y}))^{-1}dy$$

is given in terms of conformal curvature line parameters (x, y) of f by an analog of Christoffel's formula (5.3), "commutes" with the gauge transformation.

As a consequence we have proved the first part of the following

5.5.25 Lemma. *Let $F = F_0 : M^2 \to Sl(2, \mathbb{H})$ be an adapted frame for an isothermic surface $f = Fv_0\mathbb{H} : M^2 \to \mathbb{HP}^1$ with connection form $\Phi = \Phi_0$. Then*

$$\Phi_\lambda := \begin{pmatrix} \hat{\varphi} & \psi \\ \hat{\psi} + \lambda\psi^* & \varphi \end{pmatrix}, \quad where \quad \psi^* := (\psi(\tfrac{\partial}{\partial x}))^{-1}dx - (\psi(\tfrac{\partial}{\partial y}))^{-1}dy$$

in terms of conformal curvature line parameters $(x, y) : M^2 \to \mathbb{R}^2$, defines a family of integrable connection forms, $dF_\lambda = F_\lambda\Phi_\lambda$, so that $f_\lambda := F_\lambda v_0\mathbb{H}$ are the T_λ-transforms of f.

Moreover, $T_\lambda = F_\lambda F_0^{-1}$ are the Calapso transformations for f.

5.5.26 Proof. It remains to show that $T_\lambda = F_\lambda F_0^{-1}$. To verify this fact we show that the connection form

$$d(F_\lambda F_0^{-1}) = (F_\lambda F_0^{-1}) \cdot F_0(\Phi_\lambda - \Phi_0)F_0^{-1} =: (F_\lambda F_0^{-1}) \cdot \lambda F_0\Psi^*F_0^{-1}$$

of $F_\lambda F_0^{-1}$ is a[51] closed retraction form of f. It is easy to show that $F_0\Psi^*F_0^{-1}$ is a retraction form of f: Since $f = Fv_0\mathbb{H}$ and $\mathrm{im}\Psi^* = \ker\Psi^* = v_0\mathbb{H}$, we

---

[51] The scaling of the parameter is taken care of by the fact that $f_\lambda = F_\lambda v_0\mathbb{H}$ is the T_λ-transform of f; note that Ψ^* is independent of λ.

clearly have $\mathrm{im}(F_0\Psi^*F_0^{-1}) = \ker(F_0\Psi^*F_0^{-1}) = f\mathbb{H}$. It remains to show closeness,

$$d(F_0\Psi^*F_0^{-1}) = F_0(d\Psi^* + \Phi_0 \wedge \Psi^* + \Psi^* \wedge \Phi_0)F_0^{-1} = 0.$$

But this follows directly from the integrability of Φ_λ, for all λ, and the fact that $\Psi^* = \Phi_1 - \Phi_0$ is nilpotent, $\Psi^* \circ \Psi^* = 0$:

$$\begin{aligned}
0 &= \{d\Phi_1 + \Phi_1 \wedge \Phi_1\} - \{d\Phi_0 + \Phi_0 \wedge \Phi_0\} - (\Phi_1 - \Phi_0) \wedge (\Phi_1 - \Phi_0) \\
&= d(\Phi_1 - \Phi_0) + (\Phi_1 - \Phi_0) \wedge \Phi_0 + \Phi_0 \wedge (\Phi_1 - \Phi_0).
\end{aligned}$$

Thus $\tau_\lambda = \lambda F_0\Psi^*F_0^{-1}$ is a closed retraction form of f and, because $F_\lambda F_0^{-1}$ integrates τ_λ, it qualifies as the T_λ-transformation for f. ◁

5.5.27 Example. We resume our discussions from §5.2.11, §5.3.21, and §5.4.24. Thus we consider $f = \begin{pmatrix} -j\mathfrak{g} \\ 1 \end{pmatrix} \mathbb{H} : \mathbb{C} \supset U \to \mathbb{HP}^1$, where $\mathfrak{g}$ is a holomorphic (or meromorphic) function, as a totally umbilic isothermic surface. Coupling $\mathfrak{g}$ with another holomorphic function $\mathfrak{h}$ as a Christoffel partner,

$$f = -j\mathfrak{g} \quad \text{and} \quad f^* = \mathfrak{h}j,$$

provides the totally umbilic surface with "curvature lines," or a "polarization" (cf., §5.2.18), given by $q = -j(\mathfrak{g}'\mathfrak{h}'dz^2)j \simeq \mathfrak{g}'\mathfrak{h}'dz^2$ (compare §5.2.14 for the last identification). As in §5.4.24 we have

$$\tau_\lambda = \lambda \begin{pmatrix} -j & 0 \\ 0 & 1 \end{pmatrix} \begin{pmatrix} \mathfrak{g}d\mathfrak{h} & -\mathfrak{g}^2 d\mathfrak{h} \\ d\mathfrak{h} & -\mathfrak{g}d\mathfrak{h} \end{pmatrix} \begin{pmatrix} j & 0 \\ 0 & 1 \end{pmatrix} =: J^{-1}\tau_\lambda^{\mathbb{C}} J,$$

so that we can choose $T_\lambda = J^{-1}T_\lambda^{\mathbb{C}}J$ with $T_\lambda^{\mathbb{C}} : U \to Sl(2, \mathbb{C})$. As a consequence the T-transforms

$$Jf_\lambda = T_\lambda^{\mathbb{C}} \begin{pmatrix} \mathfrak{g} \\ 1 \end{pmatrix} \mathbb{H} : U \to \mathbb{CP}^1 \subset \mathbb{HP}^1$$

of $f = J^{-1} \begin{pmatrix} \mathfrak{g} \\ 1 \end{pmatrix} \mathbb{H}$ take values in the same 2-sphere as f does — we could just as well have derived this fact as a trivial consequence of our lemma in §5.5.16.

To proceed we choose more convenient homogeneous coordinates for f: Assuming that $\mathfrak{g}$ has no branch points, that is, that our surface is an immersion, we can choose a function $a = \frac{1}{\sqrt{\mathfrak{g}'}}$ so that our lift $f = J^{-1} \begin{pmatrix} \mathfrak{g} \\ 1 \end{pmatrix} a$ satisfies[52]

$$\det(Jf', Jf) = \det \begin{pmatrix} \mathfrak{g}'a + \mathfrak{g}a' & \mathfrak{g}a \\ a' & a \end{pmatrix} = a^2\mathfrak{g}' = 1.$$

---

[52] Note that $Jf : U \to \mathbb{C}^2$ is complex-valued, so that the determinant makes sense.

It is well known (or easily verified by straightforward computation) that, with this normalization, Jf satisfies Hill's equation

$$Jf'' + Jf \cdot \sigma = 0, \quad \text{where} \quad \sigma := \left(\tfrac{\mathfrak{g}''}{2\mathfrak{g}'}\right)' - \left(\tfrac{\mathfrak{g}''}{2\mathfrak{g}'}\right)^2$$

is the Schwarzian derivative of $\mathfrak{g}$. Remember that the Schwarzian derivative is invariant under postcomposition of $\mathfrak{g}$ with a Möbius transformation,[53] so that it is a well-defined notion for maps $Jf : U \to \mathbb{CP}^1$ into the 2-sphere. Conversely, rewriting Hill's equation as a first-order system

$$(Jf', Jf)' = (Jf', Jf) \begin{pmatrix} 0 & 1 \\ -\sigma & 0 \end{pmatrix} \tag{5.29}$$

shows that Jf can be uniquely reconstructed from its Schwarzian, up to Möbius transformation.

To compute how the Schwarzian changes under the T-transformation we first verify that the "proper" normalization is preserved: Because

$$d(T_\lambda f) = T_\lambda(\tau_\lambda f + df) = T_\lambda df \quad \Leftrightarrow \quad Jf'_\lambda = T_\lambda^{\mathbb{C}} Jf'$$

and $T_\lambda^{\mathbb{C}}$ takes values in $Sl(2, \mathbb{C})$, we have $\det(Jf'_\lambda, Jf_\lambda) = \det(Jf', Jf) = 1$. Taking one more derivative we obtain

$$Jf''_\lambda = T_\lambda^{\mathbb{C}}(\tau_\lambda^{\mathbb{C}}(\tfrac{\partial}{\partial z})Jf' + Jf'') = T_\lambda^{\mathbb{C}} Jf (\lambda \mathfrak{g}' \mathfrak{h}' - \sigma) = -Jf_\lambda (\sigma - \lambda \mathfrak{g}' \mathfrak{h}').$$

Thus Jf_λ satisfies Hill's equation with Schwarzian $\sigma_\lambda = \sigma_0 - \lambda \mathfrak{g}' \mathfrak{h}'$. Otherwise said, the difference between the Schwarzians[54] of f_λ and f is given by a multiple of the polarization q:

$$(\sigma_\lambda - \sigma_0) \, dz^2 = -\lambda d\mathfrak{g} \, d\mathfrak{h} \simeq -\lambda q.$$

In [48] the authors show, after introducing a notion of Schwarzian derivative for conformal immersions, that this becomes the general behavior of Calapso's transformation of isothermic surfaces. In fact, they use this behavior to describe Calapso's transformation: They prove, as a type of fundamental theorem for surfaces in Möbius geometry, that this Schwarzian and the conformal Hopf differential (cf., §5.1.10) form a complete set of invariants for surfaces in the conformal 3-sphere. When keeping the conformal Hopf differential unchanged while altering the Schwarzian by a multiple of

---

[53] This follows, for example, from the chain rule $\sigma_{\mu \circ \mathfrak{g}} = (\sigma_\mu \circ \mathfrak{g}) \cdot \mathfrak{g}'^2 + \sigma_{\mathfrak{g}}$ for the Schwarzian and the fact that the Schwarzian $\sigma_\mu = 0$ vanishes for Möbius transformations $\mu(z) = \frac{az+b}{cz+d}$.

[54] Note that the difference between two Schwarzians indeed behaves like a quadratic differential: This is a consequence of the chain rule for the Schwarzian, $\sigma_{\tilde{\mathfrak{g}} \circ w} - \sigma_{\mathfrak{g} \circ w} = ([\sigma_{\tilde{\mathfrak{g}}} - \sigma_{\mathfrak{g}}] \circ w) \cdot w'^2$.

the polarization (the holomorphic part of the conformal Hopf differential; here we need the surface to be isothermic), the integrability conditions remain satisfied so that the altered invariants determine a new (isothermic) surface (cf., §5.5.12).

Considering, as a simple example, $\mathfrak{g}(z) = z = \mathfrak{h}(z)$, we have to integrate

$$(T_\lambda^{\mathbb{C}})' = T_\lambda^{\mathbb{C}}\lambda \begin{pmatrix} z & -z^2 \\ 1 & -z \end{pmatrix} \quad \Leftrightarrow \quad (F_\lambda^{\mathbb{C}})' = F_\lambda^{\mathbb{C}} \begin{pmatrix} 0 & 1 \\ \lambda & 0 \end{pmatrix}.$$

Note the twofold possibility to interpret the system for $F_\lambda^{\mathbb{C}}$ in this case: On one hand, it is the system for the T-transformation gauged by the Euclidean frame providing the T-transforms f_λ as the image of $\binom{0}{1} H$; on the other hand, we recognize the first-order system (5.29) associated to Hill's equation. To obtain a solution for $T_\lambda^{\mathbb{C}}$ we may, according to our discussion of the relation between the Calapso and Darboux transformations in §5.5.3, just invert the fundamental system $\hat{F}_\lambda^{\mathbb{C}}$ that we determined earlier in §5.4.24:

$$T_\lambda^{\mathbb{C}}(z) = \begin{pmatrix} \cosh\sqrt{\lambda}z & \frac{1}{\sqrt{\lambda}}\sinh\sqrt{\lambda}z - z\cosh\sqrt{\lambda}z \\ \sqrt{\lambda}\sinh\sqrt{\lambda}z & \cosh\sqrt{\lambda}z - z\sqrt{\lambda}\sinh\sqrt{\lambda}z \end{pmatrix}.$$

Thus we obtain

$$f_\lambda = T_\lambda f = J^{-1} \begin{pmatrix} \frac{1}{\sqrt{\lambda}}\sinh\sqrt{\lambda}z \\ \cosh\sqrt{\lambda}z \end{pmatrix} H = \begin{pmatrix} -j\frac{1}{\sqrt{\lambda}}\tanh\sqrt{\lambda}z \\ 1 \end{pmatrix} H$$

as the T-transforms of f. Note that $f_0 = f$. A computation of the Schwarzian derivative of f_λ yields, as predicted, $\sigma_\lambda = -\lambda$.

5.5.28 Example. As a second example we reconsider the Clifford tori

$$\mathfrak{f}(x,y) = pe^{\frac{ix}{p}} + qe^{\frac{iy}{q}}j \quad \text{and} \quad \mathfrak{f}^*(x,y) = pe^{-\frac{ix}{p}} + qje^{-\frac{iy}{q}}.$$

Remember that, in §5.4.25, we restricted our attention to $\lambda \in (-\frac{1}{4q^2}, \frac{1}{4p^2})$, which provided certain reality conditions. Here we want to consider any $\lambda \in \mathbb{R}$ — in order to avoid the necessity to consider different cases, we will, in contrast to §5.4.25, write the solution of our linear system in terms of *real* functions.

Thus, we introduce functions $s_x = s_x(x)$ and $s_y = s_y(y)$ as the solutions of

$$s_x'' + (\tfrac{1}{4p^2} - \lambda)s_x = 0, \quad s_x(0) = 0, \quad s_x'(0) = 1,$$
$$s_y'' + (\tfrac{1}{4q^2} + \lambda)s_y = 0, \quad s_y(0) = 0, \quad s_y'(0) = 1.$$

Note that, as a consequence, we have the Pythagorean rules

$$s_x'^2 + (\tfrac{1}{4p^2} - \lambda)s_x^2 \equiv 1 \quad \text{and} \quad s_y'^2 + (\tfrac{1}{4q^2} + \lambda)s_y^2 \equiv 1.$$

Also, for abbreviation, we write

$$e_x = s'_x + \frac{i}{2p}s_x \quad \text{and} \quad e_y = s'_y + \frac{i}{2q}s_y.$$

For these two functions we have the relations

$$e'_x = \frac{i}{2p}e_x + \lambda s_x \quad \text{and} \quad |e_x|^2 = 1 + \lambda s_x^2,$$
$$e'_y = \frac{i}{2q}e_y - \lambda s_y \quad \text{and} \quad |e_y|^2 = 1 - \lambda s_y^2.$$

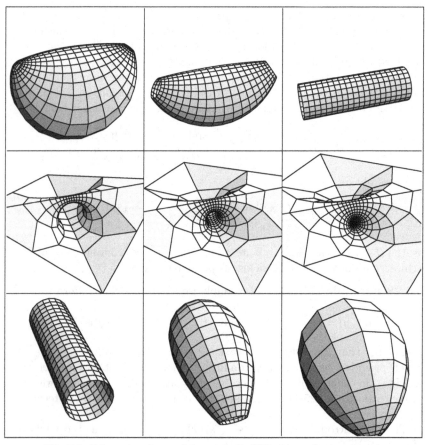

Fig. 5.7. Calapso transforms of the quadratic Clifford torus (center picture)

By using similar methods as in §5.4.25, we then determine a solution F_λ of the curved flat system

$$dF_\lambda = F_\lambda \cdot \begin{pmatrix} 0 & d\mathfrak{f} \\ \lambda d\mathfrak{f}^* & 0 \end{pmatrix}$$

in terms of these functions:

$$F_\lambda = \begin{pmatrix} e_x e_y - \lambda s_x s_y j & s_x e_y - e_x s_y j \\ e'_x e'_y - \lambda s'_x s'_y j & s'_x e'_y - e'_x s'_y j \end{pmatrix} \begin{pmatrix} e^{-(\frac{ix}{2p} + \frac{iy}{2q})} & 0 \\ 0 & i e^{\frac{ix}{2p} - \frac{iy}{2q}} \end{pmatrix}.$$

Since the Study determinant of F_λ is constant by the fact that its connection form takes values in $\mathfrak{sl}(2, \mathbb{H})$, it suffices to compute it at a single point. At the origin $(x, y) = (0, 0)$ we have

$$F_\lambda(0, 0) = \begin{pmatrix} 1 & 0 \\ -(\frac{1}{4pq} + \lambda j) & \frac{i}{2q} - \frac{i}{2p} j \end{pmatrix} \begin{pmatrix} 1 & 0 \\ 0 & i \end{pmatrix},$$

so that $[F_\lambda] \equiv 1 + \frac{1}{4p^2} + \frac{1}{4q^2} = 1 + \frac{1}{4p^2 q^2}$ since $p^2 + q^2 = 1$. Therefore, a common scaling of the F_λ's, which depends on the initial Clifford torus only, provides us with a family of curved flat frames with values in $Sl(2, \mathbb{H})$. However, to keep the formulas simple, we neglect this scaling. In fact, a rescaling of F_λ by a real constant has no effect on the T-transforms

$$f_\lambda = F_\lambda \begin{pmatrix} 0 \\ 1 \end{pmatrix} \mathbb{H} = \begin{pmatrix} s_x e_y - e_x s_y j \\ s'_x e'_y - e'_x s'_y j \end{pmatrix} \mathbb{H}.$$

To understand the geometry of the surfaces f_λ we introduce two curves of spheres

$$\mathcal{S}_x := \begin{pmatrix} 2s'_x(\frac{i}{2q} s_x - e_x j) + j & -2s_x^2 \\ -2(\frac{1}{4q^2} + \lambda)s'^2_x & -2s'_x(\frac{i}{2q} s_x - e_x j) - j \end{pmatrix},$$

$$\mathcal{S}_y := \begin{pmatrix} -2s'_y(\frac{i}{2q} s_y + e_y j) + j & 2s_y^2 \\ 2(\frac{1}{4p^2} - \lambda)s'^2_y & 2s'_y(\frac{i}{2p} s_y + e_y j) - j \end{pmatrix}.$$

First note that

$$4(\frac{1}{4q^2} + \lambda)s_x^2 s'^2_x - |2s'_x(\frac{i}{2q} s_x - e_x j) + j|^2 = 4\lambda s_x^2 s'^2_x - |1 - 2s'_x e_x|^2 = -1,$$
$$4(\frac{1}{4p^2} - \lambda)s_y^2 s'^2_y - |j - 2s'_y(\frac{i}{2q} s_x + e_x j)|^2 = -4\lambda s_y^2 s'^2_y - |1 - 2s'_y e_y|^2 = -1,$$

so that $\mathcal{S}_x$ and $\mathcal{S}_y$ really take values in the space $\mathfrak{S}(\mathbb{H}^2)$ of 2-spheres, and that $\mathcal{S}_x$ and $\mathcal{S}_y$ are symmetric with respect to the quaternionic Hermitian form S^3 given by $\begin{pmatrix} 0 & 1 \\ 1 & 0 \end{pmatrix}$, so that the 2-spheres of both curves lie in the corresponding 3-sphere (cf., §4.8.6). Further, it is a lengthy but straightforward computation to see that

$$\mathcal{S}_x f_\lambda \parallel f_\lambda \quad \text{and} \quad \mathcal{S}'_x f_\lambda \parallel f_\lambda,$$
$$\mathcal{S}_y f_\lambda \parallel f_\lambda \quad \text{and} \quad \mathcal{S}'_y f_\lambda \parallel f_\lambda.$$

This shows that f_λ envelopes the two sphere curves. As a consequence, all f_λ's are cyclides of Dupin (see §1.8.8).

Another way to verify this statement is by projecting f_λ stereographically:

$$f_\lambda \;=\; (s_x e_y - e_x s_y j)(s'_x e'_y - e'_x s'_y j)^{-1}$$

$$= \;\frac{1}{(\frac{1}{4q^2}+\lambda)s'^2_x+(\frac{1}{4p^2}-\lambda)s'^2_y}\begin{pmatrix} -\frac{1}{2q}s_x s'_x - \frac{1}{2p}s_y s'_y \\ (s'^2_x - s'^2_y) \\ \frac{1}{2p}s_x s'_x - \frac{1}{2q}s_y s'_y \end{pmatrix}.$$

By computing[55] the unit normal field of f_λ we arrive at

$$\mathfrak{n}_\lambda = \begin{pmatrix} 0 \\ 1 \\ 0 \end{pmatrix} + \frac{2}{(\frac{1}{4q^2}+\lambda)s'^2_x+(\frac{1}{4p^2}-\lambda)s'^2_y}\begin{pmatrix} -\frac{1}{2q}s'^2_y \cdot s''_x s'_x + \frac{1}{2p}s'^2_x \cdot s''_y s'_y \\ -\frac{1}{4p^2 q^2}s'^2_x s'^2_y \\ -\frac{1}{2p}s'^2_y \cdot s''_x s'_x - \frac{1}{2q}s'^2_x \cdot s''_y s'_y \end{pmatrix};$$

as a consequence,

$$\mathfrak{c}_y \;:=\; \mathfrak{n}_\lambda - 2(\tfrac{1}{4p^2} - \lambda)s'^2_y \cdot f_\lambda \;=\; \begin{pmatrix} -\frac{1}{p}s_y s'_y \\ 1 - 2s'^2_y \\ -\frac{1}{q}s_y s'_y \end{pmatrix}$$

and

$$\mathfrak{c}_x \;:=\; \mathfrak{n}_\lambda + 2(\tfrac{1}{4q^2} + \lambda)s'^2_x \cdot f_\lambda \;=\; \begin{pmatrix} \frac{1}{q}s_x s'_x \\ 1 - 2s'^2_x \\ -\frac{1}{p}s_x s'_x \end{pmatrix}$$

only depend on one parameter each. Taking derivative with respect to the other parameter shows that f_λ has principal curvatures $2(\frac{1}{4q^2} + \lambda)s'^2_x$ and $-2(\frac{1}{4p^2} - \lambda)s'^2_y$, by Rodrigues' formula. Thus the two principal curvatures are constant along their curvature lines, that is, the curvature lines are circles. Consequently, f_λ is channel in two ways by §1.8.20, confirming that we have a cyclide of Dupin (cf., §1.8.8).

Note that f_λ defines a torus (with one point at infinity[56]) if spectral parameter $\lambda \in (-\frac{1}{4q^2}, \frac{1}{4p^2})$; it is a cylinder for $\lambda = -\frac{1}{4q^2}, \frac{1}{4p^2}$; and it becomes (part of) a spindle cyclide otherwise (cf., [207]). In Figure 5.7, some Calapso transforms of the minimal Clifford torus (the surface in the middle), where $p = q = \frac{1}{\sqrt{2}}$, are shown.

Also note that the Calapso transformation changes the periods of the surface: As long as $\frac{1}{4p^2} - \lambda > 0$ or $\frac{1}{4q^2} + \lambda > 0$, the corresponding function, s_x

---

[55] Postcomposing f_λ by a rotation with angle $\alpha = \arctan(q/p)$ in the i-k-plane simplifies the computation slightly.

[56] Therefore, the fact that the Gaussian curvature of f_λ does not change sign is no contradiction to Gauss-Bonnet's theorem.

or s_y, is a scaled sine function with period $\frac{4p\pi}{\sqrt{1-4\lambda p^2}}$ or $\frac{4q\pi}{\sqrt{1+4\lambda q^2}}$, respectively. Therefore, f_λ has period $\frac{2p\pi}{\sqrt{1-4\lambda p^2}}$ or $\frac{2q\pi}{\sqrt{1+4\lambda q^2}}$ in the corresponding direction.

5.5.29 Example. A noteworthy property of the surfaces obtained in the previous example is that they all have constant mean curvature in some space of constant curvature. To conclude our discussions of the Calapso transformation we want to establish this as a general fact:

The Calapso transforms of a surface of constant mean curvature in some space form all have constant mean curvature in some space form.

In fact, we will do better than that and identify the Calapso transformation as Lawson's correspondence, up to Möbius equivalence (cf., §3.3.7): To prove this statement (again) we will imitate our argument from §3.3.7 — however, here we will see (again; cf., §5.3.21) how nicely one can mix up the (intrinsic) complex structure given in terms of complex coordinates with the quaternionic structure of the ambient space.

Thus let $f : M^2 \to S^3$ denote a surface of constant mean curvature in some space form Q_κ^3 given by $\mathcal{K} \in \mathbb{R}_1^6 \cong \mathfrak{H}(\mathbb{H}^2)$ (cf., §1.4.1 and §4.3.5). To frame f we choose the tangent plane map S as an enveloped sphere congruence (see §1.4.11), so that the second surface becomes Möbius equivalent to f (see §1.7.8); as tangential frame vectors we choose $S_1 = e^{-u}f_x$ and $S_2 = e^{-u}f_y$, where (x, y) denote conformal curvature line parameters, $I = e^{2u}(dx^2 + dy^2)$. The structure equations then read[57]

$$
\begin{aligned}
dS_1 &= (-u_y dx + u_x dy)\, S_2 + (He^u + \hat{H}e^{-u})dx\, S - \tfrac{\kappa}{2}e^u dx\, f + 2e^u dx\, \hat{f}, \\
dS_2 &= (u_y dx - u_x dy)\, S_1 + (He^u - \hat{H}e^{-u})dy\, S - \tfrac{\kappa}{2}e^u dy\, f + 2e^u dy\, \hat{f}, \\
&\text{and} \\
dS &= -(He^u + \hat{H}e^{-u})dx\, S_1 - (He^u - \hat{H}e^{-u})dy\, S_2, \\
df &= e^u dx\, S_1 \qquad\qquad + e^u dy\, S_2, \\
d\hat{f} &= -\tfrac{\kappa}{4}e^u dx\, S_1 \qquad\qquad - \tfrac{\kappa}{4}e^u dy\, S_2.
\end{aligned}
$$

Or, by interpreting f and $\hat{f}$ as homogeneous coordinates of two maps into the conformal 3-sphere $S^3 \subset \mathbb{HP}^1$, $F = (\hat{f}, f)$, and using our "translation table" in §4.5.12 with $S^3 = S_1$, $S = FS_i$, $S_1 = FS_j$, and $S_2 = FS_k$, we obtain

$$
\begin{aligned}
d\hat{f} &= \hat{f} \cdot \tfrac{i}{2}[\star du - (He^u\, dz + \hat{H}e^{-u}\, d\bar{z})j] \;+\; f \cdot \tfrac{\kappa}{4}e^u dz j, \\
df &= \hat{f} \cdot e^u dz j \;+\; f \cdot \tfrac{i}{2}[\star du - (He^u\, dz + \hat{H}e^{-u}\, d\bar{z})j],
\end{aligned}
\tag{5.30}
$$

where $z = x + iy$ and $\star$ is the Hodge operator, $\star dx = dy$ and $\star dy = -dx$. From the Codazzi equation $e^{-u}\hat{H}_z = e^u H_{\bar{z}} = 0$ we learn that $\hat{H} \equiv const$

---

[57] Remember that, in contrast to §1.7.8, we normalized $\hat{f}$ so that $\langle f, \hat{f} \rangle \equiv -\tfrac{1}{2}$ in the quaternionic setup; in this way, $\hat{f} = \tfrac{1}{2}\mathcal{K} - \tfrac{\kappa}{4}f$ instead of $\hat{f} = \tfrac{\kappa}{2}f - \mathcal{K}$.

since H is constant; therefore, to simplify later computations, we exclude the (trivial) case $\hat{H} \equiv 0$ and choose the coordinates (x, y) so that $\hat{H} \equiv 1$.

Note how the hypersphere complex $\mathcal{K} \in \mathbb{R}_1^6 \cong \mathfrak{H}(\mathbb{H}^2)$ is, in the quaternionic setup, given as a quaternionic Hermitian form by

$$\mathcal{K}(f, f) = 2, \quad \mathcal{K}(f, \hat{f}) = 0, \quad \text{and} \quad \mathcal{K}(\hat{f}, \hat{f}) = \tfrac{\kappa}{2},$$

so that $\langle \mathcal{K}, S_f \rangle = -\tfrac{1}{2}\mathcal{K}(f, f) \equiv -1$ when using (4.4) to identify vectors in $\mathbb{H}^2$ with (isotropic) quaternionic Hermitian forms.

By §5.5.25, any adapted frame can be used to determine the Calapso transforms f_λ of f. Thus we may use the above metric frame, $F = (\hat{f}, f)$, with respect to $v_\infty = \binom{1}{0}$ and $v_0 = \binom{0}{1}$ as a basis of $\mathbb{H}^2$: Writing (5.30) in matrix form and, for abbreviation, $\theta := -(He^u dz + e^{-u} d\bar{z})$, we must integrate the system

$$dF_\lambda = F_\lambda \Phi_\lambda \quad \text{with} \quad \Phi_\lambda := \begin{pmatrix} \tfrac{i}{2}(\star du + \theta j) & e^u dz j \\ (\tfrac{\kappa}{4}e^u dz - \lambda e^{-u} d\bar{z})j & \tfrac{i}{2}(\star du + \theta j) \end{pmatrix}, \tag{5.31}$$

since $\psi^* = (e^u j)^{-1} dx - (e^u k)^{-1} dy = -e^{-u}(dx - i dy)j$. Then $f_\lambda = F_\lambda v_0$. Note that, for $\lambda = 0$, (5.31) gives us the original frame $F = F_0$ up to Möbius transformation, and, from §5.5.25 again, $T_\lambda = F_\lambda F_0^{-1}$.

Next, let $\mathcal{Z} \in \mathfrak{G}(\mathbb{H}^2)$ denote the central sphere congruence of $f = f_0$; using (4.22) to identify $\mathcal{Z} \simeq Z = S^3(., \mathcal{Z}.) \in \mathfrak{H}(\mathbb{H}^2)$, we have $Z = S + HS_f$, where we identify the homogeneous coordinates of f with the quaternionic Hermitian form S_f via (4.4), as before. Note that $\langle \mathcal{K}, Z \rangle = H\langle \mathcal{K}, S_f \rangle = -H$ gives us the mean curvature H of f.

From §5.5.18 we know that $\mathcal{Z}_\lambda = T_\lambda \mathcal{Z} T_\lambda^{-1}$. In terms of Hermitian forms this means $Z_\lambda = T_\lambda Z$: If we fix T_λ to take values in S^3 (see §5.5.16), we obtain

$$T_\lambda Z = S^3(T_\lambda^{-1}., \mathcal{Z} T_\lambda^{-1}.) = (T_\lambda S^3)(., \mathcal{Z}_\lambda.) = S^3(., \mathcal{Z}_\lambda.) = Z_\lambda.$$

Now we claim that f_λ takes values in the space form given by

$$\mathcal{K}_\lambda := T_\lambda(\mathcal{K} + 2\lambda Z) :$$

Clearly, $\mathcal{K}_\lambda(f_\lambda, f_\lambda) \equiv 2$, so that we only have to show that $\mathcal{K}_\lambda \equiv const$ is constant and consequently defines a quadric $Q_{\kappa_\lambda}^3$ of constant curvature κ_λ. First note that

$$\mathcal{K}_\lambda = F_\lambda C_\lambda, \quad \text{where} \quad C_\lambda := F_0^{-1}(\mathcal{K} + 2\lambda Z)$$

is a *constant* quaternionic Hermitian form:

$$C_\lambda(v_0, v_0) = 2, \quad C_\lambda(v_0, v_\infty) = 2\lambda i, \quad C_\lambda(v_\infty, v_\infty) = \tfrac{\kappa}{2} + 2\lambda H,$$

since $Z = S + HS_f = F(S_i + HS_0)$ with $S_0 = S_{v_0}$ as in §4.5.12 (cf., (4.4)). With this knowledge we find $d\mathcal{K}_\lambda = F_\lambda \cdot \varrho(\Phi_\lambda)C_\lambda$, where ϱ denotes the Lie algebra homomorphism from (4.7). Now it is straightforward to check that the quaternionic Hermitian form $\varrho(\Phi_\lambda)C_\lambda = 0$, proving that $\mathcal{K}_\lambda$ is constant.

Thus $\mathcal{K}_\lambda$ defines a quadric $Q^3_{\kappa_\lambda}$ of constant curvature

$$\kappa_\lambda = -|\mathcal{K}_\lambda|^2 = -|\mathcal{K}|^2 - 4\lambda\langle\mathcal{K}, Z\rangle - 4\lambda^2 = \kappa + 4\lambda H - 4\lambda^2 \qquad (5.32)$$

indeed, so that f_λ takes values in $Q^3_{\kappa_\lambda}$ with constant mean curvature

$$H_\lambda = -\langle\mathcal{K}_\lambda, Z_\lambda\rangle = -\langle\mathcal{K} + 2\lambda Z, Z\rangle = H - 2\lambda. \qquad (5.33)$$

This proves our first assertion. Note that $\kappa_\lambda + H_\lambda^2 \equiv \kappa + H^2$ does not depend on λ — or, otherwise said, the ambient curvature κ_λ is a quadratic polynomial of the varying mean curvature H_λ. Every such family of Calapso transforms contains exactly one minimal surface at $\lambda = \frac{1}{2}H$, that takes values in the space of maximal ambient curvature $\kappa_{\frac{1}{2}H} = \kappa + H^2$ occurring in the family (cf., [60]). Using this surface as the "initial surface" of the family, that is, $H = H_0 = 0$, its transforms $f_{\pm\lambda}$ take values in the same quadric of constant curvature $\kappa_{\pm\lambda} = \kappa_0 - 4\lambda^2$ (up to identification via Möbius transformation) but with opposite mean curvatures $H_{\pm\lambda} = \mp 2\lambda$.

In order to see that, moreover, the T_λ-transformation is Lawson's correspondence, we have to verify that the induced metrics I_λ and the trace-free part $I\!I_\lambda - H_\lambda I_\lambda$ of the second fundamental forms of the f_λ's (as maps into the quadric $Q^3_{\kappa_\lambda}$) do not depend on λ. Here we employ our "translation table" from §4.5.12 again: Because the trace-free part of the second fundamental form is the same for *all* enveloped sphere congruences S_λ, considered as a normal field to f_λ, we do not need to change frames (to obtain the tangent plane maps of f_λ) and directly read off, from (5.31),

$$I_\lambda = e^{2u}|dz|^2 = I_0 \quad \text{and} \quad I\!I_\lambda = H_\lambda I_\lambda + \text{Re}\,dz^2 = I\!I_0 - 2\lambda I_0. \qquad (5.34)$$

Note that, in the particular case where f_0 is a minimal surface in Euclidean ambient space Q^3_0, the Calapso transforms f_λ of f_0 are also obtained by the Umehara-Yamada perturbation (see [285]): They are surfaces of constant mean curvature $H_\lambda = -2\lambda$ in hyperbolic spaces of curvature $\kappa_\lambda = -4\lambda^2$. That is, they have the mean curvature of the horospheres of the ambient hyperbolic space. For this reason we will also refer to them as "horospherical surfaces." For later reference we state the following (local) characterization of horospherical surfaces, as an obvious consequence of our discussions[58]:

An isothermic surface $f : M^2 \to S^3$ is a horospherical surface in some hyperbolic space if and only if it is (locally) the Calapso transform of a minimal surface in Euclidean space.

---

[58] Compare [253].

5.6 Bianchi's permutability theorems

We already mentioned Bianchi's outstanding contributions to the transformation theory of isothermic surfaces in the Introduction: He not only introduced Calapso's transformation, independently of Calapso, but he also was the first to elaborate on this transformation theory systematically and to study the interrelations of the various transformations in terms of permutability theorems (see [21] and [20]). For us, these permutability theorems will not only be of interest because they clarify the interrelations between the four kinds of transformation that we discussed, but they also turn out to be crucial for the transformations of minimal and constant mean curvature surfaces (cf., [153]); as well as they will be the key to our discussion of discrete isothermic surfaces (cf., [150]).

5.6.1 Notation. To make our point we will use, throughout this section, symbolic notations for the Christoffel, Goursat, and Darboux transformations. Thus we will write $C\mathfrak{f}$ for the Christoffel transform $\mathfrak{f}^*$ of an isothermic surface $\mathfrak{f}$ in Euclidean space, $G\mathfrak{f}$ or $G_\mu\mathfrak{f}$ for its Goursat transforms (we write G_μ if we wish to emphasize the Möbius transformation μ associated with the Goursat transformation), and $D_\lambda f$ for a Darboux transform of an isothermic surface f in the conformal sphere — similar to the way we write $T_\lambda f$ for the Calapso transform f_λ of f.

However, in contrast to the Calapso transforms $T_\lambda f$, the notations $D_\lambda f$, $G\mathfrak{f}$, and $C\mathfrak{f}$ are just symbolic notations.

Note that we will use the notation "T_λ" in both senses: in the concrete sense of a map into the Möbius group as well as in the abstract sense of a transformation of an isothermic surface. In the abstract sense we will (notationally) not distinguish between the transformations for different surfaces, that is, we will write $T_\lambda f$ and $T_\lambda \hat{f}$ at the same time, even though, as maps into the Möbius group, the two T_λ's are different.

5.6.2 We have already proved three such permutability theorems for the Christoffel, Goursat, and Calapso transformations, in §5.2.4, §5.3.5, and §5.5.9: Symbolically,

$$C^2 = id, \quad G_{\mu_1} G_{\mu_2} = G_{\mu_1 \circ \mu_2}, \quad \text{and} \quad T_{\lambda_1} T_{\lambda_2} = T_{\lambda_1 + \lambda_2},$$

where the first equality is up to translation and homothety, the second up to similarity, and the last equality is up to Möbius transformation — that is, the equalities have to be interpreted as equalities in the respective geometry of the transformation.

Thus we are, so far, missing a permutability theorem for two Darboux transformations. This is given by Bianchi's most famous permutability theorem in the transformation theory of isothermic surfaces, which has a similar flavor as the permutability theorems for the Bäcklund transformation of pseudospherical surfaces or for the Ribaucour transformation (cf., e.g., [28]), which is more general than the permutability theorem for the Darboux transformation.

However, before attacking this theorem, it turns out to be useful to discuss the interrelation of the Christoffel and Darboux transformations for isothermic surfaces in Euclidean space (see [21]).

5.6.3 Theorem. *Let $\mathfrak{f}$ denote an isothermic surface in Euclidean space, with Christoffel transform $\mathfrak{f}^* = C\mathfrak{f}$ and a Darboux transform $\hat{\mathfrak{f}} = D_\lambda\mathfrak{f}$. Then there is a surface $\hat{\mathfrak{f}}^*$, given algebraically by $\hat{\mathfrak{f}}^* = \mathfrak{f}^* + \frac{1}{\lambda}(\hat{\mathfrak{f}} - \mathfrak{f})^{-1}$, that is the Christoffel transform of $\hat{\mathfrak{f}}$ and a D_λ-transform of $\mathfrak{f}^*$ at the same time[59]; symbolically,*

$$D_\lambda C = C D_\lambda.$$

5.6.4 *Proof.* We already proved half of that theorem earlier without stating it: Remember that, when proving that the Darboux transform of an isothermic surface is indeed isothermic, we used the fact that (5.18) defines a Christoffel transform of $\hat{\mathfrak{f}}$. The only thing left is to show that $\hat{\mathfrak{f}}^*$ is a Darboux transform of $\mathfrak{f}^*$, that is, it satisfies the Riccati equation (5.17):

$$d\hat{\mathfrak{f}}^* = \tfrac{1}{\lambda}(\hat{\mathfrak{f}} - \mathfrak{f})^{-1}d\mathfrak{f}(\hat{\mathfrak{f}} - \mathfrak{f})^{-1} = \lambda(\hat{\mathfrak{f}}^* - \mathfrak{f}^*)d\mathfrak{f}^{**}(\hat{\mathfrak{f}}^* - \mathfrak{f}^*)$$

since $\lambda(\hat{\mathfrak{f}}^* - \mathfrak{f}^*) = (\hat{\mathfrak{f}} - \mathfrak{f})^{-1}$ and $C^2 = id$. ◁

5.6.5 In particular, this gives us a canonical scaling (and position) for the Christoffel transform of a Darboux transform $\hat{\mathfrak{f}}$ as soon as the Christoffel transform $\mathfrak{f}^*$ of $\mathfrak{f}$ is scaled (and positioned). Note that this scaling is compatible with Christoffel's formula (5.3): If (x, y) denote conformal curvature line parameters, then

$$\mathfrak{f}_x\mathfrak{f}_x^* = 1 \quad \text{and} \quad \mathfrak{f}_y\mathfrak{f}_y^* = -1 \quad \Leftrightarrow \quad \hat{\mathfrak{f}}_x\hat{\mathfrak{f}}_x^* = 1 \quad \text{and} \quad \hat{\mathfrak{f}}_y\hat{\mathfrak{f}}_y^* = -1.$$

Because the scaling of the Christoffel transform of an isothermic surface fixes a "scale" for the parameter of the Darboux (and Calapso) transformations, this "canonical" scaling of $\hat{\mathfrak{f}}^*$ provides a canonical scale for the parameter of iterated Darboux transformations. With this in mind we can now formulate Bianchi's permutability theorem for the Darboux transformation (see [19] or [21], and [106]):

---

[59] Therefore, the notation $\hat{\mathfrak{f}}^*$, which does not refer to the order of transformations, makes sense.

5.6.6 Theorem. *Let $\hat{f}_i = D_{\lambda_i} f$, $i = 1, 2$, denote two Darboux transforms of an isothermic surface f. Then there is a fourth surface $\hat{f}$ that is a D_{λ_2}-transform of $\hat{f}_1$ and a D_{λ_1}-transform of $\hat{f}_2$ at the same time; symbolically,*

$$D_{\lambda_1} D_{\lambda_2} = D_{\lambda_2} D_{\lambda_1}.$$

Moreover, corresponding points of f, $\hat{f}_1$, $\hat{f}_2$, and $\hat{f}$ are concircular and have constant (real) cross-ratio[60] $[f; \hat{f}_1; \hat{f}; \hat{f}_2] \equiv \frac{\lambda_2}{\lambda_1}$.

5.6.7 Proof. The cross-ratio condition $[f; \hat{f}_1; \hat{f}; \hat{f}_2] \equiv \frac{\lambda_2}{\lambda_1}$ uniquely determines a surface $\hat{f}$ (see (4.27)). We intend to show that this surface is a D_{λ_2}-transform of $\hat{f}_1$ and a D_{λ_1}-transform of $\hat{f}_2$. For this purpose we stereographically project the surfaces, so that the $\hat{\mathfrak{f}}_i$'s satisfy the Riccati equation (5.17), and write

$$\hat{f} = \binom{\hat{\mathfrak{f}}_1}{1} (\hat{\mathfrak{f}}_1 - \mathfrak{f})^{-1} \lambda_2 - \binom{\hat{\mathfrak{f}}_2}{1} (\hat{\mathfrak{f}}_2 - \mathfrak{f})^{-1} \lambda_1.$$

Using the formula from §4.9.8, it is now straightforward to check that this indeed defines $\hat{f}$: By writing $f = \binom{\mathfrak{f}}{1}$ and $\hat{\varphi}_i = (1, -\hat{\mathfrak{f}}_i)$, we compute

$$
\begin{aligned}
[f; \hat{f}_1; \hat{f}; \hat{f}_2] &= (\hat{\varphi}_1 f)(\hat{\varphi}_2 f)^{-1} (\hat{\varphi}_1 \hat{f})(\hat{\varphi}_2 \hat{f})^{-1} \\
&= \tfrac{\lambda_2}{\lambda_1} (\hat{\mathfrak{f}}_1 - \mathfrak{f})(\hat{\mathfrak{f}}_2 - \mathfrak{f})^{-1} (\hat{\mathfrak{f}}_1 - \hat{\mathfrak{f}}_2)(\hat{\mathfrak{f}}_1 - \mathfrak{f})^{-1} (\hat{\mathfrak{f}}_2 - \mathfrak{f})(\hat{\mathfrak{f}}_1 - \hat{\mathfrak{f}}_2)^{-1} \\
&= \tfrac{\lambda_2}{\lambda_1}.
\end{aligned}
$$

Further, by using the fact that the $\hat{\mathfrak{f}}_i$'s satisfy the Riccati equation (5.17), we find

$$d\hat{f} = \binom{\hat{\mathfrak{f}}_1}{1} (\hat{\mathfrak{f}}_1 - \mathfrak{f})^{-1} d\mathfrak{f} (\hat{\mathfrak{f}}_1 - \mathfrak{f})^{-1} \lambda_2 - \binom{\hat{\mathfrak{f}}_2}{1} (\hat{\mathfrak{f}}_2 - \mathfrak{f})^{-1} d\mathfrak{f} (\hat{\mathfrak{f}}_2 - \mathfrak{f})^{-1} \lambda_1;$$

and, with the retraction form (5.19) of the Darboux transform $\hat{f}_1$, we obtain

$$\hat{\tau}_1 \hat{f} = \binom{\hat{\mathfrak{f}}_1}{1} (\hat{\mathfrak{f}}_1 - \mathfrak{f})^{-1} d\mathfrak{f} \left((\hat{\mathfrak{f}}_2 - \mathfrak{f})^{-1} - (\hat{\mathfrak{f}}_1 - \mathfrak{f})^{-1} \right).$$

Combining these two equations, we finally derive that

$$(d + \lambda_2 \hat{\tau}_1) \hat{f} = \hat{f} \, d\mathfrak{f} \, (\hat{\mathfrak{f}}_2 - \mathfrak{f})^{-1} \parallel \hat{f},$$

---

[60] Note that this equation is symmetric in $\hat{f}_1$ and $\hat{f}_2$ by the identities for the cross-ratio shown in Figure 4.3.

so that $\hat{f}$ is indeed a D_{λ_2}-transform of $\hat{f}_1$, by §5.4.8. By symmetry, $\hat{f}$ is also a D_{λ_1}-transform of $\hat{f}_2$, and Bianchi's permutability theorem is proved. ◁

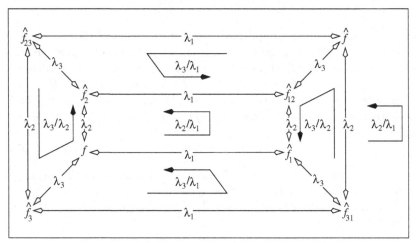

Fig. 5.8. Three Darboux transformations commute

5.6.8 A simple consequence, using the hexahedron lemma from §4.9.13, is now an extension of this permutability theorem to three (see [21]) Darboux transformations (see Figure 5.8):

$$D_{\lambda_{i(1)}} D_{\lambda_{i(2)}} D_{\lambda_{i(3)}} = D_{\lambda_1} D_{\lambda_2} D_{\lambda_3}$$

for any permutation $i : \{1, 2, 3\} \to \{1, 2, 3\}$. To verify this assertion just contemplate Figure 4.4, to obtain the scheme shown in Figure 5.8.

5.6.9 Next we tackle a permutability theorem, in its original form also due to Bianchi [20], that interrelates the Christoffel, Darboux, and Calapso transformations: Remember that, in §3.3.8, we obtained the Christoffel transformation as a limiting case of Darboux transformations. The Sym formula from §5.5.4 indicates that the same is true in our present setup.

In fact, we already proved the converse statement in §5.5.20, that is, the first part of the following theorem:

5.6.10 Theorem. *Let* $\mathfrak{f}$ *be an isothermic surface in Euclidean space with Christoffel transform* $\mathfrak{f}^*$. *Then their* T_λ-*transforms* f_λ *and* $\hat{f}_\lambda$ *can be positioned to form a* $D_{-\lambda}$-*pair in the conformal sphere.*

Conversely, if f *and* $\hat{f}$ *form a* D_λ-*pair of isothermic surfaces in the conformal sphere, then their* T_λ-*transforms form, after suitable stereographic projections, a Christoffel pair; symbolically,*[61)]

$$T_\lambda C = D_{-\lambda} T_\lambda \quad \text{and} \quad T_\lambda D_\lambda = C T_\lambda.$$

5.6.11 *Proof.* It remains to prove the second statement of the theorem.

Let $(f, \hat{f}) : M^2 \to \mathfrak{P}$ denote a D_λ-pair of isothermic surfaces in the conformal 4-sphere, and let $T_\lambda : M^2 \to Sl(2, \mathbb{H})$ be the T-transformation for f. Then we know from §5.5.3 that $T_\lambda \hat{f} \equiv const =: v_\infty \mathbb{H} \in \mathbb{HP}^1$. Let v_0 be so that (v_∞, v_0) is a basis of $\mathbb{H}^2$, and let (ν_∞, ν_0) be its pseudodual basis. By possibly replacing $\hat{f}$ by $\hat{f}(\nu_0 T_\lambda \hat{f})^{-1}$ and f by $f(\nu_\infty T_\lambda f)^{-1}$ we can achieve

$$T_\lambda \hat{f} \equiv v_\infty \quad \text{and} \quad T_\lambda f = f_\lambda = v_0 + v_\infty \mathfrak{f}_\lambda. \tag{5.35}$$

Let $\mathfrak{f}_\lambda^*$ denote the Christoffel transform of $\mathfrak{f}_\lambda$. We intend to show that $\hat{f}$ is the $T_{-\lambda}$-transform of $\mathfrak{f}_\lambda^*$. For this, remember from §5.5.9 that T_λ^{-1} is the $T_{-\lambda}$-transformation[62)] for f_λ. The $T_{-\lambda}$-transformation for $v_0 + v_\infty \mathfrak{f}_\lambda^*$ is (from §5.5.22) given by

$$T_{-\lambda}^* = T_{-\lambda}(v_\infty \nu_\infty - \lambda(v_0 + v_\infty \mathfrak{f}_\lambda)(\nu_0 - \mathfrak{f}_\lambda^* \nu_\infty)).$$

Now, as in §5.5.22, $T_{-\lambda}^*(v_0 + v_\infty \mathfrak{f}_\lambda^*) = T_{-\lambda} v_\infty = T_\lambda^{-1} v_\infty = \hat{f}$.   ◁

5.6.12 It may be a worthwhile exercise for the reader to verify that $(T_{-\lambda}^*)^{-1}$ is a T_λ-transform for $\hat{f}$, $(T_{-\lambda}^*)^{-1} = \hat{T}_\lambda$: For this, one needs to check that

$$\tfrac{1}{\lambda} T_{-\lambda}^* d(T_{-\lambda}^*)^{-1} = T_{-\lambda}^* \tau_1^*(T_{-\lambda}^*)^{-1}$$

is a closed retraction form for $\hat{f}$ (cf., §5.5.9).

5.6.13 Example. Remember that, in §5.5.29, we learned that the Calapso transformation yields, for surfaces of constant mean curvature in space forms, Lawson's correspondence. In particular, the T_λ-transforms f_λ of a minimal surface $f_0 : M^2 \to Q_0^3$ into Euclidean space, given by an isotropic quaternionic Hermitian form $\mathcal{K} = S_\infty$ that corresponds to a vector v_∞ via (4.4), have the constant mean curvature $H_\lambda = -2\lambda$ of horospheres in the hyperbolic spaces $Q_{-4\lambda^2}^3$ of constant curvature $\kappa_\lambda = -4\lambda^2$, given by the sphere complexes $\mathcal{K}_\lambda = T_\lambda(S_\infty + 2\lambda Z)$: In §5.5.29 we used this observation

---

[61)] Remember that, for the symbolic notations, we do not distinguish (notationally) between different T_λ's: As maps into the Möbius group, the four T_λ's in this formula will be different.

[62)] As before in the present section, we omit the upper indices for the Calapso transformations for different isothermic surfaces from an associated family.

to characterize surfaces that are horospherical in some hyperbolic space. Here we intend to derive a second characterization.

Namely, we know from §5.2.10 that the Christoffel transform of a minimal surface is its Gauss map $n_0 : M^2 \to S^2 \subset Q_0^3$. In particular, it takes values in a 2-sphere, and, as a consequence, all its Calapso transforms n_λ take values in certain 2-spheres (cf., §5.5.27). Thus our permutability theorem in §5.6.10 implies that a surface of constant mean curvature $H_\lambda = -2\lambda$ in $Q_{-4\lambda^2}^3$ has a totally umbilic $D_{-\lambda}$-transform.

Conversely, if an isothermic surface $f : M^2 \to S^3$ has a totally umbilic Darboux transform $n = D_\lambda f$, then their T_λ-transforms form, after suitable stereographic projection, a Christoffel pair in Euclidean 3-space. Since $T_\lambda n$ is totally umbilic, we may, without loss of generality,[63] assume that $T_\lambda n$ parametrizes part of the unit sphere in Q_0^3. Then $T_\lambda n$ is the Gauss map of $T_\lambda f$, so that $T_\lambda f$ is identified as a minimal surface (see §5.2.10). Then, since $f = T_{-\lambda}(T_\lambda f)$ by §5.5.9, f is a $T_{-\lambda}$-transform of a minimal surface in Euclidean space and therefore has constant mean curvature $H = 2\lambda$ in some hyperbolic space of curvature $\kappa = -4\lambda^2$.

To summarize[64] (cf., [253], [153]): *An isothermic surface $f : M^2 \to S^3$ is a horospherical surface in some hyperbolic space if and only if it has a totally umbilic Darboux transform.*

After having obtained this second characterization of horospherical surfaces, we want to study the geometry of the situation a bit further. First note that the central sphere congruence $Z = Z_0$ of a minimal surface $f = f_0$ in Euclidean space Q_0^3 is its tangent plane map. Thus all spheres of Z contain the point $v_\infty H$ at infinity,

$$Z(v_\infty, v_\infty) \equiv 0 \quad \Leftrightarrow \quad \mathcal{Z}v_\infty H = v_\infty H. \tag{5.36}$$

Using the curved flat frame $F_\lambda = T_\lambda F_0$ from §5.5.20, where F_0 denotes the Euclidean frame of $f = f_0$, we have

$$f_\lambda = T_\lambda f_0 = F_\lambda v_0 \quad \text{and} \quad n_\lambda = F_\lambda v_\infty = T_\lambda v_\infty.$$

Hence $\mathcal{K}_\lambda(n_\lambda, n_\lambda) = (S_\infty + 2\lambda Z)(v_\infty, v_\infty) \equiv 0$, showing that n_λ is not only totally umbilic but, more specifically, takes values in the infinity boundary

---

[63] Here we use the scaling and translation ambiguity of the Christoffel transformation — note that we exclude the case where $T_\lambda n$ becomes flat: This is the case of pairs of meromorphic functions discussed in §5.4.24. On the other hand, a totally umbilic surface clearly qualifies as a horospherical surface in many hyperbolic spaces.

[64] In the classical literature, isothermic surfaces with a totally umbilic (planar) Darboux transform sometimes also appear under the name "surfaces of Thybaut" (cf., [234] and [253]); of interest should also be the papers [280], [281], and [282] by A. Thybaut.

This characterization of horospherical surfaces as Darboux transforms of their hyperbolic Gauss maps (see below) provides a possibility to construct horospherical surfaces in this way (cf., [286] and [151]). See also [188] (cf., [22]).

of the hyperbolic space that makes f_λ a horospherical surface. Because it is also a Darboux transform of f_λ, we deduce that corresponding points of f_λ and n_λ lie on circles that intersect f_λ as well as (the infinity boundary) n_λ orthogonally, that is, they lie on hyperbolic geodesics that are orthogonal to f_λ; moreover, f_λ and n_λ induce conformally equivalent metrics. These two properties characterize the hyperbolic Gauss map of a horospherical surface[65] [42]:

The Calapso transforms $n_\lambda = T_\lambda n_0$ of the Gauss map of a minimal surface in Euclidean space provide the hyperbolic Gauss maps of the horospherical Calapso transforms of the minimal surface.

Finally, remember from §5.5.18 that the central sphere congruence of a Calapso transform $T_\lambda f$ of $f = f_0$ is given by $\mathcal{Z}_\lambda = T_\lambda \mathcal{Z} T_\lambda^{-1}$. Moreover, from §5.5.17, we know that $d\mathcal{Z}_\lambda = T_\lambda d\mathcal{Z} T_\lambda^{-1}$ since $[\tau_\lambda, \mathcal{S}] = 0$ for *any* 2-sphere congruence enveloped by f. As a consequence,

$$\mathcal{Z}_\lambda n_\lambda \parallel n_\lambda \quad \text{and} \quad (d\mathcal{Z}_\lambda) n_\lambda \parallel n_\lambda,$$

since $\mathcal{Z} v_\infty \mathbb{H} = v_\infty \mathbb{H}$: The central sphere congruence of a minimal surface is its tangent plane congruence. Thus n_λ satisfies the enveloping conditions from §4.8.15. Otherwise said[66] (cf., §3.6.4):

It is the central sphere congruence of a horospherical surface in a hyperbolic space form that provides its hyperbolic Gauss map as a Darboux transform of the surface[67]; the central sphere congruence of a horospherical surface therefore consists of horospheres[68] of the ambient hyperbolic space form.

This provides another justification for our notion of a "horospherical surface" for the surfaces of constant mean curvature H in a hyperbolic space of curvature $\kappa = -H^2$.

---

[65] Note that the hyperbolic Gauss map of a horospherical surface is a *globally* defined object. Therefore, the above characterization of horospherical surfaces by the existence of a totally umbilic Darboux transform works globally.

[66] This is, in fact, nearly a characterization of horospherical surfaces: the only surfaces for which the central sphere congruence provides a Darboux transformation are the horospherical surfaces and the minimal surfaces in Euclidean space (where the Darboux transform degenerates to the point at infinity) (see [234] and [153]).

[67] The converse of this statement is also true: If the central sphere congruence of an isothermic surface provides a Darboux transformation, then the surface is minimal in Euclidean space or horospherical in some hyperbolic space. For more details on this as well as on the notion of "isothermic surfaces of spherical type" (another way of characterizing minimal and horospherical surfaces among isothermic surfaces, via the curvature of their conformal Gauss maps), see [153].

[68] This second assertion also follows from $\frac{1}{|\mathcal{K}_\lambda|}\langle \mathcal{K}_\lambda, \mathcal{Z}_\lambda \rangle = \pm 1$: The central spheres $\mathcal{Z}_\lambda$ touch the infinity boundary of the ambient hyperbolic space.

5.6.14 Applying the above theorem from §5.6.10 twice, along with the "1-parameter group property" of the Calapso transformation, we obtain a permutability theorem for the Darboux and Calapso transformations, namely,

$$T_\lambda D_\mu T_\lambda^{-1} = T_{\lambda-\mu} T_\mu D_\mu T_\mu^{-1} T_{\lambda-\mu}^{-1} = T_{\lambda-\mu} C T_{\lambda-\mu}^{-1} = D_{\mu-\lambda}.$$

More accurately, we should argue as follows: Given a D_μ-pair of isothermic surfaces, their T_μ-transforms can be stereographically projected to yield a Christoffel pair of isothermic surfaces in Euclidean space; taking the $T_{\lambda-\mu}$-transforms of the surfaces obtained in this way yields, on one hand, the T_λ-transforms of the original surfaces, and, on the other hand, these last surfaces can be positioned to form a $D_{\mu-\lambda}$-pair of isothermic surfaces in the conformal sphere.

We want to complement this result with a discussion of the Calapso transformations that perform the mapping. Thus consider the D_μ-pair $(\hat{f}, f)$ and write the Christoffel pair, obtained as T_μ-transforms from $\hat{f}$ and f, in terms of affine coordinates as in the proof of our previous permutability theorem (see (5.35)):

$$f_\mu = T_\mu f = v_0 + v_\infty \mathfrak{f}_\mu \quad \text{and} \quad f_\mu^\perp := \nu_0 - \mathfrak{f}_\mu \nu_\infty,$$
$$\hat{f}_\mu = \hat{T}_\mu \hat{f} = v_0 + v_\infty \hat{\mathfrak{f}}_\mu \quad \text{and} \quad \hat{f}_\mu^\perp := \nu_0 - \hat{\mathfrak{f}}_\mu \nu_\infty.$$

We wish to examine the relation between $T_\lambda = T_{\lambda-\mu}^\mu T_\mu$ and $\hat{T}_\lambda = \hat{T}_{\lambda-\mu}^\mu \hat{T}_\mu$. From §5.5.22 we know that

$$\hat{T}_\mu^{-1} = T_\mu^{-1}(v_\infty \nu_\infty - \mu f_\mu \hat{f}_\mu^\perp) \quad \text{and} \quad \hat{T}_{\lambda-\mu}^\mu = T_{\lambda-\mu}^\mu(v_\infty \nu_\infty + (\lambda - \mu) f_\mu \hat{f}_\mu^\perp).$$

A straightforward computation then yields

$$
\begin{aligned}
\hat{T}_\lambda &= T_{\lambda-\mu}^\mu(v_\infty \nu_\infty + (\lambda - \mu) f_\mu \hat{f}_\mu^\perp)(-\tfrac{1}{\mu} v_\infty \nu_\infty + \hat{f}_\mu f_\mu^\perp) T_\mu \\
&= T_{\lambda-\mu}^\mu(v_\infty f_\mu^\perp + (1 - \tfrac{\lambda}{\mu}) f_\mu \nu_\infty) T_\mu \\
&= T_{\lambda-\mu}^\mu T_\mu - \tfrac{\lambda}{\mu} T_{\lambda-\mu}^\mu T_\mu f \nu_\infty T_\mu \\
&= T_\lambda(1 - \tfrac{\lambda}{\mu} f(\hat{f}^\perp f)^{-1} \hat{f}^\perp),
\end{aligned}
\tag{5.37}
$$

where we use $\nu_\infty T_\mu = (\hat{f}^\perp f)^{-1} \hat{f}^\perp$ for the last equality: This follows from our normalizations (5.35), which yield $\nu_\infty T_\mu \hat{f} \equiv 0$ and $\nu_\infty T_\mu f \equiv 1$. Note that the last expression in (5.37) is independent of choices of homogeneous coordinates.

We obtain the following "generalization[69]" of our theorem in §5.6.10:

---

[69] This "generalizes" the above theorem when considering the Christoffel transformation as a D_0-transformation. This viewpoint may be justified by sending $\lambda \to 0$ in §5.6.10 (remember that we can assume $T_0 = id$). However, the Riccati equation (5.17), or the corresponding linear system in §5.4.8, defining the Darboux transform of an isothermic surface, degenerates for $\lambda = 0$, so that any solution will be constant: As discussed in §3.3.8, the limiting surfaces have to be "blown up" to give a surface, as in §5.5.4, where the derivative with respect to the spectral parameter λ takes care of the blowing up.

5.6.15 Theorem. *Let f be an isothermic surface with D_μ-transform $\hat{f}$. Then their T_λ-transforms $T_\lambda f$ and $\hat{T}_\lambda \hat{f}$ can be positioned in the conformal sphere so that they form a $D_{\mu-\lambda}$-pair of isothermic surfaces; symbolically,*

$$T_\lambda D_\mu = D_{\mu-\lambda} T_\lambda.$$

$\hat{T}_\lambda = T_\lambda (1 - \frac{\lambda}{\mu} f (\hat{f}^\perp f)^{-1} \hat{f}^\perp)$ *is the[70] Calapso transformation of $\hat{f}$.*

5.6.16 From §5.6.10 we know that the Calapso transformation transforms Christoffel pairs into Darboux pairs, and vice versa. Next we want to address the question of whether it is possible to simultaneously stereographically project the Calapso transforms of the four surfaces satisfying the Christoffel-Darboux permutability theorem from §5.6.3 so that they satisfy this permutability scheme again, that is, whether the Calapso transformation "preserves" the Christoffel-Darboux permutability. This is indeed the case — for brevity, we state the corresponding permutability theorem[71] only graphically:

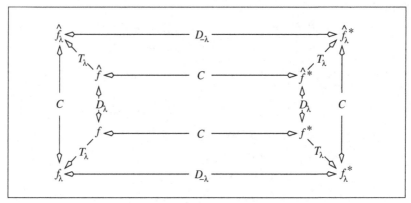

Fig. 5.9. A permutability theorem

5.6.17 *Proof.* Let f, $f^* = Cf$, $\hat{f} = D_\lambda f$, and $\hat{f}^* = CD_\lambda f$ denote four isothermic surfaces in Euclidean space satisfying the permutability scheme from §5.6.3. In particular, we then have

$$\lambda (\hat{f}^* - f^*)(\hat{f} - f) \equiv 1. \qquad (5.38)$$

As usual, we denote by $f = (v_0 + v_\infty f)\mathbb{H}$, $f^* = (v_0 + v_\infty f^*)\mathbb{H}$, and so on, the corresponding maps into the conformal sphere and let T_λ, T_λ^*, and so

---

[70] The "correct" one. Compare with T_λ^* as given in §5.5.22: Do we have $\hat{T}_\lambda \to T_\lambda^*$ as $\mu \to 0$?

[71] This permutability theorem provides an interpretation of the duality for horospherical surfaces (see [287]) in terms of the isothermic transformations (cf., [153]).

on, be the respective Calapso transformations. Further, let (v_∞, v_0) denote a basis of $I\!H^2$ with $T_\lambda \hat{f} \equiv v_\infty I\!H$, as in §5.6.11.

We already learned in §5.6.11 that a suitable choice of normalizations (5.35) yields

$$T_\lambda f(\nu_\infty T_\lambda f)^{-1} = v_0 + v_\infty \mathfrak{f}_\lambda \quad \text{and} \quad \hat{T}_\lambda \hat{f}(\nu_0 T_\lambda \hat{f})^{-1} = v_0 + v_\infty \hat{\mathfrak{f}}_\lambda$$

with $(\hat{\mathfrak{f}}_\lambda, \mathfrak{f}_\lambda)$ forming a Christoffel pair, where we chose[72] (cf., §5.6.12)

$$\hat{T}_\lambda = -\tfrac{1}{\lambda}(v_\infty \nu_\infty - \lambda(v_0 + v_\infty \hat{\mathfrak{f}}_\lambda)(\nu_0 - \mathfrak{f}_\lambda \nu_\infty))T_\lambda.$$

On the other hand, from §5.5.22, we know that T_λ^* can be adjusted so that

$$T_\lambda^* = T_\lambda(v_\infty \nu_\infty + \lambda(v_0 + v_\infty \mathfrak{f})(\nu_0 - \mathfrak{f}^* \nu_\infty)); \qquad (5.39)$$

this choice assures that $(f_\lambda^*, \mathfrak{f}_\lambda)$, where $f_\lambda^* = T_\lambda^* f^* = T_\lambda v_\infty I\!H$, form a Darboux pair.

Now, the key observation is the following: Using (5.38), we find that

$$T_\lambda^*(v_0 + v_\infty \hat{\mathfrak{f}}^*)(\hat{\mathfrak{f}} - \mathfrak{f}) = T_\lambda(v_0 + v_\infty \hat{\mathfrak{f}}) = v_\infty I\!H.$$

That is, $T_\lambda \hat{f} = T_\lambda^* \hat{f}^* \equiv const \in I\!H P^1$ define the same (constant) point in the conformal 4-sphere. Hence we can normalize f^* and $\hat{f}^*$ as in (5.35) to obtain

$$T_\lambda^* f^*(\nu_\infty T_\lambda f^*)^{-1} = v_0 + v_\infty \mathfrak{f}_\lambda^* \quad \text{and} \quad \hat{T}_\lambda^* \hat{f}^*(\nu_0 T_\lambda \hat{f}^*)^{-1} = v_0 + v_\infty \hat{\mathfrak{f}}_\lambda^*,$$

where $(\hat{\mathfrak{f}}_\lambda^*, \mathfrak{f}_\lambda^*)$ form a Christoffel pair and, as in (5.39),

$$\hat{T}_\lambda^* = -\tfrac{1}{\lambda}(v_\infty \nu_\infty - \lambda(v_0 + v_\infty \hat{\mathfrak{f}}_\lambda^*)(\nu_0 - \mathfrak{f}_\lambda^* \nu_\infty))T_\lambda^*.$$

Finally, from the permutability theorem for the Christoffel and Darboux transformations in §5.6.3, we know that $\hat{\mathfrak{f}}_\lambda = C\mathfrak{f}_\lambda$ and $\hat{\mathfrak{f}}_\lambda^* = C\mathfrak{f}_\lambda^*$ can be scaled and positioned to form a $D_{-\lambda}$-pair.   ◁

5.6.18 The author remembers that there was a very simple proof of this theorem, using the curved flat system from §5.5.20 to obtain the Calapso transforms. However, because the notes got lost, we must content ourselves with the above proof. It may be a challenge for the reader to "recover" that lost proof.

---

[72] Remember that the T-transformation for an isothermic surface is only determined up to postcomposition by a (constant) Möbius transformation.

5.6.19 Finally we want to establish a theorem that brings the Goursat transformation into the picture: In §5.6.10 we learned that proper stereographic projections of the T_λ-transforms of a D_λ-pair yields a Christoffel pair,

$$T_\lambda D_\lambda = CT_\lambda.$$

Thus, if $\hat{f}_1$ and $\hat{f}_2$ denote two (different) D_λ-transforms of an isothermic surface f, then their T_λ-transforms are the Christoffel transforms of $T_\lambda f$ after (generically different) stereographic projections. That is, $T_\lambda \hat{f}_1$ and $T_\lambda \hat{f}_2$ are, after stereographic projection, Goursat transforms of each other:

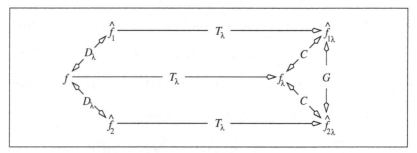

Fig. 5.10. Another permutability theorem

Reversing the argument we learn, with §5.6.10 again, that the T_λ-transforms of two isothermic surfaces that are Goursat transforms of each other are (different) $D_{-\lambda}$-transforms of one isothermic surface — namely, of the T_λ-transform of the Christoffel transform of either of the two original surfaces: Because the Calapso transformation is a transformation for Möbius equivalence classes of isothermic surfaces, it does not matter which of the two (Möbius equivalent; see §5.3.3) Christoffel transforms is used to determine the T_λ-transform.

In this sense the Goursat transformation "measures" the difference between two Darboux transformations of an isothermic surface. We conclude with the following:

5.6.20 Theorem. *The T_λ-transforms of two D_λ-transforms of an isothermic surface are, after suitable stereographic projection into Euclidean ambient space, Goursat transforms of each other.*

Conversely, the T_λ-transforms of two isothermic surfaces that are Goursat transforms of each other are $D_{-\lambda}$-transforms of one single isothermic surface.

5.6.21 Example. In this example we want to determine the Enneper cousins, that is, the Calapso transforms of the Enneper surface, illustrating various aspects of our theory. First reconsider $z \mapsto zj$ and $z \mapsto -jz$ as a

Christoffel pair of (totally umbilic) isothermic surfaces, as in §5.5.27, where we determined the Calapso transformations of

$$n := v_0 - v_\infty jz = \begin{pmatrix} -jz \\ 1 \end{pmatrix}, \quad v_0 = \begin{pmatrix} 0 \\ 1 \end{pmatrix}, \quad \text{and} \quad v_\infty = \begin{pmatrix} 1 \\ 0 \end{pmatrix},$$

to be given (with respect to (v_∞, v_0) as the basis of $I\!H^2$) by

$$T_\lambda(z) = \begin{pmatrix} -j & 0 \\ 0 & 1 \end{pmatrix} \begin{pmatrix} \cosh\sqrt{\lambda}z & \frac{1}{\sqrt{\lambda}}\sinh\sqrt{\lambda}z - z\cosh\sqrt{\lambda}z \\ \sqrt{\lambda}\sinh\sqrt{\lambda}z & \cosh\sqrt{\lambda}z - z\sqrt{\lambda}\sinh\sqrt{\lambda}z \end{pmatrix} \begin{pmatrix} j & 0 \\ 0 & 1 \end{pmatrix}.$$

Thus $n_\lambda(z) = (T_\lambda n)(z) \simeq v_0 - v_\infty j\frac{1}{\sqrt{\lambda}}\tanh(\sqrt{\lambda}z)$, as discussed in §5.5.27. Note that, by §5.5.20, the T_λ-transforms of $h(z) := v_0 + v_\infty zj$ are

$$h_\lambda(z) = T_\lambda(z)v_\infty \simeq v_0 - v_\infty j\frac{1}{\sqrt{\lambda}}\coth(\sqrt{\lambda}z)$$

since the Euclidean frame $F_0(z) = \begin{pmatrix} 1 & -jz \\ 0 & 1 \end{pmatrix}$ of n has v_∞ as an eigenvector. It is easily verified (for example, by using the Riccati equation (5.17)) that h_λ is a $D_{-\lambda}$-transform of n_λ, as predicted by our permutability theorem §5.6.10; moreover, h_λ and n_λ are Möbius equivalent — as one might expect, since they are Calapso transforms of Möbius equivalent maps, $h \simeq \begin{pmatrix} j & 0 \\ 0 & -j \end{pmatrix} n$.

Next we want to determine the master Christoffel transform of n: A fabulously complicated way[73] to do that seems to be by using the Sym formula that we derived in §5.5.4:

$$\mathfrak{F}^*(z) = (T_\lambda^{-1}\tfrac{\partial}{\partial\lambda}T_\lambda)_{\lambda=0}(z) = \begin{pmatrix} -j & 0 \\ 0 & 1 \end{pmatrix} \begin{pmatrix} \frac{1}{2}z^2 & -\frac{1}{3}z^3 \\ z & -\frac{1}{2}z^2 \end{pmatrix} \begin{pmatrix} j & 0 \\ 0 & 1 \end{pmatrix}.$$

Using v_∞ and $\nu_\infty = S^3(v_\infty, .) = (1, 0)$ to extract a Christoffel transform from $\mathfrak{F}^*$ (see §5.3.17), we recover $\nu_\infty\mathfrak{F}^*(z)v_\infty = zj$ as a Christoffel transform of $-jz$, as expected. However, using a different basis

$$\left.\begin{array}{rcl} \tilde{v}_\infty &=& v_\infty(-i) + v_0 \\ \tilde{v}_0 &=& v_\infty \end{array}\right\} \quad\leftrightarrow\quad \left\{\begin{array}{rcl} \tilde{\nu}_\infty &=& i\nu_\infty + \nu_0 \\ \tilde{\nu}_0 &=& \nu_\infty, \end{array}\right.$$

as in §5.3.21, we find the Enneper surface

$$\begin{array}{rcl} f(z) &=& \tilde{\nu}_\infty\mathfrak{F}^*(z)\tilde{v}_\infty \\ &=& -\tfrac{i}{2}(z^2 - jz^2 j) + \tfrac{1}{3}jz^3 - zj \\ &=& -i(x^2 - y^2) + j(\tfrac{1}{3}x^3 - (1 + y^2)x) + k(\tfrac{1}{3}y^3 - (1 + x^2)y) \end{array}$$

---

[73] The λ-derivative of T_λ is probably best computed in terms of the power series expansions of its components.

as a Goursat transform of zj (see §5.3.3). Writing n in terms of the new basis shows that $\mathfrak{f}$ is the Christoffel transform of (a scale translation of) the stereographic projection (5.12) of $-jz$ onto the 2-sphere:

$$n(z) = v_0 - v_\infty jz = \{\tilde{v}_0 + \tilde{v}_\infty(i - jz)^{-1}\}(i - jz) \simeq \tilde{v}_0 + \tilde{v}_\infty(i - jz)^{-1}.$$

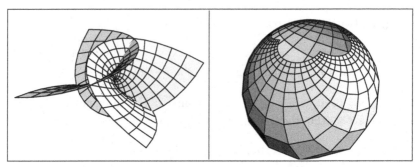

Fig. 5.11. The Enneper surface as a Goursat transform of $z \mapsto zj$

To determine the T_λ-transforms of Enneper's surface $f = \tilde{v}_0 + \tilde{v}_\infty \mathfrak{f}$, we compute

$$f_\lambda(z) = T_\lambda(z)\tilde{v}_\infty = \begin{pmatrix} -j & 0 \\ 0 & 1 \end{pmatrix} \begin{pmatrix} \frac{1}{\sqrt{\lambda}}\sinh(\sqrt{\lambda}z) - \cosh(\sqrt{\lambda}z)j(i - jz) \\ \cosh(\sqrt{\lambda}z) - \sqrt{\lambda}\sinh(\sqrt{\lambda}z)j(i - jz) \end{pmatrix}.$$

As an exercise, the reader may want to examine the effect of the above change of basis on the linear system that we have to solve to determine the Calapso transformations: It turns into a truly quaternionic system that we certainly do not want to solve.

However, from §5.5.20 we know that f_λ such obtained is the T_λ-transform of the Enneper surface f; on the other hand, from our permutability theorem in §5.6.10, we know that f_λ is another $D_{-\lambda}$-transform of n_λ, providing an explicit example for the above permutability theorem in §5.6.20. Moreover, from §5.6.13 we know that the f_λ's are horospherical surfaces in a hyperbolic space whose infinity boundary can be determined from the hyperbolic Gauss maps n_λ. Thus, stereographically projecting n_λ as before, $n_\lambda \simeq -j\frac{1}{\sqrt{\lambda}}\tanh(\sqrt{\lambda}z)$, the surfaces f_λ are horospherical in (one of) the hyperbolic half-spaces $\mathrm{Im}\,\mathbb{H} \setminus \mathbb{C}j$ (cf., §P.1.3). In Figure 5.12, the conformal deformation of the Enneper surface into its cousins is shown[74]; their hyperbolic Gauss maps are included in the bottom three pictures. Similar pictures can be obtained with Bryant's original Weierstrass-type representation from [42] (see [9]), providing evidence for our assertions from §5.5.29 and §5.6.13.

---

[74] We only show transforms for positive values of λ. By the symmetry of the Enneper surface and the fact that a negative sign of λ is compensated by interchanging the roles of the curvature directions, negative values of λ yield congruent surfaces.

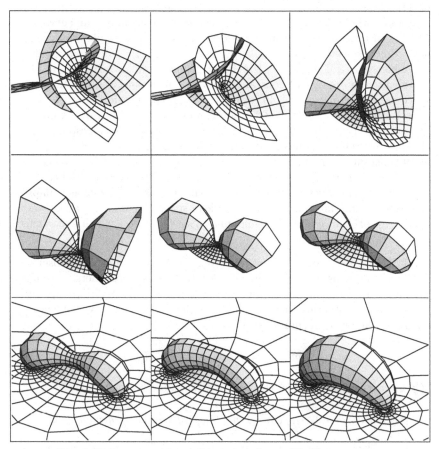

Fig. 5.12. Enneper cousins as Calapso transforms of the Enneper surface

5.7 Discrete isothermic nets and their transformations

In this final section we will discuss (one possible version of) discrete isothermic nets as discrete analogs of smooth isothermic surfaces. It turns out that, with the definition that we are going to use, the transformation theory of smooth isothermic surfaces completely carries over to the discrete setup (cf., [150] or [250]). Here, we will content ourselves with the study of the transformations and omit any discussions of whether or not the existence of these transformations can be used to characterize discrete isothermic nets, as they can in the smooth case. For partial answers to such problems the interested reader is referred to [150].

This transformation theory can then be applied to define discrete analogs of certain constant mean curvature surfaces in spaces of constant curvature — in cases where we have a characterization of the constant mean curvature surfaces in terms of isothermic transformations — as particular isothermic nets. Here we will, as an example, discuss "horospherical nets" as an analog of horospherical surfaces in hyperbolic geometry, following [150]; besides those, discrete analogs of constant mean curvature surfaces in Euclidean space have been discussed [149], using the Darboux transformation for discrete isothermic nets — this definition can clearly be carried over to other space forms using the Calapso transformation (Lawson's correspondence; cf., §5.5.29). Thus this ansatz does provide definitions for discrete analogs of all constant mean curvature surfaces in space forms that have constant mean curvature T-transforms in Euclidean space. However, for the case of horospherical nets, we have *two different* characterizations, one as Calapso transforms of minimal surfaces in Euclidean space and another as Darboux transforms of their hyperbolic Gauss maps (see §5.6.13) — which makes this case particularly interesting.

5.7.1 There are various different ways to motivate our definition of discrete isothermic nets, as a special case of discrete curvature line nets.[75] Thus a minimal requirement for a discrete net $f : \mathbb{Z}^2 \supset U \to \mathbb{HP}^1$ to qualify as a discrete isothermic net will be that the vertices of any elementary quadrilateral, or "elementary 2-cell,"

$$(f(m,n); f(m+1,n); f(m+1,n+1); f(m,n+1))$$

of f are concircular, that is, their cross-ratio has to be real:

$$[f(m,n); f(m+1,n); f(m+1,n+1); f(m,n+1)] =: q(m,n) \in \mathbb{R}.$$

In their original paper [30], A. Bobenko and U. Pinkall introduced discrete isothermic nets as the discrete analog of a conformal curvature line parametrization: Using Taylor expansion, they argue that the discrete analog of a conformal curvature line parametrization should satisfy $q \equiv -1$, that is, elementary quadrilaterals form conformal squares (see §4.9.9). Another way to motivate this definition is that it is the direct discrete (and Möbius geometric) analog of the classical definition of isothermic surfaces, as the "surfaces capable of division into *infinitesimal* squares by their curves of curvature" (cf., e.g., [64]): Using nonstandard analysis, the term "infinitesimal" can be given a rigorous meaning, and elementary considerations then show that this classical definition indeed characterizes isothermic surfaces in our sense (see [152]).

---

[75] Discrete curvature line nets will be discussed in greater detail in Chapter 8.

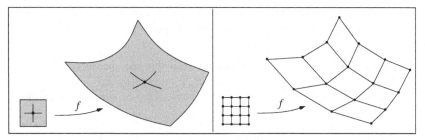

Fig. 5.13. Smooth and discrete nets

In [32], Bobenko and Pinkall gave a weakened version of the definition of a discrete isothermic net, as a discrete version of an arbitrary curvature line net on an isothermic surface, that only requires that the cross-ratios of the net's elementary quadrilaterals decompose into two single-variable real functions, $q(m, n) = \frac{a(m)}{b(n)}$. This definition can be motivated by observing the behavior of a single point when applying Bianchi's permutability theorem in §5.6.6 for the Darboux transformation repeatedly (cf., [35]). A behavior of the Calapso transformation in the discrete case that is similar to its effect on the periods of a periodic isothermic surface (cf., §5.5.28) indicates that this second definition is the "correct" definition for discrete isothermic nets:

5.7.2 Definition. *A map* $f : \mathbb{Z}^2 \supset U \to \mathbb{HP}^1$ *will be called a discrete net if the four vertices* $f(m, n)$, $f(m + 1, n)$, $f(m + 1, n + 1)$, *and* $f(m, n + 1)$ *of any elementary quadrilateral are different, that is,*

$$[f(m, n); f(m + 1, n); f(m + 1, n + 1); f(m, n + 1)] \neq 0, 1, \infty.$$

A discrete net is called a principal net, or a discrete curvature line net, if every elementary quadrilateral is a conformal rectangle, that is,

$$[f(m, n); f(m + 1, n); f(m + 1, n + 1); f(m, n + 1)] \in \mathbb{R},$$

and it will be called a (discrete) isothermic net if there are two real valued-functions $a = a(m)$ *and* $b = b(n)$ *of one (integer) variable so that the cross-ratios of its elementary quadrilaterals satisfy*

$$[f(m, n); f(m + 1, n); f(m + 1, n + 1); f(m, n + 1)] = \frac{a(m)}{b(n)}.$$

5.7.3 Note that this definition does not depend at all on $\mathbb{HP}^1$ being the target space of our net: Because the definition only refers to the cross-ratio of the (four) vertices of each elementary quadrilateral of the net, we may use this definition in any codimension, that is, with the conformal n-sphere

S^n as the ambient space of our net. Indeed, the above notion of a principal net will reappear in the context of (Ribaucour pairs of) orthogonal nets in Chapter 8.

5.7.4 Notation.   Since we will discuss stereographic projections $\mathfrak{f}$ of a discrete (isothermic) net $f = (v_0 + v_\infty \mathfrak{f})\mathbb{H}$, $\mathfrak{f} = (v_0 f)(v_\infty f)^{-1}$ as in §4.9.5, we will need to consider the "partial derivatives" of $\mathfrak{f}$ — to downsize formulas we will write, for abbreviation,

$$
\begin{aligned}
(\partial_m \mathfrak{f})(m, n) &:= \mathfrak{f}(m + 1, n) - \mathfrak{f}(m, n) \\
&= (f^\perp(m, n) v_\infty)^{-1}(f^\perp(m, n) f(m + 1, n))(v_\infty f(m + 1, n))^{-1}, \\
(\partial_n \mathfrak{f})(m, n) &:= \mathfrak{f}(m, n + 1) - \mathfrak{f}(m, n) \\
&= (f^\perp(m, n) v_\infty)^{-1}(f^\perp(m, n) f(m, n + 1))(v_\infty f(m, n + 1))^{-1},
\end{aligned}
$$

where $f^\perp = \mathbb{H}(v_0 - \mathfrak{f}v_\infty)$ denotes the "dual net" of f, with values in $(\mathbb{H}P^1)^\perp$ (see §4.9.2 and (4.26)). Note that our regularity assumption on discrete nets implies that these "partial derivatives" of $\mathfrak{f}$ do not vanish and are not equal,

$$
\partial_m \mathfrak{f}, \partial_n \mathfrak{f} \neq 0 \quad \text{and} \quad \partial_m \mathfrak{f} \neq \partial_n \mathfrak{f}.
$$

Similarly, we will use the notion of this "discrete partial derivative" for any vector space—valued maps as, for example, in the following definition. Also, to make the discussions simpler,[76] we will always assume our discrete nets and any derived objects to be defined on all of $\mathbb{Z}^2$.

As in the transformation theory of smooth isothermic surfaces, a key notion will be that of a (closed) retraction form (cf., §5.3.19):

5.7.5 Definition and Lemma.   *Let $f : \mathbb{Z}^2 \to \mathbb{H}P^1$ denote a discrete net. Then a pair $\tau = (\tau_m, \tau_n)$ of maps $\tau_m, \tau_n : \mathbb{Z}^2 \to \mathfrak{gl}(2, \mathbb{H})$ will be called a retraction form of f if*

$$
\begin{aligned}
\operatorname{im} \tau_m(m, n) &= f(m, n) \quad \text{and} \quad \ker \tau_m(m, n) = f(m + 1, n), \\
\operatorname{im} \tau_n(m, n) &= f(m, n) \quad \text{and} \quad \ker \tau_n(m, n) = f(m, n + 1).
\end{aligned}
$$

Any isothermic net $f : \mathbb{Z}^2 \to \mathbb{H}P^1$ has an integrable retraction form τ, that is, there is a map $\mathfrak{F}^* : \mathbb{Z}^2 \to \mathfrak{gl}(2, \mathbb{H})$ so that $\tau = (\partial_m \mathfrak{F}^*, \partial_n \mathfrak{F}^*)$.

5.7.6 *Proof.* Remember that, in the smooth case, the existence of a closed retraction form was ensured by the existence of a Christoffel transform $\mathfrak{f}^*$ for an arbitrary stereographic projection of f. Here we intend to proceed similarly: Define the retraction form in terms of affine coordinates for f, as

---

[76] We do not want to be obliged to discuss the "boundary points" of the domain of a discrete net, where these partial derivatives and other objects that are naturally defined on the edges (or faces) of the domain lattice may not exist.

in (5.14), and then study the integrability of the $\mathfrak{gl}(2, I\!\!H)$-valued "1-form" τ in terms of the occurring $I\!\!H$-valued discrete 1-form.

Thus assume that $\left(\begin{smallmatrix}1\\0\end{smallmatrix}\right) I\!\!H \not\subseteq f(\mathbb{Z}^2)$ and write $f = \left(\begin{smallmatrix}\mathfrak{f}\\1\end{smallmatrix}\right)$ and $f^\perp = (1, -\mathfrak{f})$. Then any retraction form τ of f must be of the form

$$\tau(m, n) = (f(m, n)\, \omega_m(m, n)\, f^\perp(m+1, n),\, f(m, n)\, \omega_n(m, n)\, f^\perp(m, n+1))$$

with some quaternionic "discrete 1-form" (ω_m, ω_n). The integrability of τ is equivalent to

$$d\tau(m, n) := \tau_n(m+1, n) + \tau_m(m, n) - \tau_m(m, n+1) - \tau_n(m, n) = 0;$$

evaluating $d\tau(m, n)$ on $\left(\begin{smallmatrix}1\\0\end{smallmatrix}\right)$ and $\left(\begin{smallmatrix}\mathfrak{f}(m+1, n+1)\\1\end{smallmatrix}\right)$, as a basis of $I\!\!H^2$, and writing the results in terms of $\left(\begin{smallmatrix}\mathfrak{f}(m, n)\\1\end{smallmatrix}\right)$ and $\left(\begin{smallmatrix}1\\0\end{smallmatrix}\right)$, the integrability of τ reduces to three (the fourth equation is trivially satisfied) quaternionic equations:

$$\begin{aligned}
0 &= \omega_n(m+1, n) + \omega_m(m, n) - \omega_m(m, n+1) - \omega_n(m, n),\\
0 &= \partial_m\mathfrak{f}(m, n) \cdot \omega_n(m+1, n) - \partial_n\mathfrak{f}(m, n) \cdot \omega_m(m, n+1),\\
0 &= \omega_m(m, n) \cdot \partial_n\mathfrak{f}(m+1, n) - \omega_n(m, n) \cdot \partial_m\mathfrak{f}(m, n+1).
\end{aligned}$$

Thus τ is integrable if and only if there is a map $\mathfrak{f}^*$ with $\omega = (\partial_m\mathfrak{f}^*, \partial_n\mathfrak{f}^*)$ that satisfies

$$\begin{aligned}
\partial_m\mathfrak{f}(m, n) \cdot \partial_n\mathfrak{f}^*(m+1, n) &= \partial_n\mathfrak{f}(m, n) \cdot \partial_m\mathfrak{f}^*(m, n+1),\\
\partial_m\mathfrak{f}^*(m, n) \cdot \partial_n\mathfrak{f}(m+1, n) &= \partial_n\mathfrak{f}^*(m, n) \cdot \partial_m\mathfrak{f}(m, n+1).
\end{aligned} \tag{5.40}$$

Taking $\mathfrak{f}^*$ to be a Christoffel transform[77] of $\mathfrak{f}$, as given in the following lemma, the two equations (5.40) are a consequence[78] of the affine versions (see (4.25)) of the cross-ratio condition,

$$\begin{aligned}
\frac{a(m)}{b(n)} &= [f(m, n); f(m+1, n); f(m+1, n+1); f(m, n+1)]\\
&= [f(m+1, n); f(m, n); f(m, n+1); f(m+1, n+1)].
\end{aligned} \tag{5.41}$$

Here the second equation follows from the cross-ratio identities in Figure 4.3.

Thus we have proved the existence of an integrable retraction form τ,

$$\begin{aligned}
\tau_m(m, n) &= (v_0 + v_\infty\mathfrak{f}(m, n))\, \partial_m\mathfrak{f}^*(m, n)(v_0 - \mathfrak{f}(m+1, n)v_\infty),\\
\tau_n(m, n) &= (v_0 + v_\infty\mathfrak{f}(m, n))\, \partial_n\mathfrak{f}^*(m, n)(v_0 - \mathfrak{f}(m+1, n)v_\infty),
\end{aligned} \tag{5.42}$$

---

[77] Note that the two equations (5.40) can be viewed as a discrete version of (5.4).

[78] Note that, for this deduction, it is crucial that the functions a and b are real-valued (so that they commute with any quaternion) and that they only depend on one variable.

for any isothermic surface as soon as we have proved the existence of a Christoffel transform, in the following lemma. ◁

5.7.7 Definition and Lemma. *Let* $\mathfrak{f} : \mathbb{Z}^2 \to \mathbb{H}$ *denote (a stereographic projection of) an isothermic net, and let* $q(m,n) = \frac{a(m)}{b(n)}$ *be a factorization of the cross-ratios of its elementary quadrilaterals with real-valued a and b. Then*

$$\partial_m \mathfrak{f}^* = a \, (\partial_m \mathfrak{f})^{-1} \quad and \quad \partial_n \mathfrak{f}^* = b \, (\partial_n \mathfrak{f})^{-1} \tag{5.43}$$

defines[79] *a second isothermic net* $\mathfrak{f}^*$ *with* $q^* = q$.

This net $\mathfrak{f}^*$ *is called a Christoffel transform of* $\mathfrak{f}$.

5.7.8 *Proof.* To check integrability, first note that the cross-ratio, given in affine coordinates as in (4.25), of an elementary quadrilateral of $\mathfrak{f}$ can be written as

$$\begin{aligned}
q(m,n) &= (\partial_m \mathfrak{f})(m,n)(\partial_n \mathfrak{f})^{-1}(m+1,n)(\partial_m \mathfrak{f})(m,n+1)(\partial_n \mathfrak{f})^{-1}(m,n), \\
&= (\partial_m \mathfrak{f})(m,n)(\partial_n \mathfrak{f})^{-1}(m,n)(\partial_m \mathfrak{f})(m,n+1)(\partial_n \mathfrak{f})^{-1}(m+1,n),
\end{aligned} \tag{5.44}$$

as in (5.41). Now the integrability is easily checked:

$$\begin{aligned}
&a(m)\{(\partial_m \mathfrak{f})^{-1}(m,n) - (\partial_m \mathfrak{f})^{-1}(m,n+1)\} \\
=\; &b(n)\{(\partial_n \mathfrak{f})^{-1}(m,n) - (\partial_n \mathfrak{f})^{-1}(m+1,n)\} \\
\Leftrightarrow\; &a(m)\{(\partial_m \mathfrak{f})(m,n+1) - (\partial_m \mathfrak{f})(m,n)\} \\
=\; &b(n)(\partial_m \mathfrak{f})(m,n)\{(\partial_n \mathfrak{f})^{-1}(m,n) - (\partial_n \mathfrak{f})^{-1}(m+1,n)\}(\partial_m \mathfrak{f})(m,n+1) \\
=\; &a(m)\{(\partial_n \mathfrak{f})(m+1,n) - (\partial_n \mathfrak{f})(m,n)\},
\end{aligned}$$

where the last equality follows from (5.44), so that integrability of $\mathfrak{f}^*$ is equivalent to the integrability of $\mathfrak{f}$.

To check that $\mathfrak{f}^*$ is an isothermic net with $q^*(m,n) = q(m,n)$, we compute:

$$\begin{aligned}
&q^*(m,n) \\
=\; &(\partial_m \mathfrak{f}^*)(m,n)(\partial_n \mathfrak{f}^*)^{-1}(m+1,n)(\partial_m \mathfrak{f}^*)(m,n+1)(\partial_n \mathfrak{f}^*)^{-1}(m,n) \\
=\; &\tfrac{a(m)}{b(n)}(\partial_m \mathfrak{f})^{-1}(m,n)\tfrac{a(m)}{b(n)}(\partial_n \mathfrak{f})(m+1,n)(\partial_m \mathfrak{f})^{-1}(m,n+1)(\partial_n \mathfrak{f})(m,n) \\
=\; &\tfrac{a(m)}{b(n)},
\end{aligned}$$

where the last equality is again obtained from (5.44). ◁

5.7.9 The above construction (5.42) of an integrable retraction form for an isothermic surface $f : \mathbb{Z}^2 \to \mathbb{HP}^1$ in terms of the Christoffel transform $\mathfrak{f}^*$

---

[79] Note that a real factorization of the cross-ratio function q is unique up to simultaneous multiplication of a and b by some real factor $\lambda \neq 0$: We recover the scaling ambiguity of the Christoffel transformation in the smooth case.

of a stereographic projection $\mathfrak{f} : \mathbb{Z}^2 \to \mathbb{H}$ of f yields, after having fixed a factorization $q = \frac{a}{b}$, a unique object that is independent of the stereographic projection.

Namely, choosing a different strerographic projection $\tilde{\mathfrak{f}} = (\tilde{\nu}_0 f)(\tilde{\nu}_\infty f)^{-1}$ for $f = v_0 + v_\infty \mathfrak{f}$ and using (4.26) yields

$$
\begin{aligned}
-\partial_m \tilde{\mathfrak{f}}(m,n) &= (f^\perp \tilde{v}_\infty)^{-1}(m+1,n)(-\partial_m \mathfrak{f})(m,n)(\tilde{\nu}_\infty f)^{-1}(m,n) \\
\Rightarrow \quad \partial_m \tilde{\mathfrak{f}}^*(m,n) &= (\tilde{\nu}_\infty f)(m,n)\,\partial_m \mathfrak{f}^*(m,n)(f^\perp \tilde{v}_\infty)(m+1,n), \\
-\partial_n \tilde{\mathfrak{f}}(m,n) &= (f^\perp \tilde{v}_\infty)^{-1}(m,n+1)(-\partial_n \mathfrak{f})(m,n)(\tilde{\nu}_\infty f)^{-1}(m,n) \\
\Rightarrow \quad \partial_n \tilde{\mathfrak{f}}^*(m,n) &= (\tilde{\nu}_\infty f)(m,n)\,\partial_n \mathfrak{f}^*(m,n)(f^\perp \tilde{v}_\infty)(m,n+1),
\end{aligned}
\tag{5.45}
$$

similar to §5.3.10. Comparison with (5.42) shows that $\tilde{\tau} = \tau$ since

$$
\tilde{v}_0 + \tilde{v}_\infty \tilde{\mathfrak{f}} = f\,(\tilde{\nu}_\infty f)^{-1} = (v_0 + v_\infty \mathfrak{f})(\tilde{\nu}_\infty f)^{-1},
$$

and similarly for $f^\perp = \nu_0 - \mathfrak{f} \nu_\infty$. Thus τ and $\mathfrak{F}^*$ are invariantly defined objects, and $\mathfrak{F}^*$ provides, as in §5.3.17, the "master Christoffel transform" for an isothermic net $f : \mathbb{Z}^2 \to \mathbb{HP}^1$: The Christoffel transform $\mathfrak{f}^*$ of any stereographic projection $\mathfrak{f} = (\nu_0 f)(\nu_\infty f)^{-1}$ of f is given by $\mathfrak{f}^* = \nu_\infty \mathfrak{F}^* v_\infty$. And, any two Christoffel transforms of (different) stereographic projections of an isothermic net f are Goursat transforms of each other — if we define, as in §5.3.3:

5.7.10 Definition. *Let* $\mathfrak{f} : \mathbb{Z}^2 \to \mathbb{H}$ *be an isothermic net and* $\mu \in M\ddot{o}b(4)$ *a Möbius transformation. Then the isothermic net* $\tilde{\mathfrak{f}} = (\mu \circ \mathfrak{f}^*)^*$ *is called a Goursat transform of* $\mathfrak{f}$.

5.7.11 For the Darboux and Calapso transformations we wish to introduce a (spectral) parameter λ: As we discussed in §5.7.7, a factorization $q = \frac{a}{b}$ of the cross-ratio function of an isothermic net is unique up to multiplication of a and b by some $\lambda \neq 0$. Since a and b enter linearly into the retraction form τ as given in (5.42), a different choice of factorization causes a rescaling of τ — just as in the smooth case (cf., §5.4.7). Thus, from now on:

We will always assume that an isothermic net $f : \mathbb{Z}^2 \to \mathbb{HP}^1$ *comes with a factorization* $q = \frac{a}{b}$ *of its cross-ratio function.*

Here we will introduce the Darboux transformation[80] a little differently from [149] and [150] — we postpone a discussion of the equivalence with the definition given in [150] as well as of the relation with our integrable retraction form τ until we have also introduced the Calapso transformation. The relevance of the factorization $q = \frac{a}{b}$ becomes obvious from the definition anyway:

---

[80] Compare [75]; see also the Darboux transformation for planar curves in [193].

5.7.12 Definition and Lemma. *Let* $f : \mathbb{Z}^2 \to \mathbb{HP}^1$, $q = \frac{a}{b}$, *be an isothermic net. Then the system*

$$
\begin{aligned}
\lambda a(m) &= [f(m,n); f(m+1,n); \hat{f}(m+1,n); \hat{f}(m,n)] \\
\lambda b(n) &= [f(m,n); f(m,n+1); \hat{f}(m,n+1); \hat{f}(m,n)]
\end{aligned}
\tag{5.46}
$$

can be solved[81] *to provide an isothermic net* $\hat{f} : \mathbb{Z}^2 \to \mathbb{HP}^1$ *with* $\hat{q} = q$.
This net $\hat{f}$ *is called a Darboux transform (or D_λ-transform) of f.*

5.7.13 *Proof.* This lemma is a direct consequence of our hexahedron lemma in §4.9.13: The claim is best verified by contemplating Figure 4.4 while taking into account the cross-ratio identities from Figure 4.3.   ◁

5.7.14 Clearly, the Darboux transformation is a Möbius invariant notion, and, by the symmetry of the defining formula (see Figure 4.3), it is involutive in the sense that, if $\hat{f}$ is a D_λ-transform of f, then f is a D_λ-transform of $\hat{f}$.

As in the smooth case, we have a $(1 + 4)$-parameter family of Darboux transforms for a given isothermic net.

Note that the hexahedron lemma also provides the existence of a "discrete sphere congruence" that is enveloped by f and $\hat{f}$. The fact of discrete isothermic nets being particular principal nets suggests that we should consider this discrete sphere congruence as a discrete Ribaucour congruence: We will come back to this viewpoint in Chapter 8.

5.7.15 In a Euclidean setup, the system (5.46) is equivalent to the Riccati-type system of difference equations

$$
\begin{aligned}
\partial_m \hat{f}(m,n) &= \lambda(\hat{f} - f)(m+1,n) \, \partial_m f^*(m,n)(\hat{f} - f)(m,n) \\
&= \lambda(\hat{f} - f)(m,n) \, \partial_m f^*(m,n)(\hat{f} - f)(m+1,n), \\
\partial_n \hat{f}(m,n) &= \lambda(\hat{f} - f)(m,n+1) \, \partial_n f^*(m,n)(\hat{f} - f)(m,n) \\
&= \lambda(\hat{f} - f)(m,n) \, \partial_n f^*(m,n)(\hat{f} - f)(m,n+1),
\end{aligned}
\tag{5.47}
$$

where the respective second equations follow from the corresponding symmetry of the cross-ratio (see Figure 4.3). This set of equations was, as a discrete version of the Riccati-type partial differential equation (5.17) in the smooth case, the starting point in [149] to introduce the Darboux transformation for discrete isothermic nets.

We now turn to the definition of the Calapso transformation for discrete isothermic nets:

---

[81] For almost all data $(\lambda, \hat{f}(m_0, n_0))$ this system yields such an isothermic net; the only problem might be the regularity of the net $\hat{f}$. However, choosing $\lambda \neq \frac{1}{a(m)}, \frac{1}{b(n)}$ for all $m, n \in \mathbb{Z}$ and the initial point $\hat{f}(m_0, n_0)$ so that none of the adjacent quadrilaterals degenerates ensures that all occurring cross-ratios are $\neq 0, 1, \infty$ — hence we obtain a *net* in our sense.

5.7.16 Definition and Lemma. *Let* $f : \mathbb{Z}^2 \to \mathbb{HP}^1$, $q = \frac{a}{b}$, *denote an isothermic net with retraction form* $\tau = (\tau_m, \tau_n)$, *and let* $\lambda \neq \frac{1}{a(m)}, \frac{1}{b(n)}$ *for all* $m, n \in \mathbb{Z}$. *Then there is a map* $T_\lambda : \mathbb{Z}^2 \to Gl(2, \mathbb{H})$ *with*

$$
\begin{aligned}
T_\lambda(m+1, n) &= T_\lambda(m, n)(1 + \lambda \tau_m(m, n)), \\
T_\lambda(m, n+1) &= T_\lambda(m, n)(1 + \lambda \tau_n(m, n)),
\end{aligned}
\tag{5.48}
$$

called a Calapso transformation (or T_λ*-transformation) for* f.

The Calapso transform $f_\lambda := T_\lambda f$ *is then isothermic with* $q_\lambda = \frac{1-\lambda b}{1-\lambda a} q$.

5.7.17 *Proof.* First note that the system (5.48) is indeed integrable since τ is an integrable retraction form for f, so that $d\tau = 0$ and $\tau \circ \tau = 0$:

$$
\begin{aligned}
&(1 + \lambda \tau_m(m, n))(1 + \lambda \tau_n(m+1, n)) - (1 + \lambda \tau_n(m, n))(1 + \lambda \tau_m(m, n+1)) \\
&= \lambda(\tau_m(m, n) + \tau_n(m+1, n) - \tau_n(m, n) - \tau_m(m, n+1)) \\
&\quad + \lambda^2(\tau_m(m, n) \circ \tau_n(m+1, n) - \tau_n(m, n) \circ \tau_m(m, n+1)) = 0.
\end{aligned}
$$

To check that T_λ can be chosen to take values in $Gl(2, \mathbb{H})$, we investigate how regularity is "transported" when integrating (5.48): With $f = v_0 + v_\infty f$ and using (5.42), we find

$$
\begin{aligned}
(1 + \lambda \tau_m(m, n)) \cdot f(m, n) &= f(m, n) \cdot (1 - \lambda a(m)) \quad \text{and} \\
(1 + \lambda \tau_m(m, n)) \cdot f(m+1, n) &= f(m+1, n), \\
(1 + \lambda \tau_n(m, n)) \cdot f(m, n) &= f(m, n) \cdot (1 - \lambda b(n)) \quad \text{and} \\
(1 + \lambda \tau_n(m, n)) \cdot f(m, n+1) &= f(m, n+1),
\end{aligned}
\tag{5.49}
$$

that is, $1 + \lambda \tau_m(m, n)$ maps the basis $(f(m, n), f(m+1, n))$ — here we need the regularity of our isothermic net f — to a basis (as long as $\lambda a \neq 1$) so that $T_\lambda(m+1, n)$ is regular as soon as $T_\lambda(m, n)$ is, and similarly for $T_\lambda(m, n+1)$.

Finally, we determine the cross-ratio of an elementary quadrilateral of the Calapso transform f_λ of f using the formula given in §4.9.8: Since (from §4.9.2) we know that $f_\lambda^\perp = (T_\lambda f)^\perp = f^\perp \circ T_\lambda^{-1}$, the relations that we derived in (5.49) give us

$$
\begin{aligned}
f_\lambda^\perp(m, n) f_\lambda(m+1, n) &= f^\perp(m, n) f(m+1, n), \\
f_\lambda^\perp(m+1, n+1) f_\lambda(m+1, n) &= f^\perp(m+1, n+1) f(m+1, n) \tfrac{1}{1-\lambda b(n)}, \\
f_\lambda^\perp(m+1, n+1) f_\lambda(m, n+1) &= f^\perp(m+1, n+1) f(m, n+1) \tfrac{1}{1-\lambda a(m)}, \\
f_\lambda^\perp(m, n) f_\lambda(m, n+1) &= f^\perp(m, n) f(m, n+1).
\end{aligned}
$$

Taking products yields the claim, $q_\lambda(m, n) = q(m, n) \frac{1-\lambda b(n)}{1-\lambda a(m)}$, and, as a consequence, f_λ is an isothermic net. ◁

5.7.18 Note that the Calapso transformation is a well-defined transformation for Möbius equivalence classes of isothermic nets, as in the smooth case:

The argument here is exactly the same as the one that provided us with the lemma in §5.5.6.

We can also prove a discrete analog of the fact that the Calapso transformation provides a second-order deformation for isothermic surfaces in the smooth case (see §5.5.11): This is another consequence of the equations (5.49). Namely, for any Calapso transform $f_\lambda = T_\lambda f$ we obtain:

$$f_\lambda(m, n+1) \parallel T_\lambda(m, n)f(m, n+1) \qquad f_\lambda(m+1, n) \parallel T_\lambda(m, n)f(m+1, n)$$

$$\nwarrow \qquad \nearrow$$

$$f_\lambda(m, n) \parallel T_\lambda(m, n)f(m, n)$$

$$\swarrow \qquad \searrow$$

$$f_\lambda(m-1, n) \parallel T_\lambda(m, n)f(m-1, n) \qquad f_\lambda(m, n-1) \parallel T_\lambda(m, n)f(m, n-1).$$

That is, any "vertex star" of f_λ consisting of the center vertex $f_\lambda(m, n)$ and its four neighbors is Möbius equivalent (via $T_\lambda(m, n)$) to the corresponding vertex star of the original isothermic net.

5.7.19 We are now prepared to study the relation between the Darboux and Calapso transformations, to obtain the discrete version of the relation discussed in §5.5.3: In this way we will prove the equivalence of our definition in §5.7.12 and that given in [150]. Thus we want to study the implications of requiring a discrete net $\hat{f} : \mathbb{Z}^2 \to \mathbb{HP}^1$ to satisfy

$$T_\lambda \hat{f} \equiv const \in \mathbb{HP}^1,$$

where T_λ is a Calapso transformation for a discrete isothermic net f. This is equivalent to

$$(T_\lambda \hat{f})(m+1, n) = T_\lambda(m, n)(1 + \lambda \tau_m(m, n))\hat{f}(m+1, n) \parallel T_\lambda(m, n)\hat{f}(m, n)$$
$$\Leftrightarrow \quad \hat{f}^\perp(m, n)(1 + \lambda \tau_m(m, n))\hat{f}(m+1, n) = 0, \quad \text{and}$$
$$(T_\lambda \hat{f})(m+1, n) = T_\lambda(m, n)(1 + \lambda \tau_n(m, n))\hat{f}(m, n+1) \parallel T_\lambda(m, n)\hat{f}(m, n)$$
$$\Leftrightarrow \quad \hat{f}^\perp(m, n)(1 + \lambda \tau_n(m, n))\hat{f}(m, n+1) = 0.$$

Writing $\hat{f} = v_0 + v_\infty \hat{\mathfrak{f}}$ and $\hat{f}^\perp = v_0 - \hat{\mathfrak{f}} v_\infty$ in terms of affine coordinates and using the form (5.42) for our retraction form τ, these equations read

$$0 = \partial_m \hat{\mathfrak{f}}(m, n) + \lambda(\mathfrak{f} - \hat{\mathfrak{f}})(m, n) \, \partial_m \mathfrak{f}^*(m, n)(\hat{\mathfrak{f}} - \mathfrak{f})(m+1, n),$$
$$0 = \partial_n \hat{\mathfrak{f}}(m, n) + \lambda(\mathfrak{f} - \hat{\mathfrak{f}})(m, n) \, \partial_n \mathfrak{f}^*(m, n)(\hat{\mathfrak{f}} - \mathfrak{f})(m, n+1),$$

that is, $T_\lambda \hat{f} \equiv const$ if and only if a stereographic projection $\hat{\mathfrak{f}}$ of $\hat{f}$ satisfies the Riccati-type system (5.47) of difference equations. Thus we have proved the following:

5.7.20 Lemma. *A discrete net* $\hat{f} : \mathbb{Z}^2 \to \mathbb{HP}^1$ *is a D_λ-transform of an isothermic net f if and only if $T_\lambda \hat{f} \equiv const$, where T_λ denotes the T_λ-transformation of f.*

5.7.21 Next we want to see that a Calapso transformation T_λ of a codimension 1 isothermic net $f : \mathbb{Z}^2 \to S^3$ can be chosen to fix S^3: This will become crucial when discussing horospherical nets as a discrete analog of horospherical surfaces. Thus we want to show that $T_\lambda S^3 \parallel S^3$, where we identify the 3-sphere with some choice of homogeneous coordinates, $S^3 \in \mathfrak{H}(\mathbb{H}^2) \cong \mathbb{R}_1^6$, for the (real) line of quaternionic Hermitian forms that defines the 3-sphere (cf., §4.3.12). To prove this fact we proceed as in the proof of §5.7.16: We show that preserving the 3-sphere is "transported" when integrating (5.48).

Thus we assume that $T_\lambda(m,n)S^3 \parallel S^3$ and consider

$$T_\lambda(m+1,n)S^3 = T_\lambda(m,n)(1+\lambda\tau_m(m,n))S^3.$$

Evaluating the quaternionic Hermitian form $(1+\lambda\tau_m(m,n))S^3$ on $f(m,n)$ and $f(m+1,n)$ as a basis of $\mathbb{H}^2$, the equations (5.49) yield

$$(1+\lambda\tau_m(m,n))S^3(f(m,n),f(m,n))$$
$$= \tfrac{1}{(1-\lambda a(m))^2}S^3(f(m,n),f(m,n)) = 0,$$

$$(1+\lambda\tau_m(m,n))S^3(f(m+1,n),f(m,n))$$
$$= \tfrac{1}{1-\lambda a(m)}S^3(f(m+1,n),f(m,n)), \quad \text{and}$$

$$(1+\lambda\tau_m(m,n))S^3(f(m+1,n),f(m+1,n))$$
$$= S^3(f(m+1,n),f(m+1,n)) = 0,$$

since $S^3(f,f) \equiv 0$ when $f : \mathbb{Z}^2 \to S^3$ takes values in S^3. Hence $T_\lambda(m+1,n)$ preserves,[82] with $T_\lambda(m,n)$ and $(1+\lambda\tau_m(m,n))$, our 3-sphere S^3. A similar computation applies to $T_\lambda(m,n+1)$, and we have proven a discrete version of §5.5.16.

As a trivial consequence, the Calapso transform f_λ of a codimension 1 isothermic net $f : \mathbb{Z}^2 \to S^3$ takes, up to Möbius transformation, values in that same 3-sphere. Moreover, the relation between the Darboux and Calapso transformations that we established in §5.7.20 provides us with a similar assertion for the Darboux transforms $\hat{f} = T_\lambda^{-1}const$ of f: Choosing $const \in S^3$, the Darboux transform $\hat{f}$ will take values in S^3 since T_λ^{-1} preserves, with T_λ, our 3-sphere.

Summarizing, we have the following:

5.7.22 Lemma. *Let $f : \mathbb{Z}^2 \to S^3$ be an isothermic net with values in some 3-sphere S^3. Then its Calapso transformation T_λ can be chosen to preserve*

---

[82] For this conclusion we need the function a to be real.

that 3-sphere. Moreover, the Calapso transform f_λ and a 3-parameter family of Darboux transforms $\hat{f} = T_\lambda^{-1}const$ then takes values in that 3-sphere.

5.7.23 Now we turn to the various permutability theorems for our transformations of discrete isothermic nets — here we will adopt the notation conventions from Section 5.6.

Clearly, the Christoffel transformation is involutive. This follows directly from the defining formulas in §5.7.7 and the fact that the Christoffel transform f^* of an isothermic net f with cross-ratios $q = \frac{a}{b}$ is isothermic with a cross-ratio $q^* = q$:

5.7.24 Theorem. *The Christoffel transformation is involutive, $C^2 = id$.*

5.7.25 This was the only fact that we used to derive that the Goursat transformations of an isothermic surface behave like the Möbius group in the smooth case (see §5.3.5). Thus we have the same "permutability" behavior in the discrete case:

5.7.26 Theorem. *The Goursat transformations act as a group on discrete isothermic nets: $G_{\mu_1} G_{\mu_2} = G_{\mu_1 \circ \mu_2}$.*

5.7.27 To derive the discrete analog of Bianchi's permutability theorem for the Darboux transformation in §5.6.6, we employ the same method as in the smooth case: Given two Darboux transforms $\hat{f}_1 = D_{\lambda_1} f$ and $\hat{f}_2 = D_{\lambda_2} f$ of an isothermic net f, we construct the fourth net $\hat{f}$ by solving the cross-ratio condition

$$[f; \hat{f}_1; \hat{f}; \hat{f}_2] = \tfrac{\lambda_2}{\lambda_1}$$

for $\hat{f}$ (see (4.27)). Now, restricting our attention to, say, the m-edges of the four nets, the hexahedron lemma tells us that

$$[\hat{f}_2(m,n); \hat{f}_2(m+1,n); \hat{f}(m+1,n); \hat{f}(m,n)] = \lambda_1 a(m) \quad \text{and}$$
$$[\hat{f}_1(m,n); \hat{f}_1(m+1,n); \hat{f}(m+1,n); \hat{f}(m,n)] = \lambda_2 a(m)$$

(see Figure 4.4). Since a similar argument works for the n-edges of the nets, we learn that $\hat{f}$ such constructed is a D_{λ_1}-transform of $\hat{f}_2$ and a D_{λ_2}-transform of $\hat{f}_1$ at the same time and, as such, is an isothermic net with a cross-ratio $\hat{q} = \hat{q}_1 = \hat{q}_2 = q$:

5.7.28 Theorem. *If $\hat{f}_i = D_{\lambda_i} f$, $i = 1, 2$, are two Darboux transforms of an isothermic net f, then there is a fourth net $\hat{f}$, satisfying $[f; \hat{f}_1; \hat{f}; \hat{f}_2] \equiv \frac{\lambda_2}{\lambda_1}$, that is a D_{λ_2}-transform of $\hat{f}_1$ and a D_{λ_1}-transform of $\hat{f}_2$ at the same time,*

$$D_{\lambda_1} D_{\lambda_2} = D_{\lambda_2} D_{\lambda_1}.$$

5.7.29 Next we want to obtain the 1-parameter group property for the Calapso transformation (compare §5.5.9): If T_{λ_1} is the T_{λ_1}-transformation for an isothermic net f and $T_{\lambda_2}^{\lambda_1}$ denotes the T_{λ_2}-transformation for the Calapso transform $f_{\lambda_1} = T_{\lambda_1} f$ of f, then $T_{\lambda_2}^{\lambda_1} \circ T_{\lambda_1}$ gives the $T_{\lambda_1+\lambda_2}$-transformation for f. Thus we want to show that $T_{\lambda_2}^{\lambda_1} \circ T_{\lambda_1}$ satisfies the system (5.48) defining the Calapso transformation with $\lambda_1 + \lambda_2$ as a parameter. Employing the fact that T_{λ_1} and $T_{\lambda_2}^{\lambda_1}$ satisfy (5.48) with retraction forms τ and τ^{λ_1} of f and f_{λ_1}, respectively, this reduces to

$$T_{\lambda_2}^{\lambda_1}(1 + \lambda_2 \tau_m^{\lambda_1}) T_{\lambda_1}(1 + \lambda_1 \tau_m) \overset{!}{=} T_{\lambda_2}^{\lambda_1} T_{\lambda_1}(1 + (\lambda_1 + \lambda_2)\tau_m) \quad \text{and}$$
$$T_{\lambda_2}^{\lambda_1}(1 + \lambda_2 \tau_n^{\lambda_1}) T_{\lambda_1}(1 + \lambda_1 \tau_n) \overset{!}{=} T_{\lambda_2}^{\lambda_1} T_{\lambda_1}(1 + (\lambda_1 + \lambda_2)\tau_n).$$

Evaluating both sides of the first equation on $f(m,n)$ and $f(m+1,n)$, as a basis of $\mathbb{H}^2$, and the second equation on $f(m,n)$ and $f(m,n+1)$, our eigenvector equations (5.49) provide the desired result since $a^{\lambda_1} = \frac{a}{1-\lambda_1 a}$, and likewise for b^{λ_1} (see §5.7.16). To ensure regularity we have to require, for all $m, n \in \mathbb{Z}$,

$$\lambda_1 \neq \tfrac{1}{a(m)} \quad \text{and} \quad \lambda_2 \neq \tfrac{1}{a^{\lambda_1}(m)} \quad \Leftrightarrow \quad \lambda_1 + \lambda_2 \neq \tfrac{1}{a(m)},$$
$$\lambda_1 \neq \tfrac{1}{b(n)} \quad \text{and} \quad \lambda_2 \neq \tfrac{1}{b^{\lambda_1}(n)} \quad \Leftrightarrow \quad \lambda_1 + \lambda_2 \neq \tfrac{1}{b(n)}.$$

Thus we have proved the following discrete analog of §5.5.9, where, for simplicity of notation, we drop the superscript of the second transformation:

5.7.30 Theorem. *If T_{λ_1} and T_{λ_2} denote the Calapso transformations for an isothermic net f and its T_{λ_1}-transform $f_{\lambda_1} = T_{\lambda_1} f$, respectively, then*

$$T_{\lambda_2} T_{\lambda_1} = T_{\lambda_1+\lambda_2}$$

is the $T_{\lambda_1+\lambda_2}$-transformation for f; here, choosing $\lambda_1, \lambda_1 + \lambda_2 \neq \frac{1}{a(m)}, \frac{1}{b(n)}$ for all $m, n \in \mathbb{Z}$ ensures that the occurring transformations are regular.

5.7.31 Now we turn to the permutability theorems that intertwine different isothermic transformations. First we address the permutability of the Christoffel and Darboux transformations given, in the smooth case, in §5.6.3. Thus let $\mathfrak{f}$ denote an isothermic net with Christoffel transform $\mathfrak{f}^*$ and D_λ-transform $\hat{\mathfrak{f}}$, and consider $\hat{\mathfrak{f}}^* := \mathfrak{f}^* + \frac{1}{\lambda}(\hat{\mathfrak{f}} - \mathfrak{f})^{-1}$. Then, employing the Riccati system (5.47) for $\hat{\mathfrak{f}}$,

$$\partial_m(\hat{\mathfrak{f}}^* - \mathfrak{f}^*)(m,n)$$
$$= -\partial_m \mathfrak{f}^*(m,n) + \lambda[\tfrac{1}{\lambda}(\hat{\mathfrak{f}} - \mathfrak{f})^{-1}](m,n)\,\partial_m \mathfrak{f}(m,n)[\tfrac{1}{\lambda}(\hat{\mathfrak{f}} - \mathfrak{f})^{-1}](m+1,n),$$
$$\partial_n(\hat{\mathfrak{f}}^* - \mathfrak{f}^*)(m,n)$$
$$= -\partial_n \mathfrak{f}^*(m,n) + \lambda[\tfrac{1}{\lambda}(\hat{\mathfrak{f}} - \mathfrak{f})^{-1}](m,n)\,\partial_n \mathfrak{f}(m,n)[\tfrac{1}{\lambda}(\hat{\mathfrak{f}} - \mathfrak{f})^{-1}](m,n+1),$$
$$(5.50)$$

that is, $\hat{\mathfrak{f}}^*$ satisfies the Riccati system — here we use $\mathfrak{f} = \mathfrak{f}^{**}$ from §5.7.24 — and is therefore a Darboux transform of $\mathfrak{f}^*$. On the other hand, using the symmetry of (5.47) in the endpoints of corresponding edges of an isothermic net and its Darboux transform, (5.50) together with the Riccati system for $\hat{\mathfrak{f}}$ yields

$$
\begin{array}{rcccc}
(\partial_m \hat{\mathfrak{f}})(\partial_m \hat{\mathfrak{f}}^*) & = & (\hat{\mathfrak{f}} - \mathfrak{f})^{-1}(\partial_m \mathfrak{f}^*)(\partial_m \mathfrak{f})(\hat{\mathfrak{f}} - \mathfrak{f}) & = & a \quad \text{and} \\
(\partial_n \hat{\mathfrak{f}})(\partial_n \hat{\mathfrak{f}}^*) & = & (\hat{\mathfrak{f}} - \mathfrak{f})^{-1}(\partial_n \mathfrak{f}^*)(\partial_n \mathfrak{f})(\hat{\mathfrak{f}} - \mathfrak{f}) & = & b,
\end{array}
$$

so that $\hat{\mathfrak{f}}^*$ is identified as the Christoffel transform of $\mathfrak{f}$ by (5.43). Thus, completely analogous to the smooth case, we have:

5.7.32 Theorem. *If $\mathfrak{f}^*$ and $\hat{\mathfrak{f}}$ denote the Christoffel transform and a D_λ-transform of an isothermic net $\mathfrak{f} : \mathbb{Z}^2 \to \mathbb{H}$, then $\hat{\mathfrak{f}}^* = \mathfrak{f}^* + \frac{1}{\lambda}(\hat{\mathfrak{f}} - \mathfrak{f})^{-1}$ is a D_λ-transform of $\mathfrak{f}^*$ and the Christoffel transform of $\hat{\mathfrak{f}}$ at the same time,*

$$
CD_\lambda = D_\lambda C.
$$

5.7.33 Most important for the study of horospherical nets will be the effect of the Calapso transformation on Christoffel or Darboux pairs, respectively. Again, a proof can be carried out entirely analogously to the smooth case in §5.6.10: There, the key observation was that, writing the discrete isothermic net $f = v_0 + v_\infty \mathfrak{f}$ and its Christoffel transform $f^* := v_0 + v_\infty \mathfrak{f}^*$ in terms of affine coordinates, the Calapso transformation T_λ^* of f^* is given by

$$
T_\lambda^* = T_\lambda(v_\infty \nu_\infty + \lambda(v_0 + v_\infty \mathfrak{f})(\nu_0 - \mathfrak{f}^* \nu_\infty)) = T_\lambda(v_\infty \nu_\infty + \lambda f f^{*\perp}). \quad (5.51)
$$

As in the smooth case, it is straightforward to verify that T_λ^* defined in this way integrates the retraction form

$$
\begin{array}{rcl}
\tau_m^*(m, n) & = & f^*(m, n)\, \partial_m \mathfrak{f}(m, n) f^{*\perp}(m + 1, n) \\
\tau_n^*(m, n) & = & f^*(m, n)\, \partial_n \mathfrak{f}(m, n) f^{*\perp}(m, n + 1)
\end{array}
$$

of f^*, given as in (5.42):

$$
\begin{array}{rcl}
\tau_m(m, n)(v_\infty \nu_\infty + \lambda f f^{*\perp}(m + 1, n)) & = & (v_\infty \nu_\infty + \lambda f f^{*\perp}(m, n))\, \tau_m^*(m, n), \\
\tau_n(m, n)(v_\infty \nu_\infty + \lambda f f^{*\perp}(m, n + 1)) & = & (v_\infty \nu_\infty + \lambda f f^{*\perp}(m, n))\, \tau_n^*(m, n).
\end{array}
$$

Now, it follows that $f_\lambda^* = T_\lambda^* f^* = T_\lambda v_\infty$, so that $T_\lambda^{-1} f_\lambda^* \equiv const.$ On the other hand, by §5.7.30, $T_\lambda^{-1} = T_{-\lambda}$ is the $T_{-\lambda}$-transform of f_λ, so that we deduce with §5.7.20 that f_λ^* is a $D_{-\lambda}$-transform of f_λ. This provides a slight variation of our proof in the smooth case, which we gave in the context of the curved flat[83] approach to the Calapso transformation in §5.5.20.

---

[83] This would also be a good point at which to discuss a discrete version of the curved flat approach to isothermic surfaces (cf., [150]). We leave it as an exercise for the reader.

For the converse, that the T_λ-transforms of a D_λ-pair of isothermic nets can be stereographically projected to give a Christoffel pair of isothermic nets in $\mathbb{H}$, we refer the reader to our proof in the smooth case (see §5.6.11): It may serve the reader as an exercise to verify the claim for the discrete case.

Anyway, we have the following permutability, as in §5.6.10:

5.7.34 Theorem. *The T_λ-transforms of the members of a Christoffel pair $(\mathfrak{f}^*, \mathfrak{f})$ of isothermic nets in a Euclidean space form, if properly positioned in the conformal 4-sphere,*[84] *a $D_{-\lambda}$-pair.*

Conversely, the T_λ-transforms of a D_λ-pair $(\hat{f}, f)$ of isothermic nets in the conformal 4-sphere form, after suitable stereographic projection, a Christoffel pair in Euclidean 4-space:

$$T_\lambda C = D_{-\lambda} T_\lambda \quad \text{and} \quad T_\lambda D_\lambda = C \, T_\lambda.$$

5.7.35 As in §5.6.15, this theorem can be "generalized" to obtain a permutability that intertwines the Darboux and Calapso transformations for different parameters, λ and μ: To prove that theorem, we only used properties of the Darboux and Calapso transformations that we also proved in the discrete case by now. Hence the proof is also valid in the current setup:

$$T_\lambda D_\mu T_\lambda^{-1} = T_{\lambda-\mu} T_\mu D_\mu T_\mu^{-1} T_{\lambda-\mu}^{-1} = T_{\lambda-\mu} C T_{\lambda-\mu}^{-1} = D_{\mu-\lambda}.$$

We also obtain the very same formula for $\hat{T}_\lambda$ as in §5.6.15: If $(\hat{f}, f)$ denotes the original D_μ-pair, then we have

$$\hat{T}_\lambda = T_\lambda (1 - \tfrac{\lambda}{\mu} f \, (\hat{f}^\perp f)^{-1} \hat{f}^\perp).$$

Similarly, our permutability scheme in Figure 5.9, showing the intertwining of the Calapso transformation and the permutability scheme for the Darboux and Christoffel transformations, also holds in the discrete setting: It shall be another exercise for the interested reader to verify this claim by carefully analyzing the proof of the smooth version.[85] This permutability scheme becomes crucial when discussing the notion of duality, introduced by M. Umehara and K. Yamada in [287], for horospherical nets (see [150]), or for (smooth) horospherical surfaces (see [153]).

Finally, our permutability theorem in §5.7.34 directly provides the discrete analog for our permutability theorem that established the effect of the

---

[84] Remember, from §5.7.18, that the Calapso transform of an isothermic net is only well defined up to Möbius transformation (cf., §5.5.6).

[85] In fact, our proof in the smooth case is just a smooth version of the proof given in [150].

Calapso transformation on different D_λ-transforms of an isothermic surface (cf., §5.6.20):

5.7.36 Theorem. *If $\hat{f}_i$, $i = 1, 2$, denote two D_λ-transforms of the same isothermic net f, then their T_λ-transforms are, after suitable stereographic projections, Goursat transforms of each other.*

Conversely, the T_λ-transforms of two discrete isothermic nets that are Goursat transforms of each other are, when properly positioned, $D_{-\lambda}$-transforms of the same isothermic net, namely, the T_λ-transform of the Christoffel transform of either of the two original nets.

5.7.37 Example. To conclude this chapter we are going to discuss a discrete analog of horospherical surfaces in hyperbolic space. Remember that we derived two different characterizations for horospherical surfaces in terms of isothermic transformations: on one hand, as Calapso transforms of minimal surfaces in Euclidean space (see §5.5.29) and, on the other hand, as Darboux transforms of their hyperbolic Gauss maps (see §5.6.13). Thus we have *two* different options for defining horospherical nets via isothermic transformations — here we want to see that both options provide the same class of isothermic nets.

First we define, with [32], discrete minimal nets via the Christoffel transformation,[86] conforming with our characterization for minimal surfaces from §5.2.10:

An isothermic net $\mathfrak{f} : \mathbb{Z}^2 \to \mathrm{Im}\,\mathbb{H}$ is called a minimal net if its Christoffel transform $\mathfrak{f}^*$ is totally umbilic, that is, it takes values in a 2-sphere.

Without loss of generality (this amounts to a choice of factorization $q = \frac{a}{b}$ of the cross-ratios of $\mathfrak{f}$) we may then assume that $\mathfrak{n} := \mathfrak{f}^* : \mathbb{Z}^2 \to S^2$ takes values in the unit sphere — this net will then be called[87] the "discrete Gauss map" of the minimal net $\mathfrak{f}$. In [30], the authors also provided a geometric characterization for these nets that corresponds to the fact that a minimal surface has opposite principal curvatures at any point.

Since the Calapso transformation of a net into a 3-sphere can be chosen to preserve that 3-sphere (see §5.7.22), it also transforms totally umbilic nets into totally umbilic nets — or, otherwise said, the Calapso transformations of a "discrete holomorphic net" are complex (cf., §5.5.27) — because those can be described as nets taking values in the 3-spheres of an elliptic sphere pencil (see Figure T.3). As a consequence, by our permutability theorem in §5.7.34, an isothermic net has a totally umbilic Darboux transform if and only if it is a Calapso transform of a discrete minimal net. Consequently,

---

[86] Note that there are other ways of defining discrete minimal nets, for example, as solutions of a variational problem, as in [223].

[87] Note that this discrete Gauss map will be a discrete holomorphic map in the sense of [193].

the two options for defining horospherical nets as a discrete analog of horo-spherical surfaces via isothermic transformations that we proposed above describe the same class of isothermic nets.

An isothermic net $f : \mathbb{Z}^2 \to S^3$ will be called horospherical if it possesses a totally umbilic D_λ-transform $n : \mathbb{Z}^2 \to S^2$ or, equivalently, if its T_λ-transform can be stereographically projected to become a minimal net.

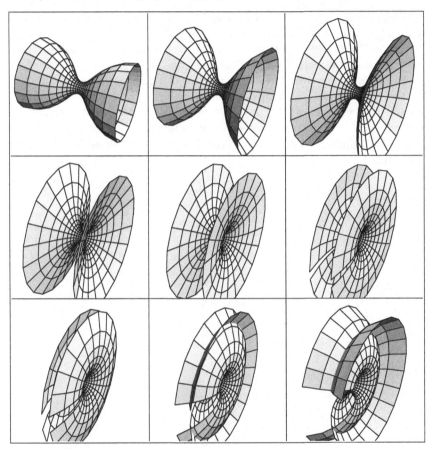

Fig. 5.14. Calapso transforms of a discrete catenoid

The totally umbilic Darboux transform n of a horospherical net provides an analog for the hyperbolic Gauss map, the "discrete hyperbolic Gauss map" of the horospherical net that determines the hyperbolic ambient space for the horospherical net. The minimal Calapso transform $f_\lambda : \mathbb{Z}^2 \to \operatorname{Im} \mathbb{H}$ of a horospherical net will be called its "minimal cousin," and the Gauss map (Christoffel transform) n_λ of this minimal cousin will be its "secondary

Gauss map": We obtain an entirely analogous situation as in the smooth case (see [42]).

Finally, note that the hyperbolic Gauss map and the parameter λ do not determine the horospherical net uniquely: There is a 3-parameter family[88] of D_λ-transforms of a given isothermic net. The effect on the minimal cousin is a Goursat transformation, as we learned from our permutability theorem in §5.7.36. On the other hand, its secondary Gauss map determines the minimal cousin uniquely, and therefore it determines the horospherical net uniquely up to Möbius transformation, that is, up to identification of the "canonical" ambient hyperbolic space. Also, this behavior conforms with the behavior of smooth horospherical surfaces.

Figure 5.14 shows the Calapso transforms of a discrete catenoid. Similar to the way in which the Enneper cousins in Figure 5.12 were determined by solving a complex version of the linear system for the Calapso transformation (see §5.6.21), here we solved the corresponding discrete system with $-j\,e^{2\pi\frac{m+in}{20}}$ and $e^{-2\pi\frac{m+in}{20}}\,j$ as a Christoffel pair of "discrete holomorphic functions."

---

[88] Corresponding, for example, to a choice of initial value for the system (5.46).

Chapter 6
A Clifford algebra model

In this chapter we make our first contact with the Clifford algebra approach
to Möbius geometry. More common than the approach that we discuss here
is a use of the Clifford algebra that, in some sense, generalizes the quaternionic model that we discussed in the preceding two chapters — a model that
uses "Vahlen matrices," that is, certain 2×2 matrices with Clifford algebra
entries (see, for example, [47], [189], or [288]). This Vahlen matrix approach
will be discussed in the following chapter — as an enhancement of the model
presented in this chapter. However, besides the fact that we will use the
material from this chapter in the following chapter on Vahlen matrices, the
model presented here is of its own interest, as the application to orthogonal
systems in Chapter 8 should show; for that reason it got its own chapter.
The model that will be presented here is, in some sense, simpler than the
Vahlen matrix approach: It is a direct reformulation of the classical model
from Chapter 1 using the Clifford algebra $A\mathbb{R}_1^{n+2}$ of the space of homogeneous coordinates $\mathbb{R}_1^{n+2}$ of the conformal n-sphere S^n (see §1.1.2). Besides
gaining an additional algebraic structure from this, the major achievement
will be that (orientation-preserving) Möbius transformations are then de-
scribed by elements of the spin group $Spin(\mathbb{R}_1^{n+2}) \subset A\mathbb{R}_1^{n+2}$. These two
features will, for example, turn out to be very useful in our discussions of
discrete analogs of "Ribaucour pairs" of triply (multiply) orthogonal systems and the related "discrete Ribaucour sphere congruences" in the first
part of Chapter 8.

The key point in developing this Clifford algebra model is to interpret
certain "pure" elements of the algebra geometrically. For this purpose we
will adopt the point of view that the underlying vector space of the Clifford
algebra is the exterior, or Grassmann, algebra for which the relation with
geometry is well known[1]; see, for example, [63], [155], or Grassmann's original expositions [130], [133] (cf., [132]; see also [303]). Then, we will replace
the exterior product[2] by the "Clifford product" and study its properties,
loosely following [117]. Prepared in this way, it will then be an easy exercise to reformulate many facts from the classical model in terms of the
Clifford algebra. In addition, we will derive a formula for the cross-ratio

---

[1] However, it is an unfortunate fact that in most modern textbooks on multilinear algebra
one can no longer find this relation to geometry — even though it was geometry that led to
its discovery, as indicated by the original term "geometric algebra."

[2] Grassmann himself already considered various products for the "geometric algebra," comprising the interior and exterior products as well as all kinds of combinations of those, including
the Clifford product [131] and quaternionic multiplication [134]; see [77] and [76].

of four points in the conformal n-sphere as a key ingredient for the theory
of discrete orthogonal nets and their Ribaucour transformations, which we
will need in Chapter 8.

The contents of this chapter are organized as follows:

Section 6.1. Some basic facts on the Grassmann algebra of a vector space
are recalled, and notations are fixed.

Section 6.2. In this section we introduce the Clifford product on the Grassmann algebra of an inner product vector space, thus giving a definition
of the Clifford algebra of this vector space. As an example, we show
that the Clifford algebra of the Euclidean plane is isomorphic to the
quaternions. Basic algebraic facts are proved, and the commutator and
anticommutator are investigated.

Section 6.3. Two involutions are introduced on the Clifford algebra: the
order involution and the Clifford conjugation. Those become important
in order to describe the (twisted adjoint) action of the Clifford group on
the underlying vector space. The representation of the Clifford group
on the orthogonal group of the vector space is used to show that any
element of the Clifford group is the product of vectors.

   Then the pin and spin groups are defined, and it is shown that the
subgroup $Spin_1^+(n + 2)$ is the (double) universal cover of the group of
orientation-preserving Möbius transformations of S^n.

Section 6.4. The Clifford dual is introduced, and its relation with the Hodge
operator is investigated. Representations for points and k-dimensional
spheres in S^n by elements of the Clifford algebra are given; the Clifford
dual is used to identify different representatives. The intersection angle
of two k-spheres in S^n is discussed.

Section 6.5. A formula for the (complex) cross-ratio of four points in the
conformal n-sphere is given and interpreted geometrically.

Section 6.6. The notion of an envelope is generalized to arbitrary dimensions, and enveloping conditions that make use of the Clifford algebra
setup are derived. The notion of a contact element is introduced.

Section 6.7. A notion of an adapted frame for a pair consisting of a map
into the conformal n-sphere and a congruence of m-spheres is introduced.
The connection form and the compatibility conditions are examined, and
the geometry of the occurring forms is discussed.

Remark. The reading of this chapter requires knowledge of the classical
model of Möbius geometry, as presented in Chapter 1. The material presented here will be used in the following chapter on Vahlen matrices.

6.1 The Grassmann algebra

First we recall a couple of facts about the Grassmann algebra, also known as the "exterior algebra" over an m-dimensional vector space V: This is a 2^m-dimensional vector space $\Lambda V = \oplus_{k=0}^m \Lambda^k V$, equipped with the "exterior multiplication" $\wedge : \Lambda V \times \Lambda V \to \Lambda V$ such that $\Lambda^0 V \cong \mathbb{R}$, $\Lambda^1 V \cong V \subset \Lambda V$ and the subspaces $\Lambda^k V \subset \Lambda V$ are spanned by $e_{i_1} \wedge \ldots \wedge e_{i_k}$, $i_1 < \ldots < i_k$, when $(e_1, \ldots, e_m)$ denotes some basis of V. Hence, $\dim \Lambda^k V = \binom{m}{k}$, so that we have $\dim \Lambda V = \sum_k \binom{m}{k} = 2^m$ indeed.

The $\wedge$-product is an alternating product, so that the product $v_1 \wedge \ldots \wedge v_k$ of k vectors $v_j = \sum_{i=1}^m v_{ij} e_i \in V$, $i = 1, \ldots, k$, can be expressed in terms of the determinants of $k \times k$ submatrices of the coefficient matrix,

$$v_1 \wedge \ldots \wedge v_k = \sum_{i_1 < \ldots < i_k} \det \begin{pmatrix} v_{i_1 1} & \cdots & v_{i_1 k} \\ \vdots & & \vdots \\ v_{i_k 1} & \cdots & v_{i_k k} \end{pmatrix} \cdot e_{i_1} \wedge \ldots \wedge e_{i_k}.$$

Note that $\Lambda^k V \wedge \Lambda^l V \subset \Lambda^{k+l} V$, which identifies ΛV as a "graded algebra."

6.1.1 The elements of $\Lambda^k V \subset \Lambda V$ are called "k-vectors," and $\mathfrak{v} \in \Lambda^k V$ is called a "pure k-vector," or "blade," if it can be decomposed as a product of k 1-vectors, $\mathfrak{v} = v_1 \wedge \ldots \wedge v_k$, $v_j \in V = \Lambda^1 V$. Note that $v_1 \wedge \ldots \wedge v_k \neq 0$ if and only if the vectors v_j are linearly independent.

6.1.2 The link to geometry is given by the following observation: Two pure k-vectors $\mathfrak{v} = v_1 \wedge \ldots \wedge v_k$, $\mathfrak{w} = w_1 \wedge \ldots \wedge w_k \in \Lambda^k V$ are linearly dependent (parallel) if and only if the k-dimensional subspaces spanned by the respective vectors coincide, $\text{span}\{v_1, \ldots, v_k\} = \text{span}\{w_1, \ldots, w_k\}$ (see [63]). In fact, this was one of the observations that led Grassmann in [130] to introduce his notion of "geometric algebra," and it is the fact that makes the Grassmann algebra a canonical object to work with in projective geometry (cf., [155]).

As a consequence, we obtain an embedding of the Grassmannian of k-planes in V into the projective space $P\Lambda^k V$ of 1-dimensional subspaces in $\Lambda^k V$. Note how the "$\wedge$-notation" that we occasionally used earlier (see, for example, Figure T.1) has now been supplied with a deeper meaning.

6.1.3 In the case of a 2-vector $\mathfrak{v} \in \Lambda^2 V$, decomposability can be detected taking an exterior product[3]: $\mathfrak{v}$ is decomposable if and only if $\mathfrak{v} \wedge \mathfrak{v} = 0$. This condition is known as the "Plücker relations"; when $V = \mathbb{R}^4$, that is, for lines in projective 3-space, the Plücker relations become one scalar equation defining the "Plücker quadric" in $P\Lambda^2 V \cong \mathbb{R}P^5$ (see [161]).

---

[3] The "only if" part is trivial since $\wedge$ is an alternating product; for an idea of proof of the "if" part see below.

6.1.4 Given an "inner product[4]" $\langle .,. \rangle$ on V, there is a canonical extension to the Grassmann algebra given by $\Lambda^k V \perp \Lambda^l V$ for $l \neq k$ and, on $\Lambda^k V$, by

$$\langle v_1 \wedge \ldots \wedge v_k, w_1 \wedge \ldots \wedge w_k \rangle = \det((\langle v_i, w_j \rangle)_{i,j=1,\ldots,m}.$$

6.1.5 In case this inner product is positive definite, the length $|\mathfrak{v}|$ of a pure k-vector $\mathfrak{v} = v_1 \wedge \ldots \wedge v_k$ is the volume of the parallelotope spanned by the vectors v_j (see [63]).

However, we will eventually use the Minkowski product on the space $\mathbb{R}_1^{n+2}$ of homogeneous coordinates of the classical model space for Möbius geometry — only being interested in the signature of the induced inner product on k-dimensional subspaces. Thus we note that, in the case of a Minkowski inner product $\langle .,. \rangle$, the sign of $|\mathfrak{v}|^2$, where $\mathfrak{v} = v_1 \wedge \ldots \wedge v_k$ is a pure k-vector, is the signature of the induced metric on span$\{v_1, \ldots, v_k\}$.

6.1.6 In the presence of a nondegenerate inner product, the space $\Lambda^2 V$ can be identified with the orthogonal algebra $\mathfrak{o}(V)$: Fixing $v_1, v_2 \in V$,

$$w \mapsto (v_1 \wedge v_2)(w) := \langle w, v_1 \rangle v_2 - \langle w, v_2 \rangle v_1$$

assigns[5] a skew-symmetric transformation to $v_1 \wedge v_2 \in \Lambda^2 V$. Because a basis of $\Lambda^2 V$ is mapped to a basis of $\mathfrak{o}(V)$, this map extends to an isomorphism between linear spaces.[6]

6.2 The Clifford algebra

Now let V be equipped with a (possibly degenerate) inner product $\langle .,. \rangle$. To obtain the associated Clifford algebra $\mathcal{A}V$, or algebra of "alternions," as it is called in some cases (cf., [243]), we equip ΛV (as a vector space) with another product:

6.2.1 Lemma and Definition. *An associative product, the Clifford product,* $\cdot : \Lambda V \times \Lambda V \to \Lambda V$, *can be uniquely defined by the scalar and vector multiplications*

$$\begin{aligned} a \cdot v = av = v \cdot a \quad &\text{for} \quad a \in \Lambda^0 V, v \in \Lambda^1 V, \\ v \cdot w = -\langle v, w \rangle + v \wedge w \quad &\text{for} \quad v, w \in \Lambda^1 V. \end{aligned} \tag{6.1}$$

---

[4] This term was apparently also coined by Grassmann: See [131] for an explanation.

[5] As an exercise, the reader should convince himself that this actually assigns a skew symmetric endomorphism to $v_1 \wedge v_2$, that is, that the endomorphism does not depend on the representation of $v_1 \wedge v_2$ as a product of two vectors.

[6] This identification can actually be used to prove that the above Plücker relations indeed characterize decomposable 2-vectors, by using normal forms of normal endomorphisms and the fact that $v_1 \wedge v_2$ is decomposable if and only if the corresponding endomorphism has rank 2.

The algebra $\mathcal{A}V = (\Lambda V, \cdot)$ is called the Clifford algebra of $(V, \langle ., . \rangle)$.

6.2.2 *Proof.* It suffices, by bilinearity, to show that (6.1) uniquely defines the products of the elements of a basis of ΛV. Thus, let $(e_1, \ldots, e_m)$ be an orthonormal basis of $(V, \langle ., . \rangle)$, that is, $\langle e_i, e_j \rangle = \varepsilon_i \delta_{ij}$ with $\varepsilon_i \in \{-1, 0, 1\}$. Now observe that, by (6.1), scalars commute with vectors while the basis vectors anticommute,

$$e_i \cdot e_j = e_i \wedge e_j = -e_j \wedge e_i = -e_j \cdot e_i$$

for $i \neq j$. Thus, using associativity, the product of any basis elements of ΛV becomes

$$\begin{aligned}(e_{i_1} \wedge \ldots \wedge e_{i_k}) \cdot (e_{j_1} \wedge \ldots \wedge e_{j_l}) &= (e_{i_1} \cdot \ldots \cdot e_{i_k}) \cdot (e_{j_1} \cdot \ldots \cdot e_{j_l}) \\ &= e_{i_1} \cdot \ldots \cdot e_{i_k} \cdot e_{j_1} \cdot \ldots \cdot e_{j_l},\end{aligned}$$

which now can be computed from (6.1) by interchanging vectors until all equal indices have been eliminated and the remaining product has been rearranged to again be one of the basis elements of ΛV.

Existence is clear — note that the expressions in (6.1) are bilinear.   ◁

6.2.3 The decomposition $\Lambda V = \oplus_{k=0}^m \Lambda^k V$, in general, no longer provides a grading for the Clifford algebra $\mathcal{A}V$, $\Lambda^k V \cdot \Lambda^l V \not\subset \Lambda^{k+l} V$. However, from the above proof we learn that $\Lambda^k V \cdot \Lambda^l V \in \oplus_{j=0}^{k+l} \Lambda^j V$, so that $\mathcal{A}V$ is a filtered algebra, $\mathcal{A}^0 V \subset \mathcal{A}^1 V \subset \ldots \subset \mathcal{A}^m V = \mathcal{A}V$ with $\mathcal{A}^k V = \oplus_{j=0}^k \Lambda^j V$.

An exception is the case when the scalar product $\langle ., . \rangle \equiv 0$ vanishes identically: In this case, the Clifford algebra becomes the Grassmann algebra, $\mathcal{A}V = \Lambda V$.

6.2.4 Example. As an example, we consider the Clifford algebra of the Euclidean plane $\mathbb{R}^2$: Choosing an orthonormal basis $(e_1, e_2) =: (j, k)$ and denoting $e_1 e_2 =: i$, we have $\mathcal{A}\mathbb{R}^2 = \mathbb{R} \oplus \mathrm{span}\{i, j, k\}$. Computing pairwise products we find

$$jk = i = -kj, \quad ki = j = -ik, \quad ij = k = -ji, \quad \text{and} \quad i^2 = j^2 = k^2 = -1.$$

Thus the Clifford algebra $\mathcal{A}\mathbb{R}^2 \cong \mathbb{H}$ is isomorphic to the field of quaternions (cf., [134], [76]).

6.2.5 In our applications we will need to compute Clifford products of nonorthogonal vectors. Thus it will be convenient to have some formulas that give the product of a vector with a pure k-vector, generalizing (6.1). The following formulas (from now on we suppress the ".") are easily verified for orthonormal (basis) vectors: Since the occurring expressions are obviously linear in each entry, this is enough to prove the lemma.

6.2.6 Lemma. *For vectors* $v_1, \ldots, v_k, w \in \Lambda^1 V$ *we have:*

$$
\begin{aligned}
w\,(v_1 \wedge \ldots \wedge v_k) &= w \wedge v_1 \wedge \ldots \wedge v_k \\
&\quad + \textstyle\sum_{j=1}^{k}(-1)^j v_1 \wedge \ldots \wedge \langle w, v_j \rangle \wedge \ldots \wedge v_k, \\
(v_1 \wedge \ldots \wedge v_k)\,w &= v_1 \wedge \ldots \wedge v_k \wedge w \\
&\quad + \textstyle\sum_{j=1}^{k}(-1)^{k-j-1} v_1 \wedge \ldots \wedge \langle w, v_j \rangle \wedge \ldots \wedge v_k.
\end{aligned} \tag{6.2}
$$

6.2.7 Thus $\Lambda^1 V \cdot \Lambda^k V \subset \Lambda^{k-1}V \oplus \Lambda^{k+1}V$, and, depending on k, the $\Lambda^{k-1}V$- and $\Lambda^{k+1}V$-parts are the symmetric and antisymmetric parts, or vice versa:

$$
\begin{aligned}
w\,(v_1 \wedge \ldots \wedge v_k) + (v_1 \wedge \ldots \wedge v_k)\,w &\in \begin{cases} \Lambda^{k+1}V & \text{for } k \equiv 0 \bmod 2, \\ \Lambda^{k-1}V & \text{for } k \equiv 1 \bmod 2, \end{cases} \\
w\,(v_1 \wedge \ldots \wedge v_k) - (v_1 \wedge \ldots \wedge v_k)\,w &\in \begin{cases} \Lambda^{k-1}V & \text{for } k \equiv 0 \bmod 2, \\ \Lambda^{k+1}V & \text{for } k \equiv 1 \bmod 2. \end{cases}
\end{aligned}
$$

Introducing, for $\mathfrak{v}, \mathfrak{w} \in \mathcal{A}V$, the symmetric and skew parts of their product,

$$
\{\mathfrak{v}, \mathfrak{w}\} := \mathfrak{v}\mathfrak{w} + \mathfrak{w}\mathfrak{v} \quad \text{and} \quad [\mathfrak{v}, \mathfrak{w}] := \mathfrak{v}\mathfrak{w} - \mathfrak{w}\mathfrak{v}, \tag{6.3}
$$

we may reformulate these properties as

$$
\{\Lambda^1 V, \Lambda^k V\} \in \begin{cases} \Lambda^{k+1}V \\ \Lambda^{k-1}V \end{cases}, \quad [\Lambda^1 V, \Lambda^k V] \in \begin{cases} \Lambda^{k-1}V & \text{for } k \equiv 0 \bmod 2 \\ \Lambda^{k+1}V & \text{for } k \equiv 1 \bmod 2. \end{cases}
$$

In particular, $\{v, w\} = -2\langle v, w \rangle$ and $[v, w] = 2v \wedge w$ for $v, w \in \Lambda^1 V$. Also we derive that, for $\mathfrak{v} \in \Lambda^k V$ and $w \in \Lambda^1 V$, we have

$$
\mathfrak{v} \wedge w = \begin{cases} \frac{1}{2}\{\mathfrak{v}, w\} & \text{if } k \equiv 0 \bmod 2, \\ \frac{1}{2}[\mathfrak{v}, w] & \text{if } k \equiv 1 \bmod 2. \end{cases}
$$

For the other part of the product we state formulas, for $k = 2, 3$, similar to the one expressing double cross-products in terms of the scalar product:

$$
\begin{aligned}
[v_1 \wedge v_2, w] &= \{w, v_2\} v_1 - \{w, v_1\} v_2 \quad \text{for } v_1, v_2, w \in \Lambda^1 V, \\
\{\mathfrak{v} \wedge v, w\} &= \{v, w\} \mathfrak{v} - [\mathfrak{v}, w] \wedge v \quad \text{for } \mathfrak{v} \in \Lambda^2 V, v, w \in \Lambda^1 V.
\end{aligned} \tag{6.4}
$$

We leave these as an exercise for the reader,[7] as well as the next, which now is a straightforward computation:

$$
\begin{aligned}
\tfrac{1}{2}(v_1 v_2 v_3 v_4 + v_4 v_3 v_2 v_1) &= v_1 \wedge v_2 \wedge v_3 \wedge v_4 \\
&\quad + \tfrac{1}{4}\{v_1, v_2\}\{v_3, v_4\} - \tfrac{1}{4}\{v_1, v_3\}\{v_2, v_4\} + \tfrac{1}{4}\{v_1, v_4\}\{v_2, v_3\}.
\end{aligned} \tag{6.5}
$$

---

[7] Hint: Prove the second for decomposable 2-vectors $\mathfrak{v} = v_1 \wedge v_2$.

Note that this sum of two quadruple products takes values in $\Lambda^0 V \oplus \Lambda^4 V$.

6.2.8 The first of the two formulas (6.4) shows that any $\mathfrak{v} \in \Lambda^2 V$ defines an endomorphism $\Lambda^1 V \ni w \mapsto [\mathfrak{v}, w] \in \Lambda^1 V$. Moreover, it is easy to check[8] that

$$0 \quad = \quad \{[\mathfrak{v}, v], w\} + \{v, [\mathfrak{v}, w]\},$$
$$[\mathfrak{v}, v \wedge w] \quad = \quad [\mathfrak{v}, v] \wedge w + v \wedge [\mathfrak{v}, w].$$

The first equation shows that the above endomorphism is skew with respect to $\{., .\}$, that is, $[\mathfrak{v}, .] \in \mathfrak{o}(V)$ when the scalar product $\{., .\} = -2\langle ., . \rangle$ is nondegenerate.

The second equation implies that $[\Lambda^2 V, \Lambda^2 V] \subset \Lambda^2 V$. Combining this with the fact that the commutator $[., .] : \mathcal{A}V \times \mathcal{A}V \to \mathcal{A}V$ makes the Clifford algebra $\mathcal{A}V$ a Lie algebra — it is straightforward to check the Jacobi identity

$$0 = [\mathfrak{v}, [\mathfrak{w}, \mathfrak{x}]] + [\mathfrak{w}, [\mathfrak{x}, \mathfrak{v}]] + [\mathfrak{x}, [\mathfrak{v}, \mathfrak{w}]]$$

for any $\mathfrak{v}, \mathfrak{w}, \mathfrak{x} \in \mathcal{A}V$ — we find that $\Lambda^2 V$ is, equipped with $[., .]$, a Lie subalgebra. In fact, since we have $[\mathfrak{v}, [\mathfrak{w}, v]] - [\mathfrak{w}, [\mathfrak{v}, v]] = [[\mathfrak{v}, \mathfrak{w}], v]$ for $\mathfrak{v}, \mathfrak{w} \in \Lambda^2 V$ and $v \in \Lambda^1 V$, the map $\Lambda^2 V \ni \mathfrak{v} \mapsto [\mathfrak{v}, .] \in \mathfrak{o}(V)$ is a Lie algebra homomorphism.

Contemplating the first equation in (6.4) for $\mathfrak{v}$ being given in terms of an orthonormal basis of V (for the nondegenerate inner product $\langle ., . \rangle$), we see that the homomorphism $\mathfrak{v} \mapsto [\mathfrak{v}, .]$ injects. Then, since $\dim \Lambda^2 V = \dim \mathfrak{o}(V)$, it is a Lie algebra isomorphism.

When comparing the first of the two formulas in (6.4) with the one in §6.1.6, all of this is indeed no surprise.

6.3 The spin group

Since we recovered the orthogonal algebra $\mathfrak{o}(V)$ as sitting inside the Clifford algebra, we might expect to find a copy of the orthogonal group in $\mathcal{A}V$: Indeed, we will recover a double cover of the (special) orthogonal group inside $\mathcal{A}V$, the (s)pin group. For a more detailed account of the covered material, see [224].

6.3.1 As a first step, we introduce two involutions on $\mathcal{A}V$:
 (i) $\mathfrak{v} \mapsto \hat{\mathfrak{v}}$: the "order involution," an automorphism of $\mathcal{A}V$, and
 (ii) $\mathfrak{v} \mapsto \bar{\mathfrak{v}}$: the "Clifford conjugation," an antiautomorphism of $\mathcal{A}V$,
both of which satisfy $\hat{v} = \bar{v} = -v$ for $v \in \Lambda^1 V = V$.

More explicitly, we can define the order involution by

$$\hat{\mathfrak{v}} \quad = \quad (-1)^k \mathfrak{v} \quad \text{for} \quad \mathfrak{v} \in \Lambda^k V.$$

---

[8] The second equation becomes easy when using $v \wedge w = \frac{1}{2}[v, w]$. In fact, it is just a special case of the Jacobi identity on $\mathcal{A}V$, given below.

Clearly, $\mathfrak{v} \mapsto \hat{\mathfrak{v}}$ such defined is linear with eigenvalues ± 1 and eigenspaces $\Lambda^k V$. Fixing an orthonormal basis $(e_1, \ldots, e_m)$ of $(V, \langle ., . \rangle)$ and computing $(\hat{.})$ of a product of basis elements,

$$(e_{i_1} \wedge \ldots \wedge e_{i_k}) \cdot (e_{j_1} \wedge \ldots \wedge e_{j_l}) \in \oplus_{i \leq k+l, \; i \equiv k+l \bmod 2} \Lambda^i V,$$

we learn that $\widehat{\mathfrak{v}\mathfrak{w}} = (-1)^{k+l} \mathfrak{v}\mathfrak{w} = \hat{\mathfrak{v}}\hat{\mathfrak{w}}$ for $\mathfrak{v} \in \Lambda^k V$ and $\mathfrak{w} \in \Lambda^l V$. As a consequence, $\mathfrak{v} \mapsto \hat{\mathfrak{v}}$ is verified to be an automorphism of $\mathcal{A}V$, that is,

$$\widehat{\mathfrak{v}\mathfrak{w}} = \hat{\mathfrak{v}}\hat{\mathfrak{w}} \quad \text{for} \quad \mathfrak{v}, \mathfrak{w} \in \mathcal{A}V.$$

The Clifford conjugation, on the other hand, has to act — since it should be an antiautomorphism, that is,

$$\overline{\mathfrak{v}\mathfrak{w}} = \bar{\mathfrak{w}}\bar{\mathfrak{v}} \quad \text{for} \quad \mathfrak{v}, \mathfrak{w} \in \mathcal{A}V$$

— on the basis elements by

$$\overline{(e_{i_1} \wedge \ldots \wedge e_{i_k})} = \overline{(e_{i_1} \cdot \ldots \cdot e_{i_k})} = (\bar{e}_{i_k} \cdot \ldots \cdot \bar{e}_{i_1})$$
$$= (-1)^k (e_{i_k} \cdot \ldots \cdot e_{i_1}) = (-1)^{\frac{k(k+1)}{2}} (e_{i_1} \wedge \ldots \wedge e_{i_k}),$$

which uniquely fixes $\mathfrak{v} \mapsto \bar{\mathfrak{v}}$ as an endomorphism:

$$\bar{\mathfrak{v}} = (-1)^{\frac{k(k+1)}{2}} \mathfrak{v} \quad \text{for} \quad \mathfrak{v} \in \Lambda^k V.$$

Note that both maps $\mathfrak{v} \mapsto \hat{\mathfrak{v}}$ and $\mathfrak{v} \mapsto \bar{\mathfrak{v}}$ are involutions, that is, $\hat{\hat{\mathfrak{v}}} = \mathfrak{v} = \bar{\bar{\mathfrak{v}}}$ for all $\mathfrak{v} \in \mathcal{A}V$.

6.3.2 Now suppose that the inner product $\langle ., . \rangle$ on V is nondegenerate. We consider the "Clifford group[9]"

$$\Gamma(V) := \{ \mathfrak{s} \in \mathcal{A}V \mid \mathfrak{s}^{-1} \text{ exists and } \mathfrak{s}V\hat{\mathfrak{s}}^{-1} \subset V \}$$

acting on V via the "twisted adjoint action" $(\mathfrak{s}, v) \mapsto r(\mathfrak{s})v := \mathfrak{s}v\hat{\mathfrak{s}}^{-1}$. Now

$$(r(\mathfrak{s})v)^2 = -(\mathfrak{s}v\hat{\mathfrak{s}}^{-1})(\widehat{\mathfrak{s}v\hat{\mathfrak{s}}^{-1}}) = -\mathfrak{s}v\hat{\mathfrak{s}}^{-1}\hat{\mathfrak{s}}\hat{v}\mathfrak{s}^{-1} = v^2,$$

showing that $\Gamma(V)$ acts on V by isometries. In particular, since $\langle ., . \rangle$ is non degenerate, each $r(\mathfrak{s})$ is an orthogonal automorphism (with respect to the linear structure on $V = \Lambda^1 V$), that is, $r : \Gamma(V) \to O(V)$.

---

[9] Clearly, $\Gamma(V)$ is closed under multiplication; since $(r(\mathfrak{s}))^{-1} = r(\mathfrak{s}^{-1})$, it is also closed under taking inverses.

Clearly, $r : \Gamma(V) \to O(V)$ is a representation of $\Gamma(V)$, that is, r is a group homomorphism, $r(\mathfrak{s}\tilde{\mathfrak{s}}) = r(\mathfrak{s}) \circ r(\mathfrak{s})$ for all $\mathfrak{s}, \tilde{\mathfrak{s}} \in \Gamma(V)$.

Next we want to determine the kernel of this representation: Thus we seek those $\mathfrak{s} \in \Gamma(V)$ that satisfy

$$\mathfrak{s}v = v\hat{\mathfrak{s}} \quad \Leftrightarrow \quad \begin{cases} [\mathfrak{s} + \hat{\mathfrak{s}}, v] = 0 \\ \{\mathfrak{s} - \hat{\mathfrak{s}}, v\} = 0 \end{cases} \tag{6.6}$$

for all $v \in \Lambda^1 V$, where the equivalence is obtained by taking $(\hat{.})$ of the equation and adding and subtracting the two equations obtained. Note that $\mathfrak{v} \mapsto \hat{\mathfrak{v}}$ has the even, respectively odd, grades of $\Lambda V = \oplus_{k=0}^{m} \Lambda^k V$ as its $+1$- and -1-eigenspaces. Since $\{\Lambda^1 V, \Lambda^k V\} \subset \Lambda^{k-1} V$ for odd k and $[\Lambda^1 V, \Lambda^k V] \subset \Lambda^{k-1} V$ for even k, we find that $\mathfrak{s} - \hat{\mathfrak{s}} = 0$ and $\mathfrak{s} + \hat{\mathfrak{s}} \in \Lambda^0 V$, by staring at (6.2).

We summarize these results in the following

6.3.3 Lemma. *The map $r : \Gamma(V) \to O(V)$, $(\mathfrak{s}, v) \mapsto r(\mathfrak{s})v = \mathfrak{s}v\hat{\mathfrak{s}}^{-1}$, is a representation of the Clifford group with kernel $\Lambda^0 V \setminus \{0\} \cong \mathbb{R} \setminus \{0\}$.*

6.3.4 Example. As an important example we examine the orthogonal transformation given by $r(s)$, where $s \in \Lambda^1 V = V$: First note that s^{-1} exists as soon as $|s|^2 = -s^2 \neq 0$ since, in this case, $s^{-1} = -\frac{1}{|s|^2}s$. Then

$$r(s)v = -svs^{-1} = \frac{1}{|s|^2}svs = \frac{1}{|s|^2}(-v\,s^2 + \{s, v\}\,s) = v - \frac{2}{|s|^2}\langle v, s\rangle\,s$$

is the reflection in the hyperplane $s^{\perp} \subset V$.

6.3.5 As for §1.3.17, we can now employ a result from linear algebra, saying that any pseudoorthogonal transformation on $\mathbb{R}_j^m$ is a composition of at most m reflections in nondegenerate hyperplanes (cf., [66]), to state that any element $\mathfrak{s} \in \Gamma(V)$ of the Clifford group is a product of at most m (where $m = \dim V$) nonisotropic *vectors* $s_i \in \Gamma(V) \cap \Lambda^1 V$. Because we assumed the inner product $\langle ., . \rangle$ on V to be nondegenerate, we have $(V, \langle ., . \rangle) \cong \mathbb{R}_j^m$ with some signature $0 \leq j \leq m$. Thus there are nonisotropic $s_i \in V$, $i \leq k \leq m$ and $s_i^2 \neq 0$, so that

$$r(\mathfrak{s}) = r(s_1) \circ \ldots \circ r(s_k) = r(s_1 \cdots s_k).$$

And, since the kernel of $r : \Gamma(V) \to O(V)$ is just $\mathbb{R} \setminus \{0\}$, we conclude that $\mathfrak{s} = \lambda s_1 \cdots s_k$ for some $\lambda \in \mathbb{R} \setminus \{0\}$. Hence, after possibly replacing s_1 by λs_1, we obtain

6.3.6 Lemma. *Any element $\mathfrak{s} \in \Gamma(V)$, $\mathfrak{s} \notin \mathbb{R} \setminus \{0\}$, of the Clifford group is the product $\mathfrak{s} = s_1 \cdots s_k$ of $k \leq m = \dim V$ vectors $s_i \in \Gamma(V)$, $1 \leq i \leq k$.*

6.3.7 With this we now find that

$$\bar{\mathfrak{s}}\mathfrak{s} = \overline{(s_1 \cdots s_k)}\,(s_1 \cdots s_k) = \bar{s}_k \cdots \bar{s}_1 \cdot s_1 \cdots s_k = \prod_{i=1}^{k} |s_i|^2 \in \mathbb{R} \setminus \{0\}. \tag{6.7}$$

Thus "normalizing" the elements of the Clifford group imposes one real condition; in this way we obtain the pin and spin groups of $(V, \langle .,. \rangle)$:

$$\begin{aligned} Pin(V) &:= \{\mathfrak{s} \in \Gamma(V) \,|\, \bar{\mathfrak{s}}\mathfrak{s} = \pm 1\}, \\ Spin(V) &:= \{\mathfrak{s} \in Pin(V) \,|\, \hat{\mathfrak{s}} = \mathfrak{s}\}. \end{aligned}$$

By §6.3.6 we can write an element $\mathfrak{s} \in Pin(V)$ as a product $\mathfrak{s} = s_1 \cdots s_k$ of vectors; from (6.7) we learn that the s_i's can be chosen in $Pin(V)$, $s_i^2 = \pm 1$. Moreover, since

$$\hat{\mathfrak{s}} = \widehat{(s_1 \cdots s_k)} = \hat{s}_1 \cdots \hat{s}_k = (-1)^k \, s_1 \cdots s_k,$$

any element $\mathfrak{s} \in Spin(V)$ is an *even* product of vectors s_i, $s_i^2 = \pm 1$. As a consequence, the corresponding orthogonal transformation $r(\mathfrak{s}) \in SO(V)$ is orientation-preserving.

We collect these results in a corollary, including the obvious consequence that the "normalization" in the definition of $Pin(V)$ has on the kernel of the restriction of the representations $r : Pin(V) \to O(V)$ of the pin group and $r : Spin(V) \to SO(V)$ of the spin group, respectively:

6.3.8 Corollary. *Any element* $\mathfrak{s} \in Pin(V)$, *or* $\mathfrak{s} \in Spin(V)$, *is the product*[10] *of* $k \le m = \dim V$ *vectors* s_i, $s_i^2 = \pm 1$, *where* $k \equiv 0 \bmod 2$ *is even when* $\mathfrak{s} \in Spin(V)$:

$$\begin{aligned} Pin(V) &= \{\textstyle\prod_{i=1}^{k} s_i \,|\, s_i \in V, \, s_i^2 = \pm 1\}, \\ Spin(V) &= \{\textstyle\prod_{i=1}^{2k} s_i \,|\, s_i \in V, \, s_i^2 = \pm 1\}. \end{aligned}$$

Moreover, the (twisted for $Pin(V)$: *r is not twisted in case of* $Spin(V)$ *where* $\hat{\mathfrak{s}} = \mathfrak{s}$) *adjoint actions*

$$\begin{aligned} r : Pin(V) \to O(V), \quad & \mathfrak{s} \mapsto [V \ni v \mapsto \mathfrak{s}v\hat{\mathfrak{s}}^{-1} \in V] \quad and \\ r : Spin(V) \to SO(V), \quad & \mathfrak{s} \mapsto [V \ni v \mapsto \mathfrak{s}v\mathfrak{s}^{-1} \in V] \end{aligned}$$

define representations of $Pin(V)$ *and* $Spin(V)$, *respectively, with kernel* ± 1.

6.3.9 Finally, we compute the Lie algebras of $Pin(V)$ and $Spin(V)$, that is, their tangent space(s) at $\mathfrak{s} = 1$. For that purpose, let $\mathfrak{s} = \mathfrak{s}(t)$, $t \in (-\varepsilon, \varepsilon)$,

---

[10] The empty product is defined to be $\prod_{i=1}^{0} s_i := 1$.

be a curve with $\mathfrak{s}(0) = 1$ in $Pin(V)$, or $Spin(V)$; let $v \in V$ denote any vector and $v(t) := r(\mathfrak{s}(t))v$, so that $v(0) = v$. Then

$$v' \;=\; \mathfrak{s}'v\,(\hat{\mathfrak{s}}(0))^{-1} - \mathfrak{s}(0)\,v\,(\hat{\mathfrak{s}}(0))^{-1}\hat{\mathfrak{s}}'(\hat{\mathfrak{s}}(0))^{-1} \;=\; \mathfrak{s}'v - v\,\hat{\mathfrak{s}}', \qquad (6.8)$$

where $(.)'$ denotes the derivative at $t = 0$. Taking $(\hat{.})$ of (6.8) and taking sum and difference of the two equations[11] we obtain

$$\{\mathfrak{s}' - \hat{\mathfrak{s}}', v\} \;=\; 0 \quad \text{and} \quad [\mathfrak{s}' + \hat{\mathfrak{s}}', v] \;=\; 2v' \in V = \Lambda^1 V.$$

Using a similar argument as the one we used to determine the kernel of r from (6.6), we deduce that $\hat{\mathfrak{s}}' = \mathfrak{s}'$ and $\mathfrak{s}' \in \mathcal{A}^2 V$. On the other hand,

$$0 \;=\; (\bar{\mathfrak{s}}\mathfrak{s})' \;=\; \bar{\mathfrak{s}}' + \mathfrak{s}',$$

so that $\mathfrak{s}'$ must be in the -1-eigenspace of $(\bar{.})$; hence $\mathfrak{s}' \in \Lambda^2 V$ by §6.3.1.

Thus we have proved that the Lie algebra of $Pin(V)$, as well as that of $Spin(V)$, is $\Lambda^2 V$. Moreover, the differential $d_1 r : \Lambda^2 V \to \mathfrak{o}(V)$ of the (twisted) adjoint representation is exactly the Lie algebra isomorphism considered in §6.2.8. As a consequence, the (identity components of the) two groups $Spin(V)$ and $SO(V)$ are locally isomorphic (see [140]). In fact, $r : Spin(V) \to SO(V)$ is a double cover of $SO(V)$ since its kernel is ± 1.

6.3.10 Theorem. *If V carries a nondegenerate inner product $\langle.,.\rangle$, then the adjoint representation*

$$r : Spin(V) \to SO(V), \quad \mathfrak{s} \mapsto r(\mathfrak{s}) = [v \ni V \mapsto \mathfrak{s}v\mathfrak{s}^{-1} \in V]$$

is a double cover of $SO(V)$ with differential

$$d_1 r : \mathfrak{spin}(V) = \Lambda^2 V \to \mathfrak{o}(V), \quad \mathfrak{v} \mapsto d_1 r(\mathfrak{v}) = [\mathfrak{v}, .].$$

6.3.11 One can say more in the cases $(V, \langle.,.\rangle) = \mathbb{R}^m_k$, $m \geq 3$ and $k \in \{0, 1\}$, which are of particular interest to us. Then the groups $Spin(\mathbb{R}^m)$ and

$$Spin^+(\mathbb{R}^m_1) := \{\mathfrak{s} \in Spin(\mathbb{R}^m_1) \mid \bar{\mathfrak{s}}\mathfrak{s} = +1\}$$

are simply connected (see [189]), so that they are, in fact, the double universal covers of the (identity components of the) respective orthogonal groups $SO(m)$ and $SO^+_1(m)$, the group of orientation- and time orientation— preserving Lorentz transformations. But $SO^+_1(n + 2)$ is just the group of orientation-preserving Möbius transformations of the conformal n-sphere[12]:

6.3.12 Theorem. *The group $Spin^+_1(n + 2) := Spin^+(\mathbb{R}^{n+2}_1)$ is the double, universal cover of the group of orientation-preserving Möbius transformations.*

---

[11] This argument becomes unnecessary in the $Spin(V)$ case since, in this case, $\hat{\mathfrak{s}}' = \mathfrak{s}'$, anyway.

[12] See §1.3.14: Here we can argue that any orientation-preserving Möbius transformations is generated by an even number of inversions (cf., §1.3.17), just as any Lorentz transformation in $SO^+_1(n + 2)$ can be generated by an even number of reflections in Minkowski hyperplanes.

6.4 Spheres and the Clifford dual

After these preparations, we now want to return to our mission: to establish a Clifford algebra model for Möbius geometry. We already saw in the previous section how the group of (orientation-preserving) Möbius transformations appears in this context via the spin group $Spin_1^+(n+2)$. Also, points and hyperspheres can be described, just as in the classical model, by isotropic and spacelike vectors. However, by our discussions on the "geometric algebra" in §6.1.2, spheres of any dimension k can now be described as decomposable spacelike $(n-k)$-vectors or, equivalently, as timelike $(k+2)$-vectors (see Figures T.1 and T.4). Before proceeding to a description of the incidence relation and the intersection angle of spheres, we want to discuss a map $\Lambda^{n-k}\mathbb{R}_1^{n+2} \to \Lambda^{k+2}\mathbb{R}_1^{n+2}$ that realizes the identification between both descriptions: the "Clifford dual."

Having this in mind, we stay for some more paragraphs with the general setup.

6.4.1 Thus let $\langle .,. \rangle$ denote a nondegenerate inner product on V, and let $(e_1, \ldots, e_m)$ be an orthonormal basis of V. We define

$$\mathrm{v} := e_1 \cdots e_m = e_1 \wedge \cdots \wedge e_m \in \Lambda^m V.$$

Clearly, v has unit length, $\bar{\mathrm{v}}\mathrm{v} = \pm 1$, where the sign depends on the signature that the inner product $\langle .,. \rangle$ has on the 1-dimensional space $\Lambda^m V$. Note that this condition determines v uniquely up to sign.

Choosing a different orthonormal basis, $\tilde{e}_i = Ae_i$ with $A \in O(V)$, we obtain

$$\tilde{\mathrm{v}} = Ae_1 \wedge \ldots \wedge Ae_m = \det A \cdot \mathrm{v} = \pm \mathrm{v}. \tag{6.9}$$

In particular, we may fix an orientation on $(V, \langle .,. \rangle)$ by choosing a sign for the "reference volume" v. Finally, we compute

$$\mathrm{v}^2 = (e_1 \cdots e_m)^2 = (-1)^{\frac{m(m-1)}{2}} \prod_{i=1}^m e_i^2 = (-1)^{\frac{m(m+1)}{2}} \bar{\mathrm{v}}\mathrm{v}. \tag{6.10}$$

6.4.2 Now let $j \mapsto i_j$ be a permutation of $\{1, \ldots, m\}$ such that $i_1 < \ldots < i_k$ and $i_{k+1} < \ldots < i_m$. Given the corresponding basis element $e_{i_1} \wedge \ldots \wedge e_{i_k}$ in $\Lambda^k V$, the product

$$\mathrm{v} \cdot (e_{i_1} \wedge \ldots \wedge e_{i_k}) = e_1 \cdots e_m \cdot e_{i_1} \cdots e_{i_k} = \pm e_{i_{k+1}} \wedge \ldots \wedge e_{i_m} \in \Lambda^{m-k} V.$$

Because we have $\mathrm{v}^2 = \pm 1$, multiplication with v yields an isomorphism $\Lambda^k V \to \Lambda^{m-k} V$ for each $k = 0, \ldots, m$:

6.4.3 Lemma and Definition. *The map* $\mathrm{v} : \Lambda^k V \to \Lambda^{m-k} V$, $\mathfrak{v} \mapsto \mathrm{v}\mathfrak{v}$, *is an isomorphism, called the Clifford dual.*

6.4.4 Example. The Clifford dual is very similar to the Hodge-$\star$ operator, or "Hodge dual," that is defined by the equation

$$\mathfrak{v} \wedge \star \mathfrak{w} = \langle \mathfrak{v}, \mathfrak{w} \rangle \, v$$

for $\mathfrak{v}, \mathfrak{w} \in \Lambda^k V$ (cf., [189]). The Hodge operator also provides isomorphisms $\star : \Lambda^k V \to \Lambda^{m-k} V$. However, in contrast to (6.10), its square becomes $\star^2 = (-1)^{k(m-k)}$ on $\Lambda^k V$ (see [298]).

As an example, we consider the Clifford algebra $\mathcal{A}\mathbb{R}^2$ of the Euclidean 2-plane (cf., §6.2.4). Choosing an orthonormal basis (e_1, e_2) of $\mathbb{R}^2$, we have $v = e_1 e_2$. Then $|v|^2 = 1$ and $v^2 = (-1)^3 = -1$. Now

$$\star 1 = v, \quad \star e_1 = e_2, \quad \star e_2 = -e_1, \quad \text{and} \quad \star v = 1;$$
$$v1 = v, \quad ve_1 = e_2, \quad ve_2 = -e_1, \quad \text{but} \quad vv = -1.$$

Thus the Clifford dual differs from the Hodge dual by a sign on some $\Lambda^k V$. It shall be left to the reader as an exercise to determine the relation in the general situation.

6.4.5 Finally, as an important step toward the identification of the two descriptions of spheres of higher codimension, we examine the effect of the Clifford dual on a nonisotropic pure k-vector $\mathfrak{v}$: First note that, since $\mathfrak{v}$ is pure, it determines a k-dimensional subspace of V, with a nondegenerate metric since $\mathfrak{v}$ is nonisotropic. Hence we can choose an orthonormal basis $(e_1, \ldots, e_k)$ of that subspace to obtain $\mathfrak{v} = \lambda e_1 \cdots e_k$ with some $\lambda \in \mathbb{R} \setminus \{0\}$. Extending that basis to a positively oriented orthonormal basis $(e_1, \ldots, e_m)$ of V, we obtain (cf., (6.9))

$$v = e_1 \cdots e_m, \quad \text{so that} \quad v\mathfrak{v} = \pm \lambda e_{k+1} \cdots e_m$$

is again a pure $(m-k)$-vector and nonisotropic since $|v\mathfrak{v}|^2 = |v|^2 |\mathfrak{v}|^2$.

6.4.6 Lemma. *Let $\mathfrak{v} \in \Lambda^k V$ be a nonisotropic pure k-vector. Then its Clifford dual $v\mathfrak{v} \in \Lambda^{m-k} V$ is a nonisotropic pure $(m-k)$-vector, with $|v\mathfrak{v}|^2 = |v|^2 |\mathfrak{v}|^2$, that represents the orthogonal complement of the subspace represented by $\mathfrak{v}$.*

6.4.7 To apply these results to Möbius geometry, let $(V, \langle ., . \rangle) = \mathbb{R}_1^{n+2}$ now be the space of homogeneous coordinates of the classical model (see §1.1.2). From Figure T.1 we read off that isotropic vectors $p \in \Lambda^1 \mathbb{R}_1^{n+2}$, $p^2 = 0$, represent the points of S^n, and spacelike k-blades $\mathfrak{s} \in \Lambda^k \mathbb{R}_1^{n+2}$, $\bar{\mathfrak{s}}\mathfrak{s} > 0$, give the $(n-k)$-dimensional spheres in the conformal n-sphere S^n. Hyperspheres are obtained as the special case of spacelike 1-vectors $s \in \Lambda^1 \mathbb{R}_1^{n+2}$, $s^2 < 0$.

Equivalently, a k-dimensional sphere in S^n can be given by the Minkowski $(k+2)$-plane spanned by $k+2$ points of the sphere in "general position," that is, by a pure timelike $(k + 2)$-vector $\tilde{\mathfrak{s}} = p_1 \wedge \ldots \wedge p_{k+2} \in \Lambda^{k+2} \mathbb{R}_1^{n+2}$. Remember that this $(k + 2)$-plane is just the orthogonal complement of the above spacelike $(n - k)$-plane, which can be interpreted as the family of all hyperspheres that contain $\mathfrak{s}$ as a subset.

The identification of both representations can be given in terms of the Clifford dual: $\tilde{\mathfrak{s}} \in \Lambda^{k+2} \mathbb{R}_1^{n+2}$ and $\mathfrak{s} = v\tilde{\mathfrak{s}} \in \Lambda^{n-k} \mathbb{R}_1^{n+2}$ represent the same k-sphere in S^n, by our lemma in §6.4.6.

6.4.8 Lemma. *The points and k-dimensional spheres in the conformal n-sphere S^n are represented by*

* *isotropic 1-vectors $p \in \Lambda^1 \mathbb{R}_1^{n+2} \setminus \{0\}$, $p^2 = 0$;*
* *spacelike pure $(n - k)$-vectors $\mathfrak{s} \in \Lambda^{n-k} \mathbb{R}_1^{n+2}$, $|\mathfrak{s}|^2 = \bar{\mathfrak{s}}\mathfrak{s} > 0$;*
  or, by their complementary
* *timelike pure $(k + 2)$-vectors $\tilde{\mathfrak{s}} = v\mathfrak{s} \in \Lambda^{k+2} \mathbb{R}_1^{n+2}$, $|\tilde{\mathfrak{s}}|^2 = |v\mathfrak{s}|^2 < 0$.*

6.4.9 Note that sphere pencils and complexes (see Figure T.1) also fit into this setup: Any sphere complex is given by a vector $\mathcal{K} \in \Lambda^1 \mathbb{R}_1^{n+2} \setminus \{0\}$ and any pure 2-vector $\mathfrak{s} \in \Lambda^2 \mathbb{R}_1^{n+2} \setminus \{0\}$ determines a sphere pencil — elliptic, parabolic, or hyperbolic, depending on the sign of $|\mathfrak{s}|^2$ (see §1.2.1).

In the elliptic and hyperbolic cases we already know by §6.4.6 that the Clifford duals $v\mathcal{K}$ and $v\mathfrak{s}$ provide pure n- and $(n + 1)$-vectors that give the orthogonal complements of the respective subspaces.

For the parabolic cases, an analog of the lemma in §6.4.6 can be proven by using a pseudo-orthogonal basis: Suppose $\mathfrak{s} \in \Lambda^{k+1} \mathbb{R}_1^{n+2} \setminus \{0\}$ is an isotropic pure $(k + 1)$-vector, $|\mathfrak{s}|^2 = 0$; then choose an orthogonal basis $(p, s_1, \ldots, s_k)$ with $p^2 = 0$ and $s_i^2 = -1$ for the corresponding subspace of $\mathbb{R}_1^{n+2}$ and extend it to a pseudo-orthonormal basis $(p, s_1, \ldots, s_k, s_{k+1}, \ldots, s_n, \hat{p})$ of $\mathbb{R}_1^{n+2}$,

$$s_i^2 = -1, \quad s_i \perp s_j \ (i \neq j), \quad p^2 = \hat{p}^2 = 0, \quad p\hat{p} + \hat{p}p = 1, \quad \text{and} \quad p, \hat{p} \perp s_i.$$

Then we have $\mathfrak{s} \parallel p \cdot s_1 \cdots s_k$ and $v = \pm(p - \hat{p})(p + \hat{p})s_1 \cdots s_n$, and we deduce

$$
\begin{aligned}
v\mathfrak{s} \quad \parallel \quad & (p\hat{p} - \hat{p}p)s_1 \cdots s_n \cdot ps_1 \cdots s_k & = \quad \pm p\hat{p}ps_{k+1} \cdots s_n \\
= \quad & \pm(p\hat{p}p + \hat{p}pp)s_{k+1} \cdots s_n & = \quad \pm ps_{k+1} \cdots s_n
\end{aligned}
$$

since $p^2 = 0$. Thus we have proved the following isotropic version of §6.4.6, which will also be of interest in the context of contact elements in the conformal n-sphere:

6.4.10 Lemma. *Let $\mathfrak{s} \neq 0$ be an isotropic pure $(k + 1)$-vector. Then its Clifford dual $v\mathfrak{s}$ is an isotropic pure $(n - k + 1)$-vector that represents the orthogonal complement of the subspace represented by $\mathfrak{s}$.*

6.4.11 Our next issue is to analyze incidence and intersection of spheres.

As in the classical model, the (unoriented) intersection angle $\alpha \in [0, \frac{\pi}{2}]$ of two hyperspheres $s, \tilde{s} \in S_1^{n+1}$ is given[13] by $\langle s, \tilde{s} \rangle^2 = \frac{1}{4}\{s, \tilde{s}\}^2 = \cos^2 \alpha$ (see Figure T.1). In particular, two hyperspheres intersect orthogonally if and only if $\langle s, \tilde{s} \rangle = -\frac{1}{2}\{s, \tilde{s}\} = 0$ or, equivalently, $s\tilde{s} = s \wedge \tilde{s} \in \Lambda^2 R_1^{n+2}$. Thus, describing a k-sphere $\mathfrak{s} = s_1 \cdots s_{n-k}$ in terms of an orthonormal basis of the corresponding subspace in R_1^{n+2} means geometrically to describe it as the orthogonal intersection of the $n - k$ hyperspheres s_i.

Now, we know that a point p lies on a k-sphere, given as the orthogonal intersection of $n - k$ hyperspheres, $\mathfrak{s} = s_1 \cdots s_{n-k}$ with $s_i \perp s_j$ for $i \neq j$, if and only if $p \perp s_i$ for all $i = 1, \ldots, n - k$. Hence, checking (6.2), we see that incidence is equivalent to $p\mathfrak{s} = p \wedge \mathfrak{s} \in \Lambda^{n-k+1} R_1^{n+2}$. In terms of the skew and symmetric parts of the product, this can be expressed as

$$\begin{aligned} [p, \mathfrak{s}] &= 0 \quad \text{for} \quad n - k \equiv 0 \bmod 2, \\ \{p, \mathfrak{s}\} &= 0 \quad \text{for} \quad n - k \equiv 1 \bmod 2. \end{aligned}$$

If, on the other hand, the sphere is given as a pure timelike $(k + 2)$-vector $\tilde{\mathfrak{s}} = v\mathfrak{s} \in \Lambda^{k+2} R_1^{n+2}$ (see §6.4.8), then the point p lies on that k-sphere if and only if p is in the corresponding Minkowski $(k + 2)$-plane, (cf., §6.4.6). Hence $p \in \tilde{\mathfrak{s}}$ lies on the k-sphere $\tilde{\mathfrak{s}}$ if and only if $p \wedge \tilde{\mathfrak{s}} = 0 \Leftrightarrow p \cdot \tilde{\mathfrak{s}} \in \Lambda^{k+1} R_1^{n+2}$.

6.4.12 Lemma. *A point $p \in R_1^{n+2}$, $p^2 = 0$, of the conformal n-sphere lies on a k-sphere $\mathfrak{s} \in \Lambda^{n-k} R_1^{n+2}$ if and only if $p\mathfrak{s} \in \Lambda^{n-k+1} R_1^{n+2}$.*

If $\mathfrak{s} = s_1 \cdots s_{n-k}$ is given as the orthogonal intersection of hyperspheres s_i, this is equivalent to saying that $p \perp s_i$ for all i.

If the sphere is given by a timelike $(k + 2)$-blade $\tilde{\mathfrak{s}} = v\mathfrak{s} \in \Lambda^{k+2} R_1^{n+2}$, then the point p lies on $\tilde{\mathfrak{s}}$ if and only if $p\tilde{\mathfrak{s}} \in \Lambda^{k+1} R_1^{n+2}$.

6.4.13 Example. We have only considered the intersection angle for hyperspheres so far — a similar ansatz that one might expect to provide a notion of intersection angle for k-spheres $\mathfrak{s}$ and $\tilde{\mathfrak{s}}$ does not work in the general case, as the following example shows.[14]

Consider two orthogonal 2-spheres $s_1, \tilde{s}_1 \in \Lambda^1 R_1^5$ in the conformal 3-sphere, and let $s_2, \tilde{s}_2 \in \Lambda^1 R_1^5$ be two other 2-spheres that both intersect s_1 and $\tilde{s}_1$ orthogonally (see Figure 6.1). Computing

$$\langle s_1 s_2, \tilde{s}_1 \tilde{s}_2 \rangle = \det \begin{pmatrix} \langle s_1, \tilde{s}_1 \rangle & \langle s_1, \tilde{s}_2 \rangle \\ \langle s_2, \tilde{s}_1 \rangle & \langle s_2, \tilde{s}_2 \rangle \end{pmatrix} = \det \begin{pmatrix} 0 & 0 \\ 0 & * \end{pmatrix} = 0,$$

---

[13] Remember that two hyperspheres intersect if and only if s and $\tilde{s}$ define an elliptic or parabolic sphere pencil, that is, if and only if $\langle s, \tilde{s} \rangle^2 \leq |s|^2 |\tilde{s}|^2$ (see §1.2.1).

[14] For a set of invariants, "generalized stationary angles," which describes the relative position of two spheres of arbitrary (possibly different) codimensions, the reader is referred to [270].

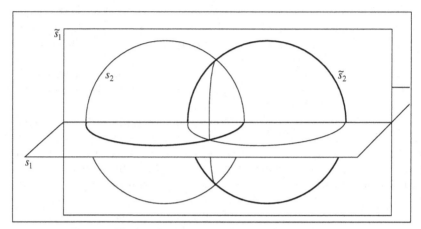

Fig. 6.1. The "intersection angle" of two circles in the 3-sphere

it becomes clear that the scalar product of $\mathfrak{s} = s_1 s_2$ and $\tilde{\mathfrak{s}} = \tilde{s}_1 \tilde{s}_2$ cannot be used (without further information) to determine the intersection angle: even though the scalar product vanishes, the two circles may not even intersect.

If, however, we *assume* the two circles to lie in a common 2-sphere, then this scalar product indeed gives the intersection angle of the two circles (as hyperspheres in that 2-sphere):

6.4.14 Lemma. *The angle α between two k-spheres $\mathfrak{s}, \tilde{\mathfrak{s}} \in \Lambda^{n-k} \mathbb{R}_1^{n+2}$ that lie in a common $(k+1)$-sphere is given by $|\mathfrak{s}|^2 |\tilde{\mathfrak{s}}|^2 \cos^2 \alpha = \langle \mathfrak{s}, \tilde{\mathfrak{s}} \rangle^2$.*

6.5 The cross-ratio

As we already discussed in Section 4.9, the (complex) cross-ratio is an important invariant of four points in the conformal n-sphere — four points always lie on some 2-sphere that we consider (after equipping it with an orientation) as a Riemann sphere so that this cross-ratio can be defined. This invariant plays a key role in discrete theories as, for example, in the theory of discrete principal or isothermic nets (cf., §5.7.2) or, more generally, in the theory of discrete orthogonal nets that we will discuss in the Chapter 8.

In our present Clifford algebra setup, the cross-ratio is given by a simple expression that we want to derive in this section — in fact, the expression that we will obtain seems simpler than in the quaternionic case, and it seems to contain more geometric information (cf., §4.9.8).

6.5.1 Contemplate (6.5): If $p_i \in L^{n+1} \subset \Lambda^1 \mathbb{R}_1^{n+2}$, $i = 1, \ldots, 4$, denote four points of the conformal n-sphere, $p_i^2 = 0$, that are in "general position," that is, not concircular, then the "imaginary part" $p_1 \wedge p_2 \wedge p_3 \wedge p_4 \neq 0$ of

$$\tfrac{1}{2}(p_1 p_2 p_3 p_4 + p_4 p_3 p_2 p_1) \in \Lambda^0 \mathbb{R}_1^{n+2} \oplus \Lambda^4 \mathbb{R}_1^{n+2}$$

determines the 2-sphere $\mathfrak{s} \in \Lambda^4 \mathbb{R}_1^{n+2}$ that contains the four points; if, on the other hand, it vanishes, then the four points lie on some circle, so that the 2-sphere containing them is no longer unique.

6.5.2 However we cut it, four points p_i always lie on a 2-sphere $S^2 \subset S^n$ (unique or not) so that there is always a Minkowski $\mathbb{R}_1^4 \subset \mathbb{R}_1^{n+2}$ that contains the four isotropic vectors $p_i \in \Lambda^1 \mathbb{R}_1^{n+2}$. Introducing an orthonormal basis (e_0, e_1, e_2, e_3), $e_i \perp e_i$ and $e_0^2 = 1 = -e_i^2$, on that Minkowski $\mathbb{R}_1^4$, we can isometrically parametrize a parabolic section (cf., §1.4.3) of its light cone by

$$\mathbb{C} \cong \mathbb{R}^2 \ni (x, y) \mapsto (1 + \tfrac{x^2+y^2}{4}, x, y, 1 - \tfrac{x^2+y^2}{4}) \in Q_0 \subset L^3 \subset \mathbb{R}_1^4.$$

With this, we compute $\{p, \tilde{p}\} = (x - \tilde{x})^2 + (y - \tilde{y})^2 = |z - \tilde{z}|^2$, where $z = x + iy$. Thus the "real part" of $2(p_1 p_2 p_3 p_4 + p_4 p_3 p_2 p_1)$ computes to

$$
\begin{aligned}
&\{p_1, p_2\}\{p_3, p_4\} - \{p_1, p_3\}\{p_2, p_4\} + \{p_1, p_4\}\{p_2, p_3\} \\
&= \ (|z_1 - z_2|^2 |z_3 - z_4|^2 - |z_1 - z_3|^2 |z_2 - z_4|^2 + |z_1 - z_4|^2 |z_2 - z_3|^2) \\
&= \ 2\mathrm{Re}\left[(z_1 - z_2)\overline{(z_2 - z_3)}(z_3 - z_4)\overline{(z_4 - z_1)}\right] \\
&= \ 2\mathrm{Re}\,\tilde{q},
\end{aligned}
$$

where $(z_1 - z_2)(z_3 - z_4) - (z_2 - z_3)(z_4 - z_1) = (z_1 - z_3)(z_2 - z_4)$ provides the second equality and $\tilde{q} := (z_1 - z_2)\overline{(z_2 - z_3)}(z_3 - z_4)\overline{(z_4 - z_1)}$. To determine the "imaginary part" (up to sign) we compute its norm:

$$|p_1 \wedge p_2 \wedge p_3 \wedge p_4|^2 = \det(-\tfrac{1}{2}|z_i - z_j|^2)_{i,j=1,\ldots,4} = \tfrac{1}{4}[(\mathrm{Re}\,\tilde{q})^2 - |\tilde{q}|^2] \leq 0.$$

Now, denoting $v_4 := e_0 e_1 e_2 e_3$, we have $|v_4|^2 = -1$, so that

$$\tfrac{1}{2}(p_1 p_2 p_3 p_4 + p_4 p_3 p_2 p_1) \ = \ \tfrac{1}{2}(\mathrm{Re}\,\tilde{q} \pm v_4 \mathrm{Im}\,\tilde{q}) \ \simeq \ \tfrac{1}{2}\tilde{q}: \qquad (6.11)$$

Note that $v_4^2 = -1$, so that $\Lambda^0 \mathbb{R}_1^4 \oplus \Lambda^4 \mathbb{R}_1^4 \cong \mathbb{C}$. The last identification in (6.11) involves a choice of orientation, $\pm v_4 \simeq \sqrt{-1}$.

6.5.3 As a consequence of (6.11) we obtain a formula for the cross-ratio of the four points:

$$\frac{z_1 - z_2}{z_2 - z_3} \frac{z_3 - z_4}{z_4 - z_1} \ = \ \frac{\tilde{q}}{|z_2 - z_3|^2 |z_4 - z_1|^2} \ \simeq \ \frac{p_1 p_2 p_3 p_4 - p_4 p_3 p_2 p_1}{\{p_2, p_3\}\{p_4, p_1\}}.$$

Clearly this last expression does not depend on the choice of homogeneous coordinates for the p_i, so that it is a conformal invariant of the four points;

the "direction" of the imaginary part $p_1 \wedge p_2 \wedge p_3 \wedge p_4$ will generally change under a conformal transformation $\mu \in Spin_1(n+2)$ though, since the 2-sphere containing the four points does not need to remain fixed.

6.5.4 Note that, on our way to this formula, we expressed the cross-ratio entirely in terms of scalar products,[15]

$$\frac{p_1p_2p_3p_4 - p_4p_3p_2p_1}{(p_2p_3 + p_3p_2)(p_4p_1 + p_1p_4)}$$
$$= \frac{\langle p_1,p_2\rangle\langle p_3,p_4\rangle - \langle p_1,p_3\rangle\langle p_2,p_4\rangle + \langle p_1,p_4\rangle\langle p_2,p_3\rangle + \sqrt{\det(\langle p_i,p_j\rangle)}}{2\langle p_1,p_4\rangle\langle p_2,p_3\rangle},$$

so that it can be given in the classical model (without referring to the Clifford multiplication). However, the geometric interpretation of the imaginary part as the 2-sphere containing the four points relies on the Clifford algebra model.

6.5.5 Lemma. *The cross-ratio of four points* $p_1, \dots, p_4 \in S^n$ *is given by*

$$[p_1; p_2; p_3; p_4] \quad = \quad \frac{p_1p_2p_3p_4 - p_4p_3p_2p_1}{(p_2p_3 + p_3p_2)(p_4p_1 + p_1p_4)}.$$

Modulo direction of the $\Lambda^4 \mathbb{R}_1^{n+2}$*-part of the Clifford product — which determines the 2-sphere that contains the four points — this expression is a Möbius invariant of the four points.*

6.6 Sphere congruences and their envelopes

Finally we turn to differential geometry: We want to formulate the conditions for a sphere congruence to be enveloped by a map. The enveloping conditions for a hypersphere congruence can be reformulated directly in terms of the Clifford algebra model (see §1.6.3):

6.6.1 Lemma. *A differentiable map* $f : M^m \to L^{n+1} \subset \Lambda^1 \mathbb{R}_1^{n+2}$ *envelopes a sphere congruence* $s : M^m \to S_1^{n+1} \subset \Lambda^1 \mathbb{R}_1^{n+2}$ *if and only if*

$$\{f, s\} = 0 \quad and \quad \{df, s\} \equiv 0.$$

6.6.2 However, we now want to extend this notion of envelope to congruences of k-spheres, where $k \geq m$: Since, geometrically, "envelope" should mean that the tangent plane of f at some point $f(p)$ is contained in the tangent plane of the sphere at that point (compare §4.7.9), $k < m$ would mean that the map f cannot be an immersion.

---

[15] Note that $\det(\langle p_i, p_j \rangle) < 0$, so that $\sqrt{\det(\langle p_i, p_j \rangle)} \in i\mathbb{R}$.

6.6.3 Definition. *A differentiable map* $f : M^m \to S^n$ *is said to envelope a congruence* $\mathfrak{s} : M^m \to \Lambda^{n-k} \mathbb{R}^{n+2}_1$ *of k-spheres,* $k \geq m$, *if, for all* $p \in M$,

$$f(p) \in \mathfrak{s}(p) \quad and \quad T_{f(p)} f(M) \subset T_{f(p)} \mathfrak{s}(p).$$

6.6.4 Note that, as in §1.6.1, we consider the tangent planes of $f(M)$ and the k-sphere $\mathfrak{s}(p) \subset S^n$ at $f(p)$ as projective planes in $\mathbb{RP}^{n+1}$, not referring to any linear structure.

We already studied the incidence of a point and a k-sphere in §6.4.12. There the main idea was to describe a k-sphere $\mathfrak{s} = s_1 \cdots s_{n-k}$ as the orthogonal intersection of $n - k$ hyperspheres s_i. With this description, the tangent plane of f at some point $f(p) \in \mathfrak{s}(p)$ will be contained in the tangent plane of $\mathfrak{s}(p)$ at $f(p)$ if and only if it is contained in all tangent planes of the spheres $s_i(p)$ (cf., §4.7.9). Combining this observation with §1.6.3, the same argument as the one that gave §6.4.12 now provides:

6.6.5 Lemma. *A differentiable map* $f : M^m \to S^n$ *envelopes a k-sphere congruence* $\mathfrak{s} : M^m \to \Lambda^{n-k} \mathbb{R}^{n+2}_1$ *if and only if*

$$\mathfrak{s}f : M \to \Lambda^{n-k+1} \mathbb{R}^{n+2}_1 \quad and \quad \mathfrak{s}df : TM \to \Lambda^{n-k+1} \mathbb{R}^{n+2}_1.$$

In case $\mathfrak{s} = s_1 \cdots s_{n-k}$ *is given in terms of* $n-k$ *congruences of orthogonally intersecting hyperspheres, this is equivalent to*

$$\{s_i, f\} = 0 \quad and \quad \{s_i, df\} \equiv 0 \quad for \quad i = 1, \ldots, n-k.$$

6.6.6 We already mentioned the term of a "contact element" in the context of §6.4.10. Suppose $f : M^m \to S^n$ is an immersion and $(v_1, \ldots, v_m)$ is a basis of $T_p M$, so that the $d_p f(v_i)$ form an orthonormal basis.[16] Then the tangent plane of f at p, as a projective m-plane in $\mathbb{RP}^{n+1}$, is given by

$$\mathfrak{t}(p) := f(p) d_p f(v_1) \cdots d_p f(v_m) \in \Lambda^{m+1} \mathbb{R}^{n+2}_1. \tag{6.12}$$

If $\mathfrak{s}(p) = (s_1 \cdots s_{n-m})(p) \in \Lambda^{n-m} \mathbb{R}^{n+2}_1$ is now an m-sphere that contains $f(p)$, then the tangent planes $\mathfrak{t}(p)$ of $f(M)$ and of $\mathfrak{s}(p)$ at $f(p)$ coincide (note that, by our assumptions, they have the same dimension) if and only

---

[16] This depends on the lift of $f : M^m \to L^{n+1} \subset \mathbb{R}^{n+2}_1$. Remember that $f : M^m \to S^n$ is an immersion if and only if any lift $f : M^m \to L^{n+1}$ induces a definite metric or, otherwise said, f and df span an $(m+1)$-dimensional subspace of the tangent space $f^\perp$ to the light cone L^{n+1} at f (see Figure T.2).

However, $d(e^u f) = e^u (df + du \cdot f)$, so that the vectors of an orthonormal basis scale with the reciprocal factor and the contact element $\mathfrak{t}$ scales with f: If $\tilde{f} = e^u f$, then $\tilde{\mathfrak{t}} = e^u \mathfrak{t}$.

if the subspaces given by $\mathfrak{t}(p)$ and $(\mathfrak{s}f)(p) \in \Lambda^{n-m+1}\mathbb{R}_1^{n+2}$ are orthogonal complements of each other. Thus, with §6.4.10, we deduce the following:

6.6.7 Lemma. *An immersion $f : M^m \to S^n$ envelopes an m-sphere congruence $\mathfrak{s} : M^m \to \Lambda^{n-m}\mathbb{R}_1^{n+2}$ if and only if $\mathfrak{t}(p) \parallel v \cdot (\mathfrak{s}f)(p)$.*

6.6.8 Geometrically, this means that $v(\mathfrak{s}f)(p) = (v\mathfrak{s}(p)) \cdot f(p)$ gives the tangent plane (or contact element) of $\mathfrak{s}(p)$ at $f(p)$.

Or, conversely, we may interpret $v\mathfrak{t}(p)$ as the family of all m-spheres that touch $f(M)$ at $f(p)$. In the case $m = n - 1$ of a hypersphere congruence enveloped by a hypersurface, $v\mathfrak{t}(p) \in \Lambda^2\mathbb{R}_1^{n+2}$ clearly defines the parabolic pencil of hyperspheres that touch $f(M)$ at $f(p)$ (see Figure T.1). In the general case, on the other hand, any choice of a spacelike $(n-m)$-dimensional subspace of the $(n - m + 1)$-dimensional space given by $v\mathfrak{t}(p)$ provides us with an m-sphere touching $f(M)$ at $f(p)$. Because the condition to be spacelike is an open condition, this leaves us with an $(n - m)$-parameter family of touching m-spheres.

For example, given one touching m-sphere $\mathfrak{s}(p) = (s_1 \cdots s_{n-m})(p)$ in terms of orthogonally intersecting hyperspheres, this $(n - m)$-parameter family of m-spheres may be parametrized by

$$(\lambda_1, \ldots, \lambda_{n-m}) \mapsto (s_1(p) + \lambda_1 f(p), \ldots, s_{n-m}(p) + \lambda_{n-m} f(p)).$$

These observations may justify the following:

6.6.9 Definition. *We will call any isotropic $(m + 1)$-blade $\mathfrak{t} \in \Lambda^{m+1}\mathbb{R}_1^{n+2}$ an m-dimensional contact element in the conformal n-sphere S^n.*

6.6.10 Note that the (unique) point $p \in S^n$ of contact of all m-spheres of a contact element $\mathfrak{t}$ is given by the intersection of the light cone L^{n+1} with the $(m + 1)$-dimensional space given by $\mathfrak{t}$. Otherwise said, it is the unique isotropic direction in that subspace.

Since we will not use the notion of contact elements much, we leave further investigations to the reader; check [66] for more information.

6.7 Frames

To conclude our presentation of this model for Möbius differential geometry, we want to briefly discuss the spinor frame equations, and how they relate to the classical frame equations.

First we slightly generalize the notion of an adapted frame from §1.7.1:

6.7.1 Definition. *Suppose we are given a map $f : M^m \to L^{n+1}$ along with an m-sphere congruence $\mathfrak{s} : M^m \to \Lambda^{n-m}\mathbb{R}_1^{n+2}$, and let $(e_0, e_1, \ldots, e_n, e_\infty)$*

denote a pseudo-orthonormal basis on $\mathbb{R}^{n+2}_1$, *that is,*

$$\langle e_i, e_j \rangle = -\tfrac{1}{2}\{e_i, e_j\} = \begin{pmatrix} 0 & 0 & -\tfrac{1}{2} \\ 0 & E_n & 0 \\ -\tfrac{1}{2} & 0 & 0 \end{pmatrix}.$$

A map $\mathfrak{z} : M^m \to Spin_1(n+2)$ *will be called an adapted frame for* $(\mathfrak{s}, f)$ *if*

$$\mathfrak{s} = s_{m+1} \cdots s_n = \mathfrak{z}e_{m+1} \cdots e_n \mathfrak{z}^{-1} \quad and \quad f = \mathfrak{z}e_0\mathfrak{z}^{-1},$$

where $s_i := \mathfrak{z}e_i\mathfrak{z}^{-1}$. *Moreover,* $\mathfrak{z}$ *will be called*
* $\mathfrak{s}$-*adapted if* $\forall p \in M : d_p s_i(T_pM) = span\{\mathfrak{z}(p)e_j\mathfrak{z}^{-1}(p) \,|\, j = 1, \ldots, m\}$
for $i = m+1, \ldots, n$, *and*
* f-*adapted if* $\forall p \in M : d_p f(T_pM) = span\{\mathfrak{z}(p)e_j\mathfrak{z}^{-1}(p) \,|\, j = 1, \ldots, m\}$.

6.7.2 In the case $m = n-1$, this is just the Clifford algebra reformulation of the definition in §1.7.1 of an $(f$-, $\mathfrak{s}$-) adapted frame for the strip $(\mathfrak{s}, f)$.

Note that, for $m < n-1$, the condition on a frame to be $\mathfrak{s}$-adapted is rather restrictive. We will discuss this in detail in a moment. However, this is the kind of frame that we will be dealing with when discussing (Ribaucour pairs of) m-orthogonal systems in Chapter 8.

6.7.3 Compatibility Conditions. Remember that, from §6.3.10, the connection form

$$\varphi := \mathfrak{z}^{-1}d\mathfrak{z} : TM \to \Lambda^2\mathbb{R}^{n+2}_1$$

of a frame $\mathfrak{z} : M \to Spin_1(n+2)$ acts on $\Lambda^1\mathbb{R}^{n+2}_1 = \mathbb{R}^{n+2}_1$ by $e_i \mapsto [\varphi, e_i]$, that is, $d(\mathfrak{z}e_i\mathfrak{z}^{-1}) = \mathfrak{z}[\varphi, e_i]\mathfrak{z}^{-1}$.

In particular, $df = \mathfrak{z}[\varphi, e_0]\mathfrak{z}^{-1}$ and $ds_i = \mathfrak{z}[\varphi, e_i]\mathfrak{z}^{-1}$ for $i = 1, \ldots, n$. Taking second derivatives, we obtain

$$\begin{aligned}
d^2(\mathfrak{z}e_i\mathfrak{z}^{-1})(x,y) \\
&= \mathfrak{z}([d\varphi(x,y), e_i] + [\varphi(x), [\varphi(y), e_i]] - [\varphi(y), [\varphi(x), e_i]])\mathfrak{z}^{-1} \\
&= \mathfrak{z}[d\varphi(x,y) + [\varphi(x), \varphi(y)], e_i]\mathfrak{z}^{-1} \\
&= \mathfrak{z}[(d\varphi + \tfrac{1}{2}[\varphi \wedge \varphi])(x,y), e_i]\mathfrak{z}^{-1}
\end{aligned}$$

when we set[17] $[\varphi \wedge \psi](x,y) := [\varphi(x), \psi(y)] - [\varphi(y), \psi(x)]$ for $x, y \in TM$. Thus the compatibility condition for the connection form φ becomes the Maurer-Cartan equation (cf., §1.7.4)

$$d\varphi + \tfrac{1}{2}[\varphi \wedge \varphi] = 0,$$

---

[17] Note that, in this context, "$\wedge$" has a different meaning than previously in this chapter: It refers to the $\wedge$-product of 1-forms rather than to the exterior product of elements of $\Lambda\mathbb{R}^{n+2}_1$.

since the adjoint representation $\Lambda^2 \mathbb{R}_1^{n+2} \ni \mathfrak{v} \mapsto [\mathfrak{v}, .] \in \mathfrak{o}_1(n+2)$ is a Lie algebra isomorphism, so that $[\mathfrak{v}, .] = 0$ implies $\mathfrak{v} = 0$ (see §6.3.10).

6.7.4 Writing this Lie algebra isomorphism $\Lambda^2 \mathbb{R}_1^{n+2} \to \mathfrak{o}_1(n+2)$ in terms of the basis $(e_0, \ldots, e_\infty)$ allows similar (the same when $m = n - 1$) interpretations of the components of φ as in §1.7.6 and §1.7.7. Thus we will only give some simple discussions and leave a detailed analysis to the reader.

First let us define the components of the connection form,[18] as in (1.1):

$$
\begin{aligned}
ds_j &= \sum_{i=1}^m s_i \omega_{ij} &+& \sum_{i=m+1}^n s_i \psi_{ij} &+& 2f\hat{\omega}_j &+& 2\hat{f}\omega_j, \\
ds_k &= -\sum_{i=1}^m s_i \psi_{ki} &+& \sum_{i=m+1}^n s_i \nu_{ik} &+& 2f\hat{\omega}_k^\perp &+& 2\hat{f}\omega_k^\perp, \\
df &= \sum_{i=1}^m s_i \omega_i &+& \sum_{i=m+1}^n s_i \omega_i^\perp &+& f\nu &+& 0, \\
d\hat{f} &= \sum_{i=1}^m s_i \hat{\omega}_i &+& \sum_{i=m+1}^n s_i \hat{\omega}_i^\perp &+& 0 &-& \hat{f}\nu,
\end{aligned}
$$
$$(6.13)$$

where $j = 1, \ldots, m$ and $k = m + 1, \ldots, n$. That is, for example, we have $\langle [\varphi, e_i], e_j \rangle = -\omega_{ij}$ for $i, j = 1, \ldots, m$ and $\langle [\varphi, e_0], e_j \rangle = \omega_j$.

The condition of §6.6.5 for f to envelope the congruence $\mathfrak{s} = s_{m+1} \cdots s_n$ of m-spheres now reads $\omega_k^\perp = 0$, and $\mathfrak{z}$ is

* f-adapted if and only if, in addition to $\omega_k^\perp = 0$, $\nu = 0$, whereas it is

* $\mathfrak{s}$-adapted if and only if, in addition to $\omega_k^\perp = 0$, $\nu_{ij} = \hat{\omega}_j^\perp = 0$.

Note that the ν_{ij}'s were not present in the codimension 1 case $m = n - 1$.

6.7.5 As mentioned above, the condition on a frame to be $\mathfrak{s}$-adapted is rather restrictive, and, in general, there will be no $\mathfrak{s}$-adapted frame for a given m-sphere congruence $\mathfrak{s}$ when $m < n - 1$.

First, in order to achieve $\omega_i^\perp = \hat{\omega}_i^\perp = 0$, the congruence has to have two envelopes f and $\hat{f}$ — in the codimension 1 case, this was guaranteed as soon as the metric induced by $\mathfrak{s}$ was spacelike (see §1.6.9).

A second condition is that the vector bundle given by $\mathfrak{s}$ has to be flat: Writing $\mathfrak{s}$ in terms of any orthonormal basis fields, $\mathfrak{s} = s_{m+1} \cdots s_n$, this is the vector bundle $\text{span}\{s_{m+1}, \cdots, s_n\}$. In terms of the above forms ν_{ij}, given in (6.13), flatness is expressed by $\varrho_{ij}^\perp := d\nu_{ij} + \sum_k \nu_{ik} \wedge \nu_{kj} = 0$. If these equations hold, then, locally, there is another choice of orthonormal frame fields that yield parallel basis fields for the bundle.

The Ricci equations — we interpret the bundle $\text{span}\{s_{m+1}, \cdots, s_n, f, \hat{f}\}$ as the "normal bundle" and its orthogonal complement, spanned by the orthonormal vector fields s_i, $i = 1, \ldots, m$, as the "tangent bundle" (see

---

[18] Note that there appear certain 2's instead of -1's as in (1.1): The reason is the different relative normalization of e_0 and e_∞, $\langle e_0, e_\infty \rangle = -\frac{1}{2}$ instead of $\langle e_0, e_\infty \rangle = 1$, as in §1.7.4.

below) — now come in four flavors[19]:

$$
\begin{aligned}
0 &= d\omega_i^\perp + \sum_{k=m+1}^{n} \nu_{ik} \wedge \omega_k^\perp + \omega_i^\perp \wedge \nu + \sum_{j=1}^{m} \psi_{ij} \wedge \omega_j, \\
0 &= d\hat\omega_i^\perp + \sum_{k=m+1}^{n} \nu_{ik} \wedge \hat\omega_k^\perp - \hat\omega_i^\perp \wedge \nu + \sum_{j=1}^{m} \psi_{ij} \wedge \hat\omega_j, \\
\tfrac{1}{2} d\nu &= \sum_{j=1}^{m} \omega_j \wedge \hat\omega_j + \sum_{k=m+1}^{n} \omega_k^\perp \wedge \hat\omega_k^\perp, \\
\varrho_{ij}^\perp &= \sum_{k=1}^{m} \psi_{ik} \wedge \psi_{jk} - 2(\omega_i^\perp \wedge \hat\omega_j^\perp + \hat\omega_i^\perp \wedge \omega_j^\perp).
\end{aligned}
\tag{6.14}
$$

If the sphere congruence $\mathfrak{s}$ has two envelopes f and $\hat f$, then, as discussed above, the forms $\omega_k^\perp = \hat\omega_k^\perp = 0$, so that the bundle given by $\mathfrak{s}$ is flat if and only if $0 = \varrho_{ij}^\perp = \sum_{k=1}^{m} \psi_{ik} \wedge \psi_{jk}$ for $i,j = m+1, \ldots, n$. Otherwise said, the bundle given by $\mathfrak{s}$ is flat if and only if the bilinear forms $\langle ds_i^T, ds_j^T \rangle$, where $ds_i^T = \sum_{j=1}^{m} \langle ds_i, s_j \rangle s_j$ denotes the "tangential part" of ds_i, are symmetric. Clearly this condition goes away in the codimension 1 case $m = n - 1$.

6.7.6 In case $\mathfrak{z}$ is f-adapted, the ω_i's and ω_{ij}'s provide the induced metric and Levi-Civita connection of $f : M^m \to L^{n+1}$, as in §1.7.7, and the ψ_{ki}'s give the Weingarten tensors with respect to the s_k as unit normal fields to f, while ω_i and $\hat\omega_i$ provide the Weingarten tensors with respect to f and $\hat f$ as normal fields, as before. The new quantities ν_{ij} give, together with the $\hat\omega_i^\perp$, the normal connection of $f : M^m \to \mathbb{R}_1^{n+2}$.

In the case of an $\mathfrak{s}$-adapted frame, on the other hand, we may interpret the components of φ in terms of any immersed section s_k, $k = m+1, \ldots, n$, of the bundle $\mathfrak{s}$, which is flat since $\nu_{ij} = 0$: Here, the ψ_{ki}'s give the induced metric of s_k and the ω_{ij}'s its Levi-Civita connection, as before. The ψ_{ji}'s, where $j = m+1, \ldots, n$ with $j \neq k$, together with the ω_i's and $\hat\omega_i$'s provide the Weingarten tensors of s_k, and its normal connection is given by ν.

6.7.7 To conclude, we determine the induced metric of an m-sphere congruence $\mathfrak{s} : M^m \to \Lambda^{n-m}\mathbb{R}_1^{n+2}$ as an immersion into $\Lambda^{n-m}\mathbb{R}_1^{n+2}$. Here it is convenient to write $\mathfrak{s}$ in terms of orthonormal basis fields using the exterior product instead of the Clifford product,[20] $\mathfrak{s} = s_{m+1} \wedge \ldots \wedge s_n$. The derivative of $\mathfrak{s}$ is then given by

$$
d\mathfrak{s} = \sum_{i=m+1}^{n} s_{m+1} \wedge \ldots \wedge ds_i \wedge \ldots \wedge s_n : TM \to \Lambda^{n-m}\mathbb{R}_1^{n+2}.
$$

Now we first note that $ds_i \perp s_i$, so that only the "tangential part" ds_i^T of ds_i counts, that is, the part orthogonal to s_j, $j = m+1, \ldots, n$. Then

$$
\langle s_i, s_{m+1} \rangle = \ldots = \langle s_i, s_{i-1} \rangle = \langle s_i, ds_i \rangle = \langle s_i, s_{i+1} \rangle = \ldots = \langle s_i, s_n \rangle = 0,
$$

---

[19] Note that the fourth type did not occur in the codimension 1 setting (see (1.4)).

[20] Remember that, for orthogonal vectors, the Clifford and exterior products coincide. Using the Clifford product, $\Lambda^{n-m-2}\mathbb{R}_1^{n+2}$-terms would occur during computations that only vanish at the end.

so that all mixed terms vanish when expanding the induced metric:

$$
\begin{aligned}
\langle d\mathfrak{s}, d\mathfrak{s} \rangle \\
= \sum_{i=m+1}^{n} \langle s_{m+1} \wedge \ldots \wedge ds_i^T \wedge \ldots \wedge s_n, s_{m+1} \wedge \ldots \wedge ds_i^T \wedge \ldots \wedge s_n \rangle \\
= \sum_{i=m+1}^{n} \langle ds_i^T, ds_i^T \rangle
\end{aligned}
$$

becomes a superposition of the metrics induced by the (orthonormal) sections s_k used to represent $\mathfrak{s}$. Expanding this expression further by using the forms from (6.13), we find $\langle d\mathfrak{s}, d\mathfrak{s} \rangle = \sum_{i=m+1}^{n} \sum_{j=1}^{m} \psi_{ij}^2$, that is, we recover a multiple of the trace form — as it should be for a Grassmannian-valued map (see [140]).

It seems less straightforward to determine the remaining geometric quantities of the m-sphere congruence $\mathfrak{s}$ from the connection form (6.13) of a frame in this general setup.

6.7.8 To Summarize. Some selected identifications from the Clifford algebra setup for Möbius geometry that we discussed in this chapter are collected in Figure T.4; compare the tables in Figure T.1 and Figure T.2 for the classical model.

Chapter 7
A Clifford algebra model: Vahlen matrices

In this chapter we discuss an enhancement of our Clifford algebra model using "Vahlen matrices," that is, certain 2×2 matrices with Clifford algebra entries. In some sense, this approach is very similar to the quaternionic approach previously discussed, or to the description of Möbius transformations by complex 2×2 matrices. However, it is more similar when considering the quaternionic model for 3-dimensional Möbius geometry or the complex approach for the Möbius geometry of the circle. But even in this case there are certain differences: For example, the Clifford algebra used for 3-dimensional Möbius geometry is $2^5 = 32$-dimensional,[1] whereas the (real) vector space of quaternionic 2×2-matrices is only 16-dimensional — thus in the quaternionic setup we have never seen the full Clifford algebra of $\mathbb{R}_1^5$. Nevertheless, there will be many similarities, as, for example, the description of Möbius transformations by fractional linear transformations on $S^n \cong \mathbb{R}^n \cup \{\infty\}$.

After Vahlen's original paper [288], this 2×2 matrix Clifford algebra approach caught the interest of various authors (see, for example, [2] and [3], or [108]); some account of the history of this approach may be found in [137]. The splitting ("conformal split") of the Clifford algebra $\mathcal{A}\mathbb{R}_1^{n+2}$ that corresponds to the 2×2 matrix representation of Möbius transformations (and of the Clifford algebra) (see the beautiful paper [154]) provides a notion of stereographic projection, as is also suggested by the fact that Möbius transformations are described as linear fractional transformations on the Euclidean n-space $\mathbb{R}^n$. This additional structure turns out to be very useful in some context (see [47], where, to the present author's knowledge, this approach to Möbius geometry has been systematically applied to submanifold geometry for the first time).

The key idea in our introduction to the Vahlen approach to Möbius geometry will be to identify the (real) vector space of 2×2 matrices with entries from the Clifford algebra $\mathcal{A}\mathbb{R}^n$ of $\mathbb{R}^n$ with the Clifford algebra $\mathcal{A}\mathbb{R}_1^{n+2}$ of the coordinate vector space $\mathbb{R}_1^{n+2}$ of the classical model — to this end, we will have to identify $\mathbb{R}_1^{n+2}$ as a subspace of the Clifford algebra 2×2 matrices. With this background it will then be rather straightforward to "translate" results from our previous considerations on the Clifford algebra model for Möbius geometry into the matrix formalism. A central result will be Vahlen's theorem on the characterization of the Clifford group $\Gamma(\mathbb{R}_1^{n+2})$. This characterization will allow us to write Möbius transformations as linear

---

[1] However, the Clifford algebra of $\mathbb{H} \cong \mathbb{R}^4$ is $2^4 = 16$-dimensional and can be identified with the quaternionic 2×2 matrices, with $\mathbb{H} \cong \{ \begin{pmatrix} 0 & q \\ -\bar{q} & 0 \end{pmatrix} \} \subset M(2 \times 2, \mathbb{H})$ (see [47]).

fractional transformations on $S^n \cong \mathbb{R}^n \cup \{\infty\}$; a discussion of the pin and spin groups as well as of Möbius involutions and spheres will then just be an inevitable consequence. A second highlight of the Vahlen matrix approach will be a representation of the space $\mathfrak{P}$ of point pairs in S^n that is just as beautiful as in the quaternionic setup — and that makes this approach so well suited for the treatment of (pairs of) isothermic surfaces of arbitrary codimension.

Before revealing these ideas we will briefly recall relevant facts from the quaternionic model in order to motivate the matrix approach, on one hand, and to make apparent differences, on the other hand.

As usual, the discussions of this chapter are complemented by a table that summarizes some identifications and results; see Figure T.5.

The contents of this chapter are organized as follows:

Section 7.1. After giving a matrix representation for the Clifford algebra of Euclidean 3-space, relevant notions and facts from the quaternionic model that we discussed in Chapter 5 are recalled: the description of 2-spheres and points in S^3 by quaternionic matrices; the Minkowski coordinate space $\mathbb{R}^5_1$ as a subspace of the quaternionic Hermitian forms on $\mathbb{H}^2$; and the spin group as consisting of even products of inversions.

Section 7.2. This section is the heart of the matter: Here we describe the coordinate Minkowski space of n-dimensional Möbius geometry in terms of 2×2 matrices with entries in the Clifford algebra of $\mathbb{R}^n$. The subspaces $\Lambda^k \mathbb{R}^{n+2}_1$ of k-vectors are identified, and the order involution and the Clifford conjugation are written in terms of 2×2 matrices.

Then Vahlen's theorem characterizing the Clifford group is proved. As a consequence, we obtain representations of the pin and spin groups in terms of 2×2 matrices.

Section 7.3. Using that the Clifford group acts by Möbius transformations on $S^n = \mathbb{R}^n \cup \{\infty\}$, an action of the Möbius group by fractional linear transformations is obtained. Inversions are discussed as an example. The incidence of a point and a sphere is discussed, and the representation of spheres is reconsidered. Finally, the space of point pairs is described as a homogeneous space in the present setup.

Section 7.4. The notions of an adapted frame for a point pair map and a single map are introduced, and the effect of the respective gauge transformations on the connection form is established. As examples, the Euclidean frame of a map into Euclidean n-space is given, and a frame for a pair of maps into Euclidean space is given. These frames will be useful later.

Section 7.5. Various topics are touched on: We show how to construct the sphere congruence that is enveloped by a given map and contains the points of another map; this result is then applied to introduce the central sphere congruence in higher codimension. Further, a criterion for a point pair map to envelope a sphere congruence is given. At the end of this section the cross-ratio of for points is computed in affine coordinates.

Remark. This chapter is merely a continuation of the previous chapter that should, therefore, be read before the present chapter. Some background on the quaternionic formalism, as discussed in Chapter 4, should be helpful but is not required.

7.1 The quaternionic model revisited

This preliminary section shall serve as a motivation for those who are familiar with the quaternionic model that we discussed earlier.[2] More specifically, our goal will be to interpret the quaternionic model (in the 3-dimensional case) in terms of our first Clifford algebra model: Embedding the coordinate Minkowski space $\mathbb{R}_1^5$ of the classical model as a subspace into the endomorphisms of $\mathbb{H}^2$, we will find an algebraic structure that very much resembles the Clifford algebra setup. In particular, the action of the Möbius group will be identified. On the other hand, we have the action of the Möbius group on $S^3 \subset \mathbb{HP}^1$: This twofold appearance of the conformal 3-sphere is one of the distinguishing features of the quaternionic model. In the Clifford algebra model that we are going to discuss in this chapter, there will be no vector space taking the role of the space $\mathbb{H}^2$ of homogeneous coordinates for $S^4 \cong \mathbb{HP}^1$; however, we will still be able to describe Möbius transformations as fractional linear transformations on $S^n \cong \mathbb{R}^n \cup \{\infty\}$ in the same way as in the quaternionic model. In this way, we will still have the twofold description of the action of the Möbius group that made the quaternionic model so powerful, especially for the treatment of point pair maps.

Thus we start by discussing the Clifford algebra of Euclidean 3-space.

7.1.1 The Clifford Algebra of $\mathbb{R}^3$. The Clifford algebra $\mathcal{A}\mathbb{R}^3$ of the (Euclidean) $\mathbb{R}^3$ is a $2^3 = 8$-dimensional associative algebra that can be obtained by introducing a product on the Grassmann (or "exterior") algebra $\Lambda\mathbb{R}^3$ of $\mathbb{R}^3$ that combines the exterior and interior products (cf., §6.2.1). Using Hodge duality[3] (cf., §6.4.4) to identify $\Lambda^2\mathbb{R}^3 \cong \Lambda^1\mathbb{R}^3$,

$$\star e_1 = e_2 \wedge e_3, \quad \star e_2 = e_3 \wedge e_1, \quad \star e_3 = e_1 \wedge e_2,$$

---

[2] Those readers who have skipped the quaternionic model may, without loosing much, skip forward to the following section.

[3] On $\Lambda^1\mathbb{R}^3$ and $\Lambda^2\mathbb{R}^3$, the Clifford dual (see §6.4.3) differs from the Hodge dual by a sign.

where (e_1, e_2, e_3) denotes some (positively oriented) orthonormal basis, the exterior product of two vectors becomes the cross-product:

$$v \wedge w = \star(v \times w)$$

for $v, w \in \mathbb{R}^3$. This suggests a close relation of the Clifford multiplication with the quaternionic multiplication: Remember from §4.1.4 that quaternionic multiplication of two vectors $v, w \in \mathbb{R}^3 \cong \mathrm{Im}\, \mathbb{H}$ can be written in terms of the scalar and cross-products,

$$vw = -\langle v, w \rangle + v \times w.$$

Indeed, the Clifford algebra $\mathcal{A}\mathbb{R}^3$ of the Euclidean $\mathbb{R}^3$ can be realized as the diagonal quaternionic 2×2 matrices (equipped with matrix multiplication as the Clifford multiplication) with $\mathbb{R}^3 \cong \Lambda^1 \mathbb{R}^3$ sitting in it as follows:

$$\Lambda^1 \mathbb{R}^3 \cong \{ \begin{pmatrix} v & 0 \\ 0 & -v \end{pmatrix} \mid v \in \mathrm{Im}\, \mathbb{H} \} \subset \{ \begin{pmatrix} q_1 & 0 \\ 0 & q_2 \end{pmatrix} \mid q_1, q_2 \in \mathbb{H} \} \cong \mathcal{A}\mathbb{R}^3.$$

The order involution $\mathfrak{v} \mapsto \hat{\mathfrak{v}}$ and the Clifford conjugation $\mathfrak{v} \mapsto \bar{\mathfrak{v}}$ on $\mathcal{A}\mathbb{R}^3$ (see §6.3.1) are then given by $\hat{\mathfrak{v}} = s^{-1} \mathfrak{v}\, s$ and $\bar{\mathfrak{v}} = \mathfrak{v}^*$, where $s = \begin{pmatrix} 0 & 1 \\ 1 & 0 \end{pmatrix}$ and $\mathfrak{v}^*$ means quaternionic conjugation and transpose. With this we can now write the Clifford group (see §6.3.2):

$$\Gamma(\mathbb{R}^3) = \{ \begin{pmatrix} q & 0 \\ 0 & \pm q \end{pmatrix} \mid q \in \mathbb{H} \setminus \{0\} \},$$

that is, $\mathfrak{v} \in \Gamma(\mathbb{R}^3)$ iff $\mathfrak{v}^{-1}$ exists and $\mathfrak{v} \in \Lambda^0 \mathbb{R}^3 \oplus \Lambda^2 \mathbb{R}^3$ or $\mathfrak{v} \in \Lambda^1 \mathbb{R}^3 \oplus \Lambda^3 \mathbb{R}^3$, as well as the groups

$$Pin(\mathbb{R}^3) = \{ \begin{pmatrix} q & 0 \\ 0 & \pm q \end{pmatrix} \mid q \in \mathbb{H}, |q|^2 = 1 \} \subset Sl(2, \mathbb{H}) \quad \text{and}$$

$$Spin(\mathbb{R}^3) = \{ \begin{pmatrix} q & 0 \\ 0 & q \end{pmatrix} \mid q \in \mathbb{H}, |q|^2 = 1 \} \subset Sl(2, \mathbb{H})$$

in terms of our quaternionic setup. By §6.3.8, every $\mathfrak{s} \in Spin(\mathbb{R}^3)$ can be decomposed into vectors, $\mathfrak{s} = \prod_{i=1}^{2k} s_i$, with $s_i \in S^2 \subset \mathbb{R}^3$. The action of each s_i on $\mathbb{R}^3$ is a reflection, as discussed in §6.3.4, since $nvn = v - 2\langle v, n \rangle n$ for $v \in \mathbb{R}^3 \cong \mathrm{Im}\, \mathbb{H}$ and $n \in S^2 = \{ q \in \mathrm{Im}\, \mathbb{H} \mid q^2 = -1 \}$.

On the other hand, we considered $Sl(2, \mathbb{H})$ as acting (by Möbius transformations) on $S^4 \cong \mathbb{H}P^1$ (see §4.4.9). Writing $\mathbb{R}^3 \cong \{ \binom{v}{1} \mid v \in \mathrm{Im}\, \mathbb{H} \}$, the above action of a vector $n \in S^2$ on $\mathbb{R}^3$ by reflection in the orthogonal 2-plane is recovered:

$$\begin{pmatrix} n & 0 \\ 0 & -n \end{pmatrix} \begin{pmatrix} v \\ 1 \end{pmatrix} \mathbb{H} = \begin{pmatrix} nvn \\ 1 \end{pmatrix} \mathbb{H} = \begin{pmatrix} v - 2\langle v, n \rangle n \\ 1 \end{pmatrix} \mathbb{H}.$$

7.1.2 Spheres in S^3. Remember that we identified 3-spheres in $S^4 \cong \mathbb{HP}^1$ with certain quaternionic Hermitian forms (see §4.3.12); thus, after fixing a basis (e_∞, e_0) of $\mathbb{H}^2$, say $e_\infty = \binom{1}{0}$ and $e_0 = \binom{0}{1}$, we may identify

$$\mathbb{HP}^1 \supset S^3 \simeq \begin{pmatrix} 0 & 1 \\ 1 & 0 \end{pmatrix} \quad \Leftrightarrow \quad S^3(e_0, e_0) = S^3(e_\infty, e_\infty) = 0, \ S^3(e_0, e_\infty) = 1,$$

Then $S^3 = \{\binom{v}{1} \mid v \in \operatorname{Im} \mathbb{H}\} \cup \{e_\infty \mathbb{H}\}$ (see §4.3.7); in particular, $0 = e_0 \mathbb{H}$ and $\infty = e_\infty \mathbb{H}$ are points in S^3.

On the other hand, we identified 2-spheres in $S^4 \cong \mathbb{HP}^1$ with (Möbius) involutions $\mathcal{S} \in \operatorname{End}(\mathbb{H}^2)$, $\mathcal{S}^2 = -id$ (see §4.8.1). By §4.8.6, $\mathcal{S} \subset S^3$ if and only if $\mathcal{S}$ is symmetric with respect to S^3. Thus, writing an endomorphism $\mathcal{S} \in \mathfrak{S}(\mathbb{H}^2)$ as a matrix with respect to (e_∞, e_0) as a basis and symmetric with respect to S^3, we have[4]

$$\mathcal{S} = \begin{pmatrix} n & 2d \\ 0 & -n \end{pmatrix} \quad \text{or} \quad \mathcal{S} = \tfrac{1}{r} \begin{pmatrix} m & |m|^2 - r^2 \\ 1 & -m \end{pmatrix} :$$

In the first case $\mathcal{S}$ is a plane in $\mathbb{R}^3 \cong \{\binom{v}{1} \mid v \in \operatorname{Im} \mathbb{H}\}$, that is, a sphere containing ∞, with unit normal $n \in S^2$ and distance $d \in \mathbb{R}$ from the origin, and in the second case $\mathcal{S}$ is a sphere with center $m \in \operatorname{Im} \mathbb{H}$ and radius r.

Also, we identified a 2-sphere $\mathcal{S} \subset S^3$ with the 3-sphere $S \in \mathfrak{H}(\mathbb{H}^2)$ that intersects S^3 orthogonally in $\mathcal{S}$. Algebraically, this identification $\mathcal{S} \leftrightarrow S$ was given by $\mathcal{S} \mapsto \mathcal{J}S^3 = S^3(.,\mathcal{S}.)$ (see (4.22)). In terms of the matrix representations given above, we obtain (cf., §4.3.10; see also §5.5.17):

$$S = \mathcal{J}S^3 \simeq \begin{pmatrix} 0 & -n \\ n & 2d \end{pmatrix} \quad \text{or} \quad S = \mathcal{J}S^3 \simeq \tfrac{1}{r} \begin{pmatrix} 1 & -m \\ m & |m|^2 - r^2 \end{pmatrix}.$$

Below we will use this identification of certain endomorphisms that are symmetric with respect to S^3 and quaternionic Hermitian forms that are orthogonal to S^3 in the sense of §4.3.5 to obtain a description of the coordinate space $\mathbb{R}^5_1$ of the classical model of 3-dimensional Möbius geometry (cf., §1.1.2). However, we will first discuss the *points* of S^3 in some detail.

7.1.3 Points in S^3. The above identification of endomorphisms $\mathcal{S}$ with 2-spheres in the conformal 3-sphere S^3, given as a quaternionic Hermitian form, can be extended to points of the conformal 3-sphere: By sending, in the affine picture, the radius of a 2-sphere $r \to 0$ (while rescaling suitably), we will obtain a point,

$$r\mathcal{S} = r \cdot \tfrac{1}{r} \begin{pmatrix} m & |m|^2 - r^2 \\ 1 & -m \end{pmatrix} \to \begin{pmatrix} m & -m^2 \\ 1 & -m \end{pmatrix} = \mathcal{P}.$$

---

[4] The interested reader may want to verify that $\mathcal{S} \in \mathfrak{S}(\mathbb{H}^2)$ and $\mathcal{S}$ is symmetric with respect to S^3 as well as the geometric interpretations given below, as an exercise (cf., §4.3.10).

More invariantly, we may identify a point $p \in S^3$ with a nontrivial 2-step nilpotent endomorphism $\mathcal{P} \in \mathfrak{N}_2(\mathbb{H}^2)$ that is symmetric with respect to S^3:

$$S^3 \ni \begin{pmatrix} v \\ 1 \end{pmatrix} \mathbb{H} \leftrightarrow \mathbb{R} \begin{pmatrix} v & -v^2 \\ 1 & -v \end{pmatrix} \in \{\mathcal{P} \in \mathfrak{N}_2(\mathbb{H}^2) \mid S^3(.,\mathcal{P}.) = S^3(\mathcal{P}.,.)\}/\mathbb{R}.$$

Namely, if $\mathcal{P}$ is 2-step nilpotent, then[5] $\mathrm{im}\mathcal{P} = \ker\mathcal{P} =: \begin{pmatrix} q \\ 1 \end{pmatrix} \mathbb{H} \in \mathbb{H}P^1$ with a suitable quaternion $q \in \mathbb{H}$ (cf., §5.3.19). Hence $\mathcal{P} = \begin{pmatrix} q\lambda & -q\lambda q \\ \lambda & -\lambda q \end{pmatrix}$ with some other quaternion λ; symmetry with respect to S^3 then yields $\lambda \in \mathbb{R}$ and $q \in \mathrm{Im}\,\mathbb{H}$, so that $\mathrm{im}\mathcal{P} \in S^3$. In fact, $S^3(.,\mathcal{P}.)$ yields, up to a real factor, the quaternionic Hermitian form that we identified in (4.4) with a point $p := (e_\infty q + e_0 1)\mathbb{H} \in \mathbb{H}P^1$:

$$S^3(e_\infty, \mathcal{P}e_\infty) = 1, \quad S^3(e_\infty, \mathcal{P}e_0) = -q, \quad \text{and} \quad S^3(e_0, \mathcal{P}e_0) = -q^2.$$

7.1.4 The Minkowski Space $\mathbb{R}_1^5$. We have already represented the elements of Möbius geometry, 2-spheres and points of S^3, by endomorphisms of $\mathbb{H}^2$; our next goal is to recover the space $\mathbb{R}_1^5$ of homogeneous coordinates of the classical model. Remember that, in §4.3.5, we identified the coordinate Minkowski space $\mathbb{R}_1^6$ of 4-dimensional Möbius geometry with the space of quaternionic Hermitian forms on $\mathbb{H}^2$, with points and hyperspheres being the unit and isotropic vectors (Hermitian forms), respectively (cf., Figure T.1). The Minkowski space $\mathbb{R}_1^5$ is now obtained as the orthogonal complement of $S^3 \in \mathbb{R}_1^6 \cong \mathfrak{H}(\mathbb{H}^2)$ — the points in S^3 are the isotropic vectors orthogonal to S^3, and the 2-spheres in S^3 can be identified with those 3-spheres that intersect S^3 orthogonally, as discussed in §4.8.8. On the other hand, *any* quaternionic Hermitian form can, in the presence of the fixed quaternionic Hermitian form $S^3 \in \mathfrak{H}(\mathbb{H}^2)$, be identified with an endomorphism that is symmetric with respect to S^3: This is the way we obtained the representations of points and 2-spheres by endomorphisms above.

To extend this description to all of $\mathbb{R}_1^5$ we first note that, in our above matrix description, any endomorphism that is symmetric with respect to $S^3 \simeq \begin{pmatrix} 0 & 1 \\ 1 & 0 \end{pmatrix}$ can be written as $\mathcal{V} = \begin{pmatrix} q & s \\ r & \bar{q} \end{pmatrix}$ with $q \in \mathbb{H}$ and $r, s \in \mathbb{R}$. The (Minkowski) scalar product (4.1) of S^3 and $S^3(.,\mathcal{V}.) \simeq \begin{pmatrix} r & \bar{q} \\ q & s \end{pmatrix}$ is then given by

$$\langle \begin{pmatrix} 0 & 1 \\ 1 & 0 \end{pmatrix}, \begin{pmatrix} r & \bar{q} \\ q & s \end{pmatrix} \rangle = \tfrac{1}{2}(q + \bar{q}).$$

---

[5] In the case where $\mathrm{im}\mathcal{P} = \ker\mathcal{P} = \begin{pmatrix} 1 \\ 0 \end{pmatrix} \mathbb{H}$, the same computation yields $\mathcal{P} = \begin{pmatrix} 0 & 1 \\ 0 & 0 \end{pmatrix}$ up to a real factor.

Hence $S^3(.,\mathcal{V}.) \perp S^3$ if and only if $q \in \text{Im}\,\mathbb{H}$, that is, if and only if

$$\mathcal{V}^2 = \begin{pmatrix} rs + q^2 & r(q + \bar{q}) \\ s(q + \bar{q}) & rs + \bar{q}^2 \end{pmatrix} \in \mathbb{R} \cdot \begin{pmatrix} 1 & 0 \\ 0 & 1 \end{pmatrix}$$

is a multiple of the identity. In fact, $\mathcal{V}^2 = -|S^3(.,\mathcal{V}.)|^2\,id$, so that we have the identification

$$\mathbb{R}_1^5 \cong \{\mathcal{V} \in End(\mathbb{H}^2)\,|\,\mathcal{V} \text{ is symmetric w.r.t. } S^3, \mathcal{V}^2 = -|S^3(.,\mathcal{V}.)|^2\,id_{\mathbb{H}^2}\}.$$

Note how this generalizes the above characterizations of 2-spheres as endomorphisms with $\mathcal{S}^2 = -1$ and points as those satisfying $\mathcal{P}^2 = 0$.

7.1.5 The Möbius Group. By §6.3.8 we know that $Spin(\mathbb{R}_1^5)$ is generated by reflections (cf., §6.3.4). With the above identification of a point $p \in S^3$ with a nilpotent endomorphism $\mathcal{P} \in \mathfrak{N}_2(\mathbb{H}^2)$, via $p = \text{im}\mathcal{P} = \ker\mathcal{P}$, we find that the action of a 2-sphere $\mathcal{S} \in \mathfrak{S}(\mathbb{H}^2)$ on $\mathcal{P}$ is given by

$$\mathcal{S}\mathcal{P}\hat{\mathcal{S}}^{-1} = \mathcal{S}\mathcal{P}\mathcal{S} = \frac{-(v-m)^2}{r^2}\begin{pmatrix} m - r^2(v-m)^{-1} & -(m - r^2(v-m)^{-1})^2 \\ 1 & -(m - r^2(v-m)^{-1}) \end{pmatrix};$$

the first equality holds because $\mathcal{S}^2 = -id$ and $\hat{\mathcal{S}} = -\mathcal{S}$, and for the second equality we have assumed that $\mathcal{S}$ and $\mathcal{P}$ are given in terms of affine coordinates as above. Thus $\mathcal{S}$ acts on the points $\mathcal{P}$ of S^3 by inversion in $\mathcal{S}$, as we expected, since a Minkowski reflection in a Minkowski hyperplane of $\mathbb{R}_1^5$ is an inversion (see Figure T.1). Similarly, the action of $\mathcal{S} = \begin{pmatrix} n & 2d \\ 0 & -n \end{pmatrix}$ is a reflection in the plane given by $\mathcal{S}$. Thus we have learned that the group

$$Spin_1^+(5) = \{\textstyle\prod_{j=1}^{2k} \mathcal{S}_j\,|\,\mathcal{S}_j \in \mathfrak{S}(\mathbb{H}^2) \cap \mathbb{R}_1^5 = S_1^4\} \subset Sl(\mathbb{H}^2) \subset End(\mathbb{H}^2)$$

sits inside the endomorphisms of $\mathbb{H}^2$ and, since $Spin_1^+(5)$ is a double cover of the Möbius group $M\ddot{o}b^+(3)$ (see §6.3.12), how the (orientation-preserving) Möbius transformations act on $\mathbb{R}_1^5 \subset End(\mathbb{H}^2)$.

7.1.6 Thus we have recovered important objects of our previously discussed Clifford algebra model in terms of the quaternionic setup, as sitting inside the endomorphisms of $\mathbb{H}^2$: the coordinate Minkowski space $\mathbb{R}_1^5$ from the classical model of 3-dimensional Möbius geometry as well as the group of orientation-preserving Möbius transformations $M\ddot{o}b^+(3)$, or its universal cover $Spin_1^+(5)$.

However, it is clear that there will be no chance to recover the Clifford algebra $\mathcal{A}(\mathbb{R}_1^5)$ itself as a subalgebra of $End(\mathbb{H}^2)$, thought of as an algebra over the reals: Just note that $\mathcal{A}(\mathbb{R}_1^5)$ is a 2^5-dimensional algebra whereas $End(\mathbb{H}^2)$ is of (real) dimension $2^2 \times 4 = 16$. Thus, in developing our Vahlen matrix approach to Möbius geometry in this chapter, we will be led by various ideas from the setup that we analyzed in this section — but we will need to suitably adjust the ideas that we elaborated on here.

7.2 Vahlen matrices and the spin group

Our first goal is to describe the Clifford algebra of the coordinate Minkowski space of n-dimensional Möbius geometry in terms of 2×2 matrices,[6] elaborating on ideas from §7.1.4: The main issue will be to express the coordinate Minkowski space $\mathbb{R}_1^{n+2}$ in terms of suitable 2×2 matrices. Clifford multiplication[7] will then be given by matrix multiplication.

After clarifying the structure of the Clifford algebra so obtained, we will characterize the Clifford, pin, and spin groups. A more comprehensive treatment of the material (also covering the more involved situation of other signatures) can be found in [189] and in [224], which served as the main source for this section.

7.2.1 Consider the space $M(2 \times 2, \mathcal{A}\mathbb{R}^n)$ of 2×2 matrices with entries from the Clifford algebra of Euclidean n-space. Clearly this is a real 2^{n+2}-dimensional vector space, in the usual way. Equipped with matrix multiplication, where multiplication of the components is by Clifford multiplication in $\mathcal{A}\mathbb{R}^n$, it becomes a noncommutative algebra with unit $1 = \begin{pmatrix} 1 & 0 \\ 0 & 1 \end{pmatrix}$. As such, it qualifies as a candidate for the Clifford algebra of an $(n+2)$-dimensional ("vector"-) subspace V of $M(2 \times 2, \mathcal{A}\mathbb{R}^n)$. Since Clifford multiplication is supposed to be matrix multiplication on $M(2 \times 2, \mathcal{A}\mathbb{R}^n)$, the elements of V have to square to a multiple of 1, and V has to generate $\mathcal{A}V$: If we choose an orthonormal basis $(\mathcal{E}_0, \ldots, \mathcal{E}_{n+1})$ of V, with respect to the scalar product obtained from squaring elements in V, then all products of k different basis vectors must provide a basis for $\Lambda^k V$, and the products of $k = 0, \ldots, n+2$ different basis vectors provide a basis of $\mathcal{A}V = \oplus_{k=0}^{n+2} \Lambda^k V$ — remember that we introduced the Clifford algebra of a vector space V as its exterior algebra, equipped with a different product that depends on the scalar product on V (see Section 6.2).

7.2.2 Lemma. $M(2 \times 2, \mathcal{A}\mathbb{R}^n) = \mathcal{A}V$ is the Clifford algebra of the $(n+2)$-dimensional Minkowski space[8]

$$V := \left\{ \begin{pmatrix} v & v_\infty \\ v_0 & -v \end{pmatrix} \mid v_0, v_\infty \in \mathbb{R}, v \in \mathbb{R}^n \right\} \subset M(2 \times 2, \mathcal{A}\mathbb{R}^n).$$

---

[6] Note that the term "endomorphism" makes no sense here since we have no (2-dimensional) vector space for an endomorphism to act on. This matrix representation of the Clifford algebra corresponds to what is called "conformal split" in [154] — this should be a useful reference, together with [155].

[7] Readers who are not familiar with Clifford algebras may wish to consult Section 6.2 for some basic information. Good references are also [189] and [224].

[8] More generally, it can be proven that $M(2 \times 2, \mathcal{A}\mathbb{R}_p^n) \cong \mathcal{A}\mathbb{R}_{p+1}^{n+2}$ (see [189]).

7.2.3 Notation. Henceforth we will not distinguish between $\mathbb{R}^{n+2}_1$ and V, and we shall write $\mathcal{A}\mathbb{R}^{n+2}_1$ for $\mathcal{A}V = M(2 \times 2, \mathcal{A}\mathbb{R}^n)$, $\Gamma(\mathbb{R}^{n+2}_1)$ for $\Gamma(V)$, and so on.

7.2.4 *Proof.* First note that, for $\mathcal{V} = \begin{pmatrix} v & v_\infty \\ v_0 & -v \end{pmatrix} \in V$, we have

$$\mathcal{V}^2 = \begin{pmatrix} v & v_\infty \\ v_0 & -v \end{pmatrix}^2 = \begin{pmatrix} v^2 + v_0 v_\infty & 0 \\ 0 & v^2 + v_0 v_\infty \end{pmatrix} = (v^2 + v_0 v_\infty) \cdot 1 \quad (7.1)$$

with $v^2 + v_0 v_\infty \in \mathbb{R}$. Since $\mathbb{R}^n \ni v \mapsto -v^2 \in \mathbb{R}$ is the quadratic form of the Euclidean scalar product on $\mathbb{R}^n$, the quadratic form $V \ni \mathcal{V} \mapsto -\mathcal{V}^2 \in \mathbb{R}$ belongs to a Minkowski scalar product, as claimed.

7.2.5 To see that V generates $M(2 \times 2, \mathcal{A}\mathbb{R}^n)$ we choose a basis that is orthonormal with respect to the Minkowski product on V:

$$\mathcal{E}_0 = \begin{pmatrix} 0 & 1 \\ 1 & 0 \end{pmatrix}, \quad \mathcal{E}_j = \begin{pmatrix} e_j & 0 \\ 0 & -e_j \end{pmatrix}, \quad \text{and} \quad \mathcal{E}_{n+1} = \begin{pmatrix} 0 & -1 \\ 1 & 0 \end{pmatrix}.$$

Because, for orthonormal vectors, the Clifford product becomes the exterior product, we can now determine $\Lambda^k V$ for $k = 0, \ldots, n + 2$: Identifying (cf., §7.1.1)

$$\Lambda^i \mathbb{R}^n \cong \left\{ \begin{pmatrix} \mathfrak{v} & 0 \\ 0 & (-1)^i \mathfrak{v} \end{pmatrix} \mid \mathfrak{v} \in \Lambda^i \mathbb{R}^n \right\},$$

and writing, for abbreviation,

$$\mathcal{E}_0 \Lambda^i \mathbb{R}^n \quad := \{ \mathcal{E}_0 \mathcal{V} \mid \mathcal{V} \in \Lambda^i \mathbb{R}^n \} \quad \cong \left\{ \begin{pmatrix} 0 & (-1)^i \mathfrak{y} \\ \mathfrak{y} & 0 \end{pmatrix} \mid \mathfrak{y} \in \Lambda^i \mathbb{R}^n \right\},$$

$$\Lambda^i \mathbb{R}^n \mathcal{E}_{n+1} := \{ \mathcal{V}\mathcal{E}_{n+1} \mid \mathcal{V} \in \Lambda^i \mathbb{R}^n \} \quad \cong \left\{ \begin{pmatrix} 0 & (-1)^{i+1} \mathfrak{x} \\ \mathfrak{x} & 0 \end{pmatrix} \mid \mathfrak{x} \in \Lambda^i \mathbb{R}^n \right\},$$

$$\mathcal{E}_0 \Lambda^i \mathbb{R}^n \mathcal{E}_{n+1} := \{ \mathcal{E}_0 \mathcal{V}\mathcal{E}_{n+1} \mid \mathcal{V} \in \Lambda^i \mathbb{R}^n \} \cong \left\{ \begin{pmatrix} \mathfrak{z} & 0 \\ 0 & (-1)^{i+1} \mathfrak{z} \end{pmatrix} \mid \mathfrak{z} \in \Lambda^i \mathbb{R}^n \right\},$$

for $i = 1, \ldots, n$, we will have

$$\Lambda^k V$$
$$= \quad \Lambda^k \mathbb{R}^n \oplus \mathcal{E}_0 \Lambda^{k-1} \mathbb{R}^n \oplus \Lambda^{k-1} \mathbb{R}^n \mathcal{E}_{n+1} \oplus \mathcal{E}_0 \Lambda^{k-2} \mathbb{R}^n \mathcal{E}_{n+1}$$
$$\cong \quad \left\{ \begin{pmatrix} \mathfrak{v} + \mathfrak{z} & \mathfrak{x} \\ \mathfrak{y} & (-1)^k(\mathfrak{v} - \mathfrak{z}) \end{pmatrix} \mid \mathfrak{v} \in \Lambda^k \mathbb{R}^n; \mathfrak{x}, \mathfrak{y} \in \Lambda^{k-1} \mathbb{R}^n; \mathfrak{z} \in \Lambda^{k-2} \mathbb{R}^n \right\}.$$
$$(7.2)$$

Hence V indeed generates $\mathcal{A}V$, $M(2 \times 2, \mathcal{A}\mathbb{R}^n) = \oplus_{k=0}^{n+2} \Lambda^k V$. ◁

7.2.6 Note that, with the setup from the previous paragraph, we can easily write the identification $\mathbb{R}^n \leftrightarrow Q^n_0$ of Euclidean n-space with a quadric of constant curvature $Q^n_0 \subset V \cong \mathbb{R}^{n+2}_1$ (see §1.4.3 and §7.1.3), namely,

$$\mathbb{R}^n \ni v \leftrightarrow \frac{1 + |v|^2}{2} \mathcal{E}_0 + \sum_{j=1}^n v_j \mathcal{E}_j + \frac{1 - |v|^2}{2} \mathcal{E}_{n+1} = \begin{pmatrix} v & -v^2 \\ 1 & -v \end{pmatrix} \in \mathbb{R}^{n+2}_1.$$

7.2.7 Next, in order to arrive at a description of the Clifford, pin, and spin groups in $\mathcal{A}V \cong \mathcal{A}\mathbb{R}_1^{n+2}$, we describe the order involution $\mathcal{V} \mapsto \hat{\mathcal{V}}$ and the Clifford conjugation $\mathcal{V} \mapsto \bar{\mathcal{V}}$ on $\mathcal{A}\mathbb{R}_1^{n+2}$ (see §6.3.1).

First we consider the order involution: On one hand, we know from §6.3.1 that we have $\hat{\mathcal{V}} = (-1)^k \mathcal{V}$ for $\mathcal{V} \in \Lambda^k \mathbb{R}_1^{n+2}$; on the other hand, we know from (7.2) that the diagonal entries of $\mathcal{V} \in \Lambda^k \mathbb{R}_1^{n+2}$ have the orders k and $k - 2$ while the off-diagonal entries have order $k - 1$. Consequently,

$$\widehat{\begin{pmatrix} a & b \\ c & d \end{pmatrix}} = \begin{pmatrix} \hat{a} & -\hat{b} \\ -\hat{c} & \hat{d} \end{pmatrix} \tag{7.3}$$

for $\mathcal{V} = \begin{pmatrix} a & b \\ c & d \end{pmatrix} \in \Lambda^k \mathbb{R}_1^{n+2}$, and hence for $\mathcal{V} \in \mathcal{A}\mathbb{R}_1^{n+2}$.

The situation is slightly more difficult for the Clifford conjugation, where we have $\bar{\mathcal{V}} = (-1)^{\frac{k(k+1)}{2}} \mathcal{V}$ for $\mathcal{V} \in \Lambda^k \mathbb{R}_1^{n+2}$ (see §6.3.1). With

$$(-1)^{\frac{(k-2)(k-1)}{2}} = -(-1)^{\frac{k(k+1)}{2}} \quad \text{and} \quad (-1)^{\frac{(k-1)k}{2}} = (-1)^k (-1)^{\frac{k(k+1)}{2}},$$

we deduce from (7.2) that

$$\begin{pmatrix} \bar{d} & \bar{b} \\ \bar{c} & \bar{a} \end{pmatrix} = (-1)^k \overline{\begin{pmatrix} a & b \\ c & d \end{pmatrix}} = \overline{\begin{pmatrix} a & b \\ c & d \end{pmatrix}} \tag{7.4}$$

for $\mathcal{V} = \begin{pmatrix} a & b \\ c & d \end{pmatrix} \in \Lambda^k \mathbb{R}_1^{n+2}$; this extends to all of $\mathcal{A}\mathbb{R}_1^{n+2} = \oplus_{k=0}^{n+2} \Lambda^k \mathbb{R}_1^{n+2}$ by linearity, as before. Using (7.3), this yields

$$\overline{\begin{pmatrix} a & b \\ c & d \end{pmatrix}} = \widehat{\overline{\begin{pmatrix} a & b \\ c & d \end{pmatrix}}} = \widehat{\begin{pmatrix} \bar{d} & \bar{b} \\ \bar{c} & \bar{a} \end{pmatrix}} = \begin{pmatrix} \hat{\bar{d}} & -\hat{\bar{b}} \\ -\hat{\bar{c}} & \hat{\bar{a}} \end{pmatrix}. \tag{7.5}$$

7.2.8 We are now prepared to establish one of the key results for the present setup, a characterization of the Clifford group $\Gamma(\mathbb{R}_1^{n+2})$ in $\mathcal{A}\mathbb{R}_1^{n+2}$. The characterizations of the pin and spin groups (see §6.3.7 and §6.3.8) will then be an easy consequence. This theorem is usually attributed to Vahlen [288] — therefore the name "Vahlen matrices" for the elements of $\Gamma(\mathbb{R}_1^{n+2})$ (cf., [189]). Our proof will loosely follow the one given by Porteous in [224], and we state it in the form given by Ahlfors [2]:

7.2.9 Theorem (Vahlen). *An element* $\mathcal{V} = \begin{pmatrix} a & b \\ c & d \end{pmatrix} \in \mathcal{A}(\mathbb{R}_1^{n+2})$ *is in the Clifford group of* $\mathbb{R}_1^{n+2}$, $\mathcal{V} \in \Gamma(\mathbb{R}_1^{n+2})$, *if and only if*
(i) $a, b, c, d \in \Gamma(\mathbb{R}^n) \cup \{0\} =: \Theta(\mathbb{R}^n)$,
(ii) $a\hat{\bar{b}}, b\hat{\bar{d}}, d\hat{\bar{c}}, c\hat{\bar{a}} \in \mathbb{R}^n$, *and*

(iii) $\Delta := a\hat{\tilde{d}} - b\hat{\tilde{c}} \in \mathbb{R} \setminus \{0\}$.

7.2.10 *Proof.* Before getting into the proof of the theorem, we collect a couple of useful facts and recall some material on the Clifford group from Chapter 6. We will also (locally) introduce some notations. Thus we want to show that the Clifford group is identical to the set $\mathfrak{V}(\mathbb{R}^n)$ of "Vahlen matrices,"

$$\Gamma(\mathbb{R}_1^{n+2}) = \mathfrak{V}(\mathbb{R}^n) := \{\mathcal{V} \in \mathcal{A}(\mathbb{R}_1^{n+2}) \,|\, \mathcal{V} \text{ satisfies (i) -- (iii)}\}.$$

In order to make the arguments a bit clearer, we introduce a third involution, the "Clifford transpose," $\mathfrak{v} \mapsto \check{\mathfrak{v}} := \hat{\bar{\mathfrak{v}}}$ on $\mathcal{A}\mathbb{R}^n$, and note that $\mathfrak{v} \mapsto \check{\mathfrak{v}}$ is an antiautomorphism, that is, $\widecheck{\mathfrak{v}\mathfrak{w}} = \check{\mathfrak{w}}\check{\mathfrak{v}}$ for $\mathfrak{v}, \mathfrak{w} \in \mathcal{A}\mathbb{R}^n$, with $\check{v} = v$ on $\mathbb{R}^n$. Also note that all three involutions of $\mathcal{A}\mathbb{R}^n$ commute — this is because they are just $\pm id$ on $\Lambda^k \mathbb{R}^n$ — and that they leave the Clifford group $\Gamma(\mathbb{R}^n)$ invariant,

$$\mathfrak{s} \in \Gamma(\mathbb{R}^n) \;\Rightarrow\; \bar{\mathfrak{s}}, \hat{\mathfrak{s}}, \check{\mathfrak{s}} \in \Gamma(\mathbb{R}^n),$$

as is easily shown with §6.3.7 in mind: By §6.3.2, we have to show that, for example, $\hat{\mathfrak{s}}$ has an inverse $\hat{\mathfrak{s}}^{-1}$ and that $\hat{\mathfrak{s}}v\mathfrak{s}^{-1} \in \mathbb{R}^n$ for any $v \in \mathbb{R}^n$. The first statement follows because, with §6.3.7, $\hat{\mathfrak{s}}^{-1} = \frac{1}{\bar{\mathfrak{s}}\mathfrak{s}}\bar{\mathfrak{s}}$, and the second follows since $\widehat{(\hat{\mathfrak{s}}v\mathfrak{s}^{-1})} = \mathfrak{s}v\hat{\mathfrak{s}}^{-1} \in \mathbb{R}^n$. For the antiautomorphisms $\bar{}$ and $\check{}$, the second statement is slightly less direct to verify: Here, $\bar{\mathfrak{s}}v\hat{\check{\mathfrak{s}}}^{-1} = -\hat{\bar{\mathfrak{s}}}^{-1}v\mathfrak{s} \in \mathbb{R}^n$ since we have $\hat{\mathfrak{s}}^{-1} \in \Gamma(\mathbb{R}^n)$.

7.2.11 With these remarks in mind, we now prove an algebraic fact that will be useful in our proof:

$$\forall a, b \in \Theta(\mathbb{R}^n) : a\tilde{b} \in \mathbb{R}^n \Leftrightarrow \bar{b}a \in \mathbb{R}^n \quad \text{and} \quad \tilde{a}b \in \mathbb{R}^n \Leftrightarrow b\bar{a} \in \mathbb{R}^n.$$

If $a = 0$ or $b = 0$, the statement is trivial, so that we may assume, without loss of generality, that $a, b \in \Gamma(\mathbb{R}^n)$. First let $v := \bar{b}a \in \mathbb{R}^n$; then

$$\mathbb{R}^n \ni bv\hat{b}^{-1} = b\bar{b}a\frac{1}{\bar{b}b}\hat{\tilde{b}} = a\tilde{b}.$$

If, on the other hand, $w := a\tilde{b} \in \mathbb{R}^n$, we compute

$$\mathbb{R}^n \ni \bar{b}w\hat{b}^{-1} = \bar{b}a,$$

since $\bar{b} \in \Gamma(\mathbb{R}^n)$ also. The second equivalence can be derived similarly. Another way[9] to obtain the second statement is by applying the first equivalence and the fact that our involutions $\check{}$ and $\bar{}$ leave $\mathbb{R}^n$ invariant:

$$\widehat{\tilde{a}b} = \bar{a}\hat{b} \in \mathbb{R}^n \Leftrightarrow \mathbb{R}^n \ni a\hat{\tilde{b}} = \bar{b}\bar{a}.$$

---

[9] Yet another way is by showing that $\tilde{a}b \in \mathbb{R}^n \Leftrightarrow a\tilde{b} \in \mathbb{R}^n$ (cf., [47] and [224]): From §6.3.6, we deduce $\tilde{a}a = a\tilde{a} = \pm \bar{a}a$; this yields $a\widehat{(\tilde{a}b)}\hat{a}^{-1} = a\bar{b}a\frac{1}{\tilde{a}a}a\tilde{a} = \pm a\tilde{b}$ and $\hat{b}^{-1}\widehat{(a\tilde{b})}b = \pm\tilde{a}b$.

As a consequence, (ii) in §7.2.9 is equivalent to $\bar{a}b, \bar{c}d \in \mathbb{R}^n$ and $a\bar{c}, b\bar{d} \in \mathbb{R}^n$. With this we are now prepared to attack the proof of the theorem:

7.2.12 First we show that $\mathfrak{V}(\mathbb{R}^n) \subset \Gamma(\mathbb{R}^n)$.

For this purpose we verify that any Vahlen matrix $\mathcal{V} \in \mathfrak{V}(\mathbb{R}^n)$ is (a) invertible and (b) acts on $\mathbb{R}_1^{n+2} \cong V$ via the twisted adjoint action (see §6.3.2).

(a) By §6.3.7, we know that the inverse $\mathcal{V}^{-1}$ of an element of the Clifford group is a multiple of its conjugate $\bar{\mathcal{V}}$. Hence it suffices to verify that $\mathcal{V}\bar{\mathcal{V}}$ is a nonzero multiple of 1; but

$$\mathcal{V}\bar{\mathcal{V}} = \begin{pmatrix} a & b \\ c & d \end{pmatrix} \begin{pmatrix} \overline{a & b} \\ c & d \end{pmatrix} = \begin{pmatrix} a\tilde{d} - b\tilde{c} & b\tilde{a} - a\tilde{b} \\ c\tilde{d} - d\tilde{c} & d\tilde{a} - c\tilde{b} \end{pmatrix} = \Delta \cdot 1 \qquad (7.6)$$

because the off-diagonal entries vanish by (ii) — V is in the $+1$-eigenspace of Clifford transpose — and $\tilde{\Delta} = \Delta \in \mathbb{R} \setminus \{0\}$ by (iii). Thus $\mathcal{V}^{-1} = \frac{1}{\Delta}\bar{\mathcal{V}}$.

(b) Here we have to check that $\mathcal{V}\mathcal{X}\hat{\mathcal{V}}^{-1} \in \mathbb{R}_1^{n+2}$ for all $\mathcal{X} \in \mathbb{R}_1^{n+2}$; since $\hat{\mathcal{V}}^{-1} = \frac{1}{\Delta}\hat{\bar{\mathcal{V}}}$, it suffices to examine

$$\begin{aligned}
\mathcal{V}\mathcal{X}\hat{\bar{\mathcal{V}}} &= \begin{pmatrix} a & b \\ c & d \end{pmatrix} \begin{pmatrix} x & x_\infty \\ x_0 & -x \end{pmatrix} \begin{pmatrix} \widehat{\overline{a & b}} \\ c & d \end{pmatrix} \\
&= \begin{pmatrix} ax\bar{d} - bx\bar{c} + x_0 b\bar{d} + x_\infty a\bar{c} & ax\bar{b} - bx\bar{a} + x_0 b\bar{b} + x_\infty a\bar{a} \\ cx\bar{d} - dx\bar{c} + x_0 d\bar{d} + x_\infty c\bar{c} & cx\bar{b} - dx\bar{a} + x_0 d\bar{b} + x_\infty c\bar{a} \end{pmatrix}.
\end{aligned}$$
$$(7.7)$$

Moreover, since the map $\mathcal{X} \mapsto \mathcal{V}\mathcal{X}\hat{\bar{\mathcal{V}}}$ is clearly (real) linear, we can break down the investigation into cases:

$x_0 = 1$, $x = 0$, $x_\infty = 0$. In this case, $\mathcal{V}\mathcal{X}\hat{\bar{\mathcal{V}}} \in \mathbb{R}_1^{n+2}$ by (i) and (ii), using our preliminary remark in §7.2.11.

$x_0 = 0$, $x = 0$, $x_\infty = 1$. Here $\mathcal{V}\mathcal{X}\hat{\bar{\mathcal{V}}} \in \mathbb{R}_1^{n+2}$, just as in the previous case.

$x_0 = 0$, x arbitrary, $x_\infty = 0$. This case is a little bit more complicated, and we consider the components in (7.7) separately. First consider an off diagonal entry, say $ax\bar{b} - bx\bar{a}$: If $a = 0$, there is nothing to show; if $a \neq 0$, we compute

$$\bar{a}(ax\bar{b} - bx\bar{a})a = \bar{a}a \cdot (x \cdot \bar{b}a - \bar{a}b \cdot x) = 2\bar{a}a \cdot \langle x, \bar{a}b \rangle \in \mathbb{R}.$$

Thus $ax\bar{b} - bx\bar{a} = 2\langle x, \bar{a}b \rangle \in \mathbb{R}$. To show that $ax\bar{d} - bx\bar{c} \in \mathbb{R}^n$, we first assume that $a \neq 0$: Then we can write

$$d = (d\tilde{a} - c\tilde{b})\tilde{a}^{-1} + c\tilde{b}\tilde{a}^{-1} = \tfrac{1}{\bar{a}a}(\Delta\hat{a} - c \cdot (\bar{b}a))$$

since $\tilde{a}^{-1} = \frac{1}{\bar{a}a}\hat{a}$ and $\tilde{b}\tilde{a} = -\bar{b}a$ because $\hat{\cdot}$ is an automorphism of $\mathcal{A}\mathbb{R}^n$ with $\mathbb{R}^n$ in its -1-eigenspace. Consequently,

$$\begin{aligned}
ax\bar{d} - bx\bar{c} &= \tfrac{1}{\bar{a}a}\{ax(\Delta\hat{\bar{a}} - (\bar{a}b)\bar{c}) - a(\bar{a}b)x\bar{c}\} \\
&= \Delta\, ax\hat{a}^{-1} + \tfrac{2}{\bar{a}a}\langle \bar{a}b, x \rangle\, a\bar{c} \in \mathbb{R}^n
\end{aligned}$$

by (i) and (ii), where we again use our remark from §7.2.11. If, on the other hand, $a = 0$, we must have $b \neq 0$ and $c = -\frac{\Delta}{b\bar{b}}\hat{b}$, so that

$$ax\bar{d} - bx\bar{c} = \Delta\, bx\hat{b}^{-1} \in \mathbb{R}^n$$

in this case also. In any case, we find that $ax\bar{d} - bx\bar{c} =: v \in \mathbb{R}^n$, so that $cx\bar{b} - dx\bar{a} = \bar{v} = -v$ and $\mathcal{V}\mathcal{X}\hat{\mathcal{V}} \in \mathbb{R}^{n+2}_1$ as sought.

Thus we have proved one inclusion.

7.2.13 To complete our proof, we show that $\Gamma(\mathbb{R}^n) \subset \mathfrak{V}(\mathbb{R}^n)$.

From §6.3.6 (and §6.3.4) we know that any element $\mathcal{V}$ of the Clifford group $\Gamma(\mathbb{R}^{n+2}_1)$ is a (nonzero) real multiple of the identity, or a product of nonisotropic vectors, $\mathcal{V} = \prod_{i=1}^{k} \mathcal{V}_i$ with $\mathcal{V}_i^2 \neq 0$. We use induction to show that any such product satisfies the properties (i) – (iii) in §7.2.9 of a Vahlen matrix.

$k = 0$, that is, $\mathcal{V} = \delta\,1$ with $\delta \neq 0$. It is a trivial exercise[10] to verify that $\mathcal{V}$ satisfies the properties of a Vahlen matrix.

$k \to k + 1$. Here we want to show that the product $\mathcal{V}\mathcal{X}$ of a Vahlen matrix $\mathcal{V} \in \mathfrak{V}(\mathbb{R}^n)$ and a vector $\mathcal{X} \in \mathbb{R}^{n+2}_1$ with $\mathcal{X}^2 \neq 0$ is a Vahlen matrix. Thus we consider

$$\begin{pmatrix} a & b \\ c & d \end{pmatrix} \begin{pmatrix} x & x_\infty \\ x_0 & -x \end{pmatrix} = \begin{pmatrix} ax + bx_0 & ax_\infty - bx \\ cx + dx_0 & cx_\infty - dx \end{pmatrix}. \tag{7.8}$$

(i) As a sample we consider $ax + bx_0$:

If $a = 0$, clearly $ax + bx_0 = bx_0 \in \Theta(\mathbb{R}^n)$.

If, on the other hand, $a \neq 0$, we write $ax + bx_0 = a(x + \frac{1}{\bar{a}a}\bar{a}b)$ — since $\bar{a}b \in \mathbb{R}^n$, we have $ax + bx_0 \in \Theta(\mathbb{R}^n)$ as the product of $a \in \Gamma(\mathbb{R}^n)$ and a vector in $\mathbb{R}^n$. Note that we have used $\bar{a}b \in \mathbb{R}^n \Leftrightarrow a\bar{b} \in \mathbb{R}^n$ from §7.2.11. The same argument works for the other entries of (7.8).

(ii) A straightforward computation yields

$$\overline{(ax + bx_0)}(ax_\infty - bx)$$
$$= (x\,\bar{a}b + \bar{a}b\,x - x_\infty\bar{a}a - x_0\bar{b}b)\,x - (x^2 + x_0x_\infty)\,\bar{a}b \in \mathbb{R}^n,$$

and likewise for $\overline{(cx + dx_0)}(cx_\infty - dx)$. Another computation gives

$$(ax + bx_0)\overline{(cx + dx_0)} = x_0(ax\bar{d} - bx\bar{c}) + x_0^2 b\bar{d} - x^2 a\bar{c} \in \mathbb{R}^n,$$

since $ax\bar{d} - bx\bar{c} \in \mathbb{R}^n$, as we saw in §7.2.12(b); by a similar computation we also find that $(ax_\infty - bx)\overline{(cx_\infty - dx)} \in \mathbb{R}^n$.

---

[10] It may be a more amusing exercise for the reader to verify that *any* diagonal matrix with real (nonvanishing diagonal) entries satisfies conditions (i) – (iii) in §7.2.9 — and then find a way to write it as a product of vectors in case it is not a multiple of the identity.

(iii) This is again a straightforward computation:

$$(ax + bx_0)(\widetilde{cx_\infty - dx}) - (ax_\infty - bx)(\widetilde{cx + dx_0}) = -\Delta \cdot \mathcal{X}^2 \in \mathbb{R} \setminus \{0\}.$$

Hence, the product in (7.8) satisfies the properties (i) – (iii) so that $\mathcal{V}\mathcal{X} \in \mathfrak{V}(\mathbb{R}^n)$ is a Vahlen matrix. This completes the proof of §7.2.9.   ◁

7.2.14 Example. A simple example of a 2×2 Clifford algebra matrix $\mathcal{V}$ that is *not* a Vahlen matrix is

$$\mathcal{V} := \begin{pmatrix} e_1 e_2 + e_3 e_4 & 0 \\ 0 & e_1 e_2 + e_3 e_4 \end{pmatrix},$$

where $e_1, e_2, e_3, e_4 \in \mathbb{R}^n$, $n \geq 4$, are four orthogonal unit vectors in $\mathbb{R}^n$: Condition (i) is not fulfilled since $e_1 e_2 + e_3 e_4$ cannot be written as a product of vectors in $\mathbb{R}^n$, and $\Delta = 2 - 2e_1 e_2 e_3 e_4 \notin \mathbb{R}$, showing that $\mathcal{V}$ also fails to satisfy (iii).

A less obvious example may be

$$\mathcal{V} := \begin{pmatrix} e_1 e_2 & e_3 \\ e_3 & e_1 e_2 \end{pmatrix}$$

with three orthogonal vectors $e_1, e_2, e_3 \in \mathbb{R}^n$, $n \geq 3$: Here, (i) and (iii) are satisfied but (ii) fails.

7.2.15 It is now an easy task to determine the pin and spin groups in $A\mathbb{R}_1^{n+2}$ according to their definitions given in §6.3.7. Since $\bar{\mathcal{V}}\mathcal{V} = \mathcal{V}\bar{\mathcal{V}}$ for $\mathcal{V} \in \Gamma(\mathbb{R}_1^{n+2})$, the condition for pin simply reads

$$\Delta = a\tilde{d} - b\tilde{c} = \pm 1,$$

as already computed in (7.6). For the spin group we must additionally have the condition $\hat{\mathcal{V}} = \mathcal{V}$, which, according to (7.3), becomes

$$\hat{a} = a, \quad \hat{d} = d, \quad \hat{b} = -b, \quad \text{and} \quad \hat{c} = -c. \tag{7.9}$$

Since, as we learned for the elements of $\Gamma(\mathbb{R}^n)$ in §6.3.6, any $a \in \Theta(\mathbb{R}^n)$ can be written as the product of a real number and a product of vectors,[11]

$$\Theta(\mathbb{R}^n) = \{a = \lambda \textstyle\prod_{i=1}^k v_i \mid \lambda \in \mathbb{R}, v_i \in \mathbb{R}^n\},$$

---

[11] Remember that the empty product is defined to be $\prod_{i=1}^0 v_i = 1$.

this tells us that a, d come from even products of vectors whereas b, c come from odd products.[12] Note that, with (7.9), we obtain

$$\Delta = a\bar{d} + b\bar{c}.$$

Summarizing, we have:

7.2.16 Corollary. *An element* $\mathcal{V} = \begin{pmatrix} a & b \\ c & d \end{pmatrix} \in \Gamma(\mathbb{R}_1^{n+2})$ *is in the pin group* $Pin(\mathbb{R}_1^{n+2})$ *if and only if* $\Delta = \pm 1$,

$$Pin(\mathbb{R}_1^{n+2}) = \{\mathcal{V} \in \Gamma(\mathbb{R}_1^{n+2}) \mid \Delta = a\tilde{d} - b\tilde{c} = \pm 1\},$$

and it is in the spin group $Spin(\mathbb{R}_1^{n+2})$ *if and only if, additionally, a and d come from even products and b and c from odd products of vectors in $\mathbb{R}^n$, that is,*

$$Spin(\mathbb{R}_1^{n+2}) = \{\mathcal{V} \in \Gamma(\mathbb{R}_1^{n+2}) \mid \hat{a} = a, \hat{b} = -b, \hat{c} = -c, \hat{d} = d, \Delta = \pm 1\}.$$

7.3 Möbius transformations and spheres

From §6.3.3 we know that the Clifford group $\Gamma(\mathbb{R}_1^{n+2})$ acts by isometries on $\mathbb{R}_1^{n+2}$ via the twisted adjoint action

$$r : \Gamma(\mathbb{R}_1^{n+2}) \to O(\mathbb{R}_1^{n+2}), \quad \mathcal{S} \mapsto [\mathbb{R}_1^{n+2} \ni \mathcal{V} \mapsto \mathcal{S}\mathcal{V}\hat{\mathcal{S}}^{-1} \in \mathbb{R}_1^{n+2}],$$

with kernel $\mathbb{R} \setminus \{0\}$. Similarly, the pin and spin groups act by (orientation-preserving) isometries on $\mathbb{R}_1^{n+2}$ with kernel ± 1 (see §6.3.8).

Interpreting $\mathbb{R}_1^{n+2}$ as the space of homogeneous coordinates for the Möbius geometry of the conformal n-sphere $S^n \cong \mathbb{R}^n \cup \{\infty\}$ (see Figure T.1), this action is by Möbius transformations on S^n (cf., §6.3.12).

Our next goal is to write this action on S^n in terms of Vahlen matrices:

7.3.1 Let $\mathcal{V} = \begin{pmatrix} v & -v^2 \\ 1 & -v \end{pmatrix} \in S^n$ (cf., §7.2.6), and let $\mathcal{S} = \begin{pmatrix} a & b \\ c & d \end{pmatrix} \in \Gamma(\mathbb{R}_1^{n+2})$. Then

$$r(\mathcal{S}) \cdot \mathcal{V} = \frac{1}{\Delta}\mathcal{S}\mathcal{V}\hat{\mathcal{S}} = \frac{(cv+d)\overline{(cv+d)}}{\Delta} \begin{pmatrix} w & -w^2 \\ 1 & -w \end{pmatrix} \simeq \begin{pmatrix} w & -w^2 \\ 1 & -w \end{pmatrix}$$

with $w := (av+b)(cv+d)^{-1}$ as soon as $cv + d \neq 0$ — if, on the other hand, $cv + d = 0$, then $r(\mathcal{S}) \cdot \mathcal{V} \simeq \begin{pmatrix} 0 & 1 \\ 0 & 0 \end{pmatrix} \simeq \infty$.

---

[12] Another way to verify this fact is by examining products of vectors in $\mathbb{R}_1^{n+2}$.

Note that $(av + b)(cv + d)^{-1} \in \mathbb{R}^n \cup \{\infty\}$ since $r(\mathcal{S}) \cdot \mathcal{V} \in \mathbb{R}_1^{n+2}$ for $\mathcal{V} \in \mathbb{R}_1^{n+2}$.

Similarly,

$$r(\mathcal{S}) \cdot \begin{pmatrix} 0 & 1 \\ 0 & 0 \end{pmatrix} = \frac{c\bar{c}}{\Delta} \begin{pmatrix} ac^{-1} & -(ac^{-1})^2 \\ 1 & -ac^{-1} \end{pmatrix} \simeq \begin{pmatrix} w & -w^2 \\ 1 & -w \end{pmatrix}$$

with $w := ac^{-1}$, that is, $r(\mathcal{S})\infty \simeq ac^{-1}$. Thus

7.3.2 Theorem. *The Clifford group* $\Gamma(\mathbb{R}_1^{n+2})$ *acts on* $S^n \cong \mathbb{R}^n \cup \{\infty\}$ *by Möbius transformations; in affine coordinates, this action is by fractional linear transformations,*

$$\left(\begin{pmatrix} a & b \\ c & d \end{pmatrix}, v \right) \mapsto \begin{pmatrix} a & b \\ c & d \end{pmatrix} \cdot v := \begin{cases} (av + b)(cv + d)^{-1} & for \quad v \in \mathbb{R}^n, \\ ac^{-1} & for \quad v = \infty. \end{cases}$$

We will also speak of $v \mapsto (av + b)(cv + d)^{-1}$ *as the Möbius transformation associated with the Vahlen matrix* $\begin{pmatrix} a & b \\ c & d \end{pmatrix}$.

Similarly, the pin and spin groups $Pin(\mathbb{R}_1^{n+2})$ *and* $Spin(\mathbb{R}_1^{n+2})$ *act on* S^n *by fractional linear transformations.*

7.3.3 Note how we used the stereographic projection $S^n \to \mathbb{R}^n$, that is, a suitable rescaling for the (lightlike) homogeneous coordinates $\mathcal{V} \in \mathbb{R}_1^{n+2}$ of a point $v \in \mathbb{R}^n$ (see §1.4.4) in order to obtain a Möbius transformation as a linear fractional transformation (cf., §7.1.3).

7.3.4 This situation is very similar to the quaternionic setup (see §4.4.5). However, in contrast to the quaternionic case, there will be no vector space that can serve as the space of homogeneous coordinates for S^n. The closest we can get to the quaternionic setup here is to identify

$$\mathbb{R}^n \ni v \leftrightarrow \begin{pmatrix} v \\ 1 \end{pmatrix} \cdot \Gamma(\mathbb{R}^n), \quad \text{and} \quad \infty \leftrightarrow \begin{pmatrix} 1 \\ 0 \end{pmatrix} \cdot \Gamma(\mathbb{R}^n)$$

and consider $\Gamma(\mathbb{R}_1^{n+2})$ as acting by matrix multiplication on these "vectors." But obviously these "vectors" do not form a vector space and $\Gamma(\mathbb{R}^n)$ is not a field, so that this "analogy" between the quaternionic setup and the present Clifford algebra setup remains formal.

Moreover, in the quaternionic setup, *every* quaternionic 2×2 matrix acted on (the vector space) $\mathbb{H}^2$ and hence on $S^4 \cong \mathbb{H}P^1$. Here it cannot only occur that a matrix with Clifford algebra entries does not provide a Möbius *transformation* of S^n (since it is not invertible), but it can also happen that it does not even provide a map $S^n \to S^n$. For example, the second 2×2 matrix from §7.2.14 does not act on $\mathbb{R}^n \cup \{\infty\}$, as the following small computation shows:

$$\begin{pmatrix} e_1 e_2 & e_3 \\ e_3 & e_1 e_2 \end{pmatrix} \begin{pmatrix} e_3 \\ 1 \end{pmatrix} = \begin{pmatrix} (e_1 - e_2)e_2 e_3 \\ (e_1 + e_2)e_2 \end{pmatrix} = \begin{pmatrix} -e_1 e_2 e_3 \\ 1 \end{pmatrix} (e_1 + e_2)e_2.$$

7.3.5 Example. As a positive example, we consider $\mathcal{S} \in \Gamma(\mathbb{R}_1^{n+2})$ to be a vector, $\mathcal{S} = \begin{pmatrix} s & s_\infty \\ s_0 & -s \end{pmatrix} \in \mathbb{R}_1^{n+2}$, of length $1 = -\mathcal{S}^2 = -(s^2 + s_0 s_\infty)$.

If $s_0 = 0$, we have $s^2 = -1$ and its action on $v \in \mathbb{R}^n$ is given by

$$\mathcal{S} v = -(sv + s_\infty)s^{-1} = svs + s_\infty s = v - 2(\langle s, v \rangle - \tfrac{1}{2}s_\infty) s,$$

which is the reflection of v in the hyperplane orthogonal to $n := s \in S^{n-1}$ and with distance $d := \tfrac{1}{2}s_\infty$ from the origin.

If, on the other hand, $s_0 \neq 0$, we obtain

$$\mathcal{S} v = (sv + s_\infty)(s_0 - sv)^{-1} = m - r^2(v - m)^{-1}$$

with $m := \frac{s}{s_0}$ and $r := \frac{1}{s_0}$ (cf., §7.1.5). In this case we obtain the reflection in the hypersphere with center $m \in \mathbb{R}^n$ and radius $|\frac{1}{s_0}| \in \mathbb{R} \setminus \{0\}$. In particular, $s = 0$ and $s_0 = -s_\infty = 1$ gives the inversion $v \mapsto \frac{v}{|v|^2}$ in the unit sphere in $\mathbb{R}^n$.

Note that, corresponding to the factorization of a Vahlen matrix in upper and lower triangular matrices, every Möbius transformation (that is not already a similarity) can be written as the composition of an inversion and a similarity: If $A = \begin{pmatrix} a & b \\ c & d \end{pmatrix} \in \Gamma(\mathbb{R}_1^{n+2})$ denotes a Vahlen matrix with $c \neq 0$, that is, the corresponding Möbius transformation is essential, we may write

$$\begin{aligned} \begin{pmatrix} a & b \\ c & d \end{pmatrix} &= \begin{pmatrix} ac^{-1}d - b & a \\ 0 & c \end{pmatrix} \begin{pmatrix} 0 & -1 \\ 1 & -m \end{pmatrix} \\ &= \begin{pmatrix} am + b & a - (am + b)m \\ 0 & c \end{pmatrix} \begin{pmatrix} m & -(1 + m^2) \\ 1 & -m \end{pmatrix}, \end{aligned} \tag{7.10}$$

where $m = -c^{-1}d \in \mathbb{R}^n$ (see §7.2.9 and §7.2.11) is the center of inversion.

7.3.6 Thus, as described in §7.1.2 for 2-spheres in S^3, we can identify hyperspheres in S^n with elements $\mathcal{S} \in \mathbb{R}_1^{n+2}$ that satisfy[13] $\mathcal{S}^2 = -1$. In this way the identification of hyperspheres with a Möbius involution that has this hypersphere as its fixed point set (cf., Section 4.8) leads us back to the

---

[13] Note that we obtain exactly the same formulas but that we can no longer interpret $\mathcal{S}$ as an endomorphism of a vector space.

classical identification of hyperspheres with the spacelike directions (or unit vectors) in our coordinate Minkowski space $\mathbb{R}_1^{n+2}$.

Thus the incidence of a point $\mathcal{V}$ and a hypersphere $\mathcal{S}$ can be detected in two ways: either, by following the classical approach, via orthogonality, or by the fact that $\mathcal{V}$ is a fixed point of the Möbius involution,

$$\mathcal{V} = \begin{pmatrix} v & -v^2 \\ 1 & -v \end{pmatrix} \perp \mathcal{S} \quad \Leftrightarrow \quad \mathcal{S}\begin{pmatrix} v \\ 1 \end{pmatrix} = \begin{pmatrix} v \\ 1 \end{pmatrix} a, \tag{7.11}$$

with a suitable element $a \in \Gamma(\mathbb{R}^n)$. Similarly, the orthogonal intersection of two hyperspheres $\mathcal{S}_1$ and $\mathcal{S}_2$ is detected by[14]

$$\{\mathcal{S}_1, \mathcal{S}_2\} = 0 \quad \Leftrightarrow \quad \mathcal{S}_1\mathcal{S}_2 = -\mathcal{S}_2\mathcal{S}_1,$$

that is, the two hyperspheres intersect orthogonally if and only if the corresponding Möbius involutions commute — remember that $Spin_1^+(n+2)$ is a *double* cover of $M\ddot{o}b(n)$ (see §6.3.12).

We collect these facts for later reference:

7.3.7 Lemma. *Any hypersphere $\mathcal{S} \subset S^n$ is encoded by a spacelike vector $\mathcal{S} \in \mathbb{R}_1^{n+2}$, $\mathcal{S}^2 < 0$; the incidence of a point $\mathcal{V} = \begin{pmatrix} v & -v^2 \\ 1 & -v \end{pmatrix} \in \mathbb{R}_1^{n+2}$ and a hypersphere $\mathcal{S} \in \mathbb{R}_1^{n+2}$ is detected by*

$$\mathcal{S}\begin{pmatrix} v \\ 1 \end{pmatrix} = \begin{pmatrix} v \\ 1 \end{pmatrix} a, \quad \text{where} \quad a \in \Gamma(\mathbb{R}^n),$$

and two hyperspheres $\mathcal{S}_1, \mathcal{S}_2 \in \mathbb{R}_1^{n+2}$ intersect orthogonally if and only if the corresponding Möbius involutions commute, $\{\mathcal{S}_1, \mathcal{S}_2\} = 0$.

7.3.8 The present description resembles very much the quaternionic setup in various aspects — despite the fact that the coordinate vector space $\mathbb{H}^2$ has no Clifford algebra counterpart.

However, the condition $\mathcal{S}^2 = -1$ on $Pin(\mathbb{R}_1^{n+2})$ also does not survive as a characterization of hyperspheres in S^n, in contrast to the case of $S^3 \subset \mathbb{H}P^1$ (cf., §7.1.2): For example, if $\mathcal{S}_1, \mathcal{S}_2 \in \mathbb{R}_1^{n+2}$ are two orthogonal hyperspheres, that is, $\mathcal{S}_i^2 = -1$ and $\{\mathcal{S}_1, \mathcal{S}_2\} = 0$, then we also have $(\mathcal{S}_1\mathcal{S}_2)^2 = -1$, but $\mathcal{S}_1\mathcal{S}_2 \in \Lambda^2\mathbb{R}_1^{n+2}$ is now a codimension 2 sphere (see §6.4.8 or Figure T.4).

More generally, the product of m orthogonal hyperspheres $\mathcal{S}_i$ provides a Möbius involution and squares to $(\prod_{i=1}^m \mathcal{S}_i)^2 = (-1)^{\frac{m(m+1)}{2}}$, so that the m-vectors corresponding to spheres of various dimensions square to -1.

---

[14] Remember that, in (6.3), we defined $[\mathfrak{v}, \mathfrak{w}] := \mathfrak{v}\mathfrak{w} - \mathfrak{w}\mathfrak{v}$ and $\{\mathfrak{v}, \mathfrak{w}\} := \mathfrak{v}\mathfrak{w} + \mathfrak{w}\mathfrak{v}$ for any two elements $\mathfrak{v}, \mathfrak{w}$ of the Clifford algebra.

On the other hand, any Möbius involution $S \in Pin(\mathbb{R}_1^{n+2})$ can be described as a product of inversions in orthogonal hyperspheres (see §1.3.19) — since the equation $S^2 = \pm 1$ characterizes such involutions, we deduce that it indeed *characterizes* spheres of certain dimensions as long as the fixed point set of the involution is not empty[15] (see §1.3.21).

Thus the condition $S^2 = -1$, or $S^2 = +1$, does not suffice to characterize the spheres of a given dimension, but, in order to control the dimension k of the sphere $S \subset S^n$, one has to require additionally that $S \in \Lambda^{n-k}\mathbb{R}_1^{n+2}$, as in §6.4.8:

7.3.9 Lemma. *Any k-sphere $S \subset S^n$ can be described as the product of $n - k$ orthogonal hyperspheres $S_i \in \mathbb{R}_1^{n+2}$, that is,*

$$S = \textstyle\prod_{i=1}^{n-k} S_i \in \Lambda^{n-k}\mathbb{R}_1^{n+2},$$

where $S_i^2 = -1$ and $\{S_i, S_j\} = 0$ for $i, j = 1, \dots, n - k$, $j \neq i$. Equivalently, a k-sphere $S \subset S^n$ can be characterized as a pure $(n - k)$-vector[16]

$$S \in \Lambda^{n-k}\mathbb{R}_1^{n+2} \quad \text{with} \quad S^2 = (-1)^{\frac{(n-k)(n-k+1)}{2}},$$

or, similarly, it can be described by its Clifford dual[17] $vS \in \Lambda^{k+2}\mathbb{R}_1^{n+2}$.

7.3.10 Now we can characterize the incidence of a point and a k-sphere S in a similar way as for hyperspheres in §7.3.7: First note that, by writing the k-sphere $S = \prod_{i=1}^{n-k} S_i$ as the orthogonal intersection of $n - k$ hyperspheres, we find that

$$S \begin{pmatrix} v \\ 1 \end{pmatrix} = \textstyle\prod_{i=1}^{n-k} S_i \begin{pmatrix} v \\ 1 \end{pmatrix} = \begin{pmatrix} v \\ 1 \end{pmatrix} \cdot \textstyle\prod_{i=0}^{n-k-1} a_{n-k-i} = \begin{pmatrix} v \\ 1 \end{pmatrix} \cdot a$$

with suitable $a_i \in \Gamma(\mathbb{R}^n)$ and $a = a_{n-k} \cdots a_1$ if $v \in S \cap \mathbb{R}^n$.

A similar argument applies to $v = \infty$ and its "coordinate vector" $\begin{pmatrix} 1 \\ 0 \end{pmatrix}$.

7.3.11 To see the converse, we note that a point $v \in \mathbb{R}^n$ is a fixed point of the Möbius involution given by S if and only if it satisfies

$$S \begin{pmatrix} v \\ 1 \end{pmatrix} \simeq \begin{pmatrix} v \\ 1 \end{pmatrix} \quad \Leftrightarrow \quad r(S) \cdot \mathcal{V} \simeq \mathcal{V},$$

---

[15] As it would be for the product of $m = n + 1$ orthogonal hyperspheres.

[16] If S is a pure $(n - k)$-vector that squares to ± 1, then $S \in Pin(\mathbb{R}_1^{n+2})$.

[17] We have $(vS)^2 = -(-1)^{\frac{(k+2)(k+2+1)}{2}}$ since the Clifford dual of a spacelike pure $(n - k)$-vector is a timelike pure $(k + 2)$-vector (cf., §6.4.6).

where $\mathcal{V} = \begin{pmatrix} v & -v^2 \\ 1 & -v \end{pmatrix} \in \mathbb{R}_1^{n+2}$ is a lightlike vector corresponding to v and $r(\mathcal{S})$
is the twisted adjoint action of $\mathcal{S}$ on $\mathbb{R}_1^{n+2}$, as in §7.3.1. With $\mathcal{S} = \prod_{i=1}^{n-k} \mathcal{S}_i$
as before, the twisted adjoint action of $r(\mathcal{S}) \in O(\mathbb{R}_1^{n+2})$ on $\mathbb{R}_1^{n+2}$ becomes
the reflection in $n - k$ orthogonal (Minkowski) hyperplanes $\mathcal{S}_i^{\perp}$,

$$r(\mathcal{S}) \cdot \mathcal{V} = \mathcal{S}_1 \cdots \mathcal{S}_{n-k} \mathcal{V} \mathcal{S}_{n-k} \cdots \mathcal{S}_1 = \mathcal{V} + \sum_{i=1}^{n-k} \{\mathcal{V}, \mathcal{S}_i\} \mathcal{S}_i,$$

where we used that $\mathcal{S}_i^2 = -1$, so that $\Delta_i = +1$. Consequently, $r(\mathcal{S})$ has
the spacelike span$\{\mathcal{S}_1, \ldots, \mathcal{S}_{n-k}\}$ as its -1 eigenspace and the Minkowski
orthogonal complement span$\{\mathcal{S}_1, \ldots, \mathcal{S}_{n-k}\}^{\perp}$ as its $+1$ eigenspace.[18] Thus,
if $\mathcal{V}$ is an isotropic eigenvector of $\mathcal{S}$, then it must be in the $+1$ eigenspace
of $r(\mathcal{S})$, and therefore (see §7.3.7 or Figure T.1) $v \in \mathcal{S}_i$ for $i = 1, \ldots, n - k$.
Consequently, $v \in \mathcal{S}$.

Again, the same arguments apply when $v = \infty$, and we have proven the
sought characterization of the incidence of a point and a k-sphere:

7.3.12 Lemma. *A point $v \in S^n$ lies on a k-sphere $\mathcal{S} \in \Lambda^{n-k}\mathbb{R}_1^{n+2}$ if and
only if v is a fixed point of the corresponding Möbius involution, that is, if
and only if*

$$\mathcal{S} \cdot v = v \quad \Leftrightarrow \quad \mathcal{S}\begin{pmatrix} v \\ 1 \end{pmatrix} = \begin{pmatrix} v \\ 1 \end{pmatrix} \cdot a \quad or \quad \mathcal{S}\begin{pmatrix} 1 \\ 0 \end{pmatrix} = \begin{pmatrix} 1 \\ 0 \end{pmatrix} \cdot a,$$

respectively, with a suitable $a \in \Gamma(\mathbb{R}^n)$.

*If $\mathcal{S} = \prod_{i=1}^{n-k} \mathcal{S}_i$ is given as the intersection of $n - k$ orthogonal hyper-
spheres, then $\mathcal{S} \cdot v = v$ if and only if $\mathcal{S}_i \cdot v = v$ for $i = 1, \ldots, n - k$.*

7.3.13 Of particular interest to us are the 0-spheres: Remember that the
space of spacelike pure n-vectors $\mathcal{S} \in \Lambda^n \mathbb{R}_1^{n+2}$, or the space of timelike pure
2-vectors $v\mathcal{S} \in \Lambda^2 \mathbb{R}_1^{n+2}$, can be interpreted as the space of point pairs in S^n
(see Figure T.4). This space naturally turned up in the theory of (Darboux
pairs of) isothermic surfaces and their spectral (Calapso) transformations
(cf., §3.3.5 and §5.5.20), and the convenient description that this space had
in the quaternionic setup (see Section 4.5) was one of the reasons why the
quaternionic model of Möbius geometry was so well suited for our investigations on isothermic surfaces and their transformations in Chapter 5.

In the present Clifford algebra setup we have a similar description of the
space of point pairs (0-spheres) as in the quaternionic setup.

7.3.14 First remember that the Lie algebra $\mathfrak{spin}(\mathbb{R}_1^{n+2}) = \Lambda^2 \mathbb{R}_1^{n+2}$ (see
§6.3.10). We already saw in (7.2) that, in our matrix version of $\mathcal{A}\mathbb{R}_1^{n+2}$,

$$\Lambda^2 \mathbb{R}_1^{n+2} = \left\{ \begin{pmatrix} r + \mathfrak{z} & v \\ w & -r + \mathfrak{z} \end{pmatrix} \mid r \in \mathbb{R}; v, w \in \mathbb{R}^n; \mathfrak{z} \in \Lambda^2 \mathbb{R}^n \right\}.$$

---

[18] Note that this argument is quite similar to the one given to prove that any Möbius involution
decomposes into inversions (see §1.3.19f).

As in Section 4.5, we naturally obtain a decomposition of $\mathfrak{spin}(\mathbb{R}_1^{n+2})$ into diagonal and off-diagonal parts,

$$\mathfrak{spin}(\mathbb{R}_1^{n+2}) = \Lambda^2 \mathbb{R}_1^{n+2} = \mathfrak{k} \oplus \mathfrak{p} := \left\{ \begin{pmatrix} r + \mathfrak{z} & 0 \\ 0 & -r + \mathfrak{z} \end{pmatrix} \right\} \oplus \left\{ \begin{pmatrix} 0 & v \\ w & 0 \end{pmatrix} \right\},$$

which is a symmetric decomposition of $\mathfrak{spin}(\mathbb{R}_1^{n+2})$,

$$[\mathfrak{k}, \mathfrak{k}] \subset \mathfrak{k}, \quad [\mathfrak{k}, \mathfrak{p}] \subset \mathfrak{p}, \quad \text{and} \quad [\mathfrak{p}, \mathfrak{p}] \subset \mathfrak{k}$$

(cf., (4.9)). Note that $\mathfrak{k}$ is the Lie algebra of the stretch-rotations of $\mathbb{R}^n$, that is, its Lie group K consists of those (orientation-preserving in the case of the spin group) Möbius transformations that fix the "base point pair" $(\infty, 0)$:

$$K = \left\{ \begin{pmatrix} r & 0 \\ 0 & \pm\frac{1}{r} \end{pmatrix} \prod_{i=1}^k \begin{pmatrix} n_i & 0 \\ 0 & -n_i \end{pmatrix} \,\middle|\, r \in \mathbb{R} \setminus \{0\}, n_i \in S^{n-1} \subset \mathbb{R}^n \right\}.$$

Thus, as in §4.5.3, the space of point pairs — or 0-spheres — in S^n is identified as a reductive homogeneous space:

7.3.15 Lemma. *The space* $\mathfrak{P} = \{(\mathcal{P}, \mathcal{Q}) \in S^n \times S^n \mid \mathcal{Q} \neq \mathcal{P}\}$ *of point pairs in* S^n *is a homogeneous space*

$$\mathfrak{P} = Spin(\mathbb{R}_1^{n+2})/K, \quad where \quad K := \left\{ \begin{pmatrix} a & 0 \\ 0 & d \end{pmatrix} \in Spin(\mathbb{R}_1^{n+2}) \right\}$$

is the isotropy group of $0, \infty \in S^n \cong \mathbb{R}^n \cup \{\infty\}$. *This isotropy group* K *consists of the rotations and homotheties[19] of* $\mathbb{R}^n$.
The tangent space of $\mathfrak{P}$ *at* $(\infty, 0) \in \mathfrak{P}$ *is* $T_{(\infty,0)}\mathfrak{P} \cong \mathfrak{p}$.

7.4 Frames and structure equations

In §4.5.8 we introduced the notion of an "adapted frame" for a point pair map in the conformal 4-sphere — choosing $\binom{1}{0}$ and $\binom{0}{1}$ as the homogeneous coordinates for a base point pair $(\infty, 0) \in \mathfrak{P}$, we could then interpret the columns of the matrix representation $F : M \to Sl(2, \mathbb{H})$ as homogeneous coordinates of the point pair map, $F = (f_\infty, f_0)$. In the current Clifford algebra setup, a similar interpretation is possible — with some restriction caused by the fact that we do not have the notion of "homogeneous coordinates" in the same sense as in the quaternionic setup.

Thus we start with a definition analogous to the one given in §4.5.8:

---

[19] Note how a homothety of $\mathbb{R}^n$ can be written as the product of two inversions in hyperspheres (with different radii) centered at $0 \in \mathbb{R}^n$.

7.4.1 Definition. *A map* $\mathcal{F} : M \to Spin(\mathbb{R}^{n+2}_1)$ *is called an adapted frame for the point pair map* $(f_\infty, f_0) : M \to \mathfrak{P}$ *if* $f_0 = \mathcal{F} \cdot 0$ *and* $f_\infty = \mathcal{F} \cdot \infty$.

7.4.2 By writing an adapted frame $\mathcal{F} : M \to Spin(\mathbb{R}^{n+2}_1)$ for a point pair map $(f_\infty, f_0) : M \to \mathfrak{P}$ as a Vahlen matrix, $\mathcal{F} = \left(\begin{smallmatrix} a & b \\ c & d \end{smallmatrix} \right)$, we obtain

$$f_\infty = ac^{-1} : M \to \mathbb{R}^n \cup \{\infty\} \quad \text{and} \quad f_0 = bd^{-1} : M \to \mathbb{R}^n \cup \{\infty\}$$

by §7.3.2 (cf., §7.2.9 and §7.2.11). Thus the "columns" of $\mathcal{F}$ give us (formally) "homogeneous coordinates" for the legs f_0 and f_∞ of our point pair map (f_∞, f_0):

$$\begin{pmatrix} a & b \\ c & d \end{pmatrix} \begin{pmatrix} 1 \\ 0 \end{pmatrix} = \begin{pmatrix} a \\ c \end{pmatrix} = \begin{pmatrix} f_\infty \\ 1 \end{pmatrix} \cdot c \quad \text{and}$$

$$\begin{pmatrix} a & b \\ c & d \end{pmatrix} \begin{pmatrix} 0 \\ 1 \end{pmatrix} = \begin{pmatrix} b \\ d \end{pmatrix} = \begin{pmatrix} f_0 \\ 1 \end{pmatrix} \cdot d,$$

where the (right) multiplication is meant to be componentwise. On the other hand, by the definition of an adapted frame for a given point pair map (f_∞, f_0), any adapted frame is of this form as soon as $f_0, f_\infty \neq \infty$:

$$\mathcal{F} = \begin{pmatrix} f_\infty \, a_\infty & f_0 \, a_0 \\ a_\infty & a_0 \end{pmatrix},$$

where $a_0, a_\infty : M \to \Gamma(\mathbb{R}^n)$ are suitable functions with values in the Clifford group of $\mathbb{R}^n$. For these functions we derive additional properties from the fact that an adapted frame has to take values in $Spin(\mathbb{R}^{n+2}_1)$: Wheeling out the conditions (i) – (iii) from §7.2.9, we find

  (i) $f_0 a_0, a_0, f_\infty a_\infty, a_\infty \in \Theta(\mathbb{R}^n)$,

 (ii) $a_0 \hat{\bar{a}}_\infty, f_\infty a_\infty \hat{\bar{a}}_0 f_0, \hat{\bar{a}}_0 f_0 a_0, \hat{\bar{a}}_\infty f_\infty a_\infty \in \mathbb{R}^n$, and

(iii) $\Delta = f_\infty a_\infty \hat{\bar{a}}_0 - f_0 a_0 \hat{\bar{a}}_\infty \in \mathbb{R} \setminus \{0\}$,

and the additional conditions from §7.2.16 read $\Delta = \pm 1$ and

$$f_\infty a_\infty = -f_\infty \hat{a}_\infty, \quad a_\infty = -\hat{a}_\infty \quad \text{and} \quad f_0 a_0 = f_0 \hat{a}_0, \quad a_0 = \hat{a}_0.$$

Thus we must have $a_0, a_\infty : M \to \Gamma(\mathbb{R}^n)$, $\hat{a}_0 = a_0$ and $\hat{a}_\infty = -a_\infty$; moreover, with $a_0 \hat{\bar{a}}_\infty = -a_0 \bar{a}_\infty \in \mathbb{R}^n$, we then deduce

$$a_0 \hat{\bar{a}}_\infty = -a_0 \bar{a}_\infty = \overline{a_0 \bar{a}_\infty} = a_\infty \bar{a}_0 = a_\infty \hat{\bar{a}}_0 \quad \text{and}$$
$$\pm 1 = \Delta = f_\infty a_\infty \hat{\bar{a}}_0 - f_0 a_0 \hat{\bar{a}}_\infty = (f_\infty - f_0) \, a_\infty \hat{\bar{a}}_0.$$

The remaining conditions are then automatically fulfilled, so that, by reformulating the last condition in terms of the Clifford conjugation instead of the Clifford transpose, we have the following:

7.4.3 Lemma. *Given a point pair map* $(f_\infty, f_0) : M \to \mathfrak{P}$, $f_0, f_\infty \neq \infty$, *any adapted frame* $\mathcal{F} : M \to Spin(\mathbb{R}_1^{n+2})$ *is of the form*

$$\mathcal{F} = \begin{pmatrix} f_\infty a_\infty & f_0 a_0 \\ a_\infty & a_0 \end{pmatrix},$$

where the functions $a_0, a_\infty : M \to \Gamma(\mathbb{R}^n)$ *satisfy* $\hat{a}_0 = a_0$ *and* $\hat{a}_\infty = -a_\infty$, *and* $(f_\infty - f_0) a_0 \bar{a}_\infty = \pm 1$.

7.4.4 Example. If $f_0, f_\infty : M \to \mathbb{R}^n$ are two maps so that $f_0(p) \neq f_\infty(p)$ for all $p \in M$, then the pair $(f_\infty, f_0) : M \to \mathfrak{P}$ qualifies as a point pair map. Choosing $a_0 \equiv 1$ and $a_\infty := (f_\infty - f_0)^{-1}$, the map

$$\mathcal{F} = \begin{pmatrix} f_\infty a_\infty & f_0 a_0 \\ a_\infty & a_0 \end{pmatrix} = \begin{pmatrix} f_\infty (f_\infty - f_0)^{-1} & f_0 \\ (f_\infty - f_0)^{-1} & 1 \end{pmatrix} : M \to Spin(\mathbb{R}_1^{n+2})$$

provides an adapted frame for (f_∞, f_0).

In fact, $\Delta = (f_0 - f_\infty)\overline{(f_\infty - f_0)^{-1}} = +1$, so that this frame takes values in $Spin_1^+(n+2) \subset Spin(\mathbb{R}_1^{n+2})$, the double universal cover of the Möbius group (see §6.3.12).

If, however, $f_\infty \equiv \infty$ and $f_0 = f : M \to \mathbb{R}^n$, we also obtain a point pair map $(\infty, f) : M \to \mathfrak{P}$. Here,

$$\mathcal{F} = \begin{pmatrix} 1 & f \\ 0 & 1 \end{pmatrix} : M \to Spin_1^+(n+2)$$

defines an adapted frame — the "Euclidean frame" of f.

7.4.5 Any two adapted frames $\mathcal{F}, \tilde{\mathcal{F}} : M \to Spin(\mathbb{R}_1^{n+2})$ of a given point pair map $(f_\infty, f_0) : M \to \mathfrak{P}$ differ by a gauge transformation

$$\mathcal{F} \mapsto \tilde{\mathcal{F}} = \mathcal{F}\mathcal{K} \quad \text{where} \quad \mathcal{K} : M \to K,$$

takes values in the isotropy group of $\mathfrak{P} = Spin(\mathbb{R}_1^{n+2})/K$ as a symmetric space, that is, in $K = \{\mathcal{K} \in Spin(\mathbb{R}_1^{n+2}) \mid \mathcal{K}\cdot 0 = 0, \mathcal{K}\cdot\infty = \infty\}$ (see §7.3.15).

7.4.6 If, on the other hand, we are interested in the geometry of a single map $f = f_0 : M \to S^n$, we may choose an arbitrary second map f_∞ to obtain a frame $\mathcal{F} : M \to Spin(\mathbb{R}_1^{n+2})$ for f (and f_∞). However, since we are not interested in this second map f_∞, we will allow the frame $\mathcal{F}$ to change in a way that only fixes its "second leg" f, that is, we are considering gauge transformations of the form

$$\mathcal{F} \mapsto \tilde{\mathcal{F}} = \mathcal{F}\mathcal{K} \quad \text{with} \quad \mathcal{K} \cdot 0 = 0$$

but $\mathcal{K} \cdot \infty$ arbitrary. In terms of Vahlen matrices, this means that

$$\mathcal{K} : M \to K_0 := \{ \begin{pmatrix} a & 0 \\ c & d \end{pmatrix} \in Spin(\mathbb{R}_1^{n+2}) \}.$$

7.4.7 Thus, here we think of the space of (single) points in S^n as the homogeneous space

$$S^n = Spin(\mathbb{R}_1^{n+2})/K_0, \quad \text{where} \quad K_0 := \{K \in Spin(\mathbb{R}_1^{n+2}) \,|\, K \cdot 0 = 0\}$$

is the isotropy group of the chosen base point $0 \in S^n$ (cf., Section 4.6). Note that this space is, as in the quaternionic setup, a nonreductive homogeneous space because there is no complement of the isotropy algebra $\mathfrak{k}_0$ that would provide a symmetric decomposition of the Lie algebra $\mathfrak{spin}(\mathbb{R}_1^{n+2})$ — in fact, S^n is a symmetric R-space (cf., §4.6.3); for more information on symmetric R-spaces in this context, consult [52].

We summarize these ideas into a definition as in §4.6.5 for the quaternionic setup:

7.4.8 Definition. *A map $\mathcal{F} : M \to Spin(\mathbb{R}_1^{n+2})$ is called an adapted frame for a map $f : M \to S^n = Spin(\mathbb{R}_1^{n+2})/K_0$ if it is a lift of f, that is, if $\mathcal{F} \cdot 0 = f$.*

7.4.9 As in all other models, key information on the differential geometry of a map $f : M \to S^n$, or of a point pair map $(f_\infty, f_0) : M \to \mathfrak{P}$, can be encoded in the structure equations $d\mathcal{F} = \mathcal{F}\Phi$ given in terms of an (adapted) frame $\mathcal{F}$, where Φ denotes the connection form of the frame. Remember how, in §6.7.4, we established a relation between Clifford algebra framings and the classical framings in Möbius geometry, using a pseudo-orthonormal basis of $\mathbb{R}_1^{n+2}$, via the twisted adjoint action of $Spin(\mathbb{R}_1^{n+2})$ on $\mathbb{R}_1^{n+2}$ and how we extracted the classical invariants from the Clifford algebra connection form. In §4.5.12 we established a similar "translation table" for the quaternionic setup. The very same procedure works in the present Clifford algebra setup — in fact, the procedure discussed in §6.7.4 applies without change in the present situation.

7.4.10 Here we restrict ourselves to studying the effect of a gauge transformation on the Clifford algebra connection form of a frame, similar to the quaternionic case (see (4.11) and (4.14)): in fact, the formulas will be exactly the same as in the quaternionic case when using Vahlen matrices.

First note that, for $\tilde{\mathcal{F}} = \mathcal{F}\mathcal{K}$, we have

$$\tilde{\Phi} = (\mathcal{F}\mathcal{K})^{-1}d(\mathcal{F}\mathcal{K}) = \mathcal{K}^{-1}\Phi\mathcal{K} + \mathcal{K}^{-1}d\mathcal{K} = \mathcal{K}^{-1}(\Phi\mathcal{K} + d\mathcal{K}).$$

Thus, if $\mathcal{K} = \begin{pmatrix} a_\infty & 0 \\ c & a_0 \end{pmatrix} : M \to K_0$ and $\Phi = \begin{pmatrix} \varphi_\infty & \psi_0 \\ \psi_\infty & \varphi_0 \end{pmatrix}$, then

$$\tilde{\Phi} = \begin{pmatrix} a_\infty^{-1} & 0 \\ -a_0^{-1}ca_\infty^{-1} & a_0^{-1} \end{pmatrix} \begin{pmatrix} (da_\infty + \varphi_\infty a_\infty) + \psi_0 c & \psi_0 a_0 \\ \psi_\infty a_\infty + (dc + \varphi_0 c) & da_0 + \varphi_0 a_0 \end{pmatrix}, \quad (7.12)$$

as in (4.14) — note that $a_0, a_\infty : M \to \Gamma(\mathbb{R}^n)$, that is, $a_0, a_\infty \neq 0$ are invertible since $\mathcal{K} : M \to K_0 \subset Spin(\mathbb{R}_1^{n+2})$ is (cf., §7.2.9).

Specializing to the case $c \equiv 0$ so that $\mathcal{K} : M \to K$ takes values in the isotropy group of $(\infty, 0) \in \mathfrak{P}$, we obtain, as in (4.11),

$$\tilde{\Phi} = \begin{pmatrix} a_\infty^{-1}da_\infty + a_\infty^{-1}\varphi_\infty a_\infty & a_\infty^{-1}\psi_0 a_0 \\ a_0^{-1}\psi_\infty a_\infty & a_0^{-1}da_0 + a_0^{-1}\varphi_0 a_0 \end{pmatrix}. \quad (7.13)$$

7.4.11 Example. First we consider the Euclidean frame $\mathcal{F} = \begin{pmatrix} 1 & f \\ 0 & 1 \end{pmatrix}$ of a map $f : M \to \mathbb{R}^n$ (see §7.4.4): The connection form of $\mathcal{F}$ then computes to

$$\Phi = \mathcal{F}^{-1}d\mathcal{F} = \begin{pmatrix} 1 & -f \\ 0 & 1 \end{pmatrix} \begin{pmatrix} 0 & df \\ 0 & 0 \end{pmatrix} = \begin{pmatrix} 0 & df \\ 0 & 0 \end{pmatrix}.$$

If we now introduce a second map $f_\infty : M \to \mathbb{R}^n$, $f_\infty(p) \neq f(p)$ for all $p \in M$, then the respective adapted frame $\tilde{\mathcal{F}}$ from §7.4.4 for the point pair map $(f_\infty, f) : M \to \mathfrak{P}$ is obtained by a gauge transformation,

$$\tilde{\mathcal{F}} = \begin{pmatrix} f_\infty(f_\infty - f)^{-1} & f \\ (f_\infty - f)^{-1} & 1 \end{pmatrix} = \begin{pmatrix} 1 & f \\ 0 & 1 \end{pmatrix} \begin{pmatrix} 1 & 0 \\ (f_\infty - f)^{-1} & 1 \end{pmatrix} = \mathcal{F}\mathcal{K}.$$

Its connection form can then be computed from (7.12):

$$\begin{aligned} \tilde{\Phi} &= \begin{pmatrix} 1 & 0 \\ -(f_\infty - f)^{-1} & 1 \end{pmatrix} \begin{pmatrix} df(f_\infty - f)^{-1} & df \\ -(f_\infty - f)^{-1}d(f_\infty - f)(f_\infty - f)^{-1} & 0 \end{pmatrix} \\ &= \begin{pmatrix} df(f_\infty - f)^{-1} & df \\ -(f_\infty - f)^{-1}df_\infty(f_\infty - f)^{-1} & -(f_\infty - f)^{-1}df \end{pmatrix}. \end{aligned}$$

7.4.12 Finally, remember from §6.7.3 that the compatibility conditions for the structure equations become the Maurer-Cartan equation for the connection form Φ of a frame,

$$d\Phi + \tfrac{1}{2}[\Phi \wedge \Phi] = 0,$$

as can be obtained by differentiating the structure equations $d\mathcal{F} = \mathcal{F}\Phi$:

$$0 = d^2\mathcal{F} = \mathcal{F} \cdot (d\Phi + \Phi \wedge \Phi) = \mathcal{F} \cdot (d\Phi + \tfrac{1}{2}[\Phi \wedge \Phi]),$$

where $[\Phi \wedge \Psi](x,y) := [\Phi(x), \Psi(y)] - [\Phi(y), \Psi(x)]$ and $[.,.]$ denotes the anti-commutator on $A\mathbb{R}_1^{n+2}$, $[\mathfrak{v}, \mathfrak{w}] = \mathfrak{v}\mathfrak{w} - \mathfrak{w}\mathfrak{v}$, that is, the Lie bracket on the Lie algebra $\Lambda^2\mathbb{R}_1^{n+2} = \mathfrak{spin}(\mathbb{R}_1^{n+2}) \cong \mathfrak{o}_1(n+2)$ (cf., §6.3.10).

7.5 Miscellaneous

As usual, some of our findings from this chapter are collected in Figure T.5.

In this final section we will touch on a couple of issues concerning envelopes of sphere congruences and on the cross-ratio of four points. However, the presented material is mainly meant to stimulate the reader to pursue these thoughts further, and most of it will no longer be used in the following chapter.

7.5.1 Remember that a differentiable map $f : M^m \to S^n$ was said to envelope a (hyper)sphere congruence $\mathcal{S}$ if

$$f(p) \in \mathcal{S}(p) \quad \text{and} \quad d_p f(T_p M) \subset T_{f(p)} \mathcal{S}(p)$$

for all $p \in M$ (see §1.6.1). In §6.6.3, we extended this notion to envelopes of congruences of k-spheres, $k \geq m$, which must satisfy the same conditions (cf., §4.7.9). In §6.6.5, we formulated a condition for a map f to envelope a k-sphere congruence; however, one might be tempted to seek a characterization analogous to §4.8.15 in the quaternionic setup after we have already proven a similar characterization for the incidence (see §7.3.12).

7.5.2 Assuming that the points of f lie on the spheres $\mathcal{S}$, that is,

$$\mathcal{S} \begin{pmatrix} f \\ 1 \end{pmatrix} = \begin{pmatrix} f \\ 1 \end{pmatrix} a \tag{7.14}$$

with $a : M^m \to \Gamma(\mathbb{R}^n)$, a way to proceed here would be to argue that f envelopes $\mathcal{S}$ if and only if any infinitesimally neighboring points $f + df$ also lie (up to higher order terms) on $\mathcal{S}$, that is,

$$\mathcal{S} \begin{pmatrix} f + df \\ 1 \end{pmatrix} = \begin{pmatrix} f + df \\ 1 \end{pmatrix} (a + \delta a),$$

where $\delta a \in \mathcal{A}(\mathbb{R}^n)$, so that $(a + \delta a) \in \Gamma(\mathbb{R}^n)$. Neglecting higher order terms and taking (7.14) into account, we would arrive at

$$\mathcal{S} \begin{pmatrix} df \\ 0 \end{pmatrix} = \begin{pmatrix} f \\ 1 \end{pmatrix} \delta a + \begin{pmatrix} df \\ 0 \end{pmatrix} a$$

as the enveloping condition. Taking, on the other hand, derivatives in (7.14), this is equivalent to

$$d\mathcal{S} \cdot \begin{pmatrix} f \\ 1 \end{pmatrix} = \begin{pmatrix} f \\ 1 \end{pmatrix} (da - \delta a), \tag{7.15}$$

thus arriving at the very same characterization as in §4.8.15.

However, we encounter a serious problem with our characterization: Our last equation (7.15) makes no sense[20] since 2×2 matrices with Clifford algebra entries do not act on S^n in general, as already discussed in §7.3.4.

7.5.3 Example. To illustrate this we consider the following sphere congruence in $S^3 \cong \mathbb{R}^3 \cup \{\infty\}$:

$$\mathcal{S}(t) = \begin{pmatrix} te_1 & -1+t^2 \\ 1 & -te_1 \end{pmatrix} \begin{pmatrix} e_2 \cos t + e_3 \sin t & 0 \\ 0 & -e_2 \cos t - e_3 \sin t \end{pmatrix},$$

where (e_1, e_2, e_3) denotes an orthonormal basis of $\mathbb{R}^3$. This is a 1-parameter family of circles of constant radius 1 in $\mathbb{R}^3$ whose centers move on the e_1-axis and whose planes rotate around the e_1-axis as the centers move. Taking its derivative at $t = 0$ yields

$$d_0\mathcal{S} = \mathcal{S}'(0)dt = \begin{pmatrix} e_1e_2 & e_3 \\ e_3 & e_1e_2 \end{pmatrix} dt,$$

that is, a 2×2 Clifford algebra matrix that does not act on S^3, as we have seen in §7.3.4. On the other hand, the sphere congruence $\mathcal{S}$ has no envelopes anyway, as is immediately clear from its geometric construction.

Allowing different speeds for the shift of the center of the circles and for the rotation of their planes, nothing changes — unless one of the two speeds is chosen to vanish. Then the sphere congruence has two envelopes: either the two lines $t \mapsto te_1 \pm e_2$, in case no rotation takes place, or the two constant maps $\pm e_1$ if the centers of the circles are 0. In either case $\mathcal{S}'$ becomes a pure 2-vector (cf., (7.2)) that is a circle itself and, therefore, acts by the reflection in that circle on S^3.

It shall be left to the reader to pursue these considerations further.

7.5.4 Now remember the condition for an immersion $f : M^m \to S^n$ to envelope an m-sphere congruence $\mathcal{S} : M^m \to \Lambda^{n-m}\mathbb{R}_1^{n+2}$ that we derived in §6.6.7 using the notion of "contact element" (see §6.6.9) — note that the dimension $\dim M = m$ of M is the same as that of the spheres. We want to use this enveloping criterion to derive a representation for an enveloped sphere congruence that contains the points of a second map[21] $\hat{f}$.

Thus let $\mathcal{F}, \hat{\mathcal{F}} : M^2 \to \mathbb{R}_1^{n+2} \subset \mathcal{A}\mathbb{R}_1^{n+2}$ denote lifts of f and $\hat{f}$; then the condition for the immersion f to envelope $\mathcal{S}$ reads

$$v\mathcal{S}\mathcal{F} = \alpha\mathcal{T}$$

---

[20] The reader may want to analyze where, in fact, we start to produce senseless equations. But, perhaps, it is possible to "repair" these problems?

[21] Note that, here, we use $\hat{\ }$ to indicate a second map, that is, in our present considerations it does *not* denote the order involution from §6.3.1 any more.

with a suitable real function α and the contact elements $\mathcal{T}$ of f. On the other hand, if we require the points $\hat{f}(p)$ of $\hat{f}$ to lie on the spheres $\mathcal{S}(p)$, then we have from §6.4.12 that

$$\hat{\mathcal{F}}\mathcal{S} - (-1)^{n-m}\mathcal{S}\hat{\mathcal{F}} = 0, \quad \text{or} \quad \hat{\mathcal{F}}(v\mathcal{S}) = (-1)^{m+1}(v\mathcal{S})\hat{\mathcal{F}} \qquad (7.16)$$

since[22] $\hat{\mathcal{F}}v = -(-1)^n v\hat{\mathcal{F}}$. From these two formulas we now learn that our sphere congruence $\mathcal{S}$ must be given by

$$v\mathcal{S} = -\frac{1}{2\langle \mathcal{F},\hat{\mathcal{F}}\rangle} v\mathcal{S}(\mathcal{F}\hat{\mathcal{F}} + \hat{\mathcal{F}}\mathcal{F}) = -\frac{\alpha}{2\langle \mathcal{F},\hat{\mathcal{F}}\rangle}(\mathcal{T}\hat{\mathcal{F}} - (-1)^m\hat{\mathcal{F}}\mathcal{T}). \qquad (7.17)$$

Note that, in any case,

$$\mathcal{T}\hat{\mathcal{F}} - (-1)^m\hat{\mathcal{F}}\mathcal{T} = \left\{ \begin{array}{ll} [\mathcal{T},\hat{\mathcal{F}}] & \text{for } m \equiv 0 \bmod 2 \\ \{\mathcal{T},\hat{\mathcal{F}}\} & \text{for } m \equiv 1 \bmod 2 \end{array} \right\} = 2\mathcal{T} \wedge \hat{\mathcal{F}} \in \Lambda^{m+2}\mathbb{R}_1^{n+2},$$

since $\mathcal{T}$ is a pure $(m+1)$-vector (cf., §6.2.7). Hence, given $\mathcal{T}$ and a second map $\hat{f}$ with $\hat{f}(p) \neq f(p)$ for all $p \in M$, so that $\mathcal{T}\hat{\mathcal{F}} - (-1)^m\hat{\mathcal{F}}\mathcal{T} \neq 0$, the equation (7.17) defines an m-sphere $\mathcal{S}$. Moreover, with §6.4.12 we can derive that $f(p) \in \mathcal{S}(p)$, since $\mathcal{T}\mathcal{F} = 0$, so that

$$v\mathcal{S}\mathcal{F} = -\frac{\alpha}{2\langle \mathcal{F},\hat{\mathcal{F}}\rangle}\mathcal{T}\hat{\mathcal{F}}\mathcal{F} = -\frac{\alpha}{2\langle \mathcal{F},\hat{\mathcal{F}}\rangle}\mathcal{T}(\hat{\mathcal{F}}\mathcal{F} + \mathcal{F}\hat{\mathcal{F}}) = \alpha\mathcal{T} : M \to \Lambda^{m+1}\mathbb{R}_1^{n+2},$$

and with §6.6.5 we deduce that $\mathcal{F}$ envelopes $\mathcal{S}$. We directly built into (7.17) that the points of $\hat{f}$ lie on the spheres $\mathcal{S}$; however, this can also be easily double checked by using (7.16).

If we additionally wish to normalize $\mathcal{S}$ "properly," $\mathcal{S}^2 = \pm 1$ as in §7.3.9, we "normalize" $\mathcal{T}$ appropriately: Let $\mathcal{T} = \mathcal{F}\mathcal{D}$, $\mathcal{D} := d\mathcal{F}(v_1)\cdots d\mathcal{F}(v_m)$ with an orthonormal basis $(v_1,\ldots,v_m)$ with respect to the metric induced by the lift $\mathcal{F}$ of f, as in (6.12), and let

$$v\mathcal{S} := \frac{1}{(\mathcal{F}\hat{\mathcal{F}}+\hat{\mathcal{F}}\mathcal{F})}(\mathcal{T}\hat{\mathcal{F}} - (-1)^m\hat{\mathcal{F}}\mathcal{T}). \qquad (7.18)$$

Then we have $\mathcal{D}^2 = (-1)^{\frac{m(m+1)}{2}} = -(-1)^{\frac{(m+2)(m+3)}{2}}$, so that

$$\begin{aligned} (\mathcal{T}\hat{\mathcal{F}} - (-1)^m\hat{\mathcal{F}}\mathcal{T})^2 &= (\mathcal{T}\hat{\mathcal{F}})^2 + (\hat{\mathcal{F}}\mathcal{T})^2 \\ &= \mathcal{F}\mathcal{D}(\hat{\mathcal{F}}\mathcal{F} + \mathcal{F}\hat{\mathcal{F}})\mathcal{D}\hat{\mathcal{F}} + \hat{\mathcal{F}}\mathcal{F}\mathcal{D}(\hat{\mathcal{F}}\mathcal{F} + \mathcal{F}\hat{\mathcal{F}})\mathcal{D} \\ &= (\mathcal{F}\hat{\mathcal{F}} + \hat{\mathcal{F}}\mathcal{F})^2\mathcal{D}^2 \\ &= -(-1)^{\frac{(m+2)(m+3)}{2}}(\mathcal{F}\hat{\mathcal{F}} + \hat{\mathcal{F}}\mathcal{F})^2 \end{aligned}$$

---

[22] To see that $\hat{\mathcal{F}}v = -(-1)^n v\hat{\mathcal{F}}$, just write $v = \mp\frac{1}{2\langle\mathcal{F},\hat{\mathcal{F}}\rangle}(\mathcal{F} - \hat{\mathcal{F}})(\mathcal{F} + \hat{\mathcal{F}})\mathcal{E}_1 \cdots \mathcal{E}_n$, where the $\mathcal{E}_i$'s form any orthonormal basis of the orthogonal complement of $\mathcal{F}$ and $\hat{\mathcal{F}}$. Then the claim follows since $\hat{\mathcal{F}}(\mathcal{F} - \hat{\mathcal{F}})(\mathcal{F} + \hat{\mathcal{F}}) = \hat{\mathcal{F}}\mathcal{F}\hat{\mathcal{F}} = -(\mathcal{F} - \hat{\mathcal{F}})(\mathcal{F} + \hat{\mathcal{F}})\hat{\mathcal{F}}$ and $\{\mathcal{E}_i, \hat{\mathcal{F}}\} = 0$.

since $\hat{\mathcal{F}}^2 = 0$ and $\mathcal{T}^2 = 0$ with $\mathcal{T}\mathcal{F} = 0$ (cf., §7.3.9 (footnote) and §6.4.6).

Thus, using the "normalized contact elements" from (6.12), the formula (7.18) provides a proper choice of homogeneous coordinates for the enveloped sphere congruence:

7.5.5 Lemma. *Let $f : M^m \to S^n$ be an immersion and $\hat{f} : M^m \to S^n$ a map so that $\hat{f}(p) \neq f(p)$ for all $p \in M$; let $\mathcal{T}$ denote the map of (normalized) contact elements of a lift $\mathcal{F}$ of f, and let $\hat{\mathcal{F}} : M^m \to \mathbb{R}_1^{n+2}$ be a lift of $\hat{f}$. Then*

$$v\mathcal{S} := \frac{1}{\langle \mathcal{F}\hat{\mathcal{F}} + \hat{\mathcal{F}}\mathcal{F}\rangle}(\mathcal{T}\hat{\mathcal{F}} - (-1)^m \hat{\mathcal{F}}\mathcal{T}) = -\frac{\mathcal{T}\wedge\hat{\mathcal{F}}}{\langle \mathcal{F},\hat{\mathcal{F}}\rangle} : M^m \to \Lambda^{m+2}\mathbb{R}_1^{n+2} \quad (7.19)$$

defines a congruence of m-spheres $\mathcal{S} : M^m \to \Lambda^{n-m}\mathbb{R}_1^{n+2}$ that is enveloped by f and contains the points of $\hat{f}$, $\hat{f}(p) \in \mathcal{S}(p)$ for all $p \in M^m$.

7.5.6 We can use this result to generalize the notion of central sphere congruence for immersions of higher codimension[23] — remember that, in §3.4.1, we introduced the notion of central sphere for a hypersurface in the conformal n-sphere in a conformally invariant way; from this definition we then derived characterizations as the "mean curvature sphere congruence" (see §3.4.4) and as the "conformal Gauss map" (see §3.4.8). Here we will generalize the mean curvature sphere characterization of the central sphere congruence to the higher codimension setup.

Thus we introduce the central sphere congruence for an m-dimensional immersion $f : M^m \to S^n \cong \mathbb{R}^n \cup \{\infty\}$ into the conformal n-sphere to consist of those touching m-spheres $\mathcal{Z}(p)$ that share the mean curvature vector with f, considered as an immersion $f : M^m \to \mathbb{R}^n$ into Euclidean space, at the point $f(p)$ of contact. That is, we will think of the sphere $\mathcal{Z}(p)$ as the sphere that lies in the affine $(m+1)$-plane that[24] contains the (affine) tangent plane of f at $f(p)$ and the normal line given by the mean curvature vector $H(p)$ of f at $f(p)$, and that has radius $r(p) = \frac{1}{|H(p)|}$ and center $m(p) = (f + \frac{1}{|H|^2}H)(p)$.

Otherwise said, we take $\mathcal{Z}(p)$ to be the m-sphere that touches f at $f(p)$ and contains $\hat{f}(p) := (f + \frac{2}{|H|^2}H)(p)$. This description allows us to use §7.5.5 in order to encode the central spheres of an immersion, in the Clifford algebra way, as spacelike pure $(n-m)$-vectors $\mathcal{Z}(p) \in \Lambda^{n-m}\mathbb{R}_1^{n+2}$ — or as their (timelike) Clifford duals $v\mathcal{Z}(p) \in \Lambda^{m+2}\mathbb{R}_1^{n+2}$. Namely, we choose lifts

$$\mathcal{F} = \begin{pmatrix} f & -f^2 \\ 1 & -f \end{pmatrix} \quad \text{and} \quad \hat{\mathcal{F}} = \begin{pmatrix} \hat{f} & -\hat{f}^2 \\ 1 & -\hat{f} \end{pmatrix} = \mathcal{F} + \frac{2}{|H|^2}\mathcal{H},$$

---

[23] Compare [241] and [107].

[24] This geometric description fails when $H(p) = 0$ — but then the central sphere should clearly be the tangent plane of f at $f(p)$. This problem will resolve itself when proceeding.

where $\mathcal{H} := \begin{pmatrix} H & -(Hf+fH)-2 \\ 0 & -H \end{pmatrix}$ (cf., §7.2.6). Then (7.19) yields

$$v\mathcal{Z} = \tfrac{1}{(\mathcal{F}\hat{\mathcal{F}}+\hat{\mathcal{F}}\mathcal{F})}(T\hat{\mathcal{F}} - (-1)^m\hat{\mathcal{F}}T) = \tfrac{1}{(\mathcal{F}\mathcal{H}+\mathcal{H}\mathcal{F})}(T\mathcal{H} - (-1)^m\mathcal{H}T)$$

since $\mathcal{F}^2 = 0$ and $T\mathcal{F} = 0$. Note that the last expression also makes sense when $H = 0$, so that $\mathcal{H} = -2 \begin{pmatrix} 0 & 1 \\ 0 & 0 \end{pmatrix} \simeq \infty$: In this case it gives the m-sphere that touches f and contains the point ∞, that is, it defines the tangent plane of f — which is its mean curvature sphere when $H = 0$.

Since we assumed $f : M^m \to \mathbb{R}^n$ to take values in Euclidean n-space, the mean curvature vector field H of f is given by $H = \tfrac{1}{m}\Delta f$, where Δ is the Laplacian with respect to the induced metric I. Hence

$$\mathcal{H} = \tfrac{1}{m}\begin{pmatrix} \Delta f & -(f\Delta f + 2I(\nabla f, \nabla f) + \Delta f\, f) \\ 0 & -\Delta f \end{pmatrix} = \tfrac{1}{m}\Delta\mathcal{F}$$

since $I(\nabla f, \nabla f) = m$. Consequently,

$$v\mathcal{Z} = \tfrac{1}{\mathcal{F}\Delta\mathcal{F}+\Delta\mathcal{F}\,\mathcal{F}}(T\Delta\mathcal{F} - (-1)^m\Delta\mathcal{F}\,T) = -\tfrac{1}{\langle\mathcal{F},\Delta\mathcal{F}\rangle}\,T \wedge \Delta\mathcal{F}. \qquad (7.20)$$

7.5.7 To see that $v\mathcal{Z}$ is a well-defined object for immersions into the conformal n-sphere, that is, $v\mathcal{Z}$ does not depend on the choice of lift $\mathcal{F}$ for f, we first note that $T \wedge d\mathcal{F}(v) = 0$ for any $v \in TM$. As a consequence,

$$T \wedge \tilde{\Delta}(e^u\mathcal{F}) = e^{-2u}T \wedge \Delta(e^u\mathcal{F}) = e^{-u}T \wedge \Delta\mathcal{F} \qquad (7.21)$$

since, by §3.4.19, $\tilde{\Delta}\varphi = e^{-2u}(\Delta\varphi + (n-2)d\varphi(\text{grad }u))$ for any function φ on M when $\tilde{\Delta}$ denotes the Laplacian with respect to the metric $\tilde{I} = e^{2u}I$ induced by the new lift $\tilde{\mathcal{F}} := e^u\mathcal{F}$ of f. Because T scales with $\mathcal{F}$, when computed with respect to an orthonormal basis as in (6.12), we obtain

$$\tilde{T} \wedge \tilde{\Delta}\tilde{\mathcal{F}} = T \wedge \Delta\mathcal{F}.$$

Since also $\langle\tilde{\mathcal{F}}, \tilde{\Delta}\tilde{\mathcal{F}}\rangle = \langle\mathcal{F}, \Delta\mathcal{F}\rangle = -m$, the representation (7.20) of the mean curvature sphere $\mathcal{Z}$ is independent of the lift $\mathcal{F}$ chosen for f.

Moreover, (7.21) shows that the Laplacian can be taken with respect to any metric that is conformally equivalent to the induced metric of the lift without any changes — the scalings caused by a conformal change of the Laplacian and by a change of lift decouple. Thus, in the 2-dimensional case $m = 2$, we may choose any conformal coordinates (x, y) to compute the central sphere congruence, without the need to scale the lift correspondingly.

7.5.8 Lemma and Definition. *Let $f : M^m \to S^n$ be an immersion into the conformal n-sphere, and let $\mathcal{F} : M^m \to \mathbb{R}_1^{n+2}$ denote any lift of f. Then*

$$v\mathcal{Z} := \tfrac{1}{2m}(T\Delta\mathcal{F} - (-1)^m\Delta\mathcal{F}\,T)$$

defines, in a conformally invariant way, an m-sphere congruence that is enveloped by f. This sphere congruence is called the central sphere congruence, or mean curvature sphere congruence, of f.

If $m = \dim M = 2$ and (x, y) denote conformal coordinates for (M^2, I), then

$$\mathbf{v}\mathcal{Z} = \tfrac{1}{4e^{4u}}\{\mathcal{F}\mathcal{F}_x\mathcal{F}_y(\mathcal{F}_{xx} + \mathcal{F}_{yy}) - (\mathcal{F}_{xx} + \mathcal{F}_{yy})\mathcal{F}\mathcal{F}_x\mathcal{F}_y\},$$

where e^{2u} is the metric factor, that is, $I = e^{2u}(dx^2 + dy^2)$ is the metric induced by the lift $\mathcal{F}$ of f.

7.5.9 As a second application of our representation (7.19) for the sphere congruence that is enveloped by a given immersion $f : M^m \to S^n$ and contains the points of a second map $\hat{f} : M^m \to S^n$, we derive a condition for two maps f and $\hat{f}$ to envelope a congruence of m-spheres — note that the dimension of M, $m = \dim M$, is the same as that of the spheres. Since we want to simultaneously obtain a formulation for the case when both maps f and $\hat{f}$ take values in Euclidean n-space (see [47]), we will assume that

$$f, \hat{f} : M^m \to \mathbb{R}^n \subset \mathbb{R}^n \cup \{\infty\} \cong S^n.$$

Thus we choose the "Euclidean lift" $\mathcal{F} = \begin{pmatrix} f & -f^2 \\ 1 & -f \end{pmatrix} : M \to \mathbb{R}_1^{n+2}$ of f into the projective light cone (see §7.2.6). Then we can express the contact element of f at $p \in M^m$ in terms of the "tangent space map" t of f:

$$T(p) = \mathcal{F}(p)d_p\mathcal{F}(v_1)\cdots d_p\mathcal{F}(v_m) = \begin{pmatrix} ft(p) & -ftf(p) \\ t(p) & -tf(p) \end{pmatrix}, \qquad (7.22)$$

where $(v_1, \ldots, v_m)$ is an orthonormal basis of T_pM for the metric induced by $f \cong \mathcal{F}$ and $t(p) := d_pf(v_1)\cdots d_pf(v_m) \in \Lambda^m\mathbb{R}^n$ represents the tangent space[25] of $f : M \to \mathbb{R}^n$ at $f(p)$.

With this relation between the contact element of f and its tangent space map we can formulate the following:

7.5.10 Lemma. *Two immersions* $f, \hat{f} : M^m \to \mathbb{R}^n$, $f \neq \hat{f}$ *pointwise, envelope an m-sphere congruence* $\mathcal{S} : M^m \to \Lambda^{n-m}\mathbb{R}_1^{n+2}$ *if and only if*

$$\hat{\mathcal{F}}T \parallel \hat{T}\mathcal{F} \quad \Leftrightarrow \quad \hat{t}\cdot(\hat{f} - f) \parallel (\hat{f} - f)\cdot t,$$

where $T, \hat{T} : M \to \Lambda^{m+1}\mathbb{R}_1^{n+2}$ *denote the contact elements[26] of* $\mathcal{F}$ *and* $\hat{\mathcal{F}}$, *and* $t, \hat{t} : M \to \Lambda^m\mathbb{R}^n$ *are the tangent plane maps of* $f, \hat{f} : M \to \mathbb{R}^n$.

---

[25] That is, the tangent space as a linear subspace of $\mathbb{R}^n$.

[26] The contact elements T and $\hat{T}$ will usually have to be computed using different orthonormal bases since the induced conformal structures of f and $\hat{f}$ will generally be different, so that an orthogonal basis $(v_1, \ldots, v_m)$ for the conformal structure of $\mathcal{F}$ will not be orthogonal for $\hat{\mathcal{F}}$.

7.5.11 *Proof.* First we note that the equivalence of

$$(\hat{\mathcal{F}}\mathcal{T})(p) \parallel (\hat{\mathcal{T}}\mathcal{F})(p) \quad \Leftrightarrow \quad (\hat{t} \cdot (\hat{f} - f))(p) \parallel ((\hat{f} - f) \cdot t)(p)$$

is a consequence of (7.22) that is obtained by straightforward computation:

$$\hat{\mathcal{F}}\mathcal{T} = \begin{pmatrix} \hat{f} & -\hat{f}^2 \\ 1 & -\hat{f} \end{pmatrix} \begin{pmatrix} ft & -ftf \\ t & -tf \end{pmatrix} = \begin{pmatrix} \hat{f}(f - \hat{f})t & -\hat{f}(f - \hat{f})tf \\ (f - \hat{f})t & -(f - \hat{f})tf \end{pmatrix}$$

$$\hat{\mathcal{T}}\mathcal{F} = \begin{pmatrix} \hat{f}\hat{t} & -\hat{f}\hat{t}\hat{f} \\ \hat{t} & -\hat{t}\hat{f} \end{pmatrix} \begin{pmatrix} f & -f^2 \\ 1 & -f \end{pmatrix} = \begin{pmatrix} \hat{f}\hat{t}(f - \hat{f}) & -\hat{f}\hat{t}(f - \hat{f})f \\ \hat{t}(f - \hat{f}) & -\hat{t}(f - \hat{f})f \end{pmatrix}.$$

A nice geometric proof for the affine criterion is given in [47] — knowing the equivalence, this proves our lemma. Here we want to complement this proof with a more algebraic proof, making use of our Clifford algebra setting and, in particular, of our formula (7.19) for the enveloped sphere congruence.

7.5.12 For this purpose we first assume that f and $\hat{f}$ do envelope a sphere congruence $\mathcal{S}$. According to §6.6.7, this means that we have

$$\mathcal{T} \parallel v\mathcal{S}\mathcal{F} \quad \text{and} \quad \hat{\mathcal{T}} \parallel v\mathcal{S}\hat{\mathcal{F}},$$

so that our claim reduces to $\hat{\mathcal{F}}v\mathcal{S}\mathcal{F} \parallel v\mathcal{S}\hat{\mathcal{F}}\mathcal{F}$. By (7.19) this holds since

$$\hat{\mathcal{F}}(\mathcal{T}\hat{\mathcal{F}} - (-1)^m\hat{\mathcal{F}}\mathcal{T})\mathcal{F} = \hat{\mathcal{F}}\mathcal{T}\hat{\mathcal{F}}\mathcal{F} = -(-1)^m(\mathcal{T}\hat{\mathcal{F}} - (-1)^m\hat{\mathcal{F}}\mathcal{T})\hat{\mathcal{F}}\mathcal{F}.$$

To see the converse we assume that $\hat{\mathcal{F}}\mathcal{T} \parallel \hat{\mathcal{T}}\mathcal{F}$, that is, there is a (real) function λ so that[27] $\hat{\mathcal{F}}\mathcal{T} = \lambda\hat{\mathcal{T}}\mathcal{F}$. Then

$$\mathcal{T}\hat{\mathcal{F}} = -\frac{1}{(f-\hat{f})^2}(\hat{\mathcal{F}}\mathcal{F} + \mathcal{F}\hat{\mathcal{F}})\mathcal{T}\hat{\mathcal{F}} = -\frac{1}{(f-\hat{f})^2}\mathcal{F}\hat{\mathcal{F}}\mathcal{T}\hat{\mathcal{F}}$$

$$= -\frac{\lambda}{(f-\hat{f})^2}\mathcal{F}\hat{\mathcal{T}}\mathcal{F}\hat{\mathcal{F}} = -\frac{\lambda}{(f-\hat{f})^2}\mathcal{F}\hat{\mathcal{T}}(\hat{\mathcal{F}}\mathcal{F} + \mathcal{F}\hat{\mathcal{F}}) = \lambda\mathcal{F}\hat{\mathcal{T}}.$$

Hence $(\mathcal{T}\hat{\mathcal{F}} - (-1)^m\hat{\mathcal{F}}\mathcal{T}) \parallel (\mathcal{F}\hat{\mathcal{T}} - (-1)^m\hat{\mathcal{T}}\mathcal{F})$, so that (7.19) yields, up to sign, the same expression for the enveloped m-sphere congruence $\mathcal{S}$.   ◁

7.5.13 Finally, we turn to the cross-ratio of four points in $\mathbb{R}^n$. Remember from §6.5.5 that the cross-ratio $[v_1; v_2; v_3; v_4]$ of four points $v_i \in \mathbb{R}^n \subset S^n$ is given by

$$[v_1; v_2; v_3; v_4] = \frac{v_1v_2v_3v_4 + v_4v_3v_2v_1}{\{v_2, v_3\}\{v_1, v_4\}} \in \Lambda^0 \mathbb{R}_1^{n+2} \oplus \Lambda^4 \mathbb{R}_1^{n+2},$$

---

[27] Note that $\hat{\mathcal{F}}\mathcal{T}, \hat{\mathcal{T}}\mathcal{F} \neq 0$ by our assumption that, pointwise, $\hat{f} \neq f$.

where $\mathcal{V}_i = \begin{pmatrix} v_i & -v_i^2 \\ 1 & -v_i \end{pmatrix} \in \mathbb{R}_1^{n+2}$ is the isotropic vector corresponding to v_i. Writing formally, as in (4.26), $\mathcal{V}_i = \begin{pmatrix} v_i \\ 1 \end{pmatrix} (1, -v_i)$ we compute

$$\{\mathcal{V}_2, \mathcal{V}_3\} = -(v_2 - v_3)^2, \quad \{\mathcal{V}_1, \mathcal{V}_4\} = -(v_1 - v_4)^2, \quad \text{and}$$

$$\mathcal{V}_1 \mathcal{V}_2 \mathcal{V}_3 \mathcal{V}_4 + \mathcal{V}_4 \mathcal{V}_3 \mathcal{V}_2 \mathcal{V}_1 = \begin{pmatrix} v_1 \mathfrak{x} + v_4 \bar{\mathfrak{x}} & -(v_1 \mathfrak{x} v_4 + v_4 \bar{\mathfrak{x}} v_1) \\ \mathfrak{x} + \bar{\mathfrak{x}} & -(\mathfrak{x} v_4 + \bar{\mathfrak{x}} v_1) \end{pmatrix}, \qquad (7.23)$$

where $\mathfrak{x} := -(v_1 - v_2)(v_2 - v_3)(v_3 - v_4)$. To simplify the argument[28] we now assume that the cross-ratio is real-valued, that is, that the four points v_i are concircular (see §6.5.1). Then the above expression simplifies[29] to

$$\mathcal{V}_1 \mathcal{V}_2 \mathcal{V}_3 \mathcal{V}_4 + \mathcal{V}_4 \mathcal{V}_3 \mathcal{V}_2 \mathcal{V}_1 = \begin{pmatrix} (v_1 - v_4)\mathfrak{x} & 0 \\ 0 & \mathfrak{x}(v_1 - v_4) \end{pmatrix} = \mathfrak{y} \cdot 1 \in \mathbb{R} \subset \mathcal{A}\mathbb{R}_1^{n+2}$$

with $\mathfrak{y} := (v_1 - v_2)(v_2 - v_3)(v_3 - v_4)(v_4 - v_1)$.

Finally, taking into account the denominator, we obtain a formula for the cross-ratio of four points $v_i \in \mathbb{R}^n$ in terms of affine coordinates (cf., [72]):

7.5.14 Lemma. *The cross-ratio of four concircular points $v_i \in \mathbb{R}^n$ is*

$$[v_1; v_2; v_3; v_4] = (v_1 - v_2)(v_2 - v_3)^{-1}(v_3 - v_4)(v_4 - v_1)^{-1} \in \mathbb{R}.$$

---

[28] It may be a worthwhile exercise for the reader to do the computations in the general case: Here it may be of profit to express the imaginary part of the (complex) cross-ratio in terms of its norm and its real part, $|\text{Im}\,[\ldots]|^2 = |[\ldots]|^2 - (\text{Re}\,[\ldots])^2$. Remember that only the length of the "imaginary" $\Lambda^4 \mathbb{R}_1^{n+2}$ part of the above expression is conformally invariant — its direction encodes the 2-sphere that contains the four points (see §6.5.5) — and the imaginary part of the complex cross-ratio was considered to be its length.

[29] In particular, we must have $\mathfrak{x} + \bar{\mathfrak{x}} = 0$; note that $\mathfrak{y} = (v_4 - v_1)\mathfrak{y}(v_4 - v_1)^{-1}$ since $\mathfrak{y}$ is real.

Chapter 8
Applications: Orthogonal systems, isothermic surfaces

This chapter comprises two relatively independent topics: As an application of the Clifford algebra formulation of the classical model that we developed in Chapter 6, we want to discuss smooth and discrete (triply) orthogonal systems and their Ribaucour transformations; and as an application of the Vahlen matrix enhancement of the Clifford algebra model that we elaborated on in Chapter 7, we will discuss (smooth) isothermic surfaces of arbitrary codimension. However, (Darboux pairs of) isothermic surfaces can be considered as a special case of (Ribaucour pairs of) orthogonal systems, which may justify treating both topics in one chapter.

Technically, an orthogonal system can be considered as an orthogonal coordinate system on $I\!\!R^n$ (or the conformal n-sphere S^n); geometrically, it is a system of n 1-parameter families of hypersurfaces so that any two hypersurfaces from different families intersect orthogonally. Triply orthogonal systems in Euclidean space have been studied intensively in classical time (see for example [17], [94], or [136]; an excellent overview is given in [248]). However, the notion of an orthogonal system is obviously conformally invariant, and important properties (as, for example, Dupin's theorem; cf. §2.4.8) can be formulated and proven in a conformal setting — remember that we already briefly touched on triply orthogonal systems in the context Guichard nets (see §2.4.4). Also, the existence of n-orthogonal systems in general position in a Riemannian manifold of dimension $n > 3$ is related to the conformal flatness of the ambient manifold (cf., §P.5.7).

More recently, relations with integrable systems methods have stimulated new interest in orthogonal systems — in Euclidean space, where the theory is governed by Lamé's equations (2.20) or (2.22) — and n-dimensional generalizations thereof (see, for example, [168] and [308]).

In §3.1.8 we introduced the notion of a Ribaucour sphere congruence; the envelopes of such a sphere congruence are usually called Ribaucour transforms of each other: They form a "Ribaucour pair" of surfaces. Our Darboux pairs of isothermic surfaces (see §3.2.5) were special cases of such Ribaucour pairs. This notion of Ribaucour pairs can be generalized to higher dimensions and other codimensions (including codimension 0), as we will discuss later. An excellent and comprehensive treatment of the Ribaucour transformation can be found in the second volume of [28], or in [106].

Bianchi's permutability theorem for the Ribaucour transformation (cf., [27]) is closely related to the integrable systems approach to orthogonal systems (cf., [123] or [187]). One possible approach to discrete orthogonal systems is via a version of this permutability theorem, as mentioned in [35].

The respective geometric definition was first proposed by A. Bobenko at a conference in 1996 (see [31]) and was taken up in various publications. The theory of discrete orthogonal systems defined using this ansatz has many features in common with the analogous smooth theory, in particular from the point of view of integrable systems (cf., [74], [104], and [166]; see also [103] and [105]).

In this chapter we will discuss Ribaucour pairs of orthogonal systems (of any dimension and codimension), taking a different approach in two ways. Because orthogonal systems as well as the Ribaucour transformation are both Möbius invariant notions, we will work in a Möbius geometric setup, in contrast to the aforementioned papers. The advantage is that certain geometric features are described with more ease; the drawback, on the other hand, is that Lamé's equations, playing a central role in many papers, are replaced by a more complicated system of differential equations.

Second — and this is where the Clifford algebra model that we just elaborated on plays a crucial role — we will define discrete orthogonal systems via a discretization of the structure equations in the smooth case. Besides the fact that this discretization procedure is very straightforward, it also provides a lot of structure that we will derive: Among other things, we will obtain a geometric description extending the aforementioned geometric definition of orthogonal systems as "circular lattices."

This was the approach that we took in [35]. In the present chapter, we will loosely follow the presentation of the material in [35], elaborating on some topics in more detail. In particular, we will give the exact discrete analog of Bianchi's permutability theorem for the Ribaucour transformation and for the Demoulin family of orthogonal systems, as it is formulated in [28].

After our discussion of smooth and discrete orthogonal systems, we will return to smooth isothermic surfaces, as an application of the Vahlen matrix approach to Möbius geometry: We will recover various results from Chapter 5 on the transformations of isothermic surfaces in arbitrary codimension — to spare the reader, the presentation will be kept rather sketchy at many places. For more details, as well as for further reading, the reader shall be pointed to F. Burstall's excellent paper [47]; in fact, our discussion in this chapter is to a large extent inspired by [47], and we will in various aspects follow the presentation there. Other good references for (transformations of) isothermic surfaces of arbitrary codimension may be [215], where a higher dimensional version of Christoffel's problem (cf., §3.3.11) is investigated; compare also [84] and [250], where the integrable geometry of isothermic surfaces in Euclidean n-space is investigated and where their transformations are treated in the spirit of [106].

Thus, in the last two sections of this chapter, we will work along the transformation theory of isothermic surfaces developed in Chapter 5, recovering all isothermic transformations in the Vahlen matrix setup for Möbius geometry. Most of the theory will turn out to be very similar, and many proofs can be carried over with only minor modifications. However, there are some differences that we will try to find out. Also, the material will be organized in a slightly different way that may provide the reader with new viewpoints. In particular, the Darboux transformation will be treated separately, after the other transformations are discussed, and their relation with curved flats will be emphasized. Besides giving an n-dimensional generalization of the transformation theory of isothermic surfaces, these two sections are also meant to provide the reader with a brief overview of the transformation theory of isothermic surfaces — therefore, any examples are omitted (nice examples, which generalize the examples of constant mean curvature surfaces given in Chapter 5, are discussed in [47]), and a systematic discussion of the permutability theorems, as in Section 5.6, shall be left for the reader as an exercise.

The contents of this chapter are organized as follows:

Section 8.1. In this preparatory section we discuss basic facts about n-orthogonal systems in the conformal n-sphere S^n: Lamé's functions are introduced and Dupin's theorem is proven. The structure equations for an n-orthogonal system are analyzed, and the geometry of coordinate submanifolds is investigated.

Section 8.2. The notions of Ribaucour sphere congruence and Ribaucour pairs of m-orthogonal systems in the conformal n-sphere are introduced. Using adapted frames, the structure equations for Ribaucour pairs of orthogonal systems are derived.

Section 8.3. The structure equations for Ribaucour pairs of m-orthogonal systems obtained in the previous section are discretized, and Ribaucour pairs of discrete m-orthogonal nets are defined via these discrete structure equations. Then the resulting geometry is analyzed: In particular, it is shown that the occurring orthogonal nets are concircular and that the nets of a Ribaucour pair "envelope" a discrete sphere congruence.

Section 8.4. In this section the discrete structure equations are interpreted geometrically. On these grounds, constructions of (Ribaucour pairs of) discrete orthogonal nets are presented. In particular, a Cauchy-type problem for discrete orthogonal nets is solved. As an example, discrete analogs of surfaces of revolution, cylinders, and cones are discussed.

Section 8.5. We conclude our investigations of Ribaucour pairs of discrete orthogonal systems with a discussion of the Demoulin family of Ribaucour transforms of a given orthogonal net — which is given by two "seed" Ribaucour transforms — and of Bianchi's permutability theorem for the Ribaucour transformation of discrete orthogonal nets.

Section 8.6. In this section we turn to smooth isothermic surfaces of an arbitrary codimension: Its Christoffel, Goursat, and Calapso transformations are introduced, similar to the way it was done in Chapter 5, and the geometry of these transformations is discussed. In particular, it is shown that the Christoffel transformation provides solutions to Christoffel's problem discussed in Chapter 3, and that the Calapso transformation yields a second-order deformation for isothermic surfaces.

A key role is played by the retraction form that can be associated to an isothermic surface, firstly because it provides a "master Christoffel transform," with values in the 2×2 Clifford algebra matrices, that encodes all possible Christoffel transforms of an isothermic surface in the conformal n-sphere and, secondly, because it yields, when interpreted as a connection form for a map into the Möbius group, the Calapso transformation for the isothermic surface.

Section 8.7. In this final section the Darboux transformation for isothermic surfaces in the conformal n-sphere will be discussed. Analyzing the geometry of Darboux pairs, we make contact with Blaschke's problem from Chapter 3, and, providing their interpretation as curved flats in the symmetric space of point pairs in S^n, we make contact with the corresponding integrable systems approach.

Remark. In this chapter the reader is expected to be familiar with the Clifford algebra approach to Möbius geometry (Chapter 6) and, for the last two sections, with the Vahlen matrix formalism (Chapter 7). Some knowledge about the transformation theory of isothermic surfaces from Chapter 3 or Chapter 5 may be helpful but should not be necessary.

8.1 Dupin's theorem and Lamé's equations

We start by deriving some fundamental facts on n-orthogonal systems in Euclidean n-space, that is, systems of n different 1-parameter families of hypersurfaces in $\mathbb{R}^n$ so that any hypersurfaces from different families intersect orthogonally. Choosing the family parameters $x_i : \mathbb{R}^n \supset V \to \mathbb{R}$, $i = 1, \ldots, n$, so that $\operatorname{grad} x_i \neq 0$, then yields locally curvilinear coordinates on $V \subset \mathbb{R}^n$: Since the gradients $\operatorname{grad} x_i$ point along the orthogonal trajectories of the families $x_i = const$, they are pairwise orthogonal and, in particular, linearly independent at any given point. Thus the inverse mapping theorem ensures that the x_i's locally form a coordinate system.

For the purposes of this chapter we will adopt the opposite viewpoint by considering the parametrization $f = (x_1, \ldots, x_n)^{-1} : \mathbb{R}^n \supset U \to \mathbb{R}^n$ as the n-orthogonal system. In fact, the condition $\frac{\partial}{\partial x_i} f \perp \frac{\partial}{\partial x_j} f$ for $i \neq j$ on f to provide an orthogonal system is obviously invariant under conformal changes of the ambient metric. Thus we define:

8.1.1 Definition. *An immersion* $f : \mathbb{R}^n \supset U \to S^n$ *into the conformal n-sphere will be called an n-orthogonal system if* $\frac{\partial}{\partial x_i} f \perp \frac{\partial}{\partial x_j} f$ *for* $i \neq j$.

8.1.2 Choosing any lift $f : U \to L^{n+1} \subset \mathbb{R}_1^{n+2}$ of an n-orthogonal system, we fix an induced metric $I = \langle df, df \rangle$ on U. The Gaussian basis fields $\frac{\partial}{\partial x_i}$ are then orthogonal with respect to that metric,

$$I(\tfrac{\partial}{\partial x_i}, \tfrac{\partial}{\partial x_j}) = \langle \tfrac{\partial}{\partial x_i} f, \tfrac{\partial}{\partial x_j} f \rangle = 0 \quad \text{for} \quad i \neq j.$$

Their lengths give the Lamé functions of the system:

8.1.3 Definition. *The positive functions* l_i *given by* $l_i^2 = I(\tfrac{\partial}{\partial x_i}, \tfrac{\partial}{\partial x_i})$ *are called the Lamé functions with respect to a lift* $f : U \to L^{n+1}$ *of an n-orthogonal system.*

8.1.4 Next we determine the Levi-Civita connection of the induced metric: By Koszul's identity we have

$$
\begin{aligned}
2I(\nabla_{\frac{\partial}{\partial x_i}} \tfrac{\partial}{\partial x_j}, \tfrac{\partial}{\partial x_k}) &= \tfrac{\partial}{\partial x_i} I(\tfrac{\partial}{\partial x_j}, \tfrac{\partial}{\partial x_k}) + \tfrac{\partial}{\partial x_j} I(\tfrac{\partial}{\partial x_i}, \tfrac{\partial}{\partial x_k}) - \tfrac{\partial}{\partial x_k} I(\tfrac{\partial}{\partial x_i}, \tfrac{\partial}{\partial x_j}) \\
&= 2(\delta_{jk} l_j \tfrac{\partial l_j}{\partial x_i} + \delta_{ik} l_i \tfrac{\partial l_i}{\partial x_j} - \delta_{ij} l_i \tfrac{\partial l_i}{\partial x_k}).
\end{aligned}
$$

In particular, $\nabla_{\frac{\partial}{\partial x_i}} \tfrac{\partial}{\partial x_j} = \tfrac{1}{l_j} \tfrac{\partial l_j}{\partial x_i} \tfrac{\partial}{\partial x_j} + \tfrac{1}{l_i} \tfrac{\partial l_i}{\partial x_j} \tfrac{\partial}{\partial x_i}$ for $i \neq j$. Writing this in terms of the normalized vector fields $v_i = \partial_i = \tfrac{1}{l_i} \tfrac{\partial}{\partial x_i}$, we obtain (cf., (2.16)),

$$\nabla_{v_j} v_i = -k_{ij} v_j \quad \text{with} \quad k_{ij} := -\tfrac{\partial_i l_j}{l_j}. \tag{8.1}$$

Because the v_i's provide unit normal vector fields for the coordinate hypersurfaces $x_i = const$, this tells us that the v_j, $j \neq i$, are curvature directions for the hypersurfaces $x_i = const$, with k_{ij} being the corresponding principal curvatures. Since the $v_j = \tfrac{1}{l_j} \tfrac{\partial}{\partial x_j}$ are, moreover, parallel to our Gaussian basis fields, we have proved the following theorem (cf., §2.4.8):

8.1.5 Theorem (Dupin). *The coordinate functions* x_j, $j \neq i$, *provide principal curvature line coordinates for the hypersurfaces* $x_i = const$.

8.1.6 Note that this situation is rather exceptional: In higher dimensions, $n \geq 4$, hypersurfaces generically do not carry principal coordinates at all.

Moreover, we learn from (8.1) that the v_{i_j}, $j = 1, \dots, m$, provide parallel normal fields for any intersection submanifold $(x_{i_1}, \dots, x_{i_m}) = (c_1, \dots, c_m)$, $m \leq n - 1$, so that the corresponding Weingarten tensor fields simultaneously diagonalize, with eigenvalues $k_{i_j i_k}$, $j = 1, \dots, m$ and $k = m+1, \dots, n$:

8.1.7 Corollary. *Any intersection submanifold* $x_{i_j} = const$, $j = 1, \dots, m$, *of an n-orthogonal system has flat normal bundle, and the remaining coordinate functions* x_{i_k}, $k = m + 1, \dots, n$, *provide curvature line coordinates.*

8.1.8 Having fixed a pseudo-orthonormal basis $(e_0, e_1, \ldots, e_n, e_\infty)$ of $\mathbb{R}^{n+2}_1$, where $e_0 e_\infty + e_\infty e_0 = 1$, we may associate a frame $\mathfrak{z} : U \to Spin_1(n+2)$ with an orthogonal system $f : U \to S^n$ satisfying

$$f = \mathfrak{z} e_0 \mathfrak{z}^{-1} \quad \text{and} \quad \partial_i f = \mathfrak{z} e_i \mathfrak{z}^{-1} =: s_i; \quad \text{let} \quad \mathfrak{z} e_\infty \mathfrak{z}^{-1} =: \hat{f}.$$

Introducing the connection form $\varphi = \mathfrak{z}^{-1} d\mathfrak{z}$, we write the structure equations as usual (see §1.7.4 and §6.7.4):

$$
\begin{aligned}
df &= \mathfrak{z} \, [\varphi, e_0] \, \mathfrak{z}^{-1} &&= \mathfrak{z}(&& \textstyle\sum_{j=1}^n \omega_j e_j &&)\mathfrak{z}^{-1} \\
ds_i &= \mathfrak{z} \, [\varphi, e_i] \, \mathfrak{z}^{-1} &&= \mathfrak{z}(2\omega_i e_\infty && - \textstyle\sum_{j=1}^n \omega_{ij} e_j && - \sigma_i e_0)\mathfrak{z}^{-1} \quad (8.2) \\
d\hat{f} &= \mathfrak{z}[\varphi, e_\infty]\mathfrak{z}^{-1} &&= \mathfrak{z}(&& - \tfrac{1}{2}\textstyle\sum_{j=1}^n \sigma_j e_j &&)\mathfrak{z}^{-1},
\end{aligned}
$$

where $\omega_i(v_j) = \delta_{ij}$ and ω_{ij} are the connection forms of the induced metric, satisfying $\omega_{ij} + \omega_{ji} = 0$; the geometry of the forms σ_i will be clarified in a moment. Thus $\varphi : TU \to \Lambda^2 \mathbb{R}^{n+2}_1$ takes the form[1]

$$\varphi = -\tfrac{1}{2}\textstyle\sum_{j=1}^n \sigma_j e_j e_0 - \tfrac{1}{4}\textstyle\sum_{j,k=1}^n \omega_{jk} e_j e_k + \textstyle\sum_{j=1}^n \omega_j e_j e_\infty. \qquad (8.3)$$

By Dupin's theorem in §8.1.5 we have $\omega_{ij} = k_{ij}\omega_j - k_{ji}\omega_i$, where k_{ij} denote the principal curvatures of the coordinate hypersurfaces $x_i = const$ (see (8.1)). Hence, when we set $k_{jj} := 0$,

$$
\begin{aligned}
\varphi &= -\tfrac{1}{2}\textstyle\sum_{i=1}^n \sigma_i e_i e_0 - \tfrac{1}{2}\textstyle\sum_{j,i=1}^n k_{ji}\omega_i e_j e_i + \textstyle\sum_{i=1}^n \omega_i e_i e_\infty \\
&= \textstyle\sum_{i=1}^n e_i \, [-\tfrac{1}{2}\sigma_i e_0 + (\tfrac{1}{2}\textstyle\sum_{j=1}^n k_{ji} e_j + e_\infty) \omega_i].
\end{aligned}
\qquad (8.4)
$$

8.1.9 The Maurer-Cartan equation $d\varphi + \tfrac{1}{2}[\varphi \wedge \varphi] = 0$ is equivalent to:

$$
\begin{aligned}
0 = d^2 f &= \left(\textstyle\sum_{i=1}^n \omega_i \wedge \sigma_i\right) f + \textstyle\sum_{i=1}^n (d\omega_i + \textstyle\sum_{j=1}^n \omega_{ij} \wedge \omega_j) s_i \\[4pt]
0 = d^2 s_i &= -(d\sigma_i + \textstyle\sum_{j=1}^n \omega_{ij} \wedge \sigma_j) f \\
&\quad - \textstyle\sum_{j=1}^n (\varrho_{ij} - \sigma_i \wedge \omega_j - \omega_i \wedge \sigma_j) s_j \qquad (8.5) \\
&\quad + 2(d\omega_i + \textstyle\sum_{j=1}^n \omega_{ij} \wedge \omega_j) \hat{f} \\[4pt]
0 = d^2 \hat{f} &= -\tfrac{1}{2}\textstyle\sum_{i=1}^n (d\sigma_i + \textstyle\sum_{j=1}^n \omega_{ij} \wedge \sigma_j) s_i + \left(\textstyle\sum_{i=1}^n \sigma_i \wedge \omega_i\right) \hat{f},
\end{aligned}
$$

where $\varrho_{ij} := d\omega_{ij} + \textstyle\sum_{k=1}^n \omega_{ik} \wedge \omega_{kj}$ denote the curvature forms of the induced metric (cf., §1.7.4). From the Ricci equation $\textstyle\sum_{i=1}^n \omega_i \wedge \sigma_i = 0$ we learn that the $(2,0)$-tensor $s(v,w) := \textstyle\sum_{i=1}^n \sigma_i(v)\omega_i(w)$ is symmetric, and, by the Gauss equations $\varrho_{ij} = \sigma_i \wedge \omega_j + \omega_i \wedge \sigma_j$, this tensor is the Schouten

---

[1] Because of the twofold skew-symmetry each term in the middle sum occurs twice.

tensor of the induced metric $I = \sum_{i=1}^{n} \omega_i^2$, while its Weyl tensor vanishes (cf., §1.7.7):

$$
\begin{aligned}
\varrho_i(v) &:= \sum_{j=1}^{n} \varrho_{ij}(v, v_j) &&= \sum_{j=1}^{n} (\sigma_i \wedge \omega_j + \omega_i \wedge \sigma_j)(v, v_j), \\
&&&= (n-2)\sigma_i(v) + (\sum_{j=1}^{n} \sigma_j(v_j))\omega_i(v), \\
\varrho &:= \sum_{i=1}^{n} \varrho_i(v_i) &&= (2n-2)\sum_{j=1}^{n} \sigma_j(v_j),
\end{aligned}
$$

where $\varrho_i(v) = ric(v, v_i)$ and $\varrho = \mathrm{tr}_I ric$ denote the Ricci forms and the scalar curvature of the induced metric I, so that

$$
\sigma_i(v) = \tfrac{1}{n-2}(\varrho_i(v) - \tfrac{\varrho}{2(n-1)}\omega_i(v)).
$$

The vanishing of the Weyl tensor is indeed no surprise, since the metric induced from a local parametrization of the conformal n-sphere clearly should be conformally flat. Moreover, the Codazzi equations $d\sigma_i + \sum_j \omega_{ij} \wedge \sigma_j = 0$ are the condition on the Schouten tensor to be a Codazzi tensor: By the theorem of Weyl-Schouten from §P.5.1, this is the condition for conformal flatness in the case of dimension $n = 3$; in higher dimensions it is a consequence of the vanishing Weyl tensor.

8.1.10 If we choose $f : U \to Q_0^n \subset L^{n+1}$ to take values in the Euclidean n-space given by $\mathcal{K}_0 := 2e_\infty$ (see §1.4.1), then $\mathfrak{z}e_\infty\mathfrak{z}^{-1} \equiv e_\infty$ (see §1.4.13). Hence $\sigma_i = 0$, so that the Gauss equations now read $\varrho_{ij} = 0$.
   Writing $\omega_{ij} = k_{ij}\omega_j - k_{ji}\omega_i$ we have

$$
\begin{aligned}
\varrho_{ij} &= dk_{ij} \wedge \omega_j - dk_{ji} \wedge \omega_i \\
&+ \sum_{l=1}^{n} k_{li}(k_{jl} - k_{ji})\omega_i \wedge \omega_l + k_{lj}(k_{ij} - k_{il})\omega_j \wedge \omega_l - k_{li}k_{lj}\omega_i \wedge \omega_j,
\end{aligned}
$$

with the principal curvatures k_{ij} of the coordinate hypersurfaces $x_i = const$ (see (8.1)). Thus the Gauss equations yield two sets of equations:

$$
\begin{aligned}
0 &= \varrho_{ij}(v_i, v_j) &&= \partial_i k_{ij} + \partial_j k_{ji} - (k_{ij}^2 + k_{ji}^2 + \sum_{l \neq i,j} k_{li}k_{lj}), \\
0 &= \varrho_{ij}(v_l, v_j) &&= \partial_l k_{ij} - k_{lj}(k_{ij} - k_{il}), \quad \text{for} \quad l \neq i, j.
\end{aligned} \tag{8.6}
$$

In the case of dimension $n = 3$, these are the classical Lamé equations (2.20) (see [248]). Using $k_{ij} = -\frac{1}{l_i l_j}\frac{\partial}{\partial x_i}l_j$ (see (8.1)), Lamé's equations (8.6) become

$$
\begin{aligned}
0 &= \tfrac{\partial}{\partial x_i}(\tfrac{1}{l_i}\tfrac{\partial l_j}{\partial x_i}) + \tfrac{\partial}{\partial x_j}(\tfrac{1}{l_j}\tfrac{\partial l_i}{\partial x_j}) + \sum_{k \neq i,j} \tfrac{1}{l_k^2}\tfrac{\partial l_i}{\partial x_k}\tfrac{\partial l_j}{\partial x_k}, \\
0 &= \tfrac{\partial}{\partial x_k}\tfrac{\partial}{\partial x_i}l_j - \tfrac{1}{l_i}\tfrac{\partial l_j}{\partial x_i}\tfrac{\partial l_i}{\partial x_k} - \tfrac{1}{l_k}\tfrac{\partial l_j}{\partial x_k}\tfrac{\partial l_k}{\partial x_i}
\end{aligned}
$$

for i, j, k pairwise distinct indices: The $\frac{1}{2}n(n-1)(2n-3)$ equations from (8.6) reduce to a system of $\frac{1}{2}n(n-1)^2$ equations. Again, in the case $n = 3$, the classical equations (2.22) are recovered.

8.2 Ribaucour pairs of orthogonal systems

We already discussed the notion of Ribaucour (hyper)sphere congruences in §3.1.8. In the case where the two envelopes f and $\hat{f}$ of a hypersphere congruence s in the conformal n-sphere are nondegenerate, that is, where the two envelopes are regular hypersurfaces in S^n, we saw that the sphere congruence is Ribaucour if and only if the curvature directions[2] on both envelopes correspond (see §3.1.6).

8.2.1 Now suppose that $s : M^{n-1} \to S^n$ is a Ribaucour sphere congruence, and let $s_1, \ldots, s_{n-1}$ denote the hypersphere congruences associated with the (common) principal directions of f and $\hat{f}$, that is, any s_i intersects s orthogonally in f and $\hat{f}$, and is orthogonal to the i-th principal direction of f and $\hat{f}$. If f (and, therefore, $\hat{f}$) carries principal coordinates $(x_1, \ldots, x_{n-1})$, then each coordinate submanifold $f|_{x_i=const}$, or $\hat{f}|_{x_i=const}$, is tangent to the congruence $ss_i|_{x_i=const}$ of $(n-2)$-spheres.

Clearly this argument carries over to any codimension: The coordinate submanifolds $(x_{i_1}, \ldots, x_{i_{k-1}}) = (c_1, \ldots, c_{k-1})$ of f as well as $\hat{f}$ are tangent to the $(n-k)$-sphere congruence $ss_{i_1} \cdots s_{i_{k-1}}|_{(x_{i_1}, \ldots, x_{i_{k-1}})=(c_1, \ldots, c_{k-1})}$, that is, the restrictions $f|_{(x_{i_1}, \ldots, x_{i_{k-1}})=(c_1, \ldots, c_{k-1})}$ and $\hat{f}|_{(x_{i_1}, \ldots, x_{i_{k-1}})=(c_1, \ldots, c_{k-1})}$ of f and $\hat{f}$ envelope this congruence of $(n-k)$-spheres (see §6.6.3).

This observation suggests a definition of Ribaucour congruence in higher codimension (compare [85]):

8.2.2 Definition. *A congruence* $\mathfrak{s} : M^m \to \Lambda^{n-k}\mathbb{R}_1^{n+2}$ *of k-spheres with two regular envelopes* $f, \hat{f} : M^m \to S^n$ *is called Ribaucour if f and $\hat{f}$ have*[3] *corresponding curvature directions.*

The two (regular) envelopes $f, \hat{f} : M^m \to S^n$ *of a Ribaucour m-sphere congruence* $\mathfrak{s} : M^m \to \Lambda^{n-m}\mathbb{R}_1^{n+2}$ *are said to form a Ribaucour pair* $(f, \hat{f})$.

8.2.3 Now remember that any of the coordinate submanifolds of an n-orthogonal system carry principal curvature line coordinates (see §8.1.7). We condense our above observation on corresponding coordinate submanifolds of a Ribaucour pair in the following definition (cf., [26]):

---

[2] We talk of curvature *directions* (not coordinates) that always exist on a hypersurface.

[3] Remember that the existence of well-defined curvature directions is equivalent to the flatness of the normal bundles of f and $\hat{f}$ (cf., Section P.6). This definition is more restrictive than it needs to be: The fact that f and $\hat{f}$ do envelope an m-sphere congruence can be used to identify their normal bundles, so that one could ask for the curvature directions with respect to corresponding normals to correspond. However, this seems unnecessarily complicated for our purposes: The above definition will be entirely satisfactory for our present discussion of (discrete) orthogonal systems.

A discussion of the general Ribaucour transformation and the respective Bianchi permutability theorem shall be given — in its "proper" setup of Lie sphere geometry — in [51].

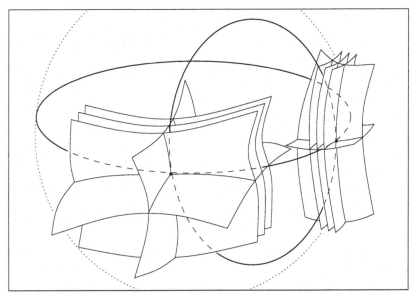

Fig. 8.1. A Ribaucour pair of triply orthogonal systems

8.2.4 Definition. *An immersion* $f : M^m \to S^n$ *is called an m-orthogonal system if it carries principal curvature line coordinates.*[4]

A pair of orthogonal systems $f, \hat{f} : M^m \to S^n$ *will be called a Ribaucour pair of orthogonal systems if any two corresponding coordinate submanifolds of dimension* $k \leq m$ *form a Ribaucour pair (of submanifolds). The orthogonal systems of a Ribaucour pair will be called Ribaucour transforms of each other.*

8.2.5 Thus for $m < n$ we regain the above notion of Ribaucour pairs with the additional demand that the corresponding curvature directions of both immersions come from coordinates: The condition on the k-dimensional coordinate submanifolds to form Ribaucour pairs is then also a consequence.

In the case $m = n$, however, the condition on two n-orthogonal systems to form a Ribaucour pair *of submanifolds* is empty — with the understanding that any orthogonal coordinate system can be considered as curvature line coordinates in codimension $n - m = 0$, any pair of orthogonal systems (that are parametrized over the same coordinate domain) would form a Ribaucour pair of submanifolds, with the (constant) enveloped Ribaucour congruence being the conformal n-sphere itself. Here the condition on any

---

[4] This definition also works for $m = n$ if, in codimension $n - m = 0$, we consider *any* orthogonal coordinate system as principal curvature line coordinates.

coordinate submanifolds to form Ribaucour pairs of submanifolds also becomes important — for $k = m - 1$, for the lower dimensions $k < m - 1$, it follows from the above observations as before.

In Figure 8.1 a Ribaucour pair of triply orthogonal systems in Euclidean 3-space is sketched. On the 2-sphere touching two of the coordinate surfaces of the system in corresponding points there are two circles that are tangent to the coordinate lines of the system, that is, to the curvature lines of the two surfaces.

8.2.6 In the case $m = 2$, the existence of principal curvature line *coordinates* is not a problem as soon as there are well-defined principal curvature *directions*: This is because any line field on a (2-dimensional) surface is integrable. Thus, in that case, the two envelopes of a Ribaucour 2-sphere congruence form a Ribaucour pair of 2-orthogonal systems without further conditions.

Note that any Darboux pair of isothermic surfaces in S^3 or S^4 (see §3.2.5 and §5.4.23) forms a Ribaucour pair of 2-orthogonal systems.

8.2.7 In some context we will weaken our assumption on the members of a Ribaucour pair being immersed: For example, it will be convenient to consider m-orthogonal systems in Euclidean n-space, coupled with the point at infinity, as a degenerate case of Ribaucour pairs.

8.2.8 After fixing our notions, we want to study the geometry of Ribaucour pairs of m-orthogonal systems a bit further. In particular, we want to derive the structure equations for suitably adapted frames: These will serve as a starting point for developing the analogous discrete theory.

Thus let $f, \hat{f} : \mathbb{R}^m \supset U \to S^n$ denote a Ribaucour pair of m-orthogonal systems, given in terms of their common (orthogonal) curvature line coordinates $(x_1, \ldots, x_m)$, and let $\mathfrak{z} : U \to Spin_1(n + 2)$ be a frame with

$$f = \mathfrak{z}e_0\mathfrak{z}^{-1}, \quad s_i := \mathfrak{z}e_i\mathfrak{z}^{-1} \quad \text{and} \quad \hat{f} = \mathfrak{z}e_\infty\mathfrak{z}^{-1},$$

where $(e_0, e_1, \ldots, e_n, e_\infty)$ denotes a pseudo-orthonormal basis of $\mathbb{R}_1^{n+2}$ as usual. By arranging the s_i, $i = m{+}1, \ldots, n$, to give the enveloped Ribaucour m-sphere congruence $\mathfrak{s}$, that is, $\mathfrak{s} = s_{m+1} \cdots s_n$, we can express the first and second fundamental forms of f and $\hat{f}$ in terms of the forms given in (6.13):

$$I = \sum_{j=1}^m \omega_j^2, \quad I\!I_i = \sum_{j=1}^m \psi_{ij}\omega_j \quad \text{and} \quad \hat{I} = \sum_{j=1}^m \hat{\omega}_j^2, \quad \hat{I\!I}_i = \sum_{j=1}^m \psi_{ij}\hat{\omega}_j.$$

Since we have common principal curvature directions for f and $\hat{f}$, given by our coordinates $(x_1, \ldots, x_m)$, we may choose (by the Cartan-Moore lemma; cf., [197]) the tangential frame to simultaneously diagonalize the Weingarten tensors of f and $\hat{f}$ so that

$$\omega_j = l_i dx_i, \quad \hat{\omega}_j = \hat{l}_j dx_j, \quad \text{and} \quad \psi_{ij} = a_{ij} dx_j.$$

From the Ricci equations (6.14) we then learn that

1. $\varrho_{ij}^{\perp} = 0$, so that the frame $\mathfrak{z}$ can be chosen to be $\mathfrak{s}$-adapted, $\nu_{ij} = 0$; and

2. $d\nu = 0$, so that f and $\hat{f}$ can be rescaled to also obtain $\nu = 0$.

Otherwise said, $s_{m+1}, \ldots, s_n, f, \hat{f}$ can be chosen to be parallel sections of the spanned vector bundle $p \mapsto \text{span}\{s_{m+1}(p), \ldots, s_n(p), f(p), \hat{f}(p)\}$. Now the common "tangential" basis fields s_i, $i = 1, \ldots, m$, point along the partial derivatives of any of these (parallel) normal sections: For $i = m + 1, \ldots, n$ and $j = 1, \ldots, m$ we have

$$\frac{\partial}{\partial x_j} f = l_j s_j, \quad \frac{\partial}{\partial x_j} \hat{f} = \hat{l}_j s_j, \quad \text{and} \quad \frac{\partial}{\partial x_j} s_i = a_{ij} s_j.$$

In particular, the frame $\mathfrak{z}$ is f-, $\hat{f}$-, and $\mathfrak{s}$-adapted in the sense of §6.7.1 at the same time:

8.2.9 Definition. *A map* $\mathfrak{z} : \mathbb{R}^m \supset U \to Spin_1(n+2)$ *will be called an adapted frame for a Ribaucour pair* $f = \mathfrak{z}e_0\mathfrak{z}^{-1}, \hat{f} = \mathfrak{z}e_\infty\mathfrak{z}^{-1} : U \to S^n$ *of m-orthogonal systems if*

$$\frac{\partial}{\partial x_j} f = l_j s_j, \quad \frac{\partial}{\partial x_j} \hat{f} = \hat{l}_j s_j, \quad \text{and} \quad \frac{\partial}{\partial x_j} s_i = a_{ij} s_j \qquad (8.7)$$

for $j = 1, \ldots, m$ *and* $i = m + 1, \ldots, n$, *where* $s_i := \mathfrak{z}e_i\mathfrak{z}^{-1}$ *and* $l_j, \hat{l}_j, a_{ij}$ *are suitable functions.*

8.2.10 Assuming that at least one of the orthogonal systems of a Ribaucour pair, say f, induces a nondegenerate Riemannian metric, that is, all $l_j \neq 0$, it is immediately clear that the enveloped Ribaucour m-sphere congruence must be given in terms of the frame fields, by $\mathfrak{s} = s_{m+1} \cdots s_n$.

8.2.11 We analyze further the structure equations (8.7) of an adapted frame for a Ribaucour pair of orthogonal systems to obtain a characterization in terms of its connection form $\varphi = \mathfrak{z}^{-1}d\mathfrak{z}$: The coordinates $(x_1, \ldots, x_m)$ of the orthogonal systems are principal coordinates for any coordinate submanifold (see §8.1.7). As in (8.1), this fact becomes apparent from the following formula,

$$\frac{\partial}{\partial x_j} s_i = -c_{ij} s_j \quad \text{with} \quad c_{ij} := l_j k_{ij} = \hat{l}_j \hat{k}_{ij}, \qquad (8.8)$$

where $1 \leq i, j \leq m$, $i \neq j$, and k_{ij} and $\hat{k}_{ij}$ are the corresponding principal curvatures, with respect to the induced metrics of f and $\hat{f}$, respectively.

Combining (8.8) with the defining equations (8.7) of an adapted frame of a Ribaucour pair, we obtain the following:

8.2.12 Lemma. $\mathfrak{z} : \mathbb{R}^m \supset U \to Spin_1(n+2)$ *is an adapted frame for a Ribaucour pair of orthogonal systems* $f = \mathfrak{z}e_0\mathfrak{z}^{-1}, \hat{f} = \mathfrak{z}e_\infty\mathfrak{z}^{-1} : U \to S^n$ *if* and only if its connection form $\varphi = \mathfrak{z}^{-1}d\mathfrak{z}$ satisfies, with suitable functions $l_j, \hat{l}_j, c_{ij}, a_{ij} : U \to \mathbb{R}$,

$$\varphi(\tfrac{\partial}{\partial x_j}) = e_j(\hat{l}_j e_0 + \tfrac{1}{2}\sum_{i=1}^{m} c_{ij}e_i + \tfrac{1}{2}\sum_{i=m+1}^{n} a_{ij}e_i + l_j e_\infty). \tag{8.9}$$

8.2.13 This lemma is easily verified from our above considerations by using that $\mathfrak{z}[\varphi, e_0]\mathfrak{z}^{-1} = df$, $\mathfrak{z}[\varphi, e_i]\mathfrak{z}^{-1} = ds_i$, and $\mathfrak{z}[\varphi, e_\infty]\mathfrak{z}^{-1} = d\hat{f}$ (see §6.7.3).

Note that, generically, an adapted frame for a Ribaucour pair is unique up to permutations of the tangential directions, *constant* orthogonal gauge transformations of the $\mathfrak{s}$-bundle $p \mapsto \mathrm{span}\{s_{m+1}, \ldots, s_n\}$ (since $\mathfrak{z}$ is $\mathfrak{s}$-adapted), and *constant* rescalings of f and $\hat{f}$ (as $\mathfrak{z}$ is an f- and $\hat{f}$-adapted frame, in the sense of §6.7.1).

8.3 Ribaucour pairs of discrete orthogonal nets

Our ansatz to obtain a discrete analog for Ribaucour pairs of orthogonal systems will be to discretize the first-order Taylor expansion of the frame equations (8.9): By §8.2.12, a map $\mathfrak{z} : \mathbb{R}^m \supset U \to Spin_1(n+2)$ is an adapted frame for a Ribaucour pair of orthogonal systems if and only if, for infinitesimal $\varepsilon \simeq 0$,

$$\mathfrak{z}(x + \varepsilon\tfrac{\partial}{\partial x_j}) \simeq \mathfrak{z}(x)(1 + \varepsilon e_j \tilde{a}_j(x)) = \mathfrak{z}(x) \cdot a_j^\varepsilon(x)e_j$$

with suitable maps[5] $a_j^\varepsilon := -(e_j + \varepsilon\tilde{a}_j) : U \to \Lambda^1 \mathbb{R}_1^{n+2}$. Obviously these structure equations can be discretized easily — here we profit from the fact that we work in the Clifford algebra since, by simply rescaling a_j^ε to take values in the Lorentz sphere S_1^{n+1}, we can achieve that these structure equations describe a map into the spin group, for finite values of ε also (cf., §6.3.12):

8.3.1 Definition. *A map* $\mathfrak{z} : \mathbb{Z}^m \supset U \to Spin_1(n+2)$ *is called an adapted frame for a Ribaucour pair* $f := \mathfrak{z}e_0\mathfrak{z}^{-1}, \hat{f} := \mathfrak{z}e_\infty\mathfrak{z}^{-1} : U \to S^n$ *of discrete orthogonal nets*[6] *if there are maps* $a_j : U \to S_1^{n+1} \subset \Lambda^1 \mathbb{R}_1^{n+2}$ *so that*

$$\mathfrak{z}(x + \delta_j) = \mathfrak{z}(x) \cdot a_j(x)e_j \tag{8.10}$$

whenever $x, x + \delta_j \in U$, *where* $\delta_j := (\delta_{1j}, \ldots, \delta_{mj}) \in \{0,1\}^m \subset \mathbb{Z}^m$.

---

[5] Note that $\tilde{a}_j \perp e_j$, that is, $e_j\tilde{a}_j = -\tilde{a}_j e_j$.

[6] Here we use the notion of a "net" since f comes with canonical coordinates (cf., Figure 5.13). Synonymously, we will use the notion of "discrete orthogonal system."

8.3.2 Starting with this rather technical definition, we want to describe the geometry of the two maps of a Ribaucour pair, seeking a more direct and more geometrical characterization.

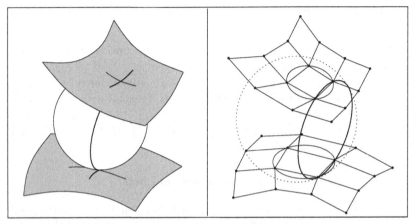

Fig. 8.2. Smooth and discrete Ribaucour pairs

First we introduce the spheres

$$s_j(x) := \mathfrak{z}(x)e_j\mathfrak{z}^{-1}(x + \delta_j) = \mathfrak{z}(x + \delta_j)e_j\mathfrak{z}^{-1}(x). \qquad (8.11)$$

Here the second equality, trivially following from $(a_je_j)^{-1} = e_ja_j$, shows that the spheres $s_j(x)$ depend symmetrically on the endpoints x and $x + \delta_j$ of the edges ("elementary 1-cells") of the domain graph U of $\mathfrak{z}$. Now

$$f(x + \delta_j) = \mathfrak{z}(x + \delta_j)e_je_0e_j\mathfrak{z}^{-1}(x + \delta_j) = s_j(x)f(x)s_j(x), \qquad (8.12)$$

and similarly for $\hat{f}$. That is, the points $f(x + \delta_j)$ and $\hat{f}(x + \delta_j)$ can be obtained from $f(x)$ and $\hat{f}(x)$, respectively, by inversion at the hypersphere $s_j(x)$ (cf., §6.3.4, or Figure T.4). In particular (cf., Figure 8.2),

8.3.3 Lemma. *The endpoints of any corresponding edges of the systems $f(U)$ and $\hat{f}(U)$ of a Ribaucour pair of discrete orthogonal systems are concircular, and the endpoint pairs of f and $\hat{f}$ do not separate on their circles.*

8.3.4 Namely, $f(x + \delta_j)$ and $\hat{f}(x + \delta_j)$ lie on the circle that contains $f(x)$ and $\hat{f}(x)$ and is orthogonal to $s_j(x)$. And, since, in the projective picture, the hyperbolic sphere pencils (cf., Figure T.1) that are associated with the endpoint pairs $f(x)$, $f(x+\delta_j)$ and $\hat{f}(x)$, $\hat{f}(x+\delta_j)$ intersect in the hypersphere $s_j(x) \in \mathbb{RP}^{n+1}_O$, the pairs do not separate.

This fact is confirmed by a computation of the cross-ratio of the four points:

$$
\begin{aligned}
[\hat{f}; f; f(.+\delta_j); \hat{f}(.+\delta_j)] &= \tfrac{3(e_\infty e_0 a_j e_0 e_\infty a_j + a_j e_\infty e_0 a_j e_0 e_\infty)3^{-1}}{2\langle a_j e_\infty a_j, e_\infty \rangle \cdot 2\langle a_j e_0 a_j, e_0 \rangle} \\
&= -\tfrac{1}{4\langle a_j, e_0 \rangle \langle a_j, e_\infty \rangle}
\end{aligned}
$$

(see Figure T.4). For this calculation it is useful to note that, since $e_0^2 = 0$, we have $e_0 a e_0 = -2\langle a, e_0 \rangle e_0$ for any $a \in \Lambda^1 \mathbb{R}_1^{n+2}$, and similarly for e_∞. Moreover, since the endpoint pairs do not separate on their circle, this cross-ratio $[\hat{f}(x); f(x); f(x+\delta_j); \hat{f}(x+\delta_j)] < 0$, that is, $\langle a_j, e_0 \rangle$ and $\langle a_j, e_\infty \rangle$ must have the same sign.

8.3.5 Generically, as long as the four points are all different, it is also possible to (geometrically, that is, up to sign) reconstruct the hypersphere $s_j(x)$ from the four points $f(x)$, $\hat{f}(x)$, $f(x+\delta_j)$, and $\hat{f}(x+\delta_j)$ as the unique hypersphere that intersects the circle of the four points orthogonally in such a way that the point pairs $f(x)$, $f(x+\delta_j)$ and $\hat{f}(x)$, $\hat{f}(x+\delta_j)$ lie symmetrically with respect to the sphere. In fact, rewriting (8.12) as

$$
f(x+\delta_j) = f(x) - 2\langle f(x), s_j(x) \rangle s_j(x),
$$

we see that $s_j(x)$ is a suitable multiple[7] of $f(x) - f(x+\delta_j)$, and similarly for $\hat{f}$.

When we are not given the "proper" homogeneous coordinates of the four points, the construction is still guaranteed by the fact that the four points are concircular, that is, their homogeneous coordinate vectors are linearly dependent, and the corresponding pairs do not separate on their circle: Consequently, the two hyperbolic sphere pencils given by $f(x)$, $f(x+\delta_j)$ and $\hat{f}(x)$, $\hat{f}(x+\delta_j)$ intersect in a point (hypersphere) $s_j(x) \in \mathbb{RP}_O^{n+1}$ of the "outer space."

8.3.6 If, however, one of the orthogonal systems of a Ribaucour pair degenerates completely, say $\hat{f} \equiv \tfrac{1}{2}\mathcal{K}_0$ is constant, then $f : U \to Q_0^n$ takes values in the Euclidean quadric given by $\mathcal{K}_0$ (see §1.4.1), and the spheres s_j are hyperplanes in that space since

$$
\hat{f}(x) = \hat{f}(x+\delta_j) = s_j(x)\hat{f}(x)s_j(x)
$$

implies $0 = \hat{f}(x)s_j(x) + s_j(x)\hat{f}(x)$, so that $s_j \perp \mathcal{K}_0$ (see §1.4.11).

---

[7] Normalizing $f(x) - f(x+\delta_j)$ to have length 1 yields $s_j(x)$, up to sign.

8.3.7 Next we wish to employ the discrete analog of the Maurer-Cartan equations from §6.7.3 for our frame $\mathfrak{z} : U \to Spin_1(n+2)$: From the structure equations (8.10) for an adapted frame of a Ribaucour pair of discrete orthogonal systems we obtain

$$\mathfrak{z}(x + \delta_i + \delta_j) = \mathfrak{z}(x + \delta_i)a_j(x + \delta_i)e_j = \mathfrak{z}(x)a_i(x)e_ia_j(x + \delta_i)e_j,$$
$$\mathfrak{z}(x + \delta_i + \delta_j) = \mathfrak{z}(x + \delta_j)a_i(x + \delta_j)e_i = \mathfrak{z}(x)a_j(x)e_ja_i(x + \delta_j)e_i.$$

Thus a necessary condition for the integrability of the structure equations are the "discrete Maurer-Cartan equations"

$$a_j(x)e_ja_i(x + \delta_j)e_i = a_i(x)e_ia_j(x + \delta_i)e_j \qquad (8.13)$$

for $i, j = 1, \ldots, m$, a discrete version of Schwarz's lemma on the independence of the second partial derivatives of the order of differentiation, that is, a discrete version of the vanishing second exterior derivative, $d^2 = 0$.

It is rather clear that the equations (8.13) are also sufficient for the integrability of the structure equations: "Integration" of the equations (8.10) is independent of the path of integration as soon as the discrete Maurer-Cartan equations hold, so that $\mathfrak{z}$ is well defined by (8.10) from the a_j, $j = 1, \ldots, m$, and an initial value.

8.3.8 Expanding the four vertices of an "elementary quadrilateral" (or "elementary 2-cell") of the orthogonal net $f : U \to S^n$ by using the structure equations, we obtain three different expressions for the vertex $f(x + \delta_i + \delta_j)$ when employing the Maurer-Cartan equations (8.13):

$$
\begin{aligned}
f(x) &= \mathfrak{z}(x)e_0\mathfrak{z}^{-1}(x),\\
f(x + \delta_i) &= \mathfrak{z}(x)a_i(x)e_0a_i(x)\mathfrak{z}^{-1}(x),\\
f(x + \delta_i + \delta_j) &= \mathfrak{z}(x)a_i(x)e_ia_j(x + \delta_i)e_0a_j(x + \delta_i)e_ia_i(x)\mathfrak{z}^{-1}(x)\\
&= \mathfrak{z}(x)a_i(x)e_ia_j(x + \delta_i)e_je_0e_ia_i(x + \delta_j)e_ja_j(x)\mathfrak{z}^{-1}(x)\\
&= \mathfrak{z}(x)a_j(x)e_ja_i(x + \delta_j)e_ie_0e_ja_j(x + \delta_i)e_ia_i(x)\mathfrak{z}^{-1}(x),\\
f(x + \delta_j) &= \mathfrak{z}(x)a_j(x)e_0a_j(x)\mathfrak{z}^{-1}(x).
\end{aligned}
$$

Using $e_0ae_0 = -2\langle a, e_0 \rangle e_0$ and the fact that e_0, e_i, and e_j are pairwise orthogonal, that is, they anti-commute, we determine the cross-ratio of this elementary quadrilateral to be (cf., Figure T.4)

$$[f(x); f(x + \delta_i); f(x + \delta_i + \delta_j); f(x + \delta_j)] = -\frac{\langle a_i(x), e_0 \rangle \langle a_i(x + \delta_j), e_0 \rangle}{\langle a_j(x), e_0 \rangle \langle a_j(x + \delta_i), e_0 \rangle}. \qquad (8.14)$$

Clearly a similar formula holds for the other orthogonal net of the Ribaucour pair. Since this cross-ratio is real, we have (cf., Figure 8.2) the following:

8.3.9 Lemma.   *The vertices of any elementary quadrilateral*

$$(f(x); f(x + \delta_i); f(x + \delta_i + \delta_j); f(x + \delta_j))$$

of an orthogonal net $f(U)$ of a Ribaucour pair are concircular.

8.3.10 This lemma generalizes to higher dimensions inductively: Suppose we know that all 2^k vertices

$$f(x + \textstyle\sum_{j=1}^{k} \varepsilon_j \delta_{i_j}), \quad \varepsilon_j \in \{0,1\},$$

of an elementary k-cell lie on a $(k-1)$-sphere. Further assume (to simplify notations) that the point $f(x + \delta_{i_{k+1}})$ is not on this $(k-1)$-sphere, so that there is a unique k-sphere containing that point as well as the original $(k-1)$-sphere. Now we know that any four points $f(x)$, $f(x + \delta_{i_j})$, $f(x + \delta_{i_{k+1}})$, and $f(x + \delta_{i_j} + \delta_{i_{k+1}})$, where $j \in \{1, \ldots, k\}$, are concircular. Since the first three points of this circle lie on our k-sphere, the whole circle lies on the sphere, and, in particular, so does the fourth point $f(x + \delta_{i_j} + \delta_{i_{k+1}})$. Repeating this argument, with one more of the $\varepsilon_{i_j} = 1$ in each step, we find that all 2^{k+1} points $f(x + \sum_{j=1}^{k+1} \varepsilon_j \delta_{i_j})$ of our elementary $(k+1)$-cell lie on the k-sphere that we just constructed.

8.3.11 Lemma. *Let f and $\hat{f}$ be a Ribaucour pair of discrete orthogonal systems. Then the vertices of any elementary k-cell of $f(U)$ and of $\hat{f}(U)$, respectively, lie on a $(k-1)$-sphere.*

8.3.12 Clearly the very same argument works for the vertices

$$f(x + \textstyle\sum_{j=1}^{k} \varepsilon_j \delta_{i_j}) \quad \text{and} \quad \hat{f}(x + \textstyle\sum_{j=1}^{k} \varepsilon_j \delta_{i_j})$$

of corresponding elementary k-cells of both nets, since we know that the endpoints of corresponding edges on both nets are concircular from §8.3.3:

8.3.13 Lemma. *The vertices of any corresponding elementary k-cells of the discrete orthogonal systems $f, \hat{f} : U \to S^n$ of a Ribaucour pair lie on a k-sphere $\mathfrak{s} \in \Lambda^{n-k} \mathbb{R}_1^{n+2}$.*

8.3.14 This provides a discrete analog of the enveloping condition from §8.2.4: Any corresponding k-dimensional subnets of the discrete orthogonal nets $f(U)$ and $\hat{f}(U)$ of a Ribaucour pair "envelope" a "discrete k-sphere congruence" $\mathfrak{s}$.

However, under certain regularity assumptions, we can do even better and explicitly determine the "enveloped" k-spheres. First note that

$$f(x + \delta_j) = \mathfrak{z}(x) a_j(x) e_0 a_j(x) \mathfrak{z}^{-1}(x) = \mathfrak{z}(x) \{e_0 - 2\langle a_j(x), e_0 \rangle \, a_j(x)\} \mathfrak{z}^{-1}(x),$$

and similarly for $\hat{f}$: Besides $f(x) = \mathfrak{z}(x) e_0 \mathfrak{z}^{-1}(x)$ and $\hat{f}(x) = \mathfrak{z}(x) e_\infty \mathfrak{z}^{-1}(x)$, the spheres $\mathfrak{s}_{i_j} = -\mathfrak{z}(x) a_{i_j}(x) \mathfrak{z}^{-1}(x)$, with $j = 1, \ldots, k$, also must be in the Minkowski $(k+2)$-space that corresponds to the k-sphere $\mathfrak{s}(x)$ as soon as one of the discrete orthogonal nets, say f, is regular in the sense that $f(x + \delta_j) \neq f(x)$ for all x and $j = 1, \ldots, m$. Assuming additionally that $f(U)$ and $\hat{f}(U)$ are disjoint, the k-sphere $\mathfrak{s}(x)$ is given by

$$v\mathfrak{s}(x) = \mathfrak{z}(x)\, e_0 \wedge a_{i_1}(x) \wedge \ldots \wedge a_{i_k}(x) \wedge e_\infty \, \mathfrak{z}^{-1}(x) \in \Lambda^{k+2} \mathbb{R}_1^{n+2}$$

(see §6.4.8). Note that, for this argument, we have used the fact that the points $f(x)$, $f(x+\delta_{i_j})$, $\hat{f}(x)$, and $\hat{f}(x+\delta_{i_j})$ lie on the k-sphere (cf., §6.4.12). For the remaining points of the elementary k-cells of $f(U)$ and $\hat{f}(U)$, incidence follows from our previous lemma.

The interested reader may want to briefly reflect on other ways of expressing $v\mathfrak{s}(x)$ using these additional incidence relations.

8.3.15 Classically, the curvature line net of a surface used to be characterized as a (coordinate) curve net that divides the surface into "infinitesimal rectangles": This fact can be justified by second order Taylor expansion of a parametrization [32] or by nonstandard analysis arguments [152].

Thus, since any elementary 2-cell of a discrete orthogonal net belonging to a Ribaucour pair is a conformal rectangle (cf., §4.9.10) by §8.3.11, it seems justified to consider the net as the discrete analog of a curvature line net. This fact fits well with the corollary to Dupin's theorem given in §8.1.7: In the sense of this notion of "discrete curvature line net," §8.3.11 is just a discrete Dupin's theorem. Combining all of these observations and facts, we obtain the following geometric description[8] for our Ribaucour pairs of discrete orthogonal systems, analogous to the definition in the smooth case (see §8.2.4):

8.3.16 Definition and Corollary. *A discrete net* $f : \mathbb{Z}^m \supset U \to S^n$ *is called a discrete curvature line net*[9] *if any elementary k-cell lies on a $(k-1)$-sphere.*

The discrete orthogonal systems of a Ribaucour pair form discrete curvature line nets. Any two corresponding k-dimensional subnets "envelope" a discrete k-sphere congruence — which we call a discrete Ribaucour congruence because its "envelopes" are discrete curvature line nets.

We will say that two discrete curvature line nets with this enveloping property satisfy the Ribaucour property, and one net will be called a Ribaucour transform of the other.[10]

---

[8] In order for this "description" to become a "characterization," we would have to show that, given the geometric configuration of the corollary, we can construct an adapted frame $\mathfrak{z}$ for the pair $(f, \hat{f})$ of discrete curvature line nets. In the following section we will learn about an ansatz that may allow us to do so.

[9] Various authors also refer to these nets as "discrete circular nets."

[10] This notion of "Ribaucour transformation" for discrete curvature line nets is involutive.

8.3.17 Note how this definition of a discrete curvature line net generalizes our definition in §5.7.2. In fact, this notion of a discrete curvature line net generalizes the notion of a discrete isothermic net, and the notion of Ribaucour transformation for a discrete curvature line net generalizes the notion of the Darboux transformation for discrete isothermic nets (cf., §5.7.12f).

8.4 Sphere constructions

We want to investigate further the geometry of the spheres s_j, given in (8.11). We already discussed their relation with the nets f and $\hat{f}$: Given an initial point pair $f(x_0)$ and $\hat{f}(x_0)$ for the Ribaucour pair of orthogonal nets, f and $\hat{f}$ can be constructed by successive reflections at the spheres s_j (see (8.12)). On the other hand, we found in §8.3.5 that, as a consequence of §8.3.3, the hyperspheres s_j can be reconstructed (up to sign) from the endpoints of corresponding edges of f and $\hat{f}$, if the four points are different. This remains true if one of the nets degenerates completely and these spheres become hyperplanes in the corresponding Euclidean n-space (cf., §8.3.6).

8.4.1 By examining (8.11) we see that the s_j's can also be used to reconstruct the frame $\mathfrak{z}$, by "integrating" the system

$$\mathfrak{z}(x + \delta_j) = -s_j(x)\mathfrak{z}(x)e_j. \tag{8.15}$$

Integrability of the system (8.15) is, as usual, obtained from the condition

$$s_i(x + \delta_j)s_j(x)\mathfrak{z}(x)e_je_i = \mathfrak{z}(x + \delta_i + \delta_j) = s_j(x + \delta_i)s_i(x)\mathfrak{z}(x)e_ie_j.$$

That is, given some initial value $\mathfrak{z}(x_0) = \mathfrak{z}_0$, the system (8.15) can be solved (uniquely) if and only if

$$0 = s_i(x + \delta_j)s_j(x) + s_j(x + \delta_i)s_i(x). \tag{8.16}$$

Since (8.15) is equivalent to (8.10) with $a_j = -\mathfrak{z}^{-1}s_{j\mathfrak{z}}$, the frame obtained by solving (8.15) is automatically an adapted frame of a Ribaucour pair of discrete orthogonal nets.

8.4.2 Lemma. *A family of maps* $s_1, \ldots, s_m : \mathbb{Z}^m \supset U \to S_1^{n+1}$ *defines an adapted frame of a Ribaucour pair via* $\mathfrak{z}(x + \delta_j) = -s_j(x)\mathfrak{z}(x)e_j$ *if and only if it satisfies* $0 = s_i(x + \delta_j)s_j(x) + s_j(x + \delta_i)s_i(x)$ *for* $i, j = 1, \ldots, m$.

8.4.3 To draw geometric insight from the integrability conditions (8.16) of (8.15), we focus on its $\Lambda^2 \mathbb{R}_1^{n+2}$ part:

$$0 = s_i(x + \delta_j) \wedge s_j(x) + s_j(x + \delta_i) \wedge s_i(x).$$

This says that $s_j(x)$, $s_i(x + \delta_j)$ and $s_i(x)$, $s_j(x + \delta_i)$ span the same sphere pencil.[11] Interpreting the spheres $s_j(x)$, and so on, as points in (the outer space of) projective $(n+1)$-space $I\!\!RP^{n+1}$ (see Figure T.4), we conclude that

8.4.4 Lemma. *Any four points (hyperspheres) $s_i(x)$, $s_j(x)$, $s_i(x+\delta_j)$, and $s_j(x + \delta_i)$ belonging to (a pair of corresponding) elementary quadrilaterals of (a Ribaucour pair of) orthogonal systems are collinear.*

8.4.5 Just as some facts we derived earlier, this lemma can be generalized to higher dimensions by induction.

For this (as well as the following) argument we can deal solely with the spheres s_j, so that we can neglect the corresponding elementary cells of the discrete orthogonal nets that they come from: Therefore, we will speak of a "k-cube" instead of the pair of "elementary k-cells" whose vertices are obtained as the images of the vertices of the k-cube.

Thus suppose we know that the $k \cdot 2^{k-1}$ spheres[12] belonging to any k-cube in U are contained in a projective $(k-1)$-plane, and consider a $(k+1)$-cube with vertices $x + \sum_{j=1}^{k+1} \varepsilon_j \delta_{i_j}$. By assumption the spheres belonging to the "base" k-cube with vertices $x + \sum_{j=1}^{k} \varepsilon_j \delta_{i_j}$ lie in a $(k-1)$-plane. Now consider all quadrilaterals consisting of the endpoints of one edge in the base cube and the two points "above" them,

$$y, \quad y + \delta_{i_j}, \quad y + \delta_{i_{k+1}}, \quad \text{and} \quad y + \delta_{i_j} + \delta_{i_{k+1}}.$$

Without loss of generality,[13] the line corresponding to one of these quadrilaterals sticks out of the $(k-1)$-plane of our base cube. This defines a k-plane. Since the sphere line of any "neighboring" quadrilateral that shares a "vertical" edge with our original quadrilateral contains two spheres that lie in that k-plane, it is entirely contained in the k-plane. Repeating this argument as often as necessary, by applying it to the "vertical neighbors"

---

[11] If $s_i(x + \delta_j) \wedge s_j(x) = 0$, we have $f(x + \delta_i + \delta_j) = f(x)$, that is, the net f (as well as $\hat{f}$) degenerates.

To verify this fact, that the four spheres belonging to a pair of elementary quadrilaterals belong to one sphere pencil, seems to be the major difficulty in obtaining the converse of our geometric description in §8.3.16. However, in case one of the maps, say $\hat{f}$, is constant (the point at infinity of a Euclidean space), so that all spheres s_j are hyperplanes (see §8.3.6) this fact is rather obvious: The four hyperplanes then intersect in an $(n-2)$-plane of that Euclidean space. Moreover, it follows from Thales's theorem that opposite angles of intersection add up to π, which makes the proof of the converse nearly complete — note that the real part of (8.16) is a condition on the intersection angles of the spheres.

[12] The spheres s_{i_j} can be associated to the *edges* of the elementary k-cube. Consequently, there are as many spheres s_{i_j} as a k-dimensional cube has edges.

[13] Otherwise, all spheres are already contained in the $(k-1)$-plane of the base cube.

of any quadrilaterals subsequently obtained, the statement extends to all spheres of the $(k + 1)$-cube that we started with.

8.4.6 Lemma. *All spheres $s_{i_j}(x)$ belonging to corresponding elementary k-cells,[14] $2 \leq k \leq m$, of the orthogonal nets $f(U)$ and $\hat{f}(U)$ of a Ribaucour pair are contained in a $(k-1)$-plane of $\mathbb{R}P^{n+1}$.*

8.4.7 Conversely, this lemma tells us how to reconstruct (geometrically, as points in $\mathbb{R}P^{n+1}$) all spheres of a $(k+1)$-cube, $k \geq 2$, if only those belonging to all k-cubes that contain one given vertex of the $(k + 1)$-cube are given: To simplify notations, assume that we are provided with all spheres that belong to the k-cubes containing $x + \sum_{j=1}^{k+1} \delta_{i_j}$. Thus we are seeking the $k + 1$ missing spheres $s_{i_j}(x)$, $j = 1, \ldots, k + 1$, of a total number of $2^k(k + 1)$ spheres that belong to the $(k + 1)$-cube.

Now, $s_{i_j}(x)$ belongs to all k-cubes that contain the i_j-edge, that is, it belongs to all k-cubes that miss out one of the other edges. In particular, there are k different k-cubes that $s_{i_j}(x)$ belongs to. For each of these k-cubes we are given[15] $(2^{k-1} - 1)k \geq k$ spheres, so that we can construct the associated $(k - 1)$-plane (in a certain k-plane) in $\mathbb{R}P^{n+1}$. The sphere $s_{i_j}(x)$ is then the intersection point of the k (generically different) $(k - 1)$-planes in $\mathbb{R}P^{n+1}$.

8.4.8 As a consequence, obtained by induction, we may reconstruct all spheres s_j of an m-cube as soon as we know the spheres belonging to all k-cubes that contain a given vertex of the m-cube, $2 \leq k \leq m$.

Because we may construct the spheres from the vertices of the elementary k-cells of a Ribaucour pair of discrete orthogonal nets $f(U)$ and $\hat{f}(U)$ (see §8.3.5), and the spheres can then be used to construct the missing points of the nets by reflections (see (8.12)), we can solve the following "Cauchy problem":

8.4.9 Theorem. *A Ribaucour pair $f, \hat{f} : \mathbb{Z}^m \supset U \to S^n$ of m-orthogonal nets can be reconstructed once all $\binom{m}{k}$ pairs of k-dimensional subnets that pass through a pair of corresponding points are given.*

8.4.10 Note that this theorem remains true if one of the m-orthogonal systems degenerates, say $\hat{f} \equiv const$. Then all provided hyperspheres are hyperplanes in the Euclidean space determined by $\infty = \hat{f}$, and the construction of the remaining hyperspheres takes place in the (parabolic) sphere complex (see Figure T.1) determined by $\hat{f} = \infty$, so that they will be hyperplanes, too. In fact, this case of a degenerate Ribaucour pair of orthogonal systems

---

[14] For $k = 1$, the statement is true but trivial (see §8.3.3).

[15] Here we need that $k \geq 2$.

can be used to establish the solution of a similar Cauchy problem for a single discrete m-orthogonal net (cf., [35]).

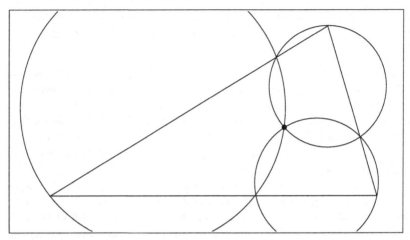

Fig. 8.3. Miguel's theorem

In the case $m = 3$, the above Cauchy problem for a discrete triply orthogonal system can be proven more directly by Miguel's theorem (see [104] or [166]): Given the seven vertices of an elementary 3-cell of a discrete curvature line net, its eighth point can be constructed uniquely. Namely, we know that the vertices of any elementary quadrilateral are concircular, and that all eight vertices lie on a 2-sphere. Hence the eighth point must be the intersection point of three circles constructed from the given points — it has to be shown that the three circles indeed intersect in one point.

Stereographically projecting the 2-sphere of the seven given points, with the point belonging to all three known 2-cells as the center of projection, this is reduced to the fact that the three circles in Figure 8.3 intersect, that is, to Miguel's theorem [12].

8.4.11 Example. As a (rather simple) example we consider 2-dimensional orthogonal systems and assume that the spheres $s_1, s_2 : \mathbb{Z}^2 \to S_1^{n+1}$ only depend on one parameter, $s_1(m, n) = s_1(m)$ and $s_2(m, n) = s_2(n)$. Then the integrability condition (8.16) reads

$$
\begin{aligned}
0 &= s_1(m, n+1)s_2(m, n) + s_2(m+1, n)s_1(m, n) \\
&= s_1(m)s_2(n) + s_2(n)s_1(m),
\end{aligned}
$$

that is, the maps s_1 and s_2 give rise, via (8.15), to an adapted frame for a Ribaucour pair of discrete orthogonal nets if and only if all the spheres $s_1(m)$ and $s_2(n)$ intersect orthogonally. If we assume that none of the two "discrete sphere curves" is constant, then at least one, say s_1, takes values in a fixed sphere pencil.

8.4.12 Knowing this we can construct certain types of (Ribaucour pairs) of 2-dimensional orthogonal nets in S^3, that is, "discrete surfaces" in terms of their "curvature lines" (cf., §1.8.5).

(i) The spheres $s_1(m)$ all belong to an elliptic sphere pencil: Then the spheres $s_1(m)$ intersect in a circle — considering one point of that circle as the point at infinity, the spheres $s_1(m)$ become planes that intersect in a line. Since the spheres $s_2(n)$ intersect all spheres $s_1(m)$ orthogonally, they intersect this line orthogonally; consequently, their centers (considered in the Euclidean space determined by the above point at infinity) lie on the line. Choosing any (pair of) initial points,[16] the discrete curve obtained by reflections in the spheres $s_2(n)$ lies in a plane containing our distinguished line; if the spheres $s_1(m)$ intersect at small and equal angles, and the plane of the "meridian curve" bisects one of these angles, then we obtain the discrete analog of a (pair of) surfaces of revolution, parametrized by curvature lines, with our line as the axis of revolution.

(ii) The spheres $s_1(m)$ all belong to a parabolic sphere pencil: That is, the spheres $s_1(m)$ all touch in one point that we consider to be the point at infinity of a Euclidean space — the spheres $s_1(m)$ then appear as parallel planes, so that the spheres $s_2(n)$ must also be planes, intersecting any plane $s_1(m)$ orthogonally. Thus a discrete surface constructed from these spheres will be a "discrete cylinder" over a discrete "meridian curve" that lies in a plane parallel to the planes $s_1(m)$. However, choosing the surface's initial point to be the point at infinity, the discrete surface completely degenerates, so that we may have a Ribaucour pair consisting of a "true" orthogonal system coupled with the point at infinity, as discussed in §8.3.6.

(iii) The spheres $s_1(m)$ all belong to a hyperbolic sphere pencil: In this case the $s_1(m)$ can be considered as concentric spheres in a Euclidean space; the spheres $s_2(n)$ then become planes that all contain the common center of the $s_1(m)$. Reflection at any such plane maps the points of any sphere in the pencil of the $s_1(m)$ onto itself. Thus our "meridian curve" becomes a spherical curve in this case, on a sphere concentric to the spheres $s_1(m)$: We obtain a (pair of) "discrete cones."

8.5 Demoulin's family and Bianchi permutability

In this section — before returning to the smooth setup — we will finally

---

[16] Or an initial value $\mathfrak{z}(0,0)$ for the frame $\mathfrak{z} : \mathbb{Z}^2 \to Spin_1(5)$.

present discrete versions of the Demoulin family of orthogonal systems (cf., [100]) and of Bianchi's permutability theorem for the Ribaucour transformation (cf., [28]), based on our geometric description in §8.3.16 of the Ribaucour transformation of discrete curvature line nets. Note that Bianchi's original formulation of the permutability theorem goes way beyond what is usually referred to as the permutability theorem in the modern integrable systems literature. However, this simpler formulation of the permutability theorem (which will also be given below) and its "higher dimensional versions" provide the same relation between discrete curvature line nets and the Ribaucour transformation of smooth orthogonal nets (cf., [35]) as the permutability theorem from §5.6.6 did for the Darboux transformation of smooth isothermic surfaces and the definition of discrete isothermic nets (see §5.7.1).

8.5.1 Theorem and Definition. *Let $f : \mathbb{Z}^m \supset U \to S^n$ be a discrete curvature line net with two Ribaucour transforms $f_0, f_1 : U \to S^n$. Then there is a 1-parameter family $(f_t)_{t \in S^1}$ of Ribaucour transforms $f_t : U \to S^n$ of f such that any point $f_t(x)$ lies on the circle $\mathfrak{c}(x)$ that is given by the three points $f(x)$, $f_0(x)$, and $f_1(x)$, $x \in U$. This family will be called the Demoulin family of f associated to f_0 and f_1.*

8.5.2 Proof. We will construct a discrete curvature line net f_t by choosing an initial point $f_t(x_0)$ and then "integrating" along the edges of f_t: Observe that the endpoints of any corresponding edges of a pair of discrete curvature lines that satisfy the Ribaucour property, in the sense of §8.3.16, have to be concircular.

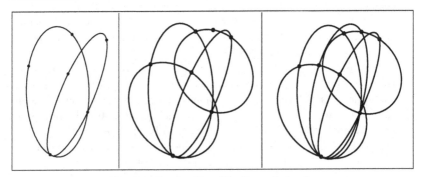

Fig. 8.4. Construction of the Demoulin family

This observation implies a construction, as indicated in Figure 8.4: Since the two circles belonging to corresponding edges of f, f_0 and f, f_1 determine a 2-sphere $\mathfrak{s}_j(x)$, which also contains the circles $\mathfrak{c}(x)$ and $\mathfrak{c}(x + \delta_j)$, the point $f_t(x + \delta_j)$ will then be given as the second (besides the point $f(x + \delta_j)$)

point[17] of intersection of the circle $\mathfrak{c}(x + \delta_j)$ and the circle $\tilde{\mathfrak{c}}_j^t(x)$ through $f(x)$, $f(x + \delta_j)$, and $f_t(x)$.

8.5.3 Now consider an elementary quadrilateral (elementary 2-cell) of our first net, $(f(x); f(x+\delta_i); f(x+\delta_i+\delta_j); f(x+\delta_j))$: Its vertices are concircular since f is a discrete curvature line net — thus there is a (generically unique) 2-sphere $\mathfrak{s}(x)$ that contains the four vertices of the original net f as well as the "initial" point $f_t(x)$. Because $\tilde{\mathfrak{c}}_j^t(x)$ contains three points of $\mathfrak{s}(x)$, it is contained in $\mathfrak{s}(x)$, $\tilde{\mathfrak{c}}_j^t(x) \subset \mathfrak{s}(x)$. On the other hand, it is also contained in the 2-sphere $\mathfrak{s}_j(x)$ that contains the two circles $\mathfrak{c}(x)$ and $\mathfrak{c}(x + \delta_j)$, so that the circle $\tilde{\mathfrak{c}}_j^t(x)$ can be described as the intersection circle of these two spheres.

Clearly the same is true for the circles $\tilde{\mathfrak{c}}_i^t(x)$, $\tilde{\mathfrak{c}}_i^t(x + \delta_j)$, and $\tilde{\mathfrak{c}}_j^t(x + \delta_i)$ that correspond to the other edges of the quadrilateral.

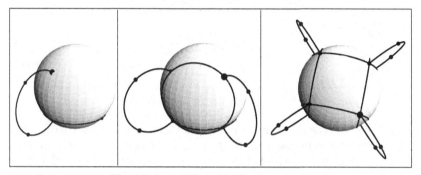

Fig. 8.5. Integrability of the Demoulin family

Thus the above construction of the points of f_t gives rise to a discrete net, that is, the "integration" is path-independent: Given the initial point $f_t(x)$, we uniquely construct $f_t(x + \delta_i)$ and $f_t(x + \delta_j)$, but there are now two possibilities to construct the point $f_t(x + \delta_i + \delta_j)$ — which give the same result since it is given as the second point (besides $f(x + \delta_i + \delta_j)$) of intersection of the sphere $\mathfrak{s}(x)$ with the circle $\mathfrak{c}(x+\delta_i+\delta_j)$ (see Figure 8.5).

Moreover, $f(x + \delta_i + \delta_j)$ lies on the circle given by $f_t(x)$, $f_t(x + \delta_i)$, and $f_t(x+\delta_j)$: Interpreting the sphere $\mathfrak{s}(x)$ as a Euclidean plane with $\infty \simeq f(x)$, this assertion is ensured by Miguel's theorem (see Figure 8.3).

8.5.4 To summarize: Thus far we have learned that a discrete net f_t can be constructed uniquely such that the endpoints of any corresponding edges of f and f_t are concircular and that the vertices of elementary quadrilaterals of f_t are concircular.

---

[17] Note that this "second point of intersection" may well coincide with $f(x + \delta_j)$: This is the degenerate situation where a point of f and its Ribaucour transform f_t coincide.

To complete the proof we just note that the fact that f_t is a discrete curvature line net (the vertices of an elementary k-cell lie on a $(k-1)$-sphere) and that it is a Ribaucour transform of f (the vertices of corresponding elementary k-cells lie on a k-sphere; see §8.3.16) can now be derived in the very same manner as our lemmas in §8.3.11 and §8.3.13 were obtained from their 1-dimensional versions.   ◁

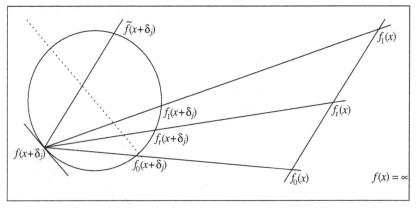

Fig. 8.6. Circle pencil of the Demoulin family

8.5.5 Note that the three circles $\tilde{\mathfrak{c}}_j^0(x)$, $\tilde{\mathfrak{c}}_j^1(x)$, and $\tilde{\mathfrak{c}}_j^t(x)$, which contain the endpoints of corresponding edges of f and f_0, of f and f_1, and of f and f_t, belong to an elliptic pencil of circles on $\mathfrak{s}_j(x)$, whereas the two circles $\mathfrak{c}(x)$ and $\mathfrak{c}(x+\delta_j)$ span a hyperbolic circle pencil on $\mathfrak{s}_j(x)$. Interpreting $\mathfrak{s}_j(x)$ as a Euclidean plane with $\infty \simeq f(x)$, the elliptic pencil $(\tilde{\mathfrak{c}}_j^t(x))_{t\in S^1}$ becomes the pencil of lines through $f(x+\delta_j) \in \mathfrak{c}(x+\delta_j) \subset \mathfrak{s}_j(x)$, and the circle $\mathfrak{c}(x)$ becomes a straight line (see Figure 8.6). Now observe that if the "initial point" $f_t(x)$ is chosen to be $f(x)$, $f_t(x) = f(x)$, then our above construction yields $f_t(x+\delta_j)$ as the (second) point of intersection of $\mathfrak{c}(x+\delta_j)$ and $\tilde{\mathfrak{c}}_j^t(x)$, that is, $f_t(x+\delta_j) = \tilde{f}(x+\delta_j)$ in Figure 8.6.

This shows that the original net f itself does not, in general, occur as a member of its Demoulin family — unless all circles $\tilde{\mathfrak{c}}_j^t(x)$ are tangent to the circles $\mathfrak{c}(x)$ and $\mathfrak{c}(x+\delta_j)$ at the same time,[18] so that the two intersection points $f(x+\delta_j)$ and $\tilde{f}(x+\delta_j)$ always coincide.

8.5.6 Also observe that the map $f_t(x) \mapsto f_t(x+\delta_j)$ preserves the cross-ratio[19]: This becomes clear by contemplating Figure 8.6. Namely, post-

---

[18] In Figure 8.6 this means that the tangent line of the circle $\mathfrak{c}(x+\delta_j)$ at $f(x+\delta_j)$ is parallel to the line $\mathfrak{c}(x)$.

[19] I am grateful to Ekki Tjaden for suggesting that fact to me.

compose the map $f_t(x) \mapsto f_t(x + \delta_j)$ by the stereographic projection of the circle $c(x + \delta_j)$ from $f(x + \delta_j)$ onto the line given by its equator 0-sphere, indicated by the dotted line in Figure 8.6. The resulting composite map is a central projection of one line, $c(x)$, onto another line, this equator line. As such, the composite map preserves the cross-ratio. On the other hand, the stereographic projection preserves, as a restriction of an inversion of the plane, that is, as a Möbius transformation, the cross-ratio (cf., §6.5.5).

As a consequence, once a third member f_∞ of the Demoulin family is determined, all other discrete curvature line nets f_t are given algebraically using the constancy of the cross-ratio function $t = [f_t; f_0; f_1; f_\infty]$:

$$f_t = t(1-t)\langle f_0, f_1\rangle \, f_\infty - (1-t)\langle f_\infty, f_1\rangle \, f_0 - t\langle f_0, f_\infty\rangle \, f_1.$$

To check that the cross-ratio $[f_t; f_0; f_1; f_\infty] = t$, we assume, without loss of generality,[20] that the relative scalings of f_∞, f_0, and f_1 are adjusted, so that

$$\langle f_1, f_0\rangle = \langle f_\infty, f_0\rangle = \langle f_1, f_\infty\rangle \equiv -1.$$

Then, using the scalar product formula for the cross-ratio given in §6.5.4, it is straightforward to compute

$$
\begin{aligned}
[f_t; f_0; f_1; f_\infty] &= \frac{\langle f_t, f_0\rangle\langle f_1, f_\infty\rangle - \langle f_t, f_1\rangle\langle f_0, f_\infty\rangle + \langle f_t, f_\infty\rangle\langle f_1, f_0\rangle}{2\langle f_t, f_\infty\rangle\langle f_1, f_0\rangle} \\
&= \frac{t^2 - (1-t)^1 + 1}{2} \\
&= t.
\end{aligned}
$$

Hence the original net f is a member of the Demoulin family if and only if the cross-ratio $[f_t; f_0; f_1; f]$ is constant.

8.5.7 We have already noted at various places that the Ribaucour transformation of discrete curvature line nets generalizes the Darboux transformation of discrete isothermic nets. Accordingly, there is a permutability theorem that generalizes the respective permutability theorem from §5.7.28 (cf., §5.6.6):

8.5.8 Theorem. *Let* $f : \mathbb{Z}^m \supset U \to S^n$ *be a discrete curvature line net with Ribaucour transforms f_0 and f_1, as in §8.5.1. Then there is a simultaneous Ribaucour transform f_{01} of f_0 and of f_1.*

Corresponding points of the four nets are concircular.[21]

---

[20] Once the relative scalings of f_0 and f_1 are chosen so that $\langle f_0, f_\infty\rangle = \langle f_1, f_\infty\rangle = -1$, a rescaling of f_∞ can be used to adjust the scalar product $\langle f_0, f_1\rangle = -a^2$.

[21] In contrast to the Darboux case (see §5.7.28), the cross-ratio of these four points will generally *not* be constant. Hence the proof given there will not carry over.

8.5.9 *Proof.* The net f_{01} can be constructed in a very similar way as the nets f_t giving the Demoulin family from §8.5.1 — in fact, as before, there will be a 1-parameter family of nets f_{01} that satisfy the permutability property: First choose an initial point $f_{01}(x)$ on the circle $\mathfrak{c}(x)$ and then "integrate" along the edges by using the fact that $f_{01}(x + \delta_j)$ has to lie on the circles $\tilde{\mathfrak{c}}^i_j(x)$, $i = 0, 1$, that are constructed from the points $f_i(x)$, $f_i(x + \delta_j)$, and $f_{01}(x)$, as well as on the circle $\mathfrak{c}(x + \delta_j)$. Since the nets f_i are Ribaucour transforms of f, these three circles lie on a 2-sphere, so that the "integrability" of the net f_{01} is a consequence of Miguel's theorem (see Figure 8.3).

The net f_{01} is, by construction, simultaneously a Ribaucour transform of f_0 as well as of f_1 as soon as it is identified as a discrete curvature line net. To see this we focus on the corresponding elementary quadrilaterals of one of the maps f_0 and f_1, say f_0, and of f_{01}, and apply Miguel's theorem once more. Since the endpoints of any corresponding edges of f_0 and f_{01} are concircular by construction, and f_0 is a discrete curvature line net, we learn that the circle through the three points $f_{01}(x)$, $f_{01}(x + \delta_i)$, and $f_{01}(x + \delta_j)$ of an elementary quadrilateral of f_{01} has $f_{01}(x + \delta_i + \delta_j)$ as a common point of intersection with the two circles $\tilde{\mathfrak{c}}^1_i(x + \delta_j)$ and $\tilde{\mathfrak{c}}^1_j(x + \delta_i)$ that are used to construct $f_{01}(x + \delta_i + \delta_j)$.

As before, the assertions about the vertices of higher dimensional elementary k-cells to lie on $(k - 1)$-spheres and on k-spheres (for the vertices of corresponding k-cells of two discrete nets) follow from the respective properties of 1-cells.   ◁

8.5.10 In our proof we could as well have taken a slightly different point of view: We could have constructed the net f_{01} by requesting it to be a Ribaucour transform of only one of the nets f_0 and f_1, say of f_0, but imposing additionally that the points of f_{01} must lie on the circles $\mathfrak{c}$ through the corresponding points of f, f_0, and f_1 — the fact that f_{01} is also a Ribaucour transform of f_1 would then have been a consequence of Miguel's theorem. But this is exactly the way we constructed the Demoulin family in §8.5.2: Note that the key fact in our proof of the existence of the Demoulin family was the existence of a 2-sphere for any edge that contains the endpoint circles $\mathfrak{c}(x)$ and $\mathfrak{c}(x + \delta_j)$. Consequently, the net f_{01} that satisfies our permutability theorem from §8.5.8 gives rise to a Demoulin family of discrete curvature line nets that all satisfy the permutability theorem. Moreover, according to §8.5.6, any four possible nets f_{01} that are Ribaucour transforms of f_0 and f_1 at the same time have constant cross-ratio.

Switching viewpoints by considering f and f_{01} as two Ribaucour transforms of f_0 (or f_1), we recognize that any net f_t of the first Demoulin family given by f_0 and f_1 is a Ribaucour transform of f and f_{01}, that is, it is a Ribaucour transform of *any* net f_{01} of the second Demoulin family.

Thus we obtain the following

8.5.11 Corollary. *Let* $f : \mathbb{Z}^m \supset U \to S^n$ *be a discrete curvature line net with Ribaucour transforms* f_0 *and* f_1, *and let* $(f_t)_{t \in S^1}$ *denote the Demoulin family of* f *associated to* f_0 *and* f_1. *Then there is a second Demoulin family* $(\hat{f}_s)_{s \in S^1}$ *that contains the original net[22]* f, *so that any net* f_t *of the first family is a Ribaucour transformation of any net* $\hat{f}_s$ *of the second family.*

The circles $\mathfrak{c}(x) = \bigcup_{t \in S^1} f_t(x)$ *and* $\hat{\mathfrak{c}}(x) = \bigcup_{s \in S^1} \hat{f}_s(x)$ *of the two Demoulin families coincide,* $\mathfrak{c}(x) = \hat{\mathfrak{c}}(x)$.

8.5.12 This last permutability theorem is the exact discrete analog of Bianchi's original permutability theorem for the Ribaucour transformation of smooth orthogonal nets (see [28] or [27]; cf., [106]).

One ansatz to prove the smooth versions of the theorems on the existence of the Demoulin family and of this permutability theorem may be to discuss a smooth limit.[23]

Another ansatz would be to use the technology discussed in §7.5.5: As in the discrete proofs, the constructions will rely on "1-dimensional versions" of the existence of the Demoulin family and of the permutability theorem, respectively, so that one will only have to deal analytically with circles (as given by (7.19)), that is, with elements in $\Lambda^0 \mathbb{R}^{n+2}_1 \oplus \cdots \oplus \Lambda^3 \mathbb{R}^{n+2}_1$. However, as of this writing, the author did not succeed in completing a proof using this ansatz — it may therefore be left to the interested reader as a challenge.

The "correct" proof, however, of the smooth versions of these two theorems should be a proof that acknowledges the invariance of the Ribaucour transformation under Lie sphere transformations. A very simple and straightforward proof in the Lie sphere geometry setup will be given in [51].

8.6 Isothermic surfaces revisited

As has been mentioned at various places, isothermic surfaces $f : M^2 \to S^n$ in the conformal n-sphere, as defined in §5.1.1, are special 2-orthogonal systems in the sense of §8.2.4 — just as (discrete) isothermic nets in the

---

[22] Clearly, f is a Ribaucour transform of f_0 and f_1 at the same time.

[23] The reader may want to investigate the problems that occur when following this ansatz: For example, the notion of "limit" involves measuring distances, but, in Möbius geometry, we only have an angle measurement at hand.

However, very recently the author was informed by A. Bobenko and D. Matthes that they had proof of an approximation result for orthogonal systems in Euclidean space, by using Clifford algebra technology, as discussed in the present chapter, and in particular the solution of the above Cauchy problem (see §8.4.9). Compare [36].

conformal n-sphere, as naturally generalized[24] from §5.7.2, are special 2-dimensional discrete curvature line nets[25] in the sense of §8.3.16. To conclude our presentation of applications of the Clifford algebra formalism we will investigate (smooth) isothermic surfaces of higher codimension: In this context, the Vahlen matrix approach will prove useful — in fact, many developments from the quaternionic treatment of isothermic surfaces in $\mathbb{HP}^1$ in Chapter 5 can be "translated" to the Vahlen matrix setup developed in Chapter 7 (compare [47]). Thus, in what follows, we will loosely follow our investigations from Chapter 5, starting with

8.6.1 The Christoffel Transformation. Remember that the Christoffel transformation is a Euclidean notion. Consequently, we consider an isothermic surface $f : M^2 \to \mathbb{R}^n$ in Euclidean n-space, and let (x, y) denote conformal curvature line parameters. The very same computations that led to §5.2.1 now provide us with an n-dimensional version of the Christoffel transformation: Since (x, y) are conformal coordinates $f_x f_x^{-1} + f_y f_x^{-1} = 0$, and as they are curvature line parameters, we have $f_{xy} = a f_x + b f_y$ with suitable functions a, b; then

$$(f_x^{-1})_y = -a f_x^{-1} - b f_x^{-1} f_y f_x^{-1} = a f_y^{-1} f_x f_y^{-1} + b f_y^{-1} = -(f_y^{-1})_x,$$

showing that the form $\psi^* := f_x^{-1} dx - f_y^{-1} dy$ is (at least locally) integrable. Moreover,

$$(f_x^{-1})_y = -a f_x^{-1} + b f_y^{-1},$$

so that (x, y) are curvature line coordinates for the corresponding (locally defined) surface f^*, $df^* = \psi^*$. Since (x, y) are clearly conformal coordinates for f^*, we obtain a new isothermic surface.

8.6.2 Theorem and Definition. *Let $f : M^2 \to \mathbb{R}^n$ be isothermic with curvature line parameters $(x, y) : M^2 \supset U \to \mathbb{R}^2$; then there is a locally defined isothermic surface f^* with*

$$df^* = f_x^{-1} dx - f_y^{-1} dy. \tag{8.17}$$

We call this surface f^* a Christoffel transform of f.

8.6.3 Clearly we have the same scaling ambiguity for f^* in the present setup as in the quaternionic setup, since (8.17) depends on a choice of conformal

---

[24] Note that the cross-ratio condition for discrete isothermic nets does not see the dimension of the ambient space: It does refers to four points of the discrete net — which always lie on some 2-sphere.

[25] At this point the reader may want to reconsider the examples discussed in §8.4.11f: The discrete analogs of surfaces of revolution, cylinders, and cones are discrete isothermic because they are invariant under a "discrete 1-parameter group" of Möbius transformations.

curvature line coordinates — which are unique up to constant rescaling, which yields a positive rescaling of f^*, and interchanging the roles of x and y, which yields a change of sign for df^*.

Also, the Christoffel transformation is involutive, $f^{**} \simeq f$, up to translation once a scaling is fixed (cf., §5.2.4).

8.6.4 However, in contrast to the quaternionic setup, (8.17) provides a somewhat "better" generalization to higher codimension of the classical notion of the Christoffel transformation than the quaternionic version of Christoffel's formula. Remember that, in §3.3.22, the Christoffel transformation was introduced as a solution to Christoffel's problem from §3.3.11, that is, as a nonsimilar surface that induces the same conformal class on the domain and has *parallel* tangent planes at corresponding points (cf., [70]). A surface f^* defined by (8.17) clearly qualifies as a Christoffel transform in this sense — in contrast to a Christoffel transform as defined by (5.3) in the quaternionic setup, where f^* has parallel tangent planes only after quaternionic conjugation (see §5.2.2).

8.6.5 As in the quaternionic case, the Christoffel transform f^* of an isothermic surface f satisfies $df^* \wedge df = 0$; taking the Clifford conjugate, we obtain

$$\overline{(df^* \wedge df)}(v,w) = \overline{df^*(v)df(w) - df^*(w)df(v)} = -(df \wedge df^*)(v,w),$$

so that, as for surfaces in Euclidean 3-space $\mathbb{R}^3 \cong \operatorname{Im} \mathbb{H}$ in the quaternionic setup, this (one) equation is symmetric in f and f^*.

Moreover, we can directly carry over the proof given in §5.2.7 in order to prove the converse: The equation $df \wedge df^* = 0$ *characterizes* Christoffel pairs of isothermic surfaces. Thus we obtain an n-dimensional version of our characterization for (Christoffel pairs of) isothermic surfaces in §5.2.6 (cf., [215]):

8.6.6 Lemma. *Two immersions* $f, f^* : M^2 \to \mathbb{R}^n$ *satisfy*

$$df \wedge df^* = 0 \tag{8.18}$$

if and only if both f *as well as* f^* *are isothermic and one is a Christoffel transform of the other, that is, they form a Christoffel pair.*

8.6.7 However, one can do better [47]: The equation $df \wedge df^* = 0$ implies that the domain is 2-dimensional, by an algebraic argument.[26]

Namely, consider two immersions $f, f^* : M^m \to \mathbb{R}^n$ of an m-dimensional manifold M^m into $\mathbb{R}^n$, let $p \in M$, and denote by $(e_1, \ldots, e_m)$ an orthonormal basis of T_pM with respect to the metric induced by f. Then (8.18) reads

$$df(e_i) \cdot df^*(e_j) = df(e_j) \cdot df^*(e_i),$$

---

[26] See [190] for a notion of a (higher dimensional) isothermic hypersurface and its implications.

364   *Applications: Orthogonal systems, isothermic surfaces*

which implies[27] that $(df^*)^\perp(e_i) = 0$ for all $i = 1, \ldots, m$, that is, the normal (with respect to f) part of df^* vanishes so that f and f^* have parallel tangent planes at p, and that, with suitable coefficients $a_{kl} \in \mathbb{R}$,

$$
\begin{aligned}
df^*(e_j) &= a_{ij}df(e_i) + a_{jj}df(e_j) \\
df^*(e_i) &= a_{ii}df(e_i) + a_{ji}df(e_j)
\end{aligned}
\right\} \quad \text{where} \quad
\left\{
\begin{aligned}
a_{ij} - a_{ji} &= 0 \\
a_{ii} + a_{jj} &= 0.
\end{aligned}
$$

Hence, if there are three distinct indices i, j, k, then, for example,

$$
a_{ij}df(e_i) + a_{jj}df(e_j) = df^*(e_j) = a_{kj}df(e_k) + a_{jj}df(e_j),
$$

so that the coefficients with mixed indices must vanish, and

$$
0 = (a_{ii} + a_{jj}) + (a_{jj} + a_{kk}) - (a_{kk} + a_{ii}) = 2a_{jj},
$$

which yields that *all* coefficients vanish, contradicting the assumption that f^* is an immersion. Consequently, we must have $m = 2$ and

$$
\begin{aligned}
df^*(e_1) &= a_{11}df(e_1) + a_{21}df(e_2) \\
df^*(e_2) &= a_{21}df(e_1) - a_{11}df(e_2)
\end{aligned}
$$

with suitable functions a_{11} and a_{21}.

Introducing conformal coordinates (x, y) for f, so that $f_x f_y + f_y f_x = 0$ and $f_x^2 = f_y^2$, and noticing that $f_x^{-1} = f_x^{-2}f_x$ and $f_y^{-1} = f_y^{-2}f_y$, we are then back to our setup in §5.2.7 to complete the proof of the more general statement in [47].

8.6.8 Also, we can recover the holomorphic quadratic differential discussed in §5.2.12 (cf., [47]): Using again conformal curvature line parameters (x, y) for an isothermic surface, such that (the differential of) its Christoffel transform is given by (8.17), the product that gave the quadratic differential in §5.2.12 now reads

$$
q := df \cdot df^* = (dx^2 - dy^2) + 2t\, dx dy, \tag{8.19}
$$

where $t := f_x f_y^{-1} = \frac{1}{|f_x|^2} f_x \wedge f_y$ is, at each point $p \in M$, a bivector that encodes the common tangent space[28] of f and f^* at p. Since, additionally,

$$
t^2 = -\frac{1}{|f_x|^4} f_x^2 f_y^2 = -1,
$$

---

[27] The $df(e_i) \cdot df(e_j)$, $i < j$, form a basis of $\Lambda^2 df(T_p M)$, and any two bivectors $df(e_i) \cdot n_1$ and $df(e_j) \cdot n_2$, where $n_1, n_2 \perp df(T_p M)$, are linearly independent.

[28] Here the tangent *space* of f (resp. f^*) is meant, that is, $t(p)$ encodes the 2-dimensional *linear subspace* $d_p f(T_p M)$ of $\mathbb{R}^n$.

we may, as before, interpret the bundle $\bigcup_{p \in M} \text{span}\{1, t(p)\}$ as a complex line bundle over $f(M)$ — note that the unit bivector t is indeed independent of the chosen coordinates, so that this bundle can be defined invariantly and hence becomes a trivial line bundle if $f(M)$ is an orientable surface. In particular, any conformal coordinates (x, y) for f can be used to compute q. Since, from (8.19), the lines $\text{Im} \, q = 0$ provide the curvature lines of f, this fact can be used to determine conformal curvature line parameters from any choice of conformal coordinates (cf., §5.2.7).

In fact, from (8.19) it becomes clear that any "Hopf differential"

$$Q_n = \langle f_{zz}, n \rangle \, dz^2 = \langle f_{xx} - f_{yy}, n \rangle \, dz^2 \simeq \langle f_{xx} - f_{yy}, n \rangle \, q,$$

where n is any unit normal field of f (cf., §5.1.8), is a real multiple of our holomorphic quadratic differential $q \simeq dz^2$.

Finally, interchanging the roles of f and f^* in (8.19) yields the Clifford conjugate $\bar{q}$ of q: This reflects the fact that f^* is equipped with the opposite orientation (cf., §5.2.14). Namely, if we compute

$$t^* = f_x^*(f_y^*)^{-1} = -f_x^{-1} f_y = -t$$

to identify $\mathcal{C} \cong \text{span}\{1, t^*\}$, then we find that the holomorphic quadratic differential does not change under the Christoffel transformation, $q^* \simeq q$.

More details on the holomorphic quadratic differential associated to an isothermic surface in $\mathbb{R}^n$ may be found in Fran Burstall's paper [47].

8.6.9 The Goursat Transformation. Clearly we can define the Goursat transformation, as in §5.3.3, for isothermic surfaces once we have a notion of Christoffel transformation[29]: Since the Christoffel transformation for isothermic surfaces depends on the Euclidean structure of the ambient space, we obtain a generically nontrivial transformation by intertwining the Christoffel transformation with Möbius transformations, $\tilde{f} = (\mu \circ f^*)^*$.

Once we know how to write Möbius transformations of $\mathbb{R}^n \cup \{\infty\}$ as fractional linear transformations by using Vahlen matrices (see §7.3.2), we may derive a similar formula for the differential $d\tilde{f}$ of a Goursat transform $\tilde{f}$ of an isothermic surface as in §5.3.10. First note that a Goursat transform $\tilde{f}$ of f given by a similarity $p \mapsto \mu(p) = (a_{11}p + a_{12})a_{22}^{-1}$ is similar to the original surface,

$$d\tilde{f} = (a_{11}f_x^{-1}a_{22}^{-1})^{-1}dx - (a_{11}(-f_y^{-1})a_{22}^{-1})^{-1}dy = d(a_{22}fa_{11}^{-1}),$$

---

[29] Remember from Section 5.3 that the Goursat transformation for isothermic surfaces generalizes the classical Goursat transformation for minimal surfaces (see [129], [167], and [213]).

and that the Goursat transforms act on a given isothermic surface like a group since the Christoffel transformation is involutive, $f^{**} \simeq f$. Consequently, we may restrict our attention to "essential" Möbius transformations

$$p \mapsto \mu(p) = -(p - m)^{-1},$$

where $m \in \mathbb{R}^n$ is the center of the inversion μ. Then we may take the same approach as in the quaternionic setup to obtain a formula for the differential of a Goursat transform of f: First express the differential of its Christoffel transform and then apply §8.6.6:

$$d\tilde{f}^* = -d(f^* - m)^{-1} = (f^* - m)^{-1} df^* (f^* - m)^{-1},$$

so that

$$d\tilde{f} = (f^* - m) \, df \, (f^* - m).$$

Again (cf., §5.3.10f), we have chosen the scaling of $\tilde{f}$ so that the holomorphic quadratic differential

$$\tilde{q} = d\tilde{f} \cdot d\tilde{f}^* = (f^* - m) \, df \cdot df^* (f^* - m)^{-1} \simeq q$$

since

$$
\begin{aligned}
\tilde{t} &= \tfrac{1}{|\tilde{f}_x|^2} \tilde{f}_x \tilde{f}_y \\
&= \tfrac{1}{(f^* - m)^4 |f_x|^2} (f^* - m) \, f_x (f^* - m)^2 f_y (f^* - m) \\
&= (f^* - m) \tfrac{1}{|f_x|^2} f_x f_y (f^* - m)^{-1} \\
&= (f^* - m) \, t \, (f^* - m)^{-1}.
\end{aligned}
$$

Summarizing, we have the following

8.6.10 Lemma. *Let $\tilde{f} = (\mu \circ f^*)^*$ denote a Goursat transform of an isothermic surface $f : M^2 \to \mathbb{R}^n$.*

If $\mu : \mathbb{R}^n \to \mathbb{R}^n$ is a similarity, then $\tilde{f} \simeq f$ is similar to the original surface.

If, on the other hand, $\mu : \mathbb{R}^n \cup \{\infty\} \to \mathbb{R}^n \cup \{\infty\}$ is an inversion at a unit sphere with center $m \in \mathbb{R}^n$, $p \mapsto \mu(p) = -(p - m)^{-1}$, then

$$d\tilde{f} = (f^* - m) \, df \, (f^* - m),$$

where the scaling of $\tilde{f}$ is chosen so that the holomorphic quadratic differentials q of f and $\tilde{q}$ of $\tilde{f}$ coincide, $\tilde{q} = d\tilde{f} \cdot d\tilde{f}^* \simeq df \cdot df^* = q$.

8.6.11 The Retraction Form. As in the quaternionic setup, we can now use this lemma to "collect" the possible Christoffel transforms of an isothermic surface $f : M^2 \to S^n$ in the conformal n-sphere into one map $\mathfrak{F}^*$ that takes values in the 2×2 Clifford algebra matrices, by switching viewpoints: As in §5.3.17, we define a "retraction form"

$$\tau := \begin{pmatrix} f\,df^* & -f\,df^*f \\ df^* & -df^*f \end{pmatrix} \qquad (8.20)$$

of $f : M^2 \to \mathbb{R}^n \subset \mathbb{R}^n \cup \{\infty\}$ that is (componentwise) integrable, $d\tau = 0$, by §8.6.6 — and then we let $\mathfrak{F}^*$ be its potential, $d\mathfrak{F}^* = \tau$.

To see that *any* Christoffel transform f^* of f, thought of as an isothermic surface in the conformal n-sphere, is obtained as an off-diagonal entry of $\mathfrak{F}^*$, we study the effect of using a Möbius transform $\mu \circ f$ of f to define τ.

First consider a similarity $f \to (a_{11}f + a_{12})a_{22}^{-1}$. Then $df^* \to a_{22}df^*a_{11}^{-1}$ so that an easy calculation yields

$$\begin{pmatrix} f\,df^* & -f\,df^*f \\ df^* & -df^*f \end{pmatrix} \to \begin{pmatrix} a_{11} & a_{12} \\ 0 & a_{22} \end{pmatrix} \begin{pmatrix} f\,df^* & -f\,df^*f \\ df^* & -df^*f \end{pmatrix} \begin{pmatrix} a_{11} & a_{12} \\ 0 & a_{22} \end{pmatrix}^{-1},$$

that is, up to a constant of integration[30] $\mathfrak{F}^* \to A\mathfrak{F}^*A^{-1}$, where A denotes the Vahlen matrix associated with μ. Note that the lower left off-diagonal entry of $A\mathfrak{F}^*A^{-1}$ indeed gives the Goursat transform $a_{22}f^*a_{11}^{-1}$ of f^*.

On the other hand, using §8.6.10, we compute the effect of taking an essential Möbius transform $f \to -(f - m)^{-1}$ of f:

$$\begin{pmatrix} f\,df^* & -f\,df^*f \\ df^* & -df^*f \end{pmatrix} \to \begin{pmatrix} 0 & -1 \\ 1 & -m \end{pmatrix} \begin{pmatrix} f\,df^* & -f\,df^*f \\ df^* & -df^*f \end{pmatrix} \begin{pmatrix} 0 & -1 \\ 1 & -m \end{pmatrix}^{-1}.$$

That is, we obtain the same answer $\mathfrak{F}^* \to A\mathfrak{F}^*A^{-1}$, also in the case of an essential Möbius transformation, and the lower left off-diagonal entry of $A\mathfrak{F}^*A^{-1}$ again yields the Christoffel transform of $-(f - m)^{-1}$. Thus, since any Möbius transformation can be decomposed into an inversion and a similarity (see (7.10)), we conclude (cf., §5.3.17 and §5.3.19) that:

8.6.12 Lemma and Definition. *Let* $f : M^2 \to \mathbb{R}^n \subset \mathbb{R}^n \cup \{\infty\}$ *be an isothermic surface with Christoffel transform* f^*. *Then the form*

$$\tau := \begin{pmatrix} f\,df^* & -f\,df^*f \\ df^* & -df^*f \end{pmatrix}$$

is called a retraction form[31] *of* f. τ *is integrable,* $d\tau = 0$, *and a potential* $\mathfrak{F}^*$ *of* τ, $d\mathfrak{F}^* = \tau$, *will be called a master Christoffel transform of* f.

---

[30] Geometrically, this constant of integration is a translation of the Christoffel transformation.

[31] Note that a rescaling of the Christoffel transform f^* results in a rescaling of τ.

If $\mu \circ f$ is a Möbius transform of f, then its retraction form is given by $A\tau A^{-1}$, where A is the Vahlen matrix associated to μ. The Christoffel transform $(\mu \circ f)^*$ of $\mu \circ f$ is the lower left off-diagonal entry of $A\mathfrak{F}^* A^{-1}$.

8.6.13 The Calapso Transformation. In fact, a retraction form τ of an isothermic surface f takes values in $\Lambda^2(\mathbb{R}_1^{n+2}) \cong \mathfrak{spin}(\mathbb{R}_1^{n+2})$ (see (7.2)). Since

$$d\tau = 0 \quad \text{and} \quad \tau \cdot \tau = 0,$$

any multiple $\tau_\lambda := \lambda\tau$, $\lambda \in \mathbb{R}$, of the retraction form τ can also be integrated into the spin-group, that is, there are (locally) maps

$$T_\lambda : M^2 \to Spin(\mathbb{R}_1^{n+2}) \quad \text{with} \quad dT_\lambda = T_\lambda \cdot \tau_\lambda.$$

Obviously the T_λ's are only well defined up to postcomposition with a (constant) Möbius transformation, that is, up to a choice of initial condition.

Replacing $f \to \mu \circ f$ by a Möbius transform of f results in $T_\lambda \to T_\lambda A^{-1}$, where A denotes the Vahlen matrix associated to μ, since $\tau \to A\tau A^{-1}$. This shows that the maps $f_\lambda := T_\lambda \cdot f$ are invariant under Möbius transformations of f, that is, the transformation $f \mapsto f_\lambda$ is a well defined transformation for Möbius equivalence classes of surfaces (cf., §5.5.6).

8.6.14 Lemma and Definition. *The retraction forms τ_λ of an isothermic surface $f : M^2 \to \mathbb{R}^n \subset \mathbb{R}^n \cup \{\infty\}$ can be integrated into the spin-group to obtain maps $T_\lambda : M^2 \to Spin(\mathbb{R}_1^{n+2})$; these maps T_λ are called the Calapso transformations, or T_λ-transformations, for f.*

The surfaces $f_\lambda := T_\lambda \cdot f$ are called the T_λ-transforms of f; the f_λ's are well defined up to Möbius transformation, and they only depend on the Möbius equivalence class of f.

8.6.15 Next we also want to understand that the Calapso transforms f_λ of f are isothermic. Remember that, in §5.5.8, we proved that fact by providing a retraction form $\tau^\lambda = \tau_1^\lambda$ for f_λ — however, we did not derive a Clifford algebra analog[32] for our lemma in §5.3.19, that is, we are lacking the "invariant description" of a retraction form as well as a statement that guarantees that a surface having such an (invariantly defined) closed retraction form is isothermic. Hence we will pursue a different strategy and derive a formula for (the differential of) a Christoffel transform f_λ^* of f_λ: Then §8.6.6 will ensure that f_λ is isothermic.

Thus let $T_\lambda = \begin{pmatrix} \tau_{11} & \tau_{12} \\ \tau_{21} & \tau_{22} \end{pmatrix} : M^2 \to Spin(\mathbb{R}_1^{n+2})$ denote a Calapso transformation for our isothermic surface $f : M^2 \to \mathbb{R}^n \subset \mathbb{R}^n \cup \{\infty\}$ and consider

---

[32] This shall be an interesting exercise: Formulate and prove a Clifford analog of §5.3.19.

the system $d\mathcal{T}_\lambda = \mathcal{T}_\lambda \cdot \tau_\lambda$ of differential equations that defines $\mathcal{T}_\lambda$:

$$\begin{pmatrix} d\tau_{11} & d\tau_{12} \\ d\tau_{21} & d\tau_{22} \end{pmatrix} = \lambda \cdot \begin{pmatrix} (\tau_{11}f + \tau_{12})df^* & -(\tau_{11}f + \tau_{12})df^*f \\ (\tau_{21}f + \tau_{22})df^* & -(\tau_{21}f + \tau_{22})df^*f \end{pmatrix}. \quad (8.21)$$

As a consequence, $d(\tau_{i1}f + \tau_{i2}) = \tau_{i1}df$, so that the differential

$$df_\lambda = \{\tau_{11} - (\tau_{11}f + \tau_{12})(\tau_{21}f + \tau_{22})^{-1}\tau_{21}\} df \, (\tau_{21}f + \tau_{22})^{-1}$$

of $f_\lambda = \mathcal{T}_\lambda \cdot f = (\tau_{11}f + \tau_{12})(\tau_{21}f + \tau_{22})^{-1}$ does not contain any derivatives of the components τ_{ij} of $\mathcal{T}_\lambda$. Then algebraic considerations — which we leave to the reader as an exercise[33] — allow us to simplify the first factor to obtain

$$\begin{aligned} df_\lambda &= \Delta \cdot (f\hat{\bar{\tau}}_{21} + \hat{\bar{\tau}}_{22})^{-1} df \, (\tau_{21}f + \tau_{22})^{-1} \\ &= \pm(-f\,\bar{\tau}_{21} + \bar{\tau}_{22})^{-1} df \, (\tau_{21}f + \tau_{22})^{-1}, \end{aligned} \quad (8.22)$$

where $(\bar{.})$ is the Clifford conjugation and $(\hat{.})$ the Clifford transpose (cf., §7.2.7), since $\Delta = \Delta(\mathcal{T}_\lambda) = \pm 1$ and $\hat{\tau}_{21} = -\tau_{21}$ and $\hat{\tau}_{22} = \tau_{22}$ when $\mathcal{T}_\lambda$ takes values in the spin-group $Spin(\mathbb{R}_1^{n+2})$ (see §7.2.16).

From (8.22) we now see what the differential of the Christoffel transform f_λ^* of f_λ must look like:

$$df_\lambda^* = \pm(\tau_{21}f + \tau_{22}) \, df^* \overline{(\tau_{21}f + \tau_{22})}. \quad (8.23)$$

Clearly, $df_\lambda \wedge df_\lambda^* = 0$; it remains to show that df_λ^* is indeed the differential of a map f_λ^*, that is, $d^2f_\lambda^* = 0$. But with $d(\tau_{21}f + \tau_{22}) = \tau_{21}df$ as obtained above from (8.21), we have

$$\begin{aligned} &d\left[(\tau_{21}f + \tau_{22}) \, df^* \overline{(\tau_{21}f + \tau_{22})}\right] \\ &= \tau_{21}df \wedge df^* \overline{(\tau_{21}f + \tau_{22})} - (\tau_{21}f + \tau_{22}) \, df^* \wedge d\bar{\tau}_{21} = 0. \end{aligned}$$

Note that we have chosen the scaling of f_λ^* so that the holomorphic quadratic differential from (8.19) does not change:

$$q_\lambda = df_\lambda \cdot df_\lambda^* = \overline{(\tau_{21}f + \tau_{22})}^{-1} q \, \overline{(\tau_{21}f + \tau_{22})} \simeq q.$$

Namely,[34] $\overline{\tau_{21}f + \tau_{22}} = |\tau_{21}f + \tau_{22}|^2(\tau_{21}f + \tau_{22})^{-1}$, so that the bivector that encodes the tangent space of f_λ is

$$t_\lambda = \tfrac{1}{|(f_\lambda)_x|^2}(f_\lambda)_x \cdot (f_\lambda)_y = \overline{(\tau_{21}f + \tau_{22})}^{-1} t \, \overline{(\tau_{21}f + \tau_{22})}$$

---

[33] Consider two cases, $\tau_{21} = 0$ and $\tau_{21} \neq 0$, and remember from §7.2.9ff that $\tau_{21}^{-1}\tau_{22} \in \mathbb{R}^n$.

[34] Here we again use the fact that $\bar{\tau}_{21}\tau_{22} \in \mathbb{R}^n$ (see §7.2.9).

by (8.22). In particular, (8.22) also shows that f_λ induces the same conformal structure on M^2 as f, so that any conformal coordinates (x, y) for f remain conformal for f_λ.

Hence the Calapso transformation preserves (conformal) curvature line parameters since the curvature lines of f_λ are given by $\operatorname{Im} q_\lambda = 0$ (see §8.6.8).

8.6.16 From (8.23) we can now also determine the retraction form τ^λ of f_λ, as given in §8.6.12:

$$
\begin{aligned}
\tau^\lambda &= \begin{pmatrix} f_\lambda df_\lambda^* & -f_\lambda df_\lambda^* f_\lambda \\ df_\lambda^* & -df_\lambda^* f_\lambda \end{pmatrix} = \begin{pmatrix} f_\lambda df_\lambda^* & f_\lambda df_\lambda^* \bar{f}_\lambda \\ df_\lambda^* & df_\lambda^* \bar{f}_\lambda \end{pmatrix} \\
&= \pm \begin{pmatrix} (\tau_{11}f + \tau_{12})df^*(\overline{\tau_{21}f + \tau_{22}}) & (\tau_{11}f + \tau_{12})df^*(\overline{\tau_{11}f + \tau_{12}}) \\ (\tau_{21}f + \tau_{22})df^*(\overline{\tau_{21}f + \tau_{22}}) & (\tau_{21}f + \tau_{22})df^*(\overline{\tau_{11}f + \tau_{12}}) \end{pmatrix} \\
&= \pm \begin{pmatrix} \tau_{11} & \tau_{12} \\ \tau_{21} & \tau_{22} \end{pmatrix} \begin{pmatrix} fdf^* & -fdf^*f \\ df^* & -df^*f \end{pmatrix} \begin{pmatrix} \bar{\tau}_{22} & \bar{\tau}_{12} \\ \bar{\tau}_{21} & \bar{\tau}_{11} \end{pmatrix} = T_\lambda \tau T_\lambda^{-1}.
\end{aligned}
$$

Thus we have recovered the formula from §5.5.8, and we can now formulate the Clifford algebra version of this lemma:

8.6.17 Lemma. *The Calapso transforms $f_\lambda = T_\lambda \cdot f$ of an isothermic surface f are isothermic with retraction forms $\tau^\lambda = T_\lambda \tau T_\lambda^{-1}$.*

8.6.18 With this lemma we are now prepared to prove that the Calapso transformations of an isothermic surface f form a 1-parameter group in the sense of §5.5.9 — in fact, the proof given there carries over to the Clifford algebra setup without changes because it only used the definition of T_λ formally and the form of τ^λ that we just derived.

Also, the proof of the fact that the Calapso transformation is a second-order deformation of an isothermic surface (see §5.5.11; cf., [62] or [203]) can directly be carried over to the Clifford algebra context — in fact, it may be even clearer in the present setup because we have the projections into Euclidean space already built into the theory in a more obvious way. Thus we leave it to the reader as an exercise to verify[35] that $T_\lambda f$ and $T_\lambda(p_0)f$ have second-order contact at some point $T_\lambda f(p_0)$: If δ denotes any linear differential operator of order at most 2, then

$$
\delta(T_\lambda f_\lambda)(p_0) = \delta(T_\lambda(p_0)f)(p_0).
$$

This fact, that the Calapso transformation is a second-order deformation, directly implies that any second order contact conditions are preserved: If

---

[35] First-order contact is already shown by (8.22); for taking a further derivative one wants to again use the fact that $d\tau_{i1}f + d\tau_{i2} = 0$. Assuming $T_\lambda(p_0) = \begin{pmatrix} 1 & 0 \\ 0 & 1 \end{pmatrix}$, without loss of generality, greatly simplifies the computation.

some 2-sphere $\mathcal{S} \in \Lambda^{n-2}\mathbb{R}_1^{n+2}$ has second-order contact with an isothermic surface in some direction(s) in $T_p M$, then its image[36)]

$$\mathcal{S}_\lambda = T_\lambda(p)\,\mathcal{S}\,T_\lambda^{-1}(p)$$

under a Calapso transformation T_λ for f will have second-order contact with f_λ in the corresponding direction(s).

8.6.19 In particular, the central sphere congruence $\mathcal{Z}$ of an isothermic surface deforms with the surface — remember that, in the codimension 1 situation, we proved this fact in §5.5.18 by using the characterization of the central sphere congruence as the conformal Gauss map (see §3.4.8). Here, using our characterization from §7.5.8 of the central sphere congruence as the mean curvature sphere congruence of f, we may directly argue with the fact that the Calapso transformation is a second-order deformation of an isothermic surface. That is, the mean curvature vectors of $T_\lambda f$ and $T_\lambda(p_0)f$ at $T_\lambda f(p_0)$ coincide — and therefore the mean curvature spheres of $T_\lambda f$ and $T_\lambda(p_0)f$ at $T_\lambda f(p_0)$ coincide.

In order to verify this fact by more computational arguments we consider the Euclidean lift $\begin{pmatrix} f & -f^2 \\ 1 & -f \end{pmatrix}$ of f and let (x,y) denote conformal coordinates. Then

$$\tfrac{\partial}{\partial x}(T_\lambda \mathcal{F} T_\lambda^{-1}) = T_\lambda(\mathcal{F}_x + [\tau_\lambda(\tfrac{\partial}{\partial x}), \mathcal{F}])T_\lambda^{-1} = T_\lambda \mathcal{F}_x T_\lambda^{-1} \qquad (8.24)$$

since $\tau_\lambda \cdot \mathcal{F} = \mathcal{F} \cdot \tau_\lambda = 0$, and likewise for $\tfrac{\partial}{\partial y}$. Consequently, the contact elements of $T_\lambda f$ become, as desired,

$$(T_\lambda \mathcal{F} T_\lambda^{-1})(T_\lambda \mathcal{F} T_\lambda^{-1})_x (T_\lambda \mathcal{F} T_\lambda^{-1})_y = T_\lambda(\mathcal{F}\mathcal{F}_x\mathcal{F}_y)T_\lambda^{-1}. \qquad (8.25)$$

Taking one more derivative, and using (8.24), we obtain

$$\begin{aligned} \tfrac{\partial^2}{\partial x^2}(T_\lambda \mathcal{F} T_\lambda^{-1}) &= T_\lambda(\mathcal{F}_{xx} + [\tau_\lambda(\tfrac{\partial}{\partial x}), \mathcal{F}_x])T_\lambda^{-1} \\ &= T_\lambda(\mathcal{F}_{xx} - 2\lambda\langle f_x, f_x^*\rangle \cdot \mathcal{F})T_\lambda^{-1}, \end{aligned}$$

where the last equality follows from the form of the retraction form τ_λ given in §8.6.12, and similarly for $\tfrac{\partial^2}{\partial y^2}$. When we now employ the formula for the central sphere congruence of a 2-dimensional surface given in §7.5.8 we find that any mean curvature sphere of $T_\lambda f$ at $T_\lambda f(p)$ is the image of

---

[36)] Remember that $T_\lambda(p) \in Spin(\mathbb{R}_1^{n+2})$ and that the action of $Spin(\mathbb{R}_1^{n+2})$ on $\mathbb{R}_1^{n+2}$ is given in exactly the presented way (see §6.3.8). Thus, writing $\mathcal{S}$ as the product of orthogonal hyperspheres, it becomes clear that the image of any sphere $\mathcal{S}$ is given by $T_\lambda(p)\mathcal{S}T_\lambda^{-1}(p)$. Another way to verify this fact is by employing §7.3.12.

the corresponding mean curvature sphere $\mathcal{Z}(p)$ at $f(p)$ under the Möbius transformation $T_\lambda(p)$: As in §5.5.18, we have

$$\mathcal{Z}_\lambda(p) = (T_\lambda \mathcal{Z} T_\lambda^{-1})(p).$$

In order to obtain the full generalization of our lemma in §5.5.18, it remains to show that all $\mathcal{Z}_\lambda$ induce the same metric on M^2 — remember that, in §6.7.7, we discussed a way to determine the induced metric of a congruence of m-spheres $\mathcal{S}$ in S^n by writing $\mathcal{S} = \prod_{j=1}^{n-m} \mathcal{S}_j$ as the product of $n - m$ (orthogonal) hyperspheres. With this ansatz, $\mathcal{Z} = \prod_{j=1}^{n-2} \mathcal{Z}_j$, it is easy to mimic the quaternionic proof: We obtain[37]

$$d\mathcal{Z}_\lambda = T_\lambda(d\mathcal{Z} + [\tau_\lambda, \mathcal{Z}])T_\lambda^{-1} = T_\lambda d\mathcal{Z} T_\lambda^{-1}.$$

Now the claim follows with §6.7.7 since $T_\lambda : M^2 \to Spin(\mathbb{R}_1^{n+2})$ acts by isometries on $\mathbb{R}_1^{n+2} \subset M(2 \times 2, \mathcal{A})$ (see §6.3.8).

8.6.20 Lemma. *Let $f : M^2 \to S^n$ be an isothermic surface with central sphere congruence $\mathcal{Z} : M^2 \to \Lambda^{n-m}\mathbb{R}_1^{n+2}$. Then $\mathcal{Z}_\lambda = T_\lambda \mathcal{Z} T_\lambda^{-1}$ are the central sphere congruences of its Calapso transforms $f_\lambda = T_\lambda f$, and all $\mathcal{Z}_\lambda$ induce the same metric on M^2.*

8.7 Curved flats and the Darboux transformation

In this final section we will discuss the Darboux transformation for isothermic surfaces in S^n. Remember from §8.6.7 that an "isothermic submanifold" — isothermic in the sense that it allows a Christoffel transformation — is necessarily 2-dimensional. The relation between Darboux pairs of isothermic surfaces and curved flats (cf., §3.3.2, [71], or [46]; see [47]) may be the "true" reason for this fact: Curved flats in the symmetric space of point pairs cannot have a higher dimension since the space of point pairs is a symmetric space of rank 2, that is, an abelian subalgebra that carries a nondegenerate metric induced by the Killing form has dimension at most 2.

8.7.1 The Darboux Transformation. In order to introduce the Darboux transformation for an isothermic immersion $f : M^2 \to S^n$, we follow the same strategy as in the quaternionic setup (see Section 5.4), by first providing an algebraic description and then discussing the underlying

---

[37] Writing, as in §5.5.17, the spheres $\mathcal{Z}_j = \begin{pmatrix} n_j & -n_j f - f n_j \\ 0 & -n_j \end{pmatrix} + a_j \begin{pmatrix} f & -f^2 \\ 1 & -f \end{pmatrix}$ in terms of tangential hyperplanes (cf., §6.6.8 and §7.3.5), it is straightforward to check that $[\tau_\lambda, \mathcal{Z}_j] = 0$.

geometry.[38] Thus we consider the equation $T_\lambda \hat{f} \equiv const$ for some map $\hat{f} : M^2 \to S^n$ into the conformal n-sphere, where T_λ is a Calapso transformation for an isothermic surface $f : M^2 \to S^n$ (cf., §5.4.8). Now remember that the spin group $Spin(\mathbb{R}_1^{n+2})$ acts by linear fractional transformations on $S^n \cong \mathbb{R}^n \cup \{\infty\}$ (see §7.3.2); to simplify the following considerations we assume that $f, \hat{f} : M^2 \to \mathbb{R}^n \subset S^n$ and investigate the above condition[39]

$$T_\lambda \hat{f} \equiv const \quad \Leftrightarrow \quad d(T_\lambda \hat{f}) = 0.$$

Writing $T_\lambda = \begin{pmatrix} \tau_{11} & \tau_{12} \\ \tau_{21} & \tau_{22} \end{pmatrix}$ and using the defining differential equation (8.21) for the Calapso transformation T_λ,

$$
\begin{aligned}
d\tau_{11} &= \lambda(\tau_{11}f + \tau_{12})\,df^*, & d\tau_{12} &= -\lambda(\tau_{11}f + \tau_{12})\,df^*f, \\
d\tau_{21} &= \lambda(\tau_{21}f + \tau_{22})\,df^*, & d\tau_{22} &= -\lambda(\tau_{21}f + \tau_{22})\,df^*f,
\end{aligned}
$$

where f^* denotes a Christoffel transform of f, we compute

$$
\begin{aligned}
d(T_\lambda \hat{f}) &= d[(\tau_{11}\hat{f} + \tau_{12})(\tau_{21}\hat{f} + \tau_{22})^{-1}] \\
&= \{a\,d\hat{f} + \lambda\,(af + b)\,df^*(\hat{f} - f)\}(\tau_{21}\hat{f} + \tau_{22})^{-1},
\end{aligned} \tag{8.26}
$$

where

$$
\begin{aligned}
a &:= \tau_{11} - (\tau_{11}\hat{f} + \tau_{12})(\tau_{21}\hat{f} + \tau_{22})^{-1}\tau_{21} &= \pm\overline{(\tau_{21}\hat{f} + \tau_{22})}^{-1}, \\
b &:= \tau_{12} - (\tau_{11}\hat{f} + \tau_{12})(\tau_{21}\hat{f} + \tau_{22})^{-1}\tau_{22} &= \mp\overline{(\tau_{21}\hat{f} + \tau_{22})}^{-1}\hat{f}.
\end{aligned} \tag{8.27}
$$

Here the first equation holds[40] since T_λ is $Spin(\mathbb{R}_1^{n+2})$-valued, as for (8.22), and the second equality is a consequence of the first since $0 = a\hat{f} + b$. Combining (8.26) and (8.27), we obtain the Clifford algebra version of our quaternionic Riccati equation (5.17) for the Darboux transformation:

$$d(T_\lambda \hat{f}) = \pm\overline{(\tau_{21}\hat{f} + \tau_{22})}^{-1}\{d\hat{f} - \lambda\,(\hat{f} - f)\,df^*(\hat{f} - f)\}(\tau_{21}\hat{f} + \tau_{22})^{-1}. \tag{8.28}$$

---

[38] As in Section 5.4, we will see that Darboux pairs envelope Ribaucour sphere congruences and induce conformally equivalent metrics: The notion of a Darboux transformation that we will define indeed generalizes the classical notion (see §3.2.5; cf., [93], [89], [98], [99], and [106]). But we will do better than there: We will show (up to a small gap for which the reader is referred to [47]) that this geometry characterizes Darboux pairs.

[39] Note that, considering the action of $Spin(\mathbb{R}_1^{n+2})$ by fractional linear transformations (or, equivalently, by Möbius transformations) we have the equivalence $T_\lambda \hat{f} \equiv const \Leftrightarrow d(T_\lambda \hat{f}) = 0$, in contrast to the quaternionic setup, where we considered $Sl(2, \mathbb{H})$ as acting on the space $\mathbb{H}^2$ of homogeneous coordinates so that $T_\lambda \hat{f} \equiv const \Leftrightarrow d(T_\lambda \hat{f}) \parallel T_\lambda \hat{f}$ (see §5.4.8).

[40] In the nontrivial case $\tau_{21} \neq 0$, we use that $\tau_{21}^{-1}\tau_{22} \in \mathbb{R}^n$ and $\tau_{11}\bar{\tau}_{22} + \tau_{12}\bar{\tau}_{21} = \pm 1$.

Thus we have the following[41]:

8.7.2 Lemma and Definition. *Let* $f : M^2 \to S^n$ *be an isothermic surface with Calapso transformation* $T_\lambda : M^2 \to Spin(\mathbb{R}_1^{n+2})$. *Then a surface* $\hat{f} : M^2 \to S^n$ *satisfying* $T_\lambda \hat{f} \equiv const$ *is called a Darboux transform of* f.

If $f, \hat{f} : M^2 \to \mathbb{R}^n \subset S^n$ *are considered as maps into Euclidean* n-*space, then* $\hat{f}$ *is a Darboux transform of* f *if and only if it satisfies the Riccati-type partial differential equation*

$$d\hat{f} = \lambda\,(\hat{f} - f)\,df^*(\hat{f} - f), \tag{8.29}$$

where f^* *denotes the Christoffel transform of* f.

8.7.3 First note that any Darboux transform $\hat{f}$ of f is an isothermic surface itself: Defining, as in (5.18),

$$\hat{f}^* := f^* + \tfrac{1}{\lambda}(\hat{f} - f)^{-1},$$

we obtain the Christoffel transform of $\hat{f}$ since

$$d\hat{f}^* = \tfrac{1}{\lambda}(\hat{f} - f)^{-1}df(\hat{f} - f)^{-1}, \tag{8.30}$$

so that $d\hat{f} \wedge d\hat{f}^* = 0$ (see §8.6.6). Obviously (8.30) is equivalent to a Riccati equation for f as a Darboux transform of $\hat{f}$; as a consequence, the Darboux transformation is involutive (cf., §5.4.12).

8.7.4 Clearly, by (8.29), f and $\hat{f}$ induce conformally equivalent metrics on M^2, and they envelope a congruence of 2-spheres by §7.5.10. In order to see that this sphere congruence is Ribaucour, we compute the holomorphic quadratic differential (see §8.6.8)

$$\hat{q} = d\hat{f} \cdot d\hat{f}^* = (\hat{f} - f)df^*df(\hat{f} - f)^{-1} = (\hat{f} - f)\,\bar{q}\,(\hat{f} - f)^{-1}.$$

Now let (x, y) denote conformal coordinates and note (cf., §5.4.14) that the bivector

$$i \simeq \hat{i} = \hat{f}_x\hat{f}_y^{-1} = -(\hat{f} - f)f_x^{-1}f_y(\hat{f} - f)^{-1} = -(\hat{f} - f)\,t\,(\hat{f} - f)^{-1}, \tag{8.31}$$

so that, under the identifications $\mathbb{C} \cong \text{span}\{1, \hat{t}\}$ and $\mathbb{C} \cong \text{span}\{1, t\}$, the holomorphic quadratic differentials of f and of $\hat{f}$ coincide, $\hat{q} \simeq q$. In particular, since $\text{Im}\,\hat{q} = \text{Im}\,q = 0$ characterizes the curvature directions of $\hat{f}$, the

---

[41] Compare [73].

curvature lines of $\hat{f}$ and f coincide and the enveloped 2-sphere congruence is Ribaucour.

Thus Darboux pairs $(\hat{f}, f) : M^2 \to S^n$ of isothermic surfaces in the conformal n-sphere provide solutions of Blaschke's problem (cf., §3.2.8).

8.7.5 Conversely, assume that two surfaces $f, \hat{f} : M^2 \to S^n$ envelope a Ribaucour congruence of 2-spheres and induce the same conformal structure on M^2. Since f and $\hat{f}$ envelope a sphere congruence, we have

$$\hat{t}(\hat{f} - f) = \pm(\hat{f} - f)t \tag{8.32}$$

by §7.5.10. Moreover, any corresponding curvature lines on both surfaces must envelope a 1-parameter family of circles because the sphere congruence is assumed to be Ribaucour: Thus, if (x, y) denote curvature line parameters for f, then we have, by §7.5.10 again,

$$\hat{f}_x(\hat{f} - f) = r_1(\hat{f} - f)f_x \quad \text{and} \quad \hat{f}_y(\hat{f} - f) = r_2(\hat{f} - f)f_y$$

with suitable functions r_1, r_2 (cf., §8.2.4).

Conformality of the induced metrics then implies $r_1^2 = r_2^2$.

In the case $r_1 = r_2$, (8.32) becomes $\hat{t}(\hat{f} - f) = (\hat{f} - f)t$ when computing t and $\hat{t}$ using the orthonormal direction fields given by $\frac{\partial}{\partial x}$ and $\frac{\partial}{\partial y}$; this does not conform with (8.31). Hence we do not find a Darboux pair in this case.[42]

In the case $r_1 = -r_2$, on the other hand, we readily check that

$$df \wedge (\hat{f} - f)^{-1}d\hat{f} = -r_1(f_x f_y + f_y f_x)(\hat{f} - f)^{-1} = 0.$$

With this we then find that $d[(\hat{f} - f)^{-1}d\hat{f}(\hat{f} - f)^{-1}] = 0$, that is, f is isothermic with a Christoffel transform f^* given by

$$df^* = (\hat{f} - f)^{-1}d\hat{f}(\hat{f} - f)^{-1} \tag{8.33}$$

(see §8.6.6). Rewriting (8.33), we obtain the Riccati equation (8.29) for $\hat{f}$, identifying $\hat{f}$ as a Darboux transform of f, which is then isothermic also.

Thus Darboux pairs of isothermic surfaces can be *characterized* by their geometric properties, that they form a Ribaucour pair and induce conformally equivalent metrics, when one additionally imposes the orientation condition (8.31) — analogous to the geometric characterization of Christoffel pairs in §3.3.22 (cf., §5.2.14 and §8.6.8).

---

[42] This case is discussed in more detail in [47]: There, the author shows that the enveloped sphere congruence is not full (cf., §3.2.5).

8.7.6 Curved Flats. Next we want to make contact with the notion of curved flats, showing that Darboux pairs yield curved flats in the symmetric space of point pairs (cf., §3.3.2 and §5.5.20).

For this purpose first consider, in $\mathcal{A}(\mathbb{R}^n)$, the linear partial differential equation

$$0 = d\hat{a} + (\hat{f} - f)^{-1} d\hat{f} \cdot \hat{a} : \qquad (8.34)$$

Since $d[(\hat{f} - f)^{-1} d\hat{f} \hat{a}] = (\hat{f} - f)^{-1} df (\hat{f} - f)^{-1} \wedge d\hat{f} \cdot \hat{a} = 0$ by (8.29) and §8.6.6, this equation is (completely) integrable, and a proper choice of an initial condition will provide us with a solution $\hat{a} : M^2 \to \Gamma(\mathbb{R}^n)$ since the coefficient differential form $(\hat{f} - f)^{-1} d\hat{f} : TM \to \Lambda^0(\mathbb{R}^n) \oplus \Lambda^2(\mathbb{R}^n)$ takes values in the Lie algebra of the Clifford group $\Gamma(\mathbb{R}^n)$. Moreover, the equation (8.34) remains unchanged under the order involution of $\mathcal{A}(\mathbb{R}^n)$ (see §6.3.1), so that $\hat{a}$ stays in the -1-eigenspace of the order involution if the initial condition is chosen there.

Next we define

$$a := [(\hat{f} - f)\hat{a}]^{-t}, \qquad (8.35)$$

where $(.)^t$ denotes the Clifford transpose[43] (cf., §7.2.9f) and $(.)^{-t}$ denotes its composition with the Clifford group inverse. It is straightforward to compute the derivative

$$da = [(\hat{f} - f)\hat{a}]^{-t} (df\hat{a})^t [(\hat{f} - f)\hat{a}]^{-t} = (\hat{f} - f)^{-1} df \cdot a \qquad (8.36)$$

of a — here we used the fact that $(\hat{f} - f)^t = (\hat{f} - f)$ and that $(.)^t$ is an antiautomorphism of $\mathcal{A}(\mathbb{R}^n)$, besides the differential equation (8.34).

Now consider $a^{-1} d\hat{a}$ and $\hat{a}^{-1} da$: Using (8.34) and (8.36), we compute

$$\begin{aligned} \hat{a}^{-1} da &= a^t df\, a, \\ a^{-1} d\hat{a} &= -a^{-1}(\hat{f} - f)^{-1} d\hat{f}(\hat{f} - f)^{-1} a^{-t} \\ &= -\lambda a^{-1} df^* a^{-t}. \end{aligned}$$

This shows that $\hat{a}^{-1} da, a^{-1} d\hat{a} : TM \to \mathbb{R}^n$ and that $\hat{a}^{-1} da \wedge a^{-1} d\hat{a} = 0$. Since, moreover, both differentials are closed, we conclude with §8.6.6 that

$$\hat{a}^{-1} da = dg \quad \text{and} \quad a^{-1} d\hat{a} = -\lambda dg^*$$

with some Christoffel pair $g, g^* : M^2 \to \mathbb{R}^n$ of isothermic surfaces.

---

[43] Here we denote the Clifford transpose by $(.)^t$ instead of $(\tilde{.})$, in order to avoid confusion with the transformations of isothermic surfaces. Remember that the Clifford transpose was introduced as the composition of the Clifford conjugation and the order involution (cf., §6.3.1), and that it could equivalently be characterized as the antiautomorphism that is the identity on $\mathbb{R}^n$. Note that the notation $(.)^{-t}$ makes sense because the Clifford transpose commutes with taking the inverse.

8.7.7 Having collected these facts, we now turn to the heart of the matter: Let

$$\mathcal{F} := \begin{pmatrix} \hat{f} & f \\ 1 & 1 \end{pmatrix} \begin{pmatrix} \hat{a} & 0 \\ 0 & a \end{pmatrix} = \begin{pmatrix} \hat{f}(\hat{f}-f)^{-1} & f \\ (\hat{f}-f)^{-1} & 1 \end{pmatrix} \begin{pmatrix} (\hat{f}-f)\hat{a} & 0 \\ 0 & a \end{pmatrix}.$$

If we have chosen $\hat{a}$ to take values in the -1-eigenspace of the order involution as discussed above, then $(\hat{f}-f)\hat{a}$ and a take values in the $+1$-eigenspace of the order involution, so that $\mathcal{F} : M^2 \to Spin(\mathbb{R}_1^{n+2})$ by (8.35) (see §7.2.16).

Thus $\mathcal{F}$ is an adapted frame for the point pair map $(\hat{f}, f) : M^2 \to \mathfrak{P}$ (see §7.4.3 and §7.4.4). It is straightforward to compute its connection form

$$\mathcal{F}^{-1}d\mathcal{F} = \begin{pmatrix} \hat{a}^{-1}\{d\hat{a} + (\hat{f}-f)^{-1}d\hat{f} \cdot \hat{a}\} & \hat{a}^{-1}(\hat{f}-f)^{-1}df\, a \\ -a^{-1}(\hat{f}-f)^{-1}d\hat{f}\, \hat{a} & a^{-1}\{da - (\hat{f}-f)^{-1}df \cdot a\} \end{pmatrix}.$$
$$(8.37)$$

Employing (8.34) and (8.36), we deduce that

$$\mathcal{F}^{-1}d\mathcal{F} = \begin{pmatrix} 0 & \hat{a}^{-1}da \\ a^{-1}d\hat{a} & 0 \end{pmatrix} = \begin{pmatrix} 0 & dg \\ -\lambda dg^* & 0 \end{pmatrix} \qquad (8.38)$$

with the aforementioned[44] Christoffel pair $g, g^* : M^2 \to \mathbb{R}^n$ (cf., §5.4.19). As a consequence (see §2.2.3 or [112]), any Darboux pair of isothermic surfaces in the conformal n-sphere forms a curved flat in the (symmetric) space of point pairs in S^n (cf., §3.3.2 and §5.5.20).

8.7.8 Conversely, if $(\hat{f}, f) : M^2 \to \mathfrak{P}$ is a curved flat, then we may choose an adapted frame $\mathcal{F} = \begin{pmatrix} \hat{f}\hat{a} & fa \\ \hat{a} & a \end{pmatrix}$ so that its connection form is off-diagonal.[45] This yields the differential equations (8.34) and (8.36) for the scaling functions a and $\hat{a}$. The Maurer-Cartan equation[46] for the connection form then shows that

$$df \wedge (\hat{f}-f)^{-1}d\hat{f}(\hat{f}-f)^{-1} = (\hat{f}-f)^{-1}d\hat{f}(\hat{f}-f)^{-1} \wedge df = 0. \qquad (8.39)$$

Since, moreover, we compute

$$d[(\hat{f}-f)^{-1}d\hat{f}(\hat{f}-f)^{-1}] = 0$$

---

[44] In fact, the shape of the connection form $\mathcal{F}^{-1}d\mathcal{F}$ implies that $a^{-1}d\hat{a}$ and $\hat{a}^{-1}da$ are closed and are the derivatives of a Christoffel pair, by the Maurer-Cartan equation.

[45] See [112] or Footnote 2.8. The interested reader may want to verify this fact in our present Clifford algebra setup, as an exercise.

[46] In fact, it is the curved flat condition $[\mathcal{F}^{-1}d\mathcal{F} \wedge \mathcal{F}^{-1}d\mathcal{F}] = 0$ that provides the equation; the closeness $d[\mathcal{F}^{-1}d\mathcal{F}] = 0$ is then a consequence.

as a consequence of (8.39), we deduce with §8.6.6 that f is isothermic with a Christoffel transform f^* satisfying

$$df^* = \tfrac{1}{\lambda}(\hat{f} - f)^{-1} d\hat{f}(\hat{f} - f)^{-1}$$

for some real constant λ; therefore, $\hat{f}$ satisfies the Riccati equation (8.29) showing that the legs f and $\hat{f}$ of our curved flat form a Darboux pair.

Thus we obtain a theorem by F. Burstall[47] [47] that characterizes Darboux pairs of isothermic surfaces in the conformal n-sphere as curved flats in the (symmetric) space of point pairs in S^n (cf., §3.3.2 and §5.5.20):

8.7.9 Theorem. *Darboux pairs* $(\hat{f}, f) : M^2 \to \mathfrak{P}$ *of isothermic surfaces in the conformal n-sphere are the curved flats in the symmetric space of point pairs.*

8.7.10 As in §5.5.25, we can determine a Calapso transformation for the isothermic surface g from the curved flat frame $\mathcal{F}$: Let $\mathcal{G} = \left(\begin{smallmatrix} 1 & g \\ 0 & 1 \end{smallmatrix}\right)$ be the Euclidean frame of g and consider $\mathcal{F}\mathcal{G}^{-1}$; then

$$\mathcal{G}\mathcal{F}^{-1}d(\mathcal{F}\mathcal{G}^{-1}) = -\lambda \begin{pmatrix} gdg^* & -gdg^*g \\ dg^* & -dg^*g \end{pmatrix},$$

identifying $\mathcal{F}\mathcal{G}^{-1}$ as the $T_{-\lambda}$-transformation for g (see §8.6.14). This shows that the original surface $f = \mathcal{F} \cdot 0 = \mathcal{F}\mathcal{G}^{-1}g$ is a Calapso transform of g. And, conversely, employing the fact that the Calapso transformations act like a 1-parameter group on a given isothermic surface (see §8.6.18), $g = T_\lambda f$ is a T_λ-transform of f.

Interchanging the roles of f and $\hat{f}$ in $\mathcal{F}$ by means of a (constant) gauge transformation $\mathcal{F} \mapsto \mathcal{F} \cdot \frac{1}{\sqrt{|\lambda|}} \left(\begin{smallmatrix} 0 & 1 \\ \lambda & 0 \end{smallmatrix}\right)$, the same line of argument applies to $\hat{f}$ and g^* in order to see that $g^* = \hat{T}_\lambda \hat{f}$ is a T_λ-transform of $\hat{f}$ (cf., §5.6.10).

8.7.11 It is a general fact that curved flats naturally come in (real) 1-parameter families, the "curved flat associated family" (see [112] or §3.3.3). In our present setup of the curved flat given by a Darboux pair of isothermic surfaces, this associated family is obtained by integrating the differential equation

$$d\mathcal{F}_\mu = \mathcal{F}_\mu \cdot \mu \begin{pmatrix} 0 & dg \\ -\lambda dg^* & 0 \end{pmatrix}$$

---

[47] In [47] the author uses the frame presented in §7.4.4 to deduce this result with more ease than is done here: For this, the frame invariance of the curved flat condition becomes important. However, to obtain the Christoffel pair $(g, g*)$ directly, we chose the framing presented here.

to obtain a family $(\mathcal{F}_\mu)_{\mu \in \mathbb{R}}$ of frames for the associated family of curved flats. Employing a gauge transformation $\mathcal{F}_\mu \to \mathcal{F}_\mu \sqrt{|\mu|} \begin{pmatrix} 1 & 0 \\ 0 & \frac{1}{\mu} \end{pmatrix}$, so that

$$\mathcal{F}_\mu^{-1} d\mathcal{F}_\mu = \begin{pmatrix} 0 & dg \\ -\lambda dg^* & 0 \end{pmatrix} \to \begin{pmatrix} 0 & dg \\ -\mu^2 \lambda dg^* & 0 \end{pmatrix},$$

we realize, by the arguments of the previous paragraph, that the legs $\hat{f}_\mu$ and f_μ of the curved flats are $T_{-\mu^2 \lambda}$-transforms of the isothermic surfaces g^* and g of the "limiting" Christoffel pair. With the 1-parameter group property of the Calapso transformation, this amounts to

$$f_\mu = T_{\lambda(1-\mu^2)} f \quad \text{and} \quad \hat{f}_\mu = \hat{T}_{\lambda(1-\mu^2)} \hat{f},$$

since $g = T_\lambda f$ and $g^* = \hat{T}_\lambda \hat{f}$ (see [47]).

In particular, the associated family $(\hat{f}_\mu, f_\mu)_{\mu \in \mathbb{R}}$ of a curved flat $(\hat{f}, f)$ consists of Calapso transforms of its two legs. Note that the surfaces $f_{\pm\mu}$ and $\hat{f}_{\pm\mu}$ are Möbius equivalent, since the respective $T_{\lambda(1-\mu^2)}$-transformations depend quadratically on μ.

8.7.12 At this point the reader will have gained enough confidence in the presented methods to continue and do his own research in the Möbius geometry of submanifolds. Thus we will end our discussions here.

Many more things could be said — for example, we could continue by revealing all the permutability theorems for the transformations of isothermic surfaces, which we discussed in Section 5.6, in the Vahlen matrix setup. However, this shall be left as an exercise for the interested reader, who has certainly noticed by now that we already touched on various permutability theorems throughout the last two sections. Of particular interest may be the permutability theorem for the Darboux transformation (cf., §5.6.6), which appears to be a special case of the permutability theorem for the Ribaucour transformation of orthogonal systems (cf., §8.5.8).

Another interesting topic to learn more about may be the implications from the relation of (Darboux pairs of) isothermic surfaces with curved flats: Integrable systems theory provides general methods that may not only be applied to (Darboux pairs of) isothermic surfaces but also to conformally flat hypersurfaces (see §2.2.13). The reader interested in this direction is pointed to F. Burstall's excellent paper [47], where he will also find references to further reading.

Further Reading

It is the author's hope that this book will not only give the reader a first introduction to the field of submanifolds in Möbius geometry, but that it will also get him interested in one or the other of the treated applications. The author chose these applications according to his personal interest and expertise — however, they seemed to be timely enough and to touch on rather different circles of problems. Obviously, this book cannot cover all of the developments in the theory since its beginnings; in fact, the author cannot claim to be able to write such a "book about everything in Möbius geometry" — there always seem to be new things to learn, and this does not only concern new developments and new technology but also old results and techniques: Again and again, the present author is astonished at how much the classical authors knew, and how much of this knowledge got lost or obscured over the years. At various times the author had the experience of seemingly obtaining new results that he discovered in the classical literature only afterwards.[1]

Anyway, it shall certainly be worthwhile to examine the classical literature and to (re)discover the results of the grand old masters of differential geometry. For the reader's convenience, various, in particular older, references are included in the Bibliography, even if they are not cited in the text: In the author's experience, it can be rather difficult to detect papers from classical times. To facilitate the decision of what is of relevance to the reader's topic of interest, references to reviews are included. However, the provided Bibliography is certainly far from being complete, and the interested reader is encouraged to also search for additional (old) literature on his own.

For this, the "Jahrbuch über die Fortschritte der Mathematik" (JFM) is certainly of great help, and it is also a worthwhile activity to investigate the "Comptes Rendus hebdomaires des séances de l'Académie des sciences."

The reader should definitely also have a look at the canonical reference [29], "Vorlesungen über Differentialgeometrie III: Differentialgeometrie der Kreise und Kugeln," by W. Blaschke and G. Thomsen, for the geometry of surfaces in Möbius geometry (and, more generally, in Lie sphere geometry; compare also [66]). The author of the present book learned Möbius geometry primarily from this book and considers it an excellent reference. Parts of the present text were strongly inspired by this book.

---

[1] Sometimes, he would even recognize the result only after deducing it by himself — some of the classical literature does not seem very easy to access, even after succeeding to obtain it.

In this context the reader should also be referred to the two papers [277] and [278] by G. Thomsen that may be considered as predecessors of [29]. Also of interest shall be a series of papers [263], [264], [265], [266], [267], [268], [269] by R. Sulanke, where a variety of problems concerning submanifolds in Möbius geometry is treated using the projective model of Möbius geometry.

Next, the author would like to point the reader's attention to the book [4], "Conformal differential geometry and its generalizations," by M. Akivis and V. Goldberg. Two remarkable features of this book are an extensive bibliography — which contains references to literature that was not available to the present author, in particular, from the Russian school — and many interesting historical notes. Besides that, the book presents generalizations of the classical theory to higher dimension and codimension, as well as to different signature: This last is relevant for the application of the theory to theoretical physics,[2] as well as establishing a relation with Lie sphere geometry. Of course there are overlaps with the presentation given in the present book — however, the text by M. Akivis and V. Goldberg differs from the present text in many respects: in its choice of topics as well as in the techniques used. Thus it shall be recommended to check this book out — even more so because it also addresses nonflat conformal geometry.

Just as Euclidean geometry can be generalized to Riemannian geometry ("nonflat Euclidean geometry"), Möbius geometry, as it was discussed in the present book, can be generalized to nonflat Möbius geometry or conformal geometry.[3] The utterly general concept of a (normal) Cartan geometry provides the framework for treating (also) nonflat Möbius geometry — in a much slicker and more conceptual way than we did in our discussions, in the Preliminaries chapter at the beginning of the present book.

To learn about this very beautiful and far-reaching approach to (not necessarily flat) Möbius geometry, the reader is referred to R. Sharpe's book [254], "Differential geometry. Cartan's generalization of Klein's Erlanger program"; an approach via connections on vector bundles[4] is presented in the excellent new paper by F. Burstall and D. Calderbank [50], "Conformal submanifold geometry," which focuses on the case of submanifolds and surfaces in conformal geometry. Readers who are interested in this approach

---

[2] Explicit discussions of important examples are given in the book.

[3] Remember that, in dimensions ≥ 3, Möbius geometry and conformal geometry are the same by Liouville's theorem (see §1.5.4).

[4] Very roughly, the basic idea can be summarized as follows: Attach to each point a classical model of Möbius geometry and identify fibers by means of a (Cartan) connection. A careful construction provides a canonical such structure on any manifold of dimension $n \geq 3$ that is equipped with a conformal structure.

to conformal geometry should definitely have a look into both references, [254] as well as [50].

Also, the author would like to draw the reader's attention to recent work on a "quaternionic function theory" that applies to the (global) Möbius geometry of surfaces in the conformal 3- and 4-spheres via the quaternionic model that we discussed in Chapter 4: Here, ideas from the theory of complex vector bundles are generalized to quaternionic vector bundles. A first announcement of the ideas was published by F. Pedit and U. Pinkall in [216], "Quaternionic analysis on Riemann surfaces and differential geometry."

Readers who are interested in this direction should also not miss the following two references: the book [49], "Conformal geometry of surfaces in S^4 and quaternions" by F. Burstall, D. Ferus, K. Leschke, F. Pedit, and U. Pinkall, and the paper [113], "Quaternionic holomorphic geometry: Plücker formula, Dirac eigenvalue estimates, and energy estimates for harmonic 2-tori" by the last four of the aforementioned authors. The first of these two texts is a good introduction to the theory and shows how the theory can be applied to global surface theory; it should complement our discussions in Chapters 4 and 5 well. The second text takes the reader to rather spectacular results: In particular, generalizations of the Riemann-Roch theorem and of the classical Plücker formula[5] for complex line bundles over a Riemann surface to quaternionic (holomorphic) line bundles over Riemann surfaces are obtained as powerful tools in the theory.

In the present author's opinion, discrete net theory, even though it may appear to be a rather exotic branch of (differential) geometry at first, promises to introduce quite interesting methodology to differential geometry — it may (re)introduce more geometric methods to differential geometry and, in this way, provide a link to the "geometric method" of classical differential geometry (cf., [152]). In order to transcend the state of pure analogy in linking discrete net theory and differential geometry, a key issue is to obtain some kind of approximation results: One asks for the existence of a corresponding discrete object that is arbitrarily close to a given differential geometric object (cf., §8.5.12). To the author's knowledge, there are currently two such approximation results available in the published literature[6]: [36] by A. Bobenko, D. Matthes, and Yu. Suris, where such an approximation result is proven for smooth and discrete surfaces of constant Gauss curvature, and [138] and [139] by Z.-X. He and O. Schramm,[7] where

---

[5] The quaternionic Plücker formula involves the Willmore energy, which relates it, for example, with the Willmore problem.

[6] As mentioned in §8.5.12, another, for orthogonal systems, can be expected to be available in the near future.

[7] This is, to the author's knowledge, the strongest available result; there are other approximation results for circle patterns, for example, the earlier result [242] by B. Rodin and D. Sullivan.

an approximation result for conformal maps is provided.

In fact, this second approximation result has a Möbius geometric flavor since the discrete objects investigated are circle packings, that is, certain maps of a planar graph into the space of circles in the complex plane. It appears that the idea of using circle packings to approximate holomorphic maps goes back to P. Koebe's 1936 paper [165], which was reconsidered by W. P. Thurston in [279]. Meanwhile, these circle packings and, more generally, circle patterns have become a topic of active research — see [33] and [34] by A. Bobenko and T. Hoffmann, [37] by these two authors with Yu. Suris, and [38] by A. Bobenko and B. Springborn, for example. The reader is also be referred to [32] by A. Bobenko and U. Pinkall: There, a definition for discrete isothermic nets, called "*S*-isothermic surfaces," is given that uses a type of spatial circle packings — this provides, in some sense, a more restrictive notion of a discrete isothermic net than the one that we discussed in Chapter 5. However, the Christoffel transformation preserves this class of *S*-isothermic nets [32], so that discrete minimal (*S*-isothermic) nets can be constructed from circle packings in the plane (cf., §5.7.37); using this construction, one gains better control over the regularity and the global behavior of the resulting net than by using arbitrary discrete isothermic nets on the 2-sphere as Gauss maps.[8]

Of course, the above is only a very small choice of a vast amount of literature in the field of conformal (sub)manifold theory. Certainly, this choice of references reflects the interest (and expertise) of the author, and it claims not to be complete in any way. However, the author hopes to have given references that introduce the reader to new and promising methods that will stimulate further development.

It remains to express my gratitude to the reader for sharing my interest in Möbius differential geometry — a wonderful field in differential geometry.

---

[8] The author is grateful to A. Bobenko and B. Springborn for the information about their corresponding recent work with T. Hoffmann.

References

1. P. Adam: *Sur les surfaces isothermiques à lignes de courbure planes dans un système ou dans les deux systèmes*, Ann. Ec. Norm. Sup. 10, 319—358 (1893; JFM 25.1186.01)

2. L. V. Ahlfors: *Möbius transformations and Clifford numbers*, Differential geometry and complex analysis, vol. dedic. H. E. Rauch, 65—73 (1985; MR 86g:20065; Zbl 569.30040); see §I.1.4, §7.0.0, §7.2.8

3. L. V. Ahlfors: *Clifford numbers and Möbius transformations in R^n*, NATO Adv. Sci. Inst. Ser. C Math. Phys. Sci. 183, 167—175 (1986; MR 88b:20074; Zbl 598.15025); see §I.1.4, §7.0.0

4. M. A. Akivis, V. V. Goldberg: *Conformal differential geometry and its generalizations*, Wiley & Sons, New York (1996; MR 98a:53023; Zbl 863.53002); see §I.0.0, §I.1.2, §P.6.3, §1.0.0, §F.0.0

5. M. A. Akivis, V. V. Goldberg: *A conformal differential invariant and the conformal rigidity of hypersurfaces*, Proc. Amer. Math. Soc. 125, 2415—2424 (1997; MR 97j:53017; Zbl 887.53030); see §P.6.3, §5.1.9

6. M. A. Akivis, V. V. Goldberg: *The Darboux mapping of canal hypersurfaces*, Beitr. Alg. Geom. 39, 395—411 (1998; MR 99g:53008; Zbl 916.53007); see §1.8.0, §1.8.16

7. M. A. Akivis, V. V. Goldberg: *Conformal and Grassmann structures*, Differ. Geom. Appl. 8, 177—203 (1998; MR 99f:53012; Zbl 924.53025); see §I.0.0

8. U. Amaldi: *Le superficie con infinite trasformazioni conformi in sè stesse*, Rom. Acc. L. Rend. (5) 10, 168—175 (1901; JFM 32.0680.01); see §1.8.6, §3.7.14

9. K. Arlt: *Eine Weierstrass-Darstellung für cmc-1-Flächen in der hyperbolischen Geometrie*, Wissensch. Hausarbeit 1. Staatsexamen, TU Berlin (2000); see §5.6.21

10. H. Aslaksen: *Quaternionic determinants*, Math. Intell. 18, 57—65 (1996; MR 97j:16028; Zbl 881.15007); see §4.0.0, §4.2.0, §4.2.5

11. M. Babich, A. Bobenko: *Willmore tori with umbilic lines and minimal surfaces in hyperbolic space*, Duke Math. J. 72, 151—185 (1993; MR 94j:53070; Zbl 820.53005); see §3.7.1, §3.7.13, §3.7.14

12. M. Berger: *Geometry I*, Universitext, Springer, Berlin (1987; MR 88a:51001a; Zbl 606.51001); see §4.9.12, §8.4.10

13. H. Bernstein: *Non-special, non-canal, isothermic tori with spherical lines of curvature*, PhD thesis, Washington Univ., St. Louis (1999) and Trans. Amer. Math. Soc. 353, 2245—2274 (2001; MR 2001k:53005; Zbl 01579449); see §5.4.25

14. L. Berwald: *Differentialinvarianten in der Geometrie. Riemannsche Mannigfaltigkeiten und ihre Verallgemeinerungen*, Encyclopaedie der mathematischen Wissenschaften III.D 11, Teubner, Leipzig (1927; JFM 53.0680.12); see §P.6.3

15. L. Bianchi: *Sulle superficie d'area minima negli spazi a curvatura costante*, Rom. Acc. L. Mem. (4) IV, 503—519 (1887; JFM 19.0833.01)

16. L. Bianchi: *Sulla teoria delle trasformazione delle superficie a curvatura costante*, Annali di Mat. 3, 185—298 (1899; JFM 30.0552.07); see §5.4.15

17. L. Bianchi: *Vorlesungen über Differentialgeometrie*, Teubner, Leipzig (1899; JFM 30.0532.03); see §2.4.13, §8.0.0

18. L. Bianchi: *Sulle superficie a linee di curvatura isoterme*, Rend. Acc. Naz. Lincei 12, 511—520 (1903; JFM 34.0654.01); see §3.0.0

19. L. Bianchi: *Il teorema di permutabilità per le trasformazioni di Darboux delle superficie isoterme*, Rend. Acc. Naz. Lincei 13, 359—367 (1904; JFM 35.0623.01); see §5.6.5

20. L. Bianchi: *Complementi alle ricerche sulle superficie isoterme*, Ann. Mat. Pura Appl. 12, 19—54 (1905; JFM 36.0675.01); see §I.1.6, §3.0.0, §5.0.0, §5.5.0, §5.6.0, §5.6.9

21. L. Bianchi: *Ricerce sulle superficie isoterme e sulla deformazione delle quadriche*, Ann. Mat. III 11, 93—157 (1905; JFM 36.0674.01); see §3.0.0, §5.0.0, §5.5.0, §5.6.0, §5.6.2, §5.6.5, §5.6.8

22. L. Bianchi: *Alcune ricerche sul rotolamento di superficie applicabili*, Palermo Rend. 38, 1—41 (1914; JFM 45.0852.02); see §5.6.13

23. L. Bianchi: *Sulle superficie a rappresentazione isoterma delle linee di curvatura come inviluppi di rotolamento*, Rend. Acc. Naz. Lincei 24, 367—377 (1915; JFM 45.0855.03)

24. L. Bianchi: *Sulla generazione, per rotolamento, delle superficie isoterme e delle superficie a rappresentazione isoterma delle linee di curvatura*, Rend. Acc. Naz. Lincei 24, 377—387 (1915; JFM 45.1367.03)

25. L. Bianchi: *Sulle superficie isoterme come superficie di rotolamento*, Rend. Acc. Naz. Lincei 24, 303—312 (1915; JFM 45.0855.02)

26. L. Bianchi: *Sulle trasformazioni di Ribaucour dei sistemi tripli ortogonali*, Rend. Acc. Naz. Lincei 24, 161—173 (1915; JFM 45.1367.02); see §8.2.3

27. L. Bianchi: *Le trasformazioni di Ribaucour dei sistemi n-pli ortogonali e il teorema generale di permutabilità*, Annali di Mat. (3) 27, 183—257 (1918; JFM 46.1123.01); see §8.0.0, §8.5.12

28. L. Bianchi: *Lezioni di geometria differenziale*, Enrico Spoerri, Pisa (1923; JFM 49.0498.06); see §5.6.2, §8.0.0, §8.5.0, §8.5.12

29. W. Blaschke: *Vorlesungen über Differentialgeometrie III: Differentialgeometrie der Kreise und Kugeln*, Grundlehren XXIX, Springer, Berlin (1929; JFM 55.0422.01); see §I.0.0, §I.1.2, §I.2.4, §1.0.0, §1.1.0, §1.2.6, §1.3.11, §1.3.22, §1.4.0, §1.4.15, §3.0.0, §3.1.0, §3.6.6, §F.0.0

30. A. Bobenko, U. Pinkall: *Discrete isothermic surfaces*, J. Reine Angew. Math. 475, 187—208 (1996; MR 97f:53004; Zbl 845.53005); see §3.0.0, §4.9.1, §5.0.0, §5.7.1, §5.7.37

31. A. Bobenko: *Discrete conformal maps and surfaces*, London Math. Soc. Lect. Note Ser. 225, 97—108 (1999; MR 2000m:53007; Zbl 1001.53001); see §8.0.0

32. A. Bobenko, U. Pinkall: *Discretization of surfaces and integrable systems*, Oxf. Lect. Ser. Math. Appl. 16, 3—58 (1999; MR 2001j:37128; Zbl 932.53004); see §5.7.1, §5.7.37, §8.3.15, §F.0.0

33. A. Bobenko, T. Hoffmann: *Conformally symmetric circle packings. A generalization of Doyle spirals*, Experimental Math. 10, 141—150 (2000; MR 2002a:52023; Zbl 987.52008); see §F.0.0

34. A. Bobenko, T. Hoffmann: *Hexagonal circle patterns and integrable systems. Patterns with constant angles*, Duke Math. J. 116, 525—566 (2003); see §F.0.0

35. A. Bobenko, U. Hertrich-Jeromin: *Orthogonal nets and Clifford algebras*, Tôhoku Math. Publ. 20, 7—22 (2001; MR 2002m:53015); see §5.7.1, §8.0.0, §8.4.10, §8.5.0

36. A. Bobenko, D. Matthes, Yu. Suris: *Nonlinear hyperbolic equations in surface theory: Integrable discretizations and approximation results*, Sfb 288 preprint 562, TU Berlin (2002) or preprint math.NA/0208042 (2002); see §8.5.12, §F.0.0

37. A. Bobenko, T. Hoffmann, Yu. Suris: *Hexagonal circle patterns and integrable systems. Patterns with the multi-ratio property and Lax equations on the regular triangular lattice*, Int. Math. Res. Notices 3 (2002; MR 2003a:52027; Zbl 993.52004); see §F.0.0

38. A. Bobenko, B. Springborn: *Variational principles for circle patterns and Koebe's theorem*, Sfb 288 preprint 545, TU Berlin (2002) or preprint math.GT/0203250 (2002); see §F.0.0

39. M. Brück, X. Du, J. Park, C.-L. Terng: *The submanifold geometries associated to Grassmannian systems*, Mem. Amer. Math. Soc. 735 (2002; MR 2002k:53103; Zbl 998.53037); see §2.2.12, §5.4.25

40. R. Bryant: *A duality theorem for Willmore surfaces*, J. Differ. Geom. 20, 23—53 (1984; MR 86j:58029; Zbl 555.53002); see §I.0.0, §3.0.0, §3.4.14, §3.5.0

41. R. Bryant, P. Griffiths: *Reduction for constrained variational problems and $\int \frac{k^2}{2} ds$*, Amer. J. Math. 108, 525—570 (1986; MR 88a:58044; Zbl 604.58022); see §3.5.1, §3.7.11, §3.7.13, §3.7.14, §3.7.18

42. R. Bryant: *Surfaces of mean curvature one in hyperbolic space*, Astérisque 154—155, 321—347 (1987; Zbl 635.53047); see §3.6.4, §4.3.4, §5.6.13, §5.6.21, §5.7.37

43. R. Bryant: *Surfaces in conformal geometry*, Proc. Symp. Pure Math. 48, 227—240 (1988; MR 89m:53102; Zbl 654.53010); see §3.5.0

44. F. E. Burstall, J. H. Rawnsley: *Twistor theory for Riemannian symmetric spaces. With applications to harmonic maps of Riemann surfaces*, Lect. Notes Math. 1424 (1990; MR 91m:58039; Zbl 699.53059); see §2.2.5

45. F. E. Burstall, E. Musso, L. Nicolodi: *Bäcklund's theorem for the Calapso equation*, Research Report CNR (1997); see §5.4.25

46. F. E. Burstall, U. Hertrich-Jeromin, F. Pedit, U. Pinkall: *Isothermic surfaces and curved flats*, Math. Z. 225, 199—209 (1997; MR 98j:53004; Zbl 902.53038); see §3.0.0, §3.3.0, §3.3.1, §3.3.4, §5.0.0, §5.1.12, §5.5.0, §8.7.0

47. F. E. Burstall: *Isothermic surfaces: Conformal geometry, Clifford algebras, and Integrable systems*, to appear, preprint math.DG/0003096 (2000); see §I.1.4, §I.1.6, §3.3.11, §5.0.0, §5.2.2, §5.5.19, §6.0.0, §7.0.0, §7.2.11, §7.5.9, §7.5.11, §8.0.0, §8.6.0, §8.6.7, §8.6.8, §8.7.0, §8.7.1, §8.7.5, §8.7.8, §8.7.11, §8.7.12

48. F. E. Burstall, F. Pedit, U. Pinkall: *Schwarzian derivatives and flows of surfaces*, Contemp. Math. 308, 39—61 (2002); see §P.6.3, §5.1.0, §5.1.9, §5.5.12, §5.5.27

49. F. E. Burstall, D. Ferus, K. Leschke, F. Pedit, U. Pinkall: *Conformal geometry of surfaces is S^4 and quaternions*, Lect. Notes Math. 1772 (2002); see §3.5.6, §4.0.0, §4.8.0, §5.4.0, §F.0.0

50. F. E. Burstall, D. M. J. Calderbank: *Conformal submanifold geometry*, manuscript (2002); see §I.0.0, §I.2.3, §P.0.0, §F.0.0

51. F. E. Burstall, U. Hertrich-Jeromin: *On the Ribaucour transformation*, manuscript (2002); see §8.2.2, §8.5.12

52. F. E. Burstall, F. Pedit, U. Pinkall: *Isothermic submanifolds of symmetric R-spaces*, in preparation; see §4.6.3, §5.4.0, §7.4.7

53. P. Calapso: *Sulle superficie a linee di curvatura isoterme*, Rend. Circ. Mat. Palermo 17, 275—286 (1903; JFM 34.0653.04); see §I.1.6, §P.6.3, §3.0.0, §3.3.19, §5.0.0, §5.5.0

54. P. Calapso: *Sugl' invarianti del gruppo delle trasformazioni conformi dello spazio*, Palermo Rend. 22, 197—213 (1906; JFM 37.0178.03); see §P.6.3

55. P. Calapso: *Sulle trasformazioni delle superficie isoterme*, Annali Mat. 24, 11—48 (1915; JFM 45.1369.01); see §I.1.6, §P.6.3, §3.0.0, §5.0.0, §5.5.0

56. R. Calapso: *Riduzione della deformazione proiettiva di una superficie R alla trasformazione C_m delle superficie isoterme*, Rend. Acc. Naz. Lincei 7, 617—626 (1928; JFM 54.0783.01); see §5.5.12

57. R. Calapso: *Una nuova trasformazione delle superficie isoterme*, Rend. Acc. Naz. Lincei 8, 287—291, 366—371 (1928; JFM 54.0790.01)

58. R. Calapso: *Sulla trasformazione delle superficie isoterme e delle superficie R*, Atti Soc. Ital. Progr. Sc. 18, 34—36 (1930; JFM 56.0602.03); see §5.5.12

59. R. Calapso: *Questioni di geometria conforme*, Atti IV. Congr. Un. Mat. Ital. 1, 287—318 (1953; MR 14:1121d; Zbl 051.38801); see §P.6.3, §5.5.0

60. D. Carfi, E. Musso: *T-transformations of Willmore isothermic surfaces*, Rend. Sem. Mat. Messina, Atti Congr. Int. P. Calapso 1998, 69—86 (2001); see §5.5.29

61. É. Cartan: *La déformation des hypersurfaces dans l'éspace conforme réell à n ≥ 5 dimensions*, Bull. Soc. Math. France 45, 57—121 (1917; JFM 46.1129.02); see §P.5.7, §P.6.3, §1.8.0, §2.0.0, §2.1.0, §2.1.6, §2.1.8, §2.3.13, §2.3.14

62. É. Cartan: *Les espaces à connexion conforme*, Ann. Soc. Pol. Math. 2, 171—221 (1923; JFM 50.0493.01); see §I.1.6, §P.6.3, §3.0.0, §5.0.0, §5.5.0, §5.5.12, §8.6.18

63. É. Cartan: *Leçons sur la géométrie des espaces de Riemann*, Gauthier-Villars, Paris (1946; MR 8:602g; Zbl 060.38101); see §6.0.0, §6.1.2, §6.1.5

64. A. Cayley: *On the surfaces divisible into squares by their curves of curvature*, Proc. London Math. Soc. 4, 8—9, 120—121 (1872; JFM 04.0364.02; JFM 04.0364.03); see §3.0.0, §5.7.1

65. A. Cayley: *Sur les surfaces divisibles en carrés par leurs courbes de courbure et sur la théorie de Dupin*, Comptes Rendus 74, 1445—1449 (1872; JFM 04.0364.01); see §P.6.5, §3.3.19

66. T. Cecil: *Lie sphere geometry. With applications to submanifolds*, Universitext, Springer, New York (1992; MR 94m:53076; Zbl 752.53003); see §1.0.0, §1.1.0, §1.2.3, §1.3.16, §1.7.11, §1.8.0, §1.8.7, §1.8.16, §6.3.5, §6.6.10, §F.0.0

67. B.-Y. Chen: *An invariant of conformal mappings*, Proc. Amer. Math. Soc. 40, 563—564 (1973; MR 47:9489; Zbl 266.53020); see §1.0.0, §3.5.0, §3.5.2

68. S.-S. Chern, J. Simons: *Characteristic forms and geometric invariants*, Ann. Math. 99, 48—69 (1974; MR 50:5811; Zbl 283.53036); see §2.0.0

69. S.-S. Chern: *On a conformal invariant of three-dimensional manifolds*, North-Holland Math. Library 34, 245—252 (1986; MR 87h:53051; Zbl 589.53011); see §2.0.0

70. E. Christoffel: *Ueber einige allgemeine Eigenschaften der Minimumsflächen*, Crelle's J. 67, 218—228 (1867); see §I.1.7, §3.0.0, §3.3.0, §3.3.10, §3.3.30, §5.2.2, §8.6.4

71. J. Cieśliński, P. Goldstein, A. Sym: *Isothermic surfaces in E^3 as soliton surfaces*, Phys. Lett., A 205, 37—43 (1995; MR 96g:53005); see §I.0.0, §3.0.0, §3.3.0, §8.7.0

72. J. Cieslinski: *The cross ratio and Clifford algebras*, Adv. Appl. Clifford Alg. 7, 133—139 (1997; MR 99b:52036; Zbl 907.15018); see §7.5.13

73. J. Cieśliński: *The Darboux-Bianchi transformation for isothermic surfaces. Classical results versus the soliton approach*, Differ. Geom. Appl. 7, 1—28 (1997; MR 98c:53003; Zbl 982.53003); see §8.7.1

74. J. Cieśliński, A. Doliwa, P. Santini: *The integrable discrete analogues of orthogonal coordinate systems are multidimensional circular lattices*, Phys. Lett., A 235, 480—488 (1997; MR 99d:58149; Zbl 969.37528); see §8.0.0

75. J. Cieśliński: *The Bäcklund transformation for discrete isothermic surfaces*, London Math. Soc. Lect. Note Ser. 255, 109—121 (1999; MR 2000f:53004; Zbl 927.37055); see §5.7.11

76. W. K. Clifford: *Applications of Grassmann's extensive algebra*, Amer. J. Math. 1, 350—358 (1878; JFM 10.0297.02); see §4.1.4, §6.0.0, §6.2.4

77. W. K. Clifford: *On the classification of geometric algebras*, unfinished paper in R. Tucker, Mathematical Papers by William Kingdon Clifford, Chelsea, New York (1968); see §I.1.4, §6.0.0

78. J. Coolidge: *Congruences and complexes of circles*, Trans. Amer. Math. Soc. 15, 107—134 (1914; JFM 45.0918.03); see §2.5.2

79. J. Coolidge: *A treatise on the circle and the sphere*, Oxford University Press, Oxford (1916; JFM 46.0921.02); see §2.2.10, §2.5.2

80. E. Cosserat: *Sur le problème de Ribaucour et sur les surfaces isothermiques*, Toulouse Bull. Mém. 3, 267—273 (1899—1900; JFM 31.0608.02); see §3.0.0, §3.1.0

81. E. Cotton: *Sur une généralisation du problème de la représentation conforme aux variétés à trois dimensions*, Comptes Rendus 125, 225—228 (1896; JFM 28.0601.03); see §P.5.0

82. E. Cotton: *Sur la représentation conforme des variétés à trois dimensions*, Comptes Rendus 127, 349—351 (1898; JFM 29.0573.03); see §P.5.0

83. A. Dahan-Dalmédico: *Mécanique et théorie des surfaces: les travaux de Sophie Germain*, Hist. Math. 14, 347—365 (1987; MR 89a:01030; Zbl 631.01013); see §3.0.0, §3.5.1

84. M. Dajczer, E. Vergasta: *Conformal hypersurfaces with the same Gauss map*, Trans. Amer. Math. Soc. 347, 2437—2456 (1995; MR 95i:53067; Zbl 832.53004); see §3.3.11, §8.0.0

85. M. Dajczer, R. Tojeiro: *An extension of the classical Ribaucour transformation*, Proc. London Math. Soc. 85, 211—232 (2002); see §8.2.1

86. G. Darboux: *Sur une classe remarquable de courbes et de surfaces algébriques et sur la théorie des imaginaires*, Mém. de Bordeaux VIII, 292—350 et IX, 1—276 (1873; JFM 05.0399.01); see §1.0.0

87. G. Darboux: *Détermination d'une classe particulière de surfaces à lignes de courbure planes dans un système et isothermes*, Comptes Rendus 96, 1202—1205, 1294—1297 (1883; JFM 15.0726.02)

88. G. Darboux: *Leçons sur la théorie générale des surfaces*, Gauthiers-Villars, Paris (1887; JFM 19.0746.02; JFM 25.1159.02); see §P.6.5, §1.0.0

89. G. Darboux: *Sur les surfaces isothermiques*, Ann. Ec. Norm. Sup. 16, 491—508 (1899; JFM 30.0555.04); see §5.0.0, §8.7.1

90. G. Darboux: *Sur une classe de surfaces isothermiques liées à la déformations des surfaces du second degré*, Comptes Rendus 128, 1483—1487 (1899; JFM 30.0555.03)

91. G. Darboux: *Sur la déformation des surfaces générales du second degré*, Comptes Rendus 128, 1264—1270 (1899; JFM 30.0555.01)

92. G. Darboux: *Sur la déformation des surfaces du second degré,* Comptes Rendus 128, 760—766, 854—859 (1899; JFM 30.0554.01; JFM 30.0554.02)

93. G. Darboux: *Sur les surfaces isothermiques,* Comptes Rendus 128, 1299—1305, 1538 (1899; JFM 30.0555.02); see §3.0.0, §3.1.0, §8.7.1

94. G. Darboux: *Leçons sur les systèmes orthogonaux et les coordonnées curvilignes,* Gauthier-Villars, Paris (1910; JFM 29.0515.03); see §8.0.0

95. P. C. Delens: *Methodes et problemes de geometries differentielles eucli- dienne et conforme,* Gauthier-Villars, Paris (1927; JFM 53.0659.01); see §I.0.0

96. A. Delgleize: *Sur les surfaces isothermiques et les surfaces de Guichard,* Bull. Acad. Roy. Belg., V. Sér. 20, 707—722 (1934; Zbl 009.37502)

97. A. Delgleize: *Sur les transformations de Ribaucour et les surfaces isothermiques,* Bull. Soc. Roy. Sci. Liège 13, 233—242 (1944; MR 7:77f; Zbl 063.01078)

98. A. Demoulin: *Sur les enveloppes de sphères dont les deux nappes se correspondent avec conservation des angles,* Comptes Rendus 141, 459—462 (1905; JFM 36.0683.01); see §3.1.0, §8.7.1

99. A. Demoulin: *Sur les surfaces isothermiques et sur une classe d'enveloppes de sphères,* Comptes Rendus 141, 1210—1212 (1905; JFM 36.0683.03); see §3.1.0, §8.7.1

100. A. Demoulin: *Sur certains couples de systèmes triple-orthogonaux,* Comptes Rendus 151, 796—800 (1910; JFM 41.0694.04); see §8.5.0

101. U. Dierkes, S. Hildebrandt, A. Küster, O. Wohlrab: *Minimal Sur- faces I,* Grundlehren 295, Springer, Berlin (1991; MR 94c:49001; Zbl 777.53012); see §5.3.0

102. M. do Carmo, M. Dajczer, F. Mercuri: *Compact conformally flat hyper- surfaces,* Trans. Amer. Math. Soc. 288, 189—203 (1985; MR 86b:53052; Zbl 554.53040); see §P.5.7, §1.8.14, §2.0.0

103. A. Doliwa, M. Manas, P. Santini: *Darboux transformation for multi- dimensional quadrilateral lattics I,* Phys. Lett., A 232, 99—105 (1997; MR 98h:58178); see §8.0.0

104. A. Doliwa, S. Manakov, P. Santini: *∂̄-reductions of the multi- dimensional quadrilateral lattice I: The multidimensional circular lattice,* Commun. Math. Phys. 196, 1—18 (1998; MR 2000a:37072; Zbl 908.35125); see §8.0.0, §8.4.10

105. A. Doliwa, P. Santini: *Geometry of discrete curves and lattices and integrable difference equations*, Oxf. Lect. Ser. Math. Appl. 16, 139—154 (1999; MR 2000k:53002; Zbl 935.53002); see §8.0.0

106. L. Eisenhart: *Transformations of surfaces*, Chelsea, New York (1962; MR 25:5455); see §3.2.6, §5.0.0, §5.6.5, §8.0.0, §8.5.12, §8.7.1

107. N. Ejiri: *Willmore surfaces with a duality in S^n*, Proc. London Math. Soc., III. Ser. 57, 383—416 (1988; MR 89h:53117; Zbl 671.53043); see §3.5.0, §7.5.6

108. J. Elstrodt, F. Grunewald, J. Mennicke: *Vahlen's groups of Clifford matrices and spin-groups*, Math. Z. 196, 369—390 (1987; MR 89b:11031; Zbl 611.20027); see §7.0.0

109. A. Enneper: *Bemerkungen über die Enveloppe einer Kugelfläche*, Gött. Nachr. 1873, 217—248 (1873; JFM 05.0409.01)

110. J.-H. Eschenburg: *Willmore surfaces and Möbius geometry*, informal notes on discussions between J. Eschenburg, U. Pinkall, and K. Voss (1988); see §3.0.0, §3.5.6

111. J. Ferrand: *The action of conformal transformations on a Riemannian manifold*, Math. Ann. 304, 277—291 (1996; MR 97c:53044; Zbl 866.53027); see §P.8.0

112. D. Ferus, F. Pedit: *Curved flats in symmetric spaces*, Manuscr. Math. 91, 445—454 (1996; MR 97k:53074; Zbl 870.53043); see §I.1.6, §2.0.0, §2.2.5, §2.2.13, §3.3.3, §5.5.0, §8.7.7, §8.7.8, §8.7.11

113. D. Ferus, K. Leschke, F. Pedit, U. Pinkall: *Quaternionic holomorphic geometry: Plücker formula, Dirac eigenvalue estimates, and energy estimates of harmonic 2-tori*, Invent. Math. 146, 507—593 (2001; MR 2003a:53057); see §I.1.3, §F.0.0

114. A. Fialkov: *The conformal theory of curves*, Trans. Amer. Math. Soc. 51, 435—501 (1942; MR 3:307e; Zbl 063.01358); see §P.6.3

115. A. Fialkov: *Conformal differential geometry of a subspace*, Trans. Amer. Math. Soc. 56, 309—433 (1944; MR 6:105c; Zbl 063.01359); see §P.6.3, §P.6.4

116. A. Fialkov: *Conformal classes of surfaces*, Amer. J. Math. 67, 583—616 (1945; MR 7:175c; Zbl 063.01360); see §P.6.3

117. W. Fiechte: *Ebene Möbiusgeometrie*, manuscript (1979); see §I.1.4, §6.0.0

118. A. Finzi: *Le ipersuperfizie a tre dimensioni che si possono rappresentare conformemente sullo spazio euclideo*, Atti. d. Veneto 62, 1049—1062 (1903; JFM 34.0668.03); see §2.0.0

119. G. Fubini: *Sulla teoria degli spazii che ammettono un gruppo conforme*, Torino Atti 38, 404—418 (1903; JFM 34.0722.03); see §P.8.0

120. G. Fubini: *Applicabilità projettiva di due superficie*, Palermo Rend. 41, 135—162 (1916; JFM 46.1098.01); see §P.6.3, §P.6.4

121. G. Fubini: *Sulle geometria di una superficie nel gruppo proiettivo e nel gruppo conforme*, Rend. Acc. L. Roma (6) 5, 373—377 (1927; JFM 53.0701.03); see §P.6.3

122. S. Gallot, D. Hulin, J. Lafontaine: *Riemannian geometry*, Universitext, Springer, Berlin (1987; MR 88k:53001; Zbl 636.53001); see §P.0.0, §P.1.3, §P.3.0, §P.3.2, §P.4.4, §1.4.10

123. E. I. Ganzha: *On approximation of solutions to some (2 + 1)-dimensional integrable systems by Bäcklund transformations*, Sib. Math. J. 41, 442—452 (2000; MR 2001j:37130; Zbl 952.37050); see §8.0.0

124. A. Garsia: *Imbeddings of some closed Riemann surfaces by canal surfaces*, Rend. Circ. Mat. Palermo 9, 313—333 (1960; MR 24:A2021; Zbl 100.17801); see §P.4.6, §1.8.0, §2.0.0

125. A. Garsia: *An imbedding of closed Riemann surfaces in Euclidean space*, Comment. Math. Helv. 25, 93—110 (1961; MR 23:A2890; Zbl 173.24001); see §P.4.6, §2.0.0

126. S. Germain: *Recherches sur la théorie des surfaces élastiques*, Courcier, Paris (1821); see §3.0.0, §3.5.1

127. S. Germain: *Remarques sur la nature, les bornes et l'etendue de la question des surfaces elastiques, et equation generale de ces surfaces*, Paris (1826); see §3.0.0

128. S. Germain: *Mémoire sur la courbure des surfaces*, Crelle's J. 7, 1—29 (1831); see §3.0.0

129. E. Goursat: *Sur un mode de transformation des surfaces minima*, Acta Math. 11, 135—186, 257—264 (1888; JFM 20.0833.02); see §5.0.0, §5.3.0, §8.6.9

130. H. Grassmann: *Die lineale Ausdehnungslehre, ein neuer Zweig der Mathematik*, Verlag Otto Wigand, Leipzig (1844); see §I.1.4, §6.0.0, §6.1.2

131. H. Grassmann: *Sur les différents genres de multiplication*, Crelle's J. 49, 123—141 (1855); see §6.0.0, §6.1.4

132. H. Grassmann: *Grundsätze der stereometrischen Multiplikation*, Crelle's J. 49, 10—20 (1855); see §6.0.0

133. H. Grassmann: *Die Ausdehnungslehre*, Verlag Th. Chr. Fr. Enslin, Berlin (1862); translated by L. Kannenberg, Extension theory, Hist. Math. 19, Amer. Math. Soc. and London Math. Soc., Rhode Island (2000; MR 2001d:01048; Zbl 953.01025); see §I.1.4, §6.0.0

134. H. Grassmann: *Der Ort der Hamilton'schen Quaternionen in der Ausdehnungslehre*, Math. Ann. 12, 375—386 (1877; JFM 09.0512.01); see §4.1.4, §6.0.0, §6.2.4

135. C. Guichard: *Sur la déformation des quadriques de révolution*, Comptes Rendus 128, 232—233 (1899; JFM 30.0552.01); see §5.0.0

136. C. Guichard: *Sur les systèmes triplement indéterminés et sur les systèmes triple orthogonaux*, Scientia 25, Gauthier-Villars, Paris (1905; JFM 36.0668.03); see §I.1.6, §2.4.3, §8.0.0

137. T. F. Havel: *Geometric algebra and Möbius sphere geometry as a basis for Euclidian invariant theory*, Invariant methods in discrete and computational geometry, Proc. Conf. Curaçao 1994, 245—256 (1995; MR 97a:15043); see §7.0.0

138. Z.-X. He, O. Schramm: *On the convergence of circle packings to the Riemann map*, Invent. Math. 125, 285—305 (1996; MR 97i:30009; Zbl 868.30010); see §F.0.0

139. Z.-X. He, O. Schramm: *The C^∞-convergence of hexagonal disk packings to the Riemann map*, Acta Math. 180, 219—245 (1998; MR 99j:52021; Zbl 913.30004); see §F.0.0

140. S. Helgason: *Differential geometry, Lie groups, and symmetric spaces*, Academic Press, New York (1978; MR 80k:53081; Zbl 451.53038); see §2.2.5, §2.2.17, §5.5.19, §6.3.9, §6.7.7

141. U. Hertrich-Jeromin, U. Pinkall: *Ein Beweis der Willmoreschen Vermutung für Kanaltori*, J. Reine Angew. Math. 430, 21—34 (1992; MR 95g:53067; Zbl 749.53040); see §3.7.0, §3.7.16, §3.7.18, §3.7.20, §3.7.22

142. U. Hertrich-Jeromin: *Über konform flache Hyperflächen in vierdimensionalen Raumformen*, PhD thesis, TU Berlin (1994; Zbl 861.53059); see §I.1.6, §2.2.13, §2.4.12

143. U. Hertrich-Jeromin: *On conformally flat hypersurfaces and Guichard's nets*, Beitr. Alg. Geom. 35, 315—331 (1994; MR 95j:53022; Zbl 820.53003); see §2.0.0, §2.3.6, §2.4.13

144. U. Hertrich-Jeromin, K. Voss: *Remarks on channel surfaces*, unpublished (1995); see §3.0.0, §3.7.0

145. U. Hertrich-Jeromin: *On conformally flat hypersurfaces, curved flats and cyclic systems*, Manuscr. Math. 91, 455—466 (1996; MR 97m:53099; Zbl 870.53007); see §2.0.0, §2.2.6, §2.2.13, §2.2.14

146. U. Hertrich-Jeromin, E. Tjaden, M. Zürcher: *On Guichard's nets and cyclic systems*, preprint FIM ETH Zürich (1997) and preprint dg-ga/9704003 (1997); see §2.0.0, §2.4.13, §2.6.1, §2.7.4

147. U. Hertrich-Jeromin, F. Pedit: *Remarks on the Darboux transform of isothermic surfaces*, Doc. Math. J. DMV 2, 313—333 (1997; MR 99k:53006; Zbl 892.53003); see §3.3.25, §5.0.0, §5.2.10, §5.4.0, §5.4.15

148. U. Hertrich-Jeromin: *Supplement on curved flats in the space of point pairs and isothermic surfaces: A quaternionic calculus*, Doc. Math. J. DMV 2, 335—350 (1997; MR 99b:53017; Zbl 892.53004); see §5.3.2, §5.5.23

149. U. Hertrich-Jeromin, T. Hoffmann, U. Pinkall: *A discrete version of the Darboux transform for isothermic surfaces*, Oxf. Lect. Ser. Math. Appl. 16, 59—81 (1999; MR 2000j:53010; Zbl 941.53010); see §4.9.14, §5.0.0, §5.4.15, §5.7.0, §5.7.11, §5.7.15

150. U. Hertrich-Jeromin: *Transformations of discrete isothermic nets and discrete cmc-1 surfaces in hyperbolic space*, Manuscr. Math. 102, 465—486 (2000; MR 2001g:53013; Zbl 979.53008); see §5.6.0, §5.7.0, §5.7.11, §5.7.19, §5.7.33, §5.7.35

151. U. Hertrich-Jeromin: *Isothermic cmc-1 cylinder*, Electr. Geom. Model 2000.09.038 (2000); see §5.6.13

152. U. Hertrich-Jeromin: *The surfaces capable of division into infinitesimal squares by their curves of curvature: A nonstandard analyis approach to classical differential geometry*, Math. Intell. 22, 54—61 (2000; MR 2001b:53003), Errata: Math. Intell. 24, 4 (2002); see §3.3.10, §5.7.1, §8.3.15, §F.0.0

153. U. Hertrich-Jeromin, E. Musso, L. Nicolodi: *Möbius geometry of surfaces of constant mean curvature 1 in hyperbolic space*, Ann. Global Anal. Geom. 19, 185—205 (2001; MR 2002a:53079); see §3.0.0, §3.3.29, §3.6.4, §5.0.0, §5.4.24, §5.6.0, §5.6.13, §5.6.16, §5.7.35

154. D. Hestenes: *The design of linear algebra and geometry*, Acta Appl. Math. 23, 65—93 (1991; MR 92i:15021; Zbl 742.51001); see §I.1.4, §7.0.0, §7.2.0

155. D. Hestenes, R. Ziegler: *Projective geometry with Clifford algebra*, Acta Appl. Math. 23, 25—63 (1991; MR 92i:15020; Zbl 735.51001); see §I.1.4, §6.0.0, §6.1.2, §7.2.0

156. G. Kamberov: *Quadratic differentials, quaternionic forms, and surfaces*, preprint dg-ga/9712011 (1997); see §5.2.11

157. G. Kamberov, F. Pedit, U. Pinkall: *Bonnet pairs and isothermic surfaces*, Duke Math. J. 92, 637—644 (1998; MR 99h:53009); see §3.0.0

158. G. Kamberov: *Quadratic differentials and surfaces shape*, Rend. Sem. Mat. Messina, Atti Congr. Int. P. Calapso 1998, 199—210 (2001); see §5.2.21

159. H. Karcher: *Embedded minimal surfaces derived from Scherk's examples*, Manuscr. Math. 62, 83—114 (1988; MR 89i:53009; Zbl 658.53006); see §5.2.21

160. F. Klein: *Vergleichende Betrachtungen über neuere geometrische Forschungen (Erlanger Programm)*, Math. Ann. 43, 63—100 (1893; JFM 25.0871.01); see §I.1.0, §I.1.2, §1.0.0, §1.4.12

161. F. Klein: *Vorlesungen über höhere Geometrie*, Grundlehren XXII, Springer, Berlin (1926; JFM 52.0624.09); see §1.0.0, §1.1.1, §1.3.13, §6.1.3

162. F. Klein: *Vorlesungen über Nicht-Euklidische Geometrie*, Grundlehren XXVI, Springer, Berlin (1928; JFM 54.0593.01); see §P.1.3, §1.0.0, §1.2.3

163. P. Klimczewski, M. Nieszporski, A. Sym: *Luigi Bianchi, Pasquale Calapso and solitons*, Rend. Sem. Mat. Messina, Atti Congr. Int. P. Calapso, Messina 1998 (2001); see §5.0.0

164. J. Knoblauch: *Ueber die Bedingung für die Isometrie der Krümmungscurven*, Crelle's J. 103, 40—43 (1888; JFM 20.0754.01); see §P.6.5, §3.3.19

165. P. Koebe: *Kontaktprobleme der konformen Abbildung*, Ber. Verh. Sächs. Akad. Leipzig 88, 141—164 (1936; Zbl 17.21701); see §F.0.0

166. B. Konopelchenko, W. Schief: *Three-dimensional integrable lattices in Euclidean spaces: Conjugacy and orthogonality*, Proc. Royal Soc. London A 454, 3075—3104 (1998; MR 2000f:53005); see §8.0.0, §8.4.10

167. W. Kreft: *Beiträge zur Goursat'schen Transformation der Minimalflächen*, PhD thesis, Westf. Wilhelms-Univ. (1908); see §5.3.0, §8.6.9

168. I. M. Krichever: *Algebraic-geometric n-orthogonal curvilinear coordinate systems and solutions of the associativity equations*, Funct. Anal. Appl. 31, 25—39 (1997; MR 98f:58049); see §8.0.0

169. W. Kühnel, U. Pinkall: *On total mean curvatures*, Q. J. Math., Oxf. II. Ser. 37, 437—447 (1986; MR 87m:53084; Zbl 627.53044); see §3.5.0

170. R. Kulkarni: *Conformal structures and Möbius structures*, Aspects Math. E 12, 1—39 (1988; MR 90f:53026; Zbl 659.53015); see §I.0.0, §P.0.0, §P.1.5, §P.8.4, §1.5.2

171. R. Kusner: *Conformal geometry and complete minimal surfaces*, Bull. Am. Math. Soc. 17, 291—295 (1987; MR 88j:53008; Zbl 634.53004); see §3.5.0

172. R. Kusner: *Comparison surfaces for the Willmore problem*, Pac. J. Math. 138, 317—345 (1989; MR 90e:53013; Zbl 643.53044); see §3.0.0, §3.5.0

173. J. Lafontaine: *Conformal geometry from the Riemannian viewpoint*, Aspects Math. E 12, 65—92 (1988; MR 90a:53022; Zbl 661.53008); see §I.0.0, §I.2.3, §P.0.0, §P.2.2, §P.3.0, §P.5.2, §2.0.0

174. V. Lalan: *Courbes isothermes sur une surface. Surfaces isothermiques*, Comptes Rendus 223, 707—709 (1946; MR 8:228d; Zbl 063.03425)

175. G. Lamé: *Sur les surfaces isothermes paraboloïdales*, Liouville J. 19, 307—318 (1874; JFM 06.0463.01); see §3.0.0

176. L. Landau, E. Lifschitz: *Lehrbuch der theoretischen Physik, Band VII. Elastizitätstheorie*, Akademie-Verlag, Berlin (1965; MR 36:2342; Zbl 166.20504); see §3.0.0, §3.5.1

177. J. Langer, D. Singer: *The total squared curvature of closed curves*, J. Differ. Geom. 20, 1—22 (1984; MR 86i:58030; Zbl 554.53013); see §3.5.1, §3.7.11, §3.7.13, §3.7.14, §3.7.16

178. J. Langer, D. Singer: *Curves in the hyperbolic plane and mean curvature of tori in 3-space*, Bull. London Math. Soc. 16, 531—534 (1984; MR 85k:53006; Zbl 554.53014); see §3.5.1, §3.7.11, §3.7.22

179. D. Laugwitz: *Differentialgeometrie*, Teubner, Stuttgart (1960; MR 22:7061; Zbl 094.16106); see §1.5.5

180. P. Li, S. T. Yau: *A new conformal invariant and its applications to the Willmore conjecture and the first eigenvalue of compact surfaces*, Invent. Math. 69, 269—291 (1982; MR 84f:53049; Zbl 503.53042); see §3.5.0, §3.7.22

181. H. Li, C.-P. Wang, F. Wu: *A Möbius characterization of Veronese surfaces in S^n*, Math. Ann. 319, 707—714 (2001; MR 2002b:53098); see §P.6.5

182. L. Lichtenstein: *Beweis des Satzes, dass jedes hinreichend kleine, im wesentlichen stetig gekrümmte, singularitätenfreie Flächenstück auf einen Teil einer Ebene zusammenhängend und in den kleinsten Teilen ähnlich abgebildet werden kann*, Abh. preuss. Akad. Wissensch., Phys.-Math. Cl., 1911, Anhang (1912; JFM 42.0710.01); see §P.4.6

183. L. Lichtenstein: *Zur Theorie der konformen Abbildung nichtanalytis-cher, singularitätenfreier Flächenstücke auf ebene Gebiete*, Krak. Anz. 1916, 192—217 (1916; JFM 46.0547.01); see §P.4.6

184. R. v. Lilienthal: *Die auf einer Fläche gezogenen Kurven*, Encyclopaedie der mathematischen Wissenschaften III.D 3, Teubner, Leipzig (1902; JFM 33.0633.04); see §3.0.0, §5.3.0

185. H. Linsenbarth: *Abhandlungen über das Gleichgewicht und die Schwingungen der ebenen elastischen Kurven von Jakob Bernoulli (1691, 1694, 1695) und Leonhard Euler (1744)*, Ostwald's Klassiker 175, Wilhelm Engelmann, Leipzig (1910); see §3.0.0

186. R. Lipowsky: *Kooperatives Verhalten von Membranen*, Phys. Bl. 52, 555—560 (1996); see §3.5.1

187. Q. P. Liu, M. Manas: *Vectorial Ribaucour transformations for the Lamé equations*, J. Phys. A, Math. Gen. 31, L193-L200 (1998; MR 99c:58178; Zbl 922.58088); see §8.0.0

188. L. Lopes de Lima, P. Roitman: *CMC-1 surfaces in hyperbolic 3-space using the Bianchi-Calò method*, An. Acad. Brasil. Ciênc. 74, 19—24 (2002; MR 2003a:53008); see §5.0.0, §5.6.13

189. P. Lounesto: *Clifford algebras and spinors*, London Math. Soc. Lect. Note Ser. 239, Cambridge Univ. Press, Cambridge (1997; MR 98k:15045; Zbl 887.15029); see §I.2.3, §6.0.0, §6.3.11, §6.4.4, §7.2.0, §7.2.2, §7.2.8

190. Y. Lumiste, E. Vyal'yas: *Isothermal hypersurfaces and three-dimensional Dupin-Mannheim cyclides*, Mat. Zametki 41, 731—740 (1987; MR 88j:53007; Zbl 628.53013); see §2.0.0, §8.6.7

191. H. v. Mangoldt: *Anwendung der Differential- und Integralrechnung auf Kurven und Flächen*, Encyclopaedie der mathematischen Wissenschaften III.D 1, 2, Teubner, Leipzig (1902; JFM 33.0633.03); see §3.0.0

192. C. McCune: *Rational minimal surfaces*, Q. J. Math. 52, 329—354 (2001; MR 2003a:53010); see §5.2.11

193. I. McIntosh, U. Hertrich-Jeromin, P. Norman, F. Pedit: *Periodic discrete conformal maps*, J. Reine Angew. Math. 534, 129—153 (2001; MR 2002f:52021; Zbl 986.37063); see §5.7.11, §5.7.37

194. F. Mercuri: *Conformally flat immersions*, Note Mat. 9, 85—99 (1989; MR 93d:53025; Zbl 795.53058); see §2.0.0

195. F. Mercuri, M. H. Noronha: *Conformal flatness, cohomogeneity one, and hypersurfaces of revolution*, Differ. Geom. Appl. 9, 243—249 (1998; MR 99j:53077; Zbl 943.53037); see §2.0.0

196. I. M. Mladenov: *Generalization of Goursat's transformation of the minimal surfaces*, C. R. Acad. Bulg. Sci. 52, 23—26 (1999; MR 2000k:53007; Zbl 947.53005); see §5.3.0

197. J. Moore: *Isometric immersions of space forms in space forms*, Pac. J. Math. 40, 157—166 (1972; MR 46:4442; Zbl 238.53033); see §2.2.13, §8.2.8

198. J. Moore: *Conformally flat submanifolds of Euclidean space*, Math. Ann. 225, 89—97 (1977; MR 55:4048; Zbl 322.53027); see §2.0.0

199. J. Moore: *On Conformal immersions of space forms*, Lect. Notes Math. 838, 203—210 (1980; MR 82k:53081; Zbl 437.53042); see §2.0.0, §2.2.0, §2.2.14

200. A. Moretti: *Konforme Abbildungen in Riemannschen Räumen*, Diplom thesis, ETH Zürich (1975); see §P.5.7

201. P. B. Mucha: *Isothermic surfaces of k-th order*, Proc. First Non-Orthodox School on Nonlinearity and Geometry, Luigi Bianchi Days, Sept. 21—28, 1995, Polish Scientific Publishers PWN, Warsaw (1998; MR 2001h:53006; Zbl 941.53008); see §3.0.0

202. E. Musso: *Willmore surfaces in the four-sphere*, Ann. Global Anal. Geom. 8, 21—41 (1990; MR 92g:53059; Zbl 705.53028); see §3.5.0

203. E. Musso: *Deformazione di superficie nello spazio di Möbius*, Rend. Ist. Mat. Univ. Trieste 27, 25—45 (1995; MR 97i:53013); see §3.0.0, §5.0.0, §5.5.0, §5.5.12, §8.6.18

204. E. Musso: *Isothermic surfaces in Euclidean space*, L. Cordero, E. García-Río, Proc. Workshop Recent Topics in Differential Geometry, July 1997, Publ. Dep. Geom. Topología, Univ. Santiago de Compostela 89, 219—235 (1998; Zbl 924.53007); see §5.1.0, §5.1.12, §5.5.12

205. E. Musso, L. Nicolodi: *Willmore canal surfaces in Euclidean space*, Rend. Ist. Mat. Univ. Trieste 31, 177—202 (1999; MR 2001f:53014); see §3.7.0, §3.7.13

206. E. Musso, L. Nicolodi: *Special isothermic surfaces and solitons*, Contemp. Math. 288, 129—148 (2001; MR 2003a:53017); see §5.0.0

207. E. Musso, L. Nicolodi: *Darboux transforms of Dupin surfaces*, Banach Center Publ. 57, 135—154 (2002); see §5.4.25, §5.5.28

208. S. Nishikawa, Y. Maeda: *Conformally flat hypersurfaces in a conformally flat Riemannian manifold*, Tôhoku Math. J. 26, 159—168 (1974; MR 49:3730; Zbl 278.53018); see §2.0.0

209. B. O'Neill: *Semi-Riemannian geometry. With applications to relativity*, Academic Press, New York (1983; MR 85f:53002; Zbl 531.53051); see §I.2.3, §2.2.1

210. K. Ogura: *On the differential geometry of inversion*, Tôhoku Math. J. 9, 216—223 (1916; JFM 46.1085.03); see §P.6.3

211. A. R. Özbek: *Über Weingarten'sche Isothermflächen auf denen die Kurven festen Krümmungsmasses eine zueinander geodätisch parallele Kurvenschar bilden*, Bull. Tech. Univ. Istanbul 14, 31—44 (1961; MR 28:5385; Zbl 101.14101); see §5.0.0

212. H. Pabel: *Geometrisches über Minimalflächen, insbesondere Scharen assoziierter und derivierter Flächen*, Fak. Math. Bayr. Julius-Maximilians Univ. (1986; Zbl 623.53001); see §5.3.0

213. H. Pabel: *Deformation von Minimalflächen*, Geometrie und ihre Anwendungen, 107—139 (1994; MR 95k:53014; Zbl 966.53008); see §5.3.0, §8.6.9

214. R. Palais, C.-L. Terng: *Critical point theory and submanifold geometry*, Lect. Notes Math. 1353, Springer, New York (1988; MR 90c:53143; Zbl 658.49001); see §2.0.0, §2.7.3

215. B. Palmer: *Isothermic surfaces and the Gauss map*, Proc. Amer. Math. Soc. 104, 876—884 (1988; MR 90a:53077; Zbl 692.53003); see §3.3.11, §5.2.2, §8.0.0, §8.6.5

216. F. Pedit, U. Pinkall: *Quaternionic analysis on Riemann surfaces and differential geometry*, Doc. Math., J. DMV, Extra Vol. ICM Berlin 1998, vol. II, 389—400 (1998; MR 2000c:53053; Zbl 910.53042); see §F.0.0

217. U. Pinkall: *Dupin'sche Hyperflächen*, PhD thesis, Freiburg i. Br. (1981); see §1.8.7

218. U. Pinkall: *Vortrag über Konformminimalflächen*, manuscript (1983); see §I.0.0, §3.5.1, §3.7.13, §3.7.14, §3.7.18

219. U. Pinkall: *Hopf tori in S^3*, Invent. Math. 81, 379—386 (1985; MR 86k:53075; Zbl 585.53051); see §3.5.0, §3.7.11

220. U. Pinkall: *Inequalities of Willmore type for submanifolds*, Math. Z. 193, 241—246 (1986; MR 88a:53053; Zbl 602.53039); see §3.5.0

221. U. Pinkall, I. Sterling: *Willmore surfaces*, Math. Intell. 9, 38—43 (1987; MR 88f:53009; Zbl 616.53049); see §3.5.0

222. U. Pinkall: *Compact conformally flat hypersurfaces*, Aspects Math. E 12, 217—236 (1988; MR 90a:53071; Zbl 657.53030); see §P.5.7, §2.0.0

223. K. Polthier, W. Rossman: *Discrete constant mean curvature surfaces and their index*, J. Reine Angew. Math. 549, 47—77 (2002); see §5.7.37

224. I. R. Porteous: *Clifford algebras and the classical groups*, Cambridge Stud. Adv. Math. 50, Cambridge Univ. Press, Cambridge (1995; MR 97c:15046; Zbl 855.15019); see §6.3.0, §7.2.0, §7.2.8, §7.2.11

225. G. Preissler: *Möbius-Differentialgeometrie von Hyperflächen*, PhD thesis, Stuttgart (1996; MR 99c:53067; Zbl 894.53024); see §P.6.3, §3.5.0

226. G. Preissler: *On a generalization of Willmore surfaces for hypersurfaces*, Result. Math. 35, 314—324 (1999; MR 2000f:53079; Zbl 959.53033); see §3.5.0

227. O. Pylarinos: *Sur les surfaces-W isothermiques*, Ann. Mat. Pura Appl., IV. Ser. 74, 37—59 (1966; MR 34:6657; Zbl 144.20502); see §5.0.0

228. O. Pylarinos: *Sur les surfaces-W isothermiques à lignes de courbure sphériques dans les deux systèmes*, Praktika Akad. Athen 42, 195—215 (1967; MR 38:616; Zbl 244.53002); see §5.0.0

229. O. Pylarinos: *Sur les surfaces isothermiques à lignes de courbure sphériques dans les deux systèmes*, Bull. Sci. Math. 92, 17—39 (1968; MR 38:617; Zbl 157.51402)

230. O. Pylarinos: *Sur la représentation des surfaces les unes sur les autres avec parallélisme des trièdres principaux*, Praktika Akad. Athen 44, 187—213 (1970; MR 49:1333; Zbl 242.53006); see §3.3.11

231. H.-B. Rademacher: *Conformal and isometric immersions of conformally flat Riemannian manifolds into spheres and Euclidean spaces*, Aspects Math. E 12, 192—216 (1988; MR 90c:53159; Zbl 665.53049); see §2.0.0

232. L. Raffy: *Surfaces doublement cylindrées et des surfaces isothermiques*, Comptes Rendus 128, 285—288 (1899; JFM 30.0548.01)

233. L. Raffy: *Sur la recherche des surfaces isothermiques*, Comptes Rendus 140, 1672—1674 (1905; JFM 36.0672.04); see §3.3.19

234. L. Raffy: *Recherches sur les surfaces isothermiques*, Ann. Ec. Norm. Sup. 22, 397—439 (1905; JFM 36.0673.01); see §3.6.4, §5.6.13

235. L. Raffy: *Remarques sur la recherche des surfaces isothermiques*, Comptes Rendus 143, 874—877 (1906; JFM 37.0630.01); see §3.3.19

236. L. Raffy: *Recherches sur les surfaces isothermiques*, Ann. Ec. Norm. Sup. 23, 387—428 (1906; JFM 37.0628.02); see §3.3.19

237. K. Reich: *Die Geschichte der Differentialgeometrie von Gauss bis Riemann (1828—1868)*, Arch. Hist. Exact Sci. 11, 273—382 (1973; MR 58:15971, Zbl 329.01008); see §3.0.0

238. K. Reinbeck: *Ueber diejenigen Flächen, auf welche die Flächen zweiten Grades durch parallele Normalen conform abgebildet werden*, PhD thesis, Göttingen (1886; JFM 18.0798.02); see §5.2.21

239. T. Reye: *Synthetische Geometrie der Kugeln und linearen Kugelsysteme. Mit einer Einleitung in die analytische Geometrie der Kugelsysteme*, Teubner, Leipzig (1879; JFM 11.0439.02); see §1.0.0

240. J. Richter: *Conformal maps of a Riemann surface into the space of quaternions*, PhD thesis, TU Berlin (1997; Zbl 896.53005); see §3.5.0

241. M. Rigoli: *The conformal Gauss map of submanifolds in the Möbius space*, Ann. Global Anal. Geom. 5, 97—116 (1987; MR 89e:53083; Zbl 642.53014); see §3.5.0, §7.5.6

242. B. Rodin, D. Sullivan: *The convergence of circle packings to the Riemann mapping*, J. Differ. Geom. 26, 349—360 (1987; MR 90c:30007; Zbl 694.30006); see §F.0.0

243. B. Rosenfeld: *Geometry of Lie groups*, Kluwer Acad. Publ., Dordrecht (1997; MR 98i:53002; Zbl 867.53002); see §6.2.0

244. R. Rothe: *Untersuchungen über die Theorie der isothermen Flächen*, PhD thesis, Berlin (1897; JFM 28.0545.01); see §P.6.5, §3.3.19

245. R. Rothe: *Sur la transformation de M. Darboux et l'équation fondamentale des surfaces isothermiques*, Comptes Rendus 143, 543—546 (1906; JFM 37.0629.02)

246. R. Rothe: *Sur les surfaces isothermiques*, Comptes Rendus 143, 578—581 (1906; JFM 37.0629.03); see §P.6.5, §3.3.19, §5.5.12

247. R. Rothe: *Über die Inversion einer Fläche und die konforme Abbildung zweier Flächen aufeinander mit Erhaltung der Krümmungslinien*, Math. Ann. 72, 57—77 (1912; JFM 43.0692.03), Erratum in Math. Ann. 73, 229 (1913); see §P.6.3, §P.6.5, §3.3.19

248. E. Salkowski: *Dreifach orthogonale Flächensysteme*, Encyclopaedie der mathematischen Wissenschaften III.D 9, Teubner, Leipzig (1902; JFM 48.0783.10); see §2.3.10, §2.4.7, §2.4.11, §8.0.0, §8.1.10

249. P. Samuel: *Correspondance conforme de deux surfaces à plans tangents parallèles*, Ann. Univ. Lyon, Sect. A. 5, 19—29 (1942; MR 8:486c; Zbl 063.06681); see §3.3.11

250. W. K. Schief: *Isothermic surfaces in spaces of arbitrary dimension: Integrability, discretization and Bäcklund transformations. A discrete Calapso equation*, Stud. Appl. Math. 106, 85—137 (2001; MR 2002k:37140); see §5.7.0, §8.0.0

251. M. Schmidt: *A proof of the Willmore conjecture*, preprint math.DG/0203224 (2002); see §3.5.0

252. J. Schouten: *Über die konforme Abbildung n-dimensionaler Mannigfaltigkeiten mit quadratischer Maßbestimmung auf eine Mannigfaltigkeit mit Euklidischer Maßbestimmung*, Math. Z. 11, 58—88 (1921; JFM 48.0857.02); see §P.5.0, §P.5.7, §P.7.4, §2.1.6

253. M. Servant: *Sur quelques applications de la géométrie non euclidienne*, Comptes Rendus 131, 827—830 (1900; JFM 31.0474.02); see §3.6.4, §5.5.29, §5.6.13

254. R. Sharpe: *Differential geometry: Cartan's generalization of Klein's Erlanger program*, Graduate Texts in Math. 166, Springer, New York (1997; Zbl 876.53001); see §I.0.0, §I.1.2, §I.2.3, §P.0.0, §1.0.0, §F.0.0

255. H. Sievert: *Ueber die Centralflächen der Enneper'schen Flächen constanten Krümmungsmasses*, Tübingen (1886; JFM 18.0739.02)

256. L. Simon: *Existence of Willmore surfaces*, Proc. Cent. Math. Anal. Aust. Natl. Univ. 10, 187—216 (1986; MR 87k:53006; Zbl 598.49029); see §3.5.0

257. G. Springer: *Introduction to Riemann surfaces*, Chelsea, New York (1981; MR 19:1169g; Zbl 501.30039); see §P.8.2

258. H. Steller: *Die Goursat-Transformation für Minimalflächen*, Wissensch. Hausarbeit 1. Staatsexamen, TU Berlin (2001); see §5.3.0

259. I. Sterling, H.C. Wente: *Existence and classification of constant mean curvature multibubbletons of finite and infinite type*, Indiana Univ. Math. J. 42, 1239—1266 (1993; MR 95a:53015; Zbl 803.53009); see §5.4.25

260. K. Strubecker: *Differentialgeometrie II—III*, Walter de Gruyter, Sammlg. Göschen, Berlin (1958—59; MR 20:4273 and MR 21:878; Zbl 082.36704); see §P.1.3, §P.4.10, §P.7.5

261. D. J. Struik: *Outline of a history of differential geometry II*, Isis 20, 161—191 (1933; Zbl 007.38806); see §3.0.0

262. E. Study: *Ein Seitenstück zur Theorie der linearen Transformationen einer komplexen Veränderlichen, Teile I—IV*, Math. Z. 18, 55—86, 201—229 (1923; JFM 49.0075.01) and 21, 45—71, 174—194 (1924; JFM 51.0588.03); see §I.1.3, §4.0.0, §4.2.0, §4.3.0, §4.8.0

263. C. Schiemangk, R. Sulanke: *Submanifolds of the Moebius space*, Math. Nachr. 96, 165—183 (1980; MR 82d:53017; Zbl 484.53008); see §F.0.0

264. R. Sulanke: *Submanifolds of the Möbius space II: Frenet formulas and curves of constant curvature*, Math. Nachr. 100, 235—247 (1981; MR 83d:53046; Zbl 484.53009); see §F.0.0

265. R. Sulanke: *Submanifolds of the Möbius space III: The analogue of O. Bonnet's theorem for hypersurfaces*, Tensor, New Ser. 38, 311—317 (1982; MR 87m:53066; Zbl 511.53060); see §P.6.3, §F.0.0

266. R. Sulanke: *Submanifolds of the Moebius space IV: Conformal invariants of immersions into spaces of constant curvature*, Potsdamer Forsch., Reihe B 43, 21—26 (1984; Zbl 643.53013); see §P.6.3, §F.0.0

267. R. Sulanke: *Submanifolds of the Moebius space V: Homogeneous surfaces in the Möbius space S^3*, Colloq. Math. Soc. Janos Bolyai 46, 1141—1154 (1988; MR 90e:53067; Zbl 643.53014); see §1.8.6, §F.0.0

268. Ch. Dittrich, R. Sulanke: *Submanifolds of the Moebius space VI: Characterization of the homogeneous tori*, Lect. Notes Math. 1410, 121—127 (1989; MR 91g:53054; Zbl 691.53043); see §1.8.6, §F.0.0

269. R. Sulanke: *Submanifolds of the Moebius space VII: On channel surfaces*, Geometry, Proc. 3rd Congr., Thessaloniki/Greece 1991, 410—419 (1992; MR 93m:53049; Zbl 759.53037); see §F.0.0

270. R. Sulanke: *Möbius invariants for pairs of spheres (S_1^m, S_2^l) in the Möbius space S^n*, Beitr. Alg. Geom. 41, 233—246 (2000; MR 2001b:53012); see §6.4.13

271. Y. Suyama: *Explicit representation of compact conformally flat hypersurfaces*, Tôhoku Math. J. 50, 179—196 (1998; MR 99e:53083; Zbl 951.53045); see §2.0.0

272. Y. Suyama: *Conformally flat hypersurfaces in Euclidean 4-space*, Nagoya Math. J. 158, 1—42 (2000; MR 2001h:53079); see §2.0.0

273. Y. Suyama: *Conformally flat hypersurfaces in Euclidean 4-space II*, preprint (2001); see §2.0.0, §2.4.13, §2.4.14

274. P. Tait: *On orthogonal isothermal surfaces*, Trans. Edinburgh 27, 105—123 (1872; JFM 07.0462.02); see §3.0.0

275. T. Takasu: *Differentialgeometrien in den Kugelräumen, Bd. I*, Tagaido Publ. Co., Kyoto, and Hafner Publ. Co., New York (1938; Zbl 019.04401); see §I.0.0, §1.0.0

276. P. Tapernoux: *Willmore-Kanalflächen*, Diplom thesis, ETH Zürich (1991); see §3.0.0, §3.5.1, §3.7.0, §3.7.13

277. G. Thomsen: *Über konforme Geometrie I: Grundlagen der konformen Flächentheorie*, Hamb. Math. Abh. 3, 31—56 (1923; JFM 49.0530.02); see §P.6.3, §1.0.0, §F.0.0

278. G. Thomsen: *Über konforme Geometrie II: Über Kreisscharen und Kurven in der Ebene und über Kugelscharen und Kurven im Raum*, Hamb. Math. Abh. 4, 117—147 (1926; JFM 51.0585.03); see §P.6.3, §1.0.0, §F.0.0

279. W. P. Thurston: *The geometry and topology of 3-manifolds*, electronic version 1.1 (2002); see §F.0.0

280. A. Thybaut: *Sur la déformation du paraboloïde et sur quelques problèmes qui s'y rattachent*, Ann. Ec. Norm. Sup. 14, 45—98 (1897; JFM 28.0566.04); see §5.0.0, §5.6.13

281. A. Thybaut: *Sur les surfaces isothermiques et la déformation du paraboloïde*, Comptes Rendus 128, 1274—1276 (1899; JFM 30.0556.02); see §3.6.4, §5.0.0, §5.6.13

282. A. Thybaut: *Sur les surfaces isothermiques*, Comptes Rendus 131, 932—935 (1900; JFM 31.0607.04); see §3.6.4, §5.6.13

283. A. Tresse: *Sur les invariants différentiells d'une surface par rapport aux transformations conformes de l'espace*, Comptes Rendus 114, 948—950 (1892; JFM 24.0131.06); see §P.6.3

284. G. Tzitzéica: *Géométrie différentielle projective des réseaux*, Cultura Nationala, Bucharest (1924); see §5.0.0, §5.1.0, §5.1.4

285. M. Umehara, K. Yamada: *A parametrization of the Weierstrass formulae and perturbation of complete minimal surfaces in $\mathbb{R}^3$ into the hyperbolic 3-space*, J. Reine Angew. Math. 432, 93—116 (1992; MR 94e:54004; Zbl 757.53033); see §5.3.21, §5.5.29

286. M. Umehara, K. Yamada: *Surfaces of constant mean curvature c in $H^3(-c^2)$ with prescribed hyperbolic Gauss map*, Math. Ann. 304, 203—224 (1996; MR 97b:53017; Zbl 841.53050); see §5.6.13

287. M. Umehara, K. Yamada: *A Duality on cmc-1 surfaces in hyperbolic space and a hyperbolic analogue of the Osserman inequality*, Tsukuba J. Math. 21, 229—237 (1997; MR 99e:53012); see §5.6.16, §5.7.35

288. K. Th. Vahlen: *Über Bewegungen und complexe Zahlen*, Math. Ann. 55, 585—593 (1902; JFM 33.0721.01); see §I.1.4, §6.0.0, §7.0.0, §7.2.8

289. L. van Hemmen: *Theoretische Membranphysik: der Formenreichtum der Vesikel*, lecture notes, winter semester 2000/01; see §3.0.0, §3.5.1

290. E. Vergasta: *Conformal immersions with the same Gauss map*, An. Acad. Bras. Ciênc. 59, 145—147 (1987; MR 89e:53005; Zbl 644.53004); see §3.3.11

291. E. S. Vergasta: *Conformal deformations preserving the Gauss map*, Pac. J. Math. 156, 359—369 (1992; MR 93j:53077; Zbl 761.53004); see §3.3.11

292. L. Verstraelen, G. Zafindratafa: *Some comments on conformally flat submanifolds*, Geometry and Topology III, 131—147 (1991; MR 96c:53088; Zbl 727.53028); see §2.0.0

293. M. E. Vessiot: *Contribution à la géométrie conforme. Théorie des surfaces*, Bull. Soc. Math. France 54, 139—179 (1926; JFM 52.0764.01) and 55, 39—79 (1927; JFM 53.0700.04); see §3.7.4

294. K. Voss: *Variation of curvature integrals*, Results Math. 20, 789—796 (1991; MR 92m:53105; Zbl 753.53004); see §3.5.9

295. K. Voss: *Canal Willmore tori*, talk given at the Oberwolfach geometry Meeting (1991); see §3.7.0

296. C.-P. Wang: *Möbius geometry for hypersurfaces in S^4*, Nagoya Math. J. 139, 1—20 (1995; MR 96k:53087; Zbl 863.53034); see §P.6.3, §2.3.5

297. C.-P. Wang: *Möbius geometry of submanifolds in S^n*, Manuscr. Math. 96, 517—534 (1998; Zbl 912.53012); see §P.6.3, §P.6.5

298. F. Warner: *Foundations of differentiable manifolds and Lie groups*, Graduate Texts in Math. 94, Springer, New York (1983; MR 84k:58001; Zbl 516.58001); see §P.4.8, §4.4.8, §6.4.4

299. J. Weingarten: *Über die Differentialgleichung der Oberflächen, welche durch ihre Krümmungslinien in unendlich kleine Quadrate getheilt werden können*, Sitzungsber. königl. preuß. Akad. Wiss. 1883 (2.Teil) 1163—1166 (1883; JFM 15.0635.01); see §P.6.5, §3.0.0, §3.3.19

300. H. Weyl: *Reine Infinitesimalgeometrie*, Math. Z. 2, 384—411 (1918); see §P.5.0

301. H. Weyl: *Zur Infinitesimalgeometrie: Einordnung der projektiven und konformen Auffassung*, Nachr. königl. Ges. Wiss. Göttingen, Math.-Phys. Kl. 1921, 99—112 (1921; JFM 48.0844.04); see §P.5.0

302. J. H. White: *A global invariant of conformal mappings in space*, Proc. Am. Math. Soc. 38, 162—164 (1973; MR 48:2954; Zbl 256.53008); see §3.5.0, §3.5.2

303. A. N. Whitehead: *A treatise on universal algebra with applications*, Hafner Publ. Co., New York (1960); see §6.0.0

304. J. Wilker: *The Quaternion formalism for Möbius groups in four or fewer dimensions*, Lin. Alg. Appl. 190, 99—136 (1993; MR 94c:20086; Zbl 786.51005); see §4.0.0, §4.8.0

305. H. Willgrod: *Ueber Flächen, welche sich durch ihre Krümmungslinien in unendlich kleine Quadrate theilen lassen*, PhD thesis, Göttingen (1883; JFM 15.0634.02); see §5.0.0

306. T. Willmore: *Note on embedded surfaces*, An. Sti. Univ. Al. I. Cuza Iasi, N. Ser., Sect. Ia 11B, 493—496 (1965; MR 34:1940; Zbl 171.20001); see §I.0.0, §3.0.0, §3.5.0

307. T. Willmore: *Surfaces in conformal geometry*, Ann. Global Anal. Geom. 18, 255—264 (2000; MR 2001i:53099; Zbl 978.53025); see §3.0.0, §3.5.0

308. V. Zakharov: *Description of the n-orthogonal curvilinear coordinate systems and Hamiltonian integrable systems of hydrodynamic type. Part I: Integration of the Lamé equations*, Duke Math. J. 94, 103—139 (1998; MR 99e:58109; Zbl 963.37068); see §8.0.0

Index

Printed in the United States
By Bookmasters